Ubiquitin and the Biology of the Cell

Ubiquitin and the Biology of the Cell

Edited by

Jan-Michael Peters
Research Institute of Molecular Pathology
Vienna, Austria

J. Robin Harris
Institute of Zoology
University of Mainz
Mainz, Germany

and

Daniel Finley
Harvard Medical School
Boston, Massachusetts

PLENUM PRESS • NEW YORK AND LONDON

Library of Congress Cataloging in Publication Data

Ubiquitin and the biology of the cell / edited by Jan-Michael Peters, J. Robin Harris, and Daniel Finley.
p. cm.
Includes bibliographical references and index.
ISBN 0-306-45649-4
1. Ubiquitin. I. Peters, Jan-Michael. II. Harris, James R. III. Finley, Daniel.
QP552.U24U253 1998
572′.6—dc21 98-18550
CIP

Front cover: Structures of 20S and 26S proteasome complexes. Crystal structures of the 20S proteasome from the archaebacterium *Thermoplasma acidophilum* (top) and the yeast *Saccharomyces cerevisiae* (bottom) are shown as space-filling representations. The molecules colored in yellow represent cocrystallized proteasome inhibitor molecules (acetyl-leucyl-leucyl-norleucinal). The 26S proteasome complex purified from oocytes of the frog *Xenopus laevis* (center) is shown as a correlation average obtained from negatively stained electron micrographs. For further details see Groll *et al., Nature* **386**, 463 – 471 (1997); Löwe *et al., Science* **268**, 533 – 539 (1995); Peters *et al., J. Mol. Biol.* **234**, 932 – 937 (1993); and Chapters 5 and 6 in this volume. The image of the *Thermoplasma* proteasome was kindly provided by Dr. Andrei Lupas and Dr. Wolfgang Baumeister. The other two images are reproduced with kind permission of Macmillan Magazines Ltd. and of Academic Press Ltd.

Back cover: Yeast cells expressing K48R mutant ubiquitin in place of wild-type ubiquitin. Photo courtesy of V. Chau and D. Finley.

ISBN 0-306-45649-4

A Division of Plenum Publishing Corporation
233 Spring Street, New York, N.Y. 10013

http://www.plenum.com

10 9 8 7 6 5 4 3 2 1

Printed in the United States of America

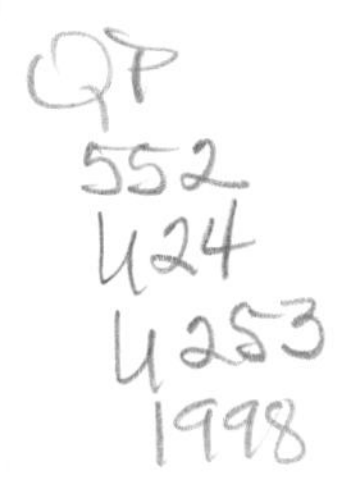

Contributors

Wolfgang Baumeister Molekulare Strukturbiologie, Max-Planck-Institute für Biochemie, D-82152 Martinsried, Germany
Jochen Beninga Department of Cell Biology, Harvard Medical School, Boston, Massachusetts 02115
C. Byrd Division of Biology, California Institute of Technology, Pasadena, California 91125
Zhijian J. Chen ProScript, Inc., Cambridge, Massachusetts 02139
Philip Coffino Department of Microbiology and Immunology and of Medicine, University of California, San Francisco, San Francisco, California 94143
I. V. Davydov Division of Biology, California Institute of Technology, Pasadena, California 91125
Raymond J. Deshaies Division of Biology, California Institute of Technology, Pasadena, California 91125
R. J. Dohmen Institut für Mikrobiologie, Heinrich-Heine-Universität, D-40225 Düsseldorf, Germany
F. Du Division of Biology, California Institute of Technology, Pasadena, California 91125
M. Ghislain Unité de Biochimie Physiologique, Université catholique de Louvain, B-1348 Louvain-la-Neuve, Belgium
Alfred Goldberg Department of Cell Biology, Harvard Medical School, Boston, Massachusetts 02115
M. Gonzalez NICHD, National Institutes of Health, Bethesda, Maryland 20892-2725
S. Grigoryev Department of Biology, University of Massachusetts, Amherst, Massachusetts 01003
Avram Hershko Unit of Biochemistry, The B. Rappaport Faculty of Medi-

cine and The Rappaport Institute for Research in the Medical Sciences, Technion–Israel Institute of Technology, Haifa 31096, Israel

Mark Hochstrasser Department of Biochemistry and Molecular Biology, University of Chicago, Chicago, Illinois 60637

Peter M. Howley Department of Pathology, Harvard Medical School, Boston, Massachusetts 02115

Jon M. Huibregtse Department of Molecular Biology and Biochemistry, Rutgers University, Piscataway, New Jersey 08855

Stefan Jentsch Zentrum für Molekulare Biologie der Universität, Heidelberg, D-69120 Heidelberg, Germany

E. S. Johnson The Rockefeller University, New York, New York 10021-6399

N. Johnsson Max-Delbrück-Laboratorium, D-50829 Cologne, Germany

J. A. Johnston Division of Biology, California Institute of Technology, Pasadena, California 91125

Randall W. King Department of Cell Biology, Harvard Medical School, Boston, Massachusetts 02115

Ron R. Kopito Department of Biological Sciences, Stanford University, Stanford, California 94305-5020

Daniel Kornitzer Department of Molecular Microbiology, The B. Rappaport Faculty of Medicine, Technion– Israel Institute of Technology, Haifa 31096, Israel

Y. T. Kwon Division of Biology, California Institute of Technology, Pasadena, California 91125

Michael Landon Department of Biochemistry, University of Nottingham Medical School, Queen's Medical Centre, Nottingham NG7 2UH, United Kingdom

F. Lévy Ludwig Institute for Cancer Research, CH-1066 Epalinges, Switzerland

O. Lomovskaya Microcide, Inc., Mountain View, California 94043

James Lowe Department of Clinical Laboratory Sciences, University of Nottingham Medical School, Queen's Medical Centre, Nottingham NG7 2UH, United Kingdom

Andrei Lupas Molekulare Strukturbiologie, Max-Planck-Institute für Biochemie, D-82152 Martinsried, Germany

K. Madura Department of Biochemistry, UMDNJ–Johnson Medical School, Piscataway, New Jersey 08854

Carl G. Maki Department of Pathology, Harvard Medical School, Boston, Massachusetts 02115

Tom Maniatis Department of Molecular and Cellular Biology, Harvard University, Cambridge, Massachusetts 02138

R. John Mayer Department of Biochemistry, University of Nottingham Medical School, Queen's Medical Centre, Nottingham NG7 2UH, United Kingdom

I. Ota Department of Chemistry and Biochemistry, University of Colorado, Boulder, Colorado 80309-0215

Jan-Michael Peters Department of Cell Biology, Harvard Medical School, Boston, Massachusetts 02115; *present address:* Research Institute of Molecular Pathology, A-1030 Vienna, Austria

Cecile M. Pickart Department of Biochemistry, Johns Hopkins University, Baltimore, Maryland 21205

Martin Rechsteiner Department of Biochemistry, University of Utah School of Medicine, Salt Lake City, Utah 84132

T. Rümenapf Institut für Virologie, Fachbereich Veterinärmedizin, 35392 Giessen, Germany

Martin Scheffner Deutsches Krebsforschungszentrum, Angewandte Tumorvirologie, D-69120 Heidelberg, Germany

T. E. Shrader Department of Biochemistry, Albert Einstein College of Medicine, Bronx, New York 10461

Susan Smith Zentrum für Molekulare Biologie der Universität Heidelberg, 69120 Heidelberg, Germany; *present address*: Mitotix, Inc. Cambridge, Massachusetts 02139

T. Suzuki Division of Biology, California Institute of Technology, Pasadena, California 91125

G. Turner Division of Biology, California Institute of Technology, Pasadena, California 91125

A. Varshavsky Division of Biology, California Institute of Technology, Pasadena, California 91125

P. R. H. Waller Testa, Hurwitz & Thibeault, Boston, Massachusetts 02110

A. Webster Division of Biology, California Institute of Technology, Pasadena, California 91125

Keith D. Wilkinson Department of Biochemistry, Emory University, Atlanta, Georgia 30083

Y. Xie Division of Biology, California Institute of Technology, Pasadena, California 91125

Preface

The last several years have been a landmark period in the ubiquitin field. The breadth of ubiquitin's roles in cell biology was first sketched, and the importance of ubiquitin-dependent proteolysis as a regulatory mechanism gained general acceptance. The many strands of work that led to this new perception are recounted in this book. A consequence of this progress is that the field has grown dramatically since the first book on ubiquitin was published almost a decade ago [M. Rechsteiner (ed.), *Ubiquitin*, Plenum Press, 1988]. In this span, students of the cell cycle, transcription, signal transduction, protein sorting, neuropathology, cancer, virology, and immunology have attempted to chart the role of ubiquitin in their particular experimental systems, and this integration of the field into cell biology as a whole continues at a remarkable pace. We hope that for active researchers in the field as well as for newcomers and those on the fence, this book will prove helpful for its breadth, historical perspective, and practical tips.

Structural data are now available on many of the components of the ubiquitin pathway. The structures have provided basic insights into the unusual biochemical mechanisms of ubiquitination and proteasome-mediated proteolysis. Because high-speed computer graphics can convey structures more effectively than print media, we have supplemented the figures of the book with a Worldwide Web site that can display the structures in a flexible, viewer-controlled format. For many of the chapters, a visit to this Web site will help to substantiate and reinforce points drawn in the text, and allow each individual to get to know those aspects of each structure that are of most interest to that person. The site will be updated as additional structures are solved.

This book was made possible by the enthusiasm of almost 50 authors who generously devoted their time to the preparation of the various chapters, and many others have contributed by providing unpublished observations and helpful comments. Our thanks to all of them. We are especially grateful to Joanna Lawrence at

Plenum Press in London, who has guided the production of this book and the accompanying Worldwide Web site, and to Markus Rohrwild and Bob Freeman at Harvard Medical School in Boston, who assembled this site.

You can visit our associated Web sites for the latest protein structures of the ubiquitin system:

http://cbweb.med.harvard.edu/ubiquitin/
http://www.at.embnet.org/mirror/ubiquitin/

Jan-Michael Peters

Vienna, Austria

J. Robin Harris

Mainz, Germany

Daniel Finley

Boston, Massachusetts

Contents

Chapter 1
The Ubiquitin System: Past, Present, and Future Perspectives

Avram Hershko

Chapter 2
Polyubiquitin Chains

Cecile M. Pickart

Chapter 5
The 20 S Proteasome

Andrei Lupas and Wolfgang Baumeister

Chapter 6
The 26 S Proteasome

Martin Rechsteiner

Chapter 7

Function of the Proteasome in Antigen Presentation

Jochen Beninga and Alfred L. Goldberg

Chapter 8
The N-End Rule Pathway

A. Varshavsky, C. Byrd, I. V. Davydov, R.J. Dohmen, F. Du,
M. Ghislain, M. Gonzalez, S. Grigoryev, E. S. Johnson, N. Johnsson,
J. A. Johnston, Y. T. Kwon, F. Lévy, O. Lomovskaya, K. Madura,
I. Ota, T. Rümenapf, T. E. Shrader, T. Suzuki, G. Turner,
P. R. H. Waller, A. Webster, and Y. Xie

Chapter 9

Ubiquitin-Dependent Degradation of Transcription Regulators

Mark Hochstrasser and Daniel Kornitzer

Chapter 10
Role of the Ubiquitin–Proteasome Pathway in NF-κB Activation

Zhijian J. Chen and Tom Maniatis

Chapter 11
Ubiquitination of the p53 Tumor Suppressor

Jon M. Huibregtse, Carl G. Maki, and Peter M. Howley

Chapter 12
Cell Cycle Control by Ubiquitin-Dependent Proteolysis

Jan-Michael Peters, Randall W. King, and Raymond J. Deshaies

Chapter 13
Ubiquitination of Integral Membrane Proteins and Proteins in the Secretory Pathway

Ron R. Kopito

Chapter 14
Degradation of Ornithine Decarboxylase

Philip Coffino

Chapter 15
Ubiquitin and the Molecular Pathology of Human Disease

R. John Mayer, Michael Landon, and James Lowe

Ubiquitin and the Biology of the Cell

CHAPTER 1

The Ubiquitin System

Past, Present, and Future Perspectives

Avram Hershko

1. INTRODUCTION

The purpose of this chapter is threefold: (1) to provide the uninitiated reader a brief overview on the functions and mechanisms of the ubiquitin system, as an introduction to topics described in detail in the different chapters of this book; (2) to give to both the uninitiated and to the expert, but relatively newcomer ubiquitinologist a historical perspective on the discovery of the ubiquitin system, because much can be learned from the way in which discoveries were made; and (3) to point out some challenges that workers in the ubiquitin field, and those who may stumble into it from other areas, may face in the future. These different subjects are discussed in light of my personal experience with the ubiquitin proteolytic system.

Avram Hershko • Unit of Biochemistry, The B. Rappaport Faculty of Medicine and The Rappaport Institute for Research in the Medical Sciences, Technion–Israel Institute of Technology, Haifa 31096, Israel.

Ubiquitin and the Biology of the Cell, edited by Peters *et al.* Plenum Press, New York, 1998.

2. DISCOVERY OF THE ROLE OF UBIQUITIN IN PROTEIN BREAKDOWN

Ubiquitin was first isolated by Goldstein *et al.* (1975) from the thymus and was originally thought to be a thymic hormone; however, it was found to be present in all tissues and organisms, and hence its name. Subsequently, it was found by Goldknopf and Busch (1977) to be a part of an unusual branched protein, in which the C-terminus of ubiquitin is linked to the ε-amino group of a lysine residue of histone 2A. The functions of ubiquitin remained totally obscure, however, until it emerged, quite unexpectedly, in our studies on the mechanisms of intracellular protein degradation. I have been interested in the question of how proteins are degraded in cells since my work, in the laboratory of Gordon Tomkins, on the degradation of tyrosine aminotransferase in hepatoma cells (Hershko and Tomkins, 1971). The dynamic turnover of cellular proteins (Schoenheimer, 1942) and the important functions of selective protein degradation in the control of the levels of specific cellular proteins (Schimke and Doyle, 1970) had been recognized for a long time, but the underlying mechanisms remained unknown. An unusual proteolytic machinery was suggested by observations indicating that intracellular protein degradation requires cellular energy (Simpson, 1953; Hershko and Tomkins, 1971). It was assumed by some investigators that energy is required to obtain specificity, but it is doubtful whether anybody (certainly, not this author) thought of a machinery as complex as the ubiquitin proteolytic system turned out to be. To elucidate the mechanism, a cell-free system was required, in which the enzyme components can be isolated and characterized by direct biochemical methods. An ATP-dependent proteolytic system in reticulocyte lysates was first established by Etlinger and Goldberg (1977), and then was analyzed by our biochemical fractionation–reconstitution studies. In this work I was greatly helped by Aaron Ciechanover, who was then my graduate student, and by collaboration with Irwin Rose, who hosted me in his laboratory at Fox Chase Cancer Center in Philadelphia for a sabbatical year in 1977–78 and many times afterwards. Initially, reticulocyte lysates were fractionated on DEAE-cellulose into two crude fractions: Fraction 1, which is not adsorbed, and Fraction 2, which contains all proteins adsorbed to the resin and eluted with high salt. The original aim of this fractionation had been to remove hemoglobin (present in Fraction 1), but we found that Fraction 2 lost most ATP-dependent proteolytic activity, as compared with crude lysates. However, ATP-stimulated proteolytic activity could be reconstituted by the addition of Fraction 1 to Fraction 2. The active component in Fraction 1 was a small, heat-stable protein (Ciechanover *et al.*, 1978). It was first called APF-1 (ATP-dependent proteolysis factor 1). The identity of APF-1 with ubiquitin was found by others (Wilkinson *et al.*, 1980), following our discovery of its conjugation to proteins, as described below.

The purification of ubiquitin from Fraction 1 allowed the elucidation of the mode of its action in the proteolytic system. At first we thought that it might be an activator or a regulatory subunit of some protease or other enzyme component of the system present in Fraction 2. To examine such a possibility, purified ubiquitin was radiolabeled and incubated with crude Fraction 2 in the presence or absence of ATP. A dramatic ATP-dependent binding of [^{125}I]ubiquitin to high-molecular-weight proteins was observed by gel filtration chromatography (Ciechanover *et al.*, 1980). We were greatly surprised, however, to find that a covalent amide linkage was formed, as shown by the stability of the "complex" to treatments with acid, alkali, hydroxylamine, or boiling with SDS and mercaptoethanol. Only then did we consider the possibility that ubiquitin may be linked to protein substrates (rather than to an enzyme of the system), thus providing a marking mechanism for selective protein breakdown. In support of this interpretation, we found that ubiquitin is conjugated to proteins that are good substrates of the ATP-dependent proteolytic system (Hershko *et al.*, 1980). On the conjugation of ubiquitin to a protein substrate, multiple bands were observed in SDS–polyacrylamide gel electrophoresis, which consisted of several molecules of ubiquitin linked to a single molecule of the protein. Based on these findings, we proposed that the ligation of ubiquitin to protein marks it for degradation by a protease that specifically recognizes ubiquitinylated proteins (Hershko *et al.*, 1980).

3. IDENTIFICATION OF THE ENZYMATIC REACTIONS OF THE UBIQUITIN PATHWAY

Following the discovery of ubiquitin ligation to protein substrates, we tried to identify and characterize the different enzymes of this system in the order of their action in the proteolytic pathway. An outline of the main enzymatic steps of this pathway, based on such studies, is shown in Fig. 1. We first described the ubiquitin-activating enzyme, E1, that carries out the ATP-dependent activation of the C-terminal glycine residue of ubiquitin prior to ligation. E1 catalyzes a two-step reaction sequence, in which ubiquitin adenylate is first formed, followed by the transfer of activated ubiquitin to a thiol site of E1 (Ciechanover *et al.*, 1981). Next, we found that two further enzymes are required for ubiquitin-protein ligation: a ubiquitin-carrier protein, E2, which accepts activated ubiquitin from E1 by transacylation and then transfers it to the protein substrate in a reaction requiring a ubiquitin-protein ligase, E3 (Hershko *et al.*, 1983). E3 enzymes apparently have an important function in the selection of proteins suitable for degradation. We have studied in detail one species of ubiquitin-protein ligase, E3α. The enzyme binds specific protein substrates (Hershko *et al.*, 1986) and E2 (Reiss *et al.*, 1989) and allows the transfer of ubiquitin from E2 to amino groups of the

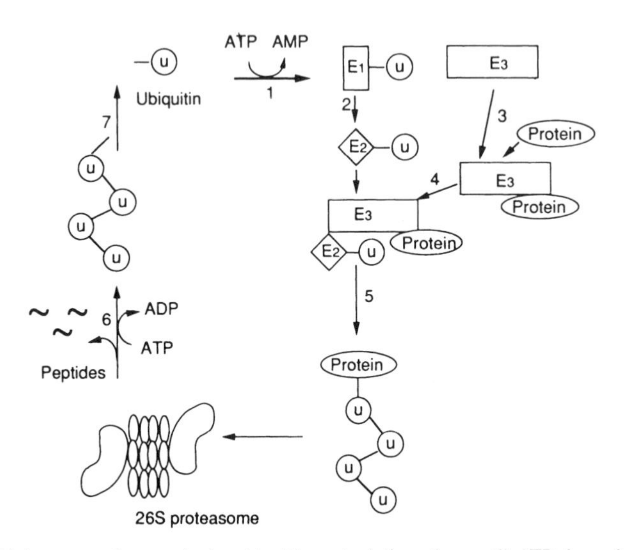

Figure 1. Main enzymatic steps in the ubiquitin proteolytic pathway. (1) ATP-dependent activation of ubiquitin with the formation of a high-energy thiolester bond with E1. (2) Transfer of activated ubiquitin to thiolester with E2. (3) Binding of the protein substrate to a specific ubiquitin ligase, E3. (4) Formation of an intermediary complex between E3, the protein substrate, and E2–ubiquitin. (5) Ligation of ubiquitin to a lysine residue of the protein substrate, followed by the formation of a multiubiquitin chain. (6) ATP-dependent degradation of ubiquitinylated protein to small peptides by the 26 S proteasome complex. (7) Recycling of free ubiquitin by the action of isopeptidases. Reproduced from Hershko (1996), with permission.

protein substrate. One signal in protein substrates recognized by E3α is the identity of the N-terminal amino acid residue. We found that E3α contains distinct binding sites for basic and bulky-hydrophobic N-terminal amino acid residues (Reiss *et al.*, 1988; Reiss and Hershko, 1990), providing an explanation for the "N-end rule" discovered by Varshavsky and co-workers (Bachmair *et al.*, 1986). Other E3 enzymes, with other E2 partners, recognize other signals in specific proteins (see below).

By the action of such ubiquitinylating enzyme systems, ubiquitin is linked to proteins by isopeptide linkages between the C-terminal glycine residue of ubiquitin and ε-amino groups of lysine residues in substrate proteins. In addition, the formation of multiubiquitin chains has been shown to occur (Hershko and Heller, 1985), in which a molecule of ubiquitin is usually linked to lysine-48 (Chau *et al.*, 1989) of another ubiquitin, which in turn is linked to the protein substrate (see Chapter 2). Such multiubiquitinylated proteins are degraded by a large 26 S protease complex, now called the 26 S proteasome, discovered by Rechsteiner and colleagues (Hough *et al.*, 1986). The 26 S complex requires ATP for its proteolytic action (Hough *et al.*, 1986) and has ATPase activity (Armon *et al.*, 1990). We found that the 26 S proteasome is formed by the ATP-dependent assembly of three components, called conjugate-degrading factors CF-1, CF-2, and CF-3 (Ganoth *et al.*, 1988). One of these, CF-3, was identified as the previously known "multicatalytic protease" or the 20 S proteasome (Eytan *et al.*, 1989; Driscoll and Goldberg, 1990), confirming the suggestion that the 20 S proteasome is the catalytic core of the 26 S complex (Hough *et al.*, 1986). The structure and mode of action of the 20 and 26 S proteasomes are currently being investigated (see below).

The last step in the ubiquitin proteolytic pathway is the regeneration of free and reutilizable ubiquitin, a process carried out by ubiquitin-C-terminal hydrolases or isopeptidases. We have studied two ubiquitin-C-terminal hydrolases that are involved in the regeneration of free ubiquitin at the final stages of protein degradation. One is an enzyme called isopeptidase T, which preferentially cleaves ubiquitin-Lys48-ubiquitin linkages in multiubiquitin chains (Hadari *et al.*, 1992). Its main function appears to be the disassembly of multiubiquitin chain remnants following the proteolysis of the protein moiety of multiubiquitin-protein conjugates by the 26 S proteasome. Isopeptidase T does not cleave the linkage between ubiquitin and Lys residues of the protein substrate. Such linkages are cleaved by another ubiquitin-C-terminal hydrolase activity which is associated with the 26 S proteasome complex (Eytan *et al.*, 1993). The release of free ubiquitin by the latter isopeptidase is stimulated by ATP. The role of ATP in hydrolase action, and the mode of the integration of the protease and ubiquitin-C-terminal hydrolase activities of the 26 S proteasome complex, remain to be elucidated.

Although the above studies provided a general outline of the main biochemical events of the ubiquitin pathway, this was undoubtedly only a first glimpse of a vastly complex, highly specific and regulated system. Some of the current progress in this field on the functions and mechanisms of the ubiquitin system and the major problems that remain to be elucidated, are discussed below.

4. CURRENT STATE OF KNOWLEDGE ON MECHANISMS AND FUNCTIONS OF THE UBIQUITIN SYSTEM

Recent progress in our knowledge of enzyme components, natural substrates, and cellular roles of the ubiquitin system is described in various chapters of this book. Some selected cases are discussed here briefly.

4.1. Enzyme Components

It now seems that in most organisms there is a single El enzyme, but there are multiple species of E2 and E3 enzymes. There is much more information at present on E2s than on E3s, presumably reflecting the small size and relative ease of cloning of E2 genes. In yeast, the genes of 13 different E2 enzymes (also called ubiquitin conjugating enzymes or *UBC*s) have been cloned (see Chapter 3). Mutations in the various E2 genes show that some of them have distinct functions. For example, *UBC2/RAD6* is involved in DNA repair and N-end rule degradation, *UBC3/CDC34* is required for G1→S cell cycle transition, and *UBC4* and *UBC5* are involved in the degradation of many short-lived normal and abnormal proteins. Because of these specific functions, it has been assumed by some workers that certain E2 enzymes recognize certain specific protein substrates. However, specific E2–protein interactions have not been demonstrated. It seems more plausible that a certain E2 binds to a specific E3, which, in turn, binds a protein substrate having a specific recognition signal. This seems to be the case with E3α, which binds specifically $E2_{14kDa}$, the mammalian homologue of Ubc2/Rad6 (Reiss *et al.*, 1989; Dohmen *et al.*, 1991). It appears, however, that certain species of E2 may bind to more than one E3, as disruption of the gene for Ubrl (the yeast homologue of E3α) has no influence on DNA repair, in contrast to the phenotype of UBC2/RAD6 mutants.

Available information on E3 enzymes is still very scanty. There is no sequence similarity between the two E3 enzymes that have been cloned thus far, Ubrl and E6-associated protein (E6-AP), suggesting that different families of ubiquitin-protein ligases may exist. E6-AP, a protein involved in ubiquitin ligation to tumor suppressor p53, is also different in that it binds and ubiquitinylates p53 only in the presence of an auxiliary protein, the oncogenic papillomavirus E6 protein (Scheffner *et al.*, 1993; see Chapter 11). It remains to be seen whether other auxiliary proteins exist that play roles in the recognition of certain cellular protein substrates by E6-AP or by other ubiquitin ligases homologous to E6-AP (Huibregtse *et al.*, 1995).

There has been considerable progress on the structure and function of the 20 and 26 S proteasome complexes (reviewed in Coux *et al.*, 1996; Chapters 5 and 6). In the archaebacterium *Thermoplasma acidophilum*, a 20 S proteasome particle was found that has a cylindrical structure very similar to that of eukaryotic proteasomes, but it is composed of only two types of subunits, termed α and β. These are arranged in four rings of seven subunits, so that the two outer rings are composed of α subunits and the two inner rings of β subunits. X-ray crystallography showed that a narrow tunnel traverses the four rings; it widens to a central cavity between the two inner rings (Lowe *et al.*, 1995). This central cavity contains all of the catalytic sites, as shown by the crystal structure with bound peptide aldehyde inhibitor. The catalytically active amino acid residue was identified as the N-terminal threonine of each β subunit of the *Thermoplasma* proteasome. It appears reasonable to assume that eukaryotic 20 S proteasomes have a similar structure of greater complexity, as they are also built of four rings of seven subunits, but they contain seven different α-type and seven different β-type subunits and they possess at least five different proteolytic activities against peptide substrates (reviewed in Coux *et al.*, 1996; Chapter 5).

There is as yet no X-ray structure of the 26 S proteasome complex, but electron microscopy showed that it is built of a central 20 S proteasome cylindrical structure, to both sides of which are attached cap-shaped complexes (Peters, 1994; see Chapter 6). In this dumbbell-shaped complex, the two cap-shaped structures are attached to the core 20 S particle in opposite orientations. When dissociated from the 26 S proteasomes, these cap structures have been purified as a 19 S complex, designated "regulatory complex," "ball," or "PA700" (Chapter 6). The 19 S regulatory complex presumably contains CF-1 and CF-2 that had been isolated earlier from ATP-depleted reticulocytes (Ganoth *et al.*, 1988), as it can be assembled with the 20 S proteasome in the presence of ATP to form the 26 S proteasome. A great number of subunits of the 19 S regulatory complex have been cloned, including several ATPase subunits (see Chapter 6) and a subunit that binds multiubiquitin chains (Deveraux *et al.*, 1994). This multiubiquitin-binding subunit is not essential in yeast (Van Nocker *et al.*, 1996), so it is possible that several different subunits may bind ubiquitinylated proteins to the 26 S proteasome. A reasonable model for the mode of action of the 26 S proteasome complex appears to be that ubiquitinylated proteins first bind to multiubiquitin-binding subunits and then are unfolded and translocated to the protease catalytic sites in a process that requires the energy of ATP hydrolysis (see Chapters 5 and 6). The location of the protease catalytic sites within the central cavity may ensure that only properly bound, unfolded, and translocated proteins are attacked by the proteasome. This and other hypotheses on the mode of action of the 26 S proteasome remain to be investigated.

A surprisingly large number of genes encoding ubiquitin-C-terminal hydrolases (also called ubiquitin-specific proteases or Ubps) have been identified (see Chapter 4). It has been suggested that certain Ubps may act specifically on ubiquitin conjugates of specific proteins, based on observations that some Ubps appear to have specific functions in growth control (Papa and Hochstrasser, 1993; Zhu *et al.*, 1996) or in development (Huang *et al.*, 1995). However, the targets of these Ubps have not been identified and there is no evidence at present indicating the selectivity of a ubiquitin deconjugating enzyme to a certain protein substrate.

4.2. Some Cellular Proteins Degraded by the Ubiquitin System

Several chapters of this book describe recent information on a variety of important regulatory proteins whose levels are controlled by ubiquitin-mediated degradation. These include many cell cycle regulatory proteins (Chapter 12 and see below), p53 tumor suppressor (Chapter 11), the transcriptional regulator NF-κB and its inhibitor IκBα (Chapter 10), many transcription factors (Chapter 9), and the *mos* protooncogene (Nishisawa *et al.*, 1993). It may be expected that the list of important cellular protein substrates of the ubiquitin system will rapidly expand in the near future, as the programmed degradation of short-lived regulatory proteins appears to be a recurring theme in temporally controlled processes. This seemingly wasteful way of using "disposable" protein regulators may be essential to ensure irreversibility. Good examples for this regulatory strategy are some regulators of the cell cycle. The cell division cycle is driven by oscillations in the activities of cyclin-dependent kinases (CDKs). The rise in activities of the different CDKs is initiated by their association with their respective cyclin subunits, whereas the subsequent inactivation of CDKs is caused by the rapid degradation of cyclins. The activities of CDKs are also controlled by specific protein inhibitors, and some of these CDK inhibitors are also degraded at specific stages of the cell cycle (reviewed in Chapter 12). It thus seems that specific protein degradation is involved in both positive and negative regulation of CDK activity. Other important regulators of the cell cycle, such as the phosphatase cdc25 (Nefsky and Beach, 1996), are also synthesized and degraded in a cell cycle-specific manner.

Several years ago, I became interested in the problem of the degradation of cyclin B (a mitotic cyclin), because of its high selectivity, intricate regulation, and important roles in the cell cycle. Cyclin B is the positive regulatory subunit of protein kinase cdc2 (CDK1). The active cyclin B–cdc2 complex causes entry of cells into mitosis, and after a short lag, activates the system that degrades its cyclin B subunit. Here again, I have used biochemical approaches to study the mechanisms and regulation of cyclin B degradation. In this case, the quest for a cell-free system led me to *Spisula solidissima*, a large clam that produces large amounts of oocytes, suitable for biochemical work. A cell-free system from fertilized clam oocytes, which reproduces faithfully cell cycle stage-specific degradation of mitotic cyclins, was established by Luca and Ruderman (1989). Using this cell-free system, we showed that the degradation of mitotic cyclins is inhibited by methylated ubiquitin, a specific inhibitor of multiubiquitin chain formation (Hershko *et al.*, 1991). At the same time, Glotzer *et al.* (1991) showed that in extracts of *Xenopus* eggs, cyclin B is ligated to ubiquitin in mitosis when it is degraded, but not in the interphase, when it is stable. The latter authors also showed that a conserved sequence of RXXLXXIXN, termed the "destruction box," is required for the ubiquitinylation and degradation of mitotic cyclins (see Chapter 12). The cumulative evidence from both studies indicated that cyclin B is degraded by the ubiquitin system.

In work carried out in part in collaboration with Joan Ruderman, we fractionated the clam oocyte system. We found that in addition to E1, two novel components are involved in cyclin B-ubiquitin ligation. These are a cyclin-selective ubiquitin-carrier protein E2-C, and a cyclin-ubiquitin ligase activity associated with particulate material (Hershko *et al.*, 1994). E2-C from clam oocytes was recently purified, cloned, and shown to be a novel Ubc family member (Aristarkhov *et al.*, 1996). A *Xenopus* homologue of E2-C was recently cloned by Yu *et al.* (1996). Although E2-C is selective for cyclin degradation, it is constitutively active, and is not regulated in the cell cycle. On the other hand, the activity of particle-associated cyclin-ubiquitin ligase is regulated: It is inactive in the interphase, but becomes active at the end of mitosis (Hershko *et al.*, 1994). We have dissociated cyclin-ubiquitin ligase activity from particles by extraction with high salt and the enzyme was partially purified by ammonium sulfate fractionation and glycerol density gradient centrifugation. It has an unusually large size of about 1500 kDa. The soluble enzyme from interphase oocytes could be activated *in vitro* by active cdc2–cyclin B complex, but only following a time lag. This preparation also retained selectivity for mitotic cyclins that contain intact "destruction box" recognition determinants. We called the particle that contains cyclin-ubiquitin ligase activity the *cyclosome*, to indicate its central role in the control of cyclin B levels in the cell cycle (Sudakin *et al.*, 1995). The conversion of the cyclosome to the active form is apparently related to its phosphorylation, as the active, mitotic form of the cyclosome is inactivated by treatment with an okadaic acid-sensitive phosphatase (Lahav-Baratz *et al.*, 1995). A similar complex was isolated by King *et al.* (1995) from extracts of *Xenopus* eggs. They named the same particle the *anaphase promoting complex* (APC). Subunits of the cyclosome/APC particle are strongly conserved in evolution and are required for exit from mitosis in various eukaryotes (reviewed in Chapter 12).

Figure 2 summarizes our present working hypothesis on the regulation of

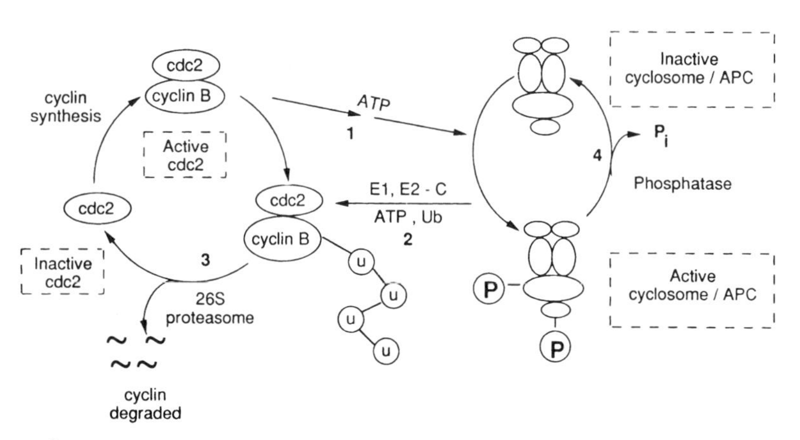

Figure 2. Proposed sequence of events in cell cycle-regulated degradation of cyclin B. (1) The active cdc2–cyclin B protein kinase converts the cyclosome/APC particle to the active, phosphorylated form at the end of mitosis. This process includes a time lag and may involve a cascade of phosphorylation reactions. (2) The active cyclosome complex ligates ubiquitin to cyclin B, in a process involving the action of El and E2-C. (3) Ubiquitinylated cyclin B is rapidly degraded by a constitutively acting 26 S proteasome and thus cdc2 is converted to the inactive, monomeric form. (4) The cyclosome complex is converted back to the inactive, interphase form by phosphatase action. Reproduced from Hershko (1996), with permission.

cyclin degradation in the embryonic cell cycle. The active cdc2–cyclin B protein kinase complex triggers the activation of the cyclosome/APC particle at the end of mitosis. Activation is related to the phosphorylation of some protein(s) of the cyclosome/APC particle. However, it is not necessary to assume that cdc2 directly phosphorylates the cyclosome, as there is a time lag in the activation of the cyclosome, which may indicate the involvement of a cascade of protein kinases. This time lag appears to have an important role in preventing the premature activation of cdc2 before the end of mitosis. The active form of the cyclosome complex ubiquitinylates cyclin B (in concert with E2-C and E1), and ubiquitinylated cyclin B is rapidly degraded by a constitutively acting 26 S proteasome. The degradation of cyclin B inactivates cdc2 and allows exit from mitosis. Subsequently, the cyclosome complex is converted back to the inactive, interphase form by the action of a phosphatase. An essential feature in this model is the interplay between protein phosphorylation and specific protein degradation in cell cycle control. Many processes in this system, especially those involved in the activation of the cyclosome by cdc2, remain to be elucidated. The reason for the large size of the cyclosome/APC particle, the mode of its action on other "destruction box"-containing proteins that are degraded at different stages of exit from mitosis (see Chapter 12), and possible additional functions of this complex in cell cycle control, remain to be investigated. It should be noted that some aspects of this model may be true only for the relatively simple embryonic cell cycle and not for somatic cells or yeasts, where the cyclin degradation system remains active until the end of G1 (Amon *et al.*, 1994; Brandeis and Hunt, 1996).

The mechanism described above for the regulation of cyclin degradation, the phosphorylation of a specific ubiquitin ligase complex, is certainly one of several regulatory mechanisms. In other cases, the phosphorylation of the substrate protein appears to be required for its ubiquitin-mediated degradation. Thus, the degradation of the yeast G1 cyclins, Cln3 (Yaglom *et al.*, 1995) and Cln2 (Lanker *et al.*, 1996), requires their phosphorylation prior to ubiquitinylation and degradation. In both cases, phosphorylation is on Ser or Thr residues within PEST sequences, regions that have been noted to occur at high frequency in rapidly degraded proteins (Rogers *et al.*, 1986). The machinery that ubiquitinylates G1 cyclins is completely different from the cyclosome system involved in the degradation of mitotic cyclins and requires the yeast proteins Cdc53, Grr1, Skp1 and the E2 protein Cdc34 (see Chapter 12). The binding of phosphorylated Cln2 to Cdc53 has been demonstrated (Willems *et al.*, 1996), but it is not yet clear whether binding is direct, or is mediated by another protein that binds to Cdc53. Phosphorylation of the protein substrate also appears to be involved in the degradation of the mammalian G1 cyclin, cyclin E (Won and Reed, 1996; Clurman *et al.*, 1996), and of the yeast CDK inhibitors Sic1 (Schwob *et al.*, 1994; Schneider *et al.*, 1996) and Far1 (McKinney and Cross, 1995). Likewise, the ubiquitinylation and degradation of IκBα is initiated by signal-induced phosphorylation of two specific Ser residues (reviewed in Chapter 10). The ubiquitin ligases that recognize the phosphorylated forms of these proteins have yet to be discovered. In other cases, such as with the *mos* protooncogene, the removal of phosphate from a specific Ser residue is required for ubiquitinylation and degradation (Nishisawa *et al.*, 1993). It thus seems that a variety of strategies have evolved to regulate the degradation of different cellular proteins.

5. CHALLENGES FOR THE FUTURE

Among the large gaps in our knowledge, notable is our ignorance of different types of ubiquitin ligases (E3s), as opposed to the important roles of such enzymes in the selectivity and regulation of ubiquitin-mediated protein degradation. Because of the large variety of specifically degraded proteins,which are presumably recognized by different E3 enzymes, it may be expected that the number of E3 enzymes will greatly exceed that of different E2 proteins. At present, only the yeast E3α (UBR1, Bartel *et al.*, 1990) and a limited number of E6-AP-like E3 enzymes (Huibregtse *et al.*, 1995; Nefsky and Beach, 1996) have been cloned. Because there is no homology between these two types of E3 enzymes, or between these and known subunits of the cyclosome/APC particle, or the proteins involved in the ubiquitinylation of G1 cyclins or of Sic1 (see Chapter 12): it is possible that many different families of E3 enzymes exist and will be identified only by the difficult process of enzyme purification and protein microsequencing. Equally important is the identification of signals in proteins that are presumably recognized by the different E3 enzymes. To date, the only well-characterized recognition signal is the "N-end rule" (see Chapter 8), which, however, appears to have very limited cellular functions. The "destruction box" signal in mitotic cyclins and the cyclosome/APC particle that acts on these proteins have been identified (see above), but it is not known whether a specific subunit of the cyclosome is responsible for the binding of the destruction box region. It also remains to be seen whether the phosphorylation of some proteins that is required for their ubiquitinylation (see above) is directly recognized by some E3 enzymes, or indirectly, such as by changing protein structure in a way that exposes another recognition determinant. Indirect interaction mediated by auxiliary proteins, such as in the case of the p53/E6/E6-AP complex, may also play a more widespread role. In all cases, the identification of a degradation signal and of the E3-like or auxiliary proteins that recognize this signal have to be done in parallel, to provide a definite answer to the selectivity problem. The mode of action of the 26 S proteasome complex, the roles of ATP in its action and assembly are formidable problems for the future, as is its action on some nonubiquitinylated proteins, as in the case of the ornithine decarboxylase–antizyme complex (Murakami *et al.*, 1992; see Chapter

14). The cellular roles of different ubiquitin-C-terminal hydrolases (Ubps) and their possible selectivity for certain types of proteins or multiubiquitin chains, remain to be elucidated.

Based on developments of the past few years, it may be predicted that the ubiquitin field will continue to expand rapidly, mainly by three types of approaches. One is the realization that the ubiquitin-mediated degradation of key regulatory proteins has important roles in temporally or spatially controlled processes, such as the cell cycle, circadian rhythms, development, and differentiation. In such cases, the challenge will be to find out how these regulatory proteins are degraded in a specific and controlled manner, as seen in the case of the cell cycle stage-specific degradation of cyclin B. A second way by which investigators from other fields may enter the ubiquitin system is by cloning a gene having an interesting function and then finding, by sequence homology, that the gene belongs to a certain family of enzymes of the ubiquitin system. Thus were, for example, the DNA repair gene, *RAD6* (Jentsch *et al.*, 1987) and *Drosophila bendless* (Muralidhar and Thomas, 1993), involved in *Drosophila* central nervous system function, identified as E2 enzymes. The product of *Drosophila hyperplastic discs* is an E3 enzyme of the E6-AP family (Mansfield *et al.*, 1994; Huibregtse *et al.*, 1995), whereas those of *fat facets* gene (Huang *et al.*, 1995), a growth-inducible gene (Zhu *et al.*, 1996), and a protein associated with the yeast protein SIR4 involved in the silencing of transcription (Moazed and Johnson, 1996), are ubiquitin-C-terminal hydrolases. In such cases, the challenge is to discover the cellular substrates of these gene products, and how the ubiquitinylation of these substrates is involved in their degradation. A third approach is the use of specific inhibitors, which implicated the involvement of the ubiquitin system in basic processes such as antigen presentation (see Chapter 7) and apoptosis (Grimm *et al.*, 1996; Sadoul *et al.*, 1996).

Finally, it should be noted that another important area for future research is the elucidation of some ubiquitinylation processes that do not lead to degradation by the 26 S proteasome. The function of the monoubiquitinylated derivative of histone 2A, the first protein shown to be a ubiquitin conjugate (Goldknopf and Busch, 1977), still remains a mystery, as the histone is not degraded. Another interesting process is the polyubiquitinylation of many cell surface receptors in response to ligand binding (see Chapter 13). In several cases, such as with the Ste2p receptor of the mating pheromone α-factor in *S. cerevisiae*, it has been conclusively shown that polyubiquitinylation leads to endocytosis and thus to degradation in the vacuole (Hicke and Riezman, 1996). It is noteworthy that mechanisms evolved to target some ubiquitinylated proteins to degradation either by the 26 S protease or by the lysosome/vacuole; it remains to be elucidated how polyubiquitinylation of receptors leads to their endocytosis. In another case, ubiquitin ligation was shown to be required for the phosphorylation of IκBα, which, in turn, is required for the ubiquitin-mediated degradation of this protein

(Chen *et al.*, 1996). The target of this ubiquitinylation reaction (a specific protein kinase?) and the mechanism of its activation by ubiquitinylation, remain to be identified.

6. CONCLUDING REMARKS

The main lesson that can be learned from this brief discussion of the past and present developments in the ubiquitin field is the importance of both biochemistry and molecular genetics for significant progress in biological research. The ubiquitin-mediated proteolytic system was discovered by a classical biochemical approach of fractionation of a cell-free system, purification and characterization of enzyme components, and reconstitution of the system from purified components. These findings were subsequently confirmed and greatly extended by molecular genetics. Without the biochemistry, the mode of action of the products of the multitude of genes of the components of the ubiquitin system, identified by cloning or by genome projects, would still be a complete mystery. On the other hand, molecular genetics is needed to uncover the great variety of basic cellular functions of the ubiquitin system. It has been stated in some of the many mini-reviews on this subject that until recently the ubiquitin proteolytic system was thought mainly to be a "garbage disposal" for the removal of damaged or abnormal proteins. This statement is certainly not true for those of us who have always appreciated the importance of selective protein degradation in the control of the levels of specific cellular proteins. It is true, however, that without molecular genetics, we would know much less about processes in which ubiquitin-mediated protein degradation has essential functions. It may thus be hoped that the combined biochemical–molecular genetic approach may lead to further significant advances in the ubiquitin field in the near future.

Acknowledgments. I wish to thank all present and past members of my laboratory who contributed to the work described in this chapter. Recent work from my laboratory was supported by grants from the United States–Israel Binational Science Foundation, the Israel Science Foundation, and the Rappaport Institute.

7. REFERENCES

Amon, A., Irniger, S., and Nasmyth, K., 1994, Closing the cell cycle circle in yeast: G2 cyclin proteolysis initiated at mitosis persists until the activation of G1 cyclins in the next cycle, *Cell* **77:**1037–1050.

Aristarkhov, A., Eytan, E., Moghe, A., Admon, A., Hershko, A., and Ruderman, J. V., 1996, E2-C, a cyclin-selective ubiquitin-carrier protein required for the destruction of mitotic cyclins, *Proc. Natl. Acad. Sci. USA* **93:**4294–4299.

Armon, T., Ganoth, D., and Hershko, A., 1990, Assembly of the 26S complex that degrades proteins ligated to ubiquitin is accompanied by the formation of ATPase activity, *J. Biol. Chem.* **265:** 20723–20726.

Bachmair, A., Finley, D., and Varshavsky, A., In vivo half-life of a protein is a function of its amino-terminal residue, *Science* **234:**179–186.

Bartel, B., Wunning, I., and Varshavsky, A., 1990, The recognition component of the N-end rule pathway, *EMBO J.* **9:**3179–3189.

Brandeis, M., and Hunt, T., 1996, The proteolysis of mitotic cyclins in mammalian cells persists from the end of mitosis until the onset of S phase, *EMBO J.* **15:**5280–5289.

Chau, V., Tobias, J. W., Bachmair, A., Mariott, D., Ecker, D. J., Gonda, D. K., and Varshavsky, A., 1989, A multiubiquitin chain is confined to a specific lysine in a targeted short-lived protein, *Science* **243:**1576–1583.

Chen, Z. J., Parent, L., and Maniatis, T., 1996, Site-specific phosphorylation of IκBα by a novel ubiquitination-dependent protein kinase activity, *Cell* **84:**853–862.

Ciechanover, A., Hod, Y., and Hershko, A., 1978, A heat-stable polypeptide component of an ATP-dependent proteolytic system from reticulocytes, *Biochem. Biophys. Res. Commun.* **81:**1100–1105.

Ciechanover, A., Heller, H., Elias, S., Haas, A. L., and Hershko, A., 1980, ATP-dependent conjugation of reticulocyte proteins with the polypeptide required for protein degradation, *Proc. Natl. Acad. Sci. USA* **77:**1365–1368.

Ciechanover, A., Heller, H., Katz-Etzion, R., and Hershko, A., 1981, Activation of the heat-stable polypeptide of the ATP-dependent proteolytic system, *Proc. Natl. Acad. Sci. USA* **78:**761–765.

Clurman, B. E., Sheaff, R. J., Thress, K., Groudine, M., and Roberts, J. M., 1996, Turnover of cyclin E by the ubiquitin-proteasome pathway is regulated by cdk2 binding and cyclin phosphorylation, *Genes Dev.* **10:**1979–1990.

Coux, O., Tanaka, K., and Goldberg, A. L., 1996, Structure and functions of the 20S and 26S proteasomes, *Annu. Rev. Biochem.* **65:**801–847.

Deveraux, Q., Ustrell, V., Pickart, C., and Rechsteiner, M., 1994, A 26S protease subunit that binds ubiquitin conjugates, *J. Biol. Chem.* **269:**7059–7061.

Dohmen, J., Madura, K., Bartel, B., and Varshavsky, A., 1991, The N-end rule is mediated by the UBC2(RAD6) ubiquitin-conjugating enzyme, *Proc. Natl. Acad. Sci. USA* **88:**7351–7355.

Driscoll, J., and Goldberg, A. L., 1990, The proteasome (multicatalytic protease) is a component of the 1500-kDa complex which degrades proteins conjugated to ubiquitin, *J. Biol. Chem.* **265:**4789–4792.

Etlinger, J. D., and Goldberg, A. L., 1977, A soluble ATP-dependent proteolytic system responsible for the degradation of abnormal proteins in reticulocytes, *Proc. Natl. Acad. Sci. USA* **74:**54–58.

Eytan, E.,Ganoth, D., Armon, T., and Hershko, A., 1989, ATP-dependent incorporation of 20S protease into the 26S complex that degrades proteins conjugated to ubiquitin, *Proc. Natl. Acad. Sci. USA* **86:**7751–7755.

Eytan, E., Armon, T., Heller, H., Beck, S., and Hershko, A., 1993, Ubiquitin-C-terminal hydrolase activity associated with the 26S protease complex, *J. Biol. Chem.* **268:**4668–4674.

Ganoth, D., Leshinsky, E., Eytan, E., and Hershko, A., 1988, A multicomponent system that degrades proteins conjugated to ubiquitin. Resolution of factors and evidence for ATP-dependent complex formation, *J. Biol. Chem.* **263:**12412–12419.

Glotzer, M., Murray, A. W., and Kirschner, M. W., 1991, Cyclin is degraded by the ubiquitin pathway, *Nature* **349:**132–138.

Goldknopf, I. L., and Busch, H., 1977, Isopeptide linkage between nonhistone and histone 2A polypeptides of chromosomal conjugate protein A24, *Proc. Natl. Acad. Sci. USA* **74:**864–868.

Goldstein, G., Steed, M., Hammerling, U., Boyse, E. A., Schlesinger, D. H., and Niall, H. D., 1975, Isolation of a polypeptide that has lymphocyte-differentiating properties and is probably represented universally in living cells, *Proc. Natl. Acad. Sci. USA* **72:**11–15.

Grimm, L. M., Goldberg, A. L., Poirier, L. M., Schwartz, L. M., and Osborne, B. A., 1996, Proteasomes play an essential role in thymocyte apoptosis, *EMBO J.* **15:**3835–3844.

Hadari, T., Warms, J. V. B., Rose, I. A., and Hershko, A., 1992, A ubiquitin-C-terminal isopeptidase that acts on polyubiquitin chains. Role in protein degradation, *J. Biol. Chem.* **267:**719–727.

Hershko, A., 1996, Lessons from the discovery of the ubiquitin system, *Trends Biochem. Sci.* **21:** 445–449.

Hershko, A., and Heller, H., 1985, Occurrence of a polyubiquitin structure in ubiquitin-protein conjugates, *Biochem. Biophys. Res. Commun.* **128:**1079–1086.

Hershko, A., and Tomkins, G. M., 1971, Studies on the degradation of tyrosine aminotransferase in hepatoma cells in culture. Influence of the composition of the medium and adenosine triphosphate dependence, *J. Biol. Chem.* **246:**710–714.

Hershko, A.,Ciechanover, A., Heller, H., Haas, A. L., and Rose, I. A., 1980, Proposed role of ATP in protein breakdown: Conjugation of proteins with multiple chains of the polypeptide of ATP-dependent proteolysis, *Proc. Natl. Acad. Sci. USA* **77:**1783–1786.

Hershko, A., Heller, H., Elias, S., and Ciechanover, A.,1983, Components of ubiquitin-protein ligase system: Resolution, affinity purification and role in protein breakdown, *J. Biol. Chem.* **258:**8206–8214.

Hershko, A., Heller, H., Eytan, E., and Reiss, Y., 1986, The protein binding site of the ubiquitin-protein ligase system, *J. Biol. Chem.* **261:**11992–11999.

Hershko, A., Ganoth, D., Pehrson, J., Palazzo, R. E., and Cohen, L. H., 1991, Methylated ubiquitin inhibits cyclin degradation in clam oocyte extracts, *J. Biol. Chem.* **266:**16376–16379.

Hershko, A., Ganoth, D., Sudakin, V., Dahan, A., Cohen, L. H., Luca, F. C., Ruderman, J. V., and Eytan, E., 1994, Components of a system that ligates cyclin to ubiquitin and their regulation by protein kinase cdc2, *J. Biol. Chem.* **269:**4940–4946.

Hicke, L., and Riezman, H., 1996, Ubiquitination of yeast plasma membrane receptor signals its ligand-stimulated endocytosis, *Cell* **84:**277–287.

Hough, R., Pratt, G., and Rechsteiner, M., 1986, Ubiquitin-lysozyme conjugates. Identification and characterization of an ATP-dependent protease from rabbit reticulocyte lysates, *J. Biol. Chem.* **261:**2400–2408.

Huang, Y., Baker, R. T., and Fischer-Vize, J. A., 1995, Control of cell fate by a deubiquitinating enzyme encoded by the *fat facets* gene, *Science* **270:**1828–1831.

Huibregtse, J. M., Scheffner, M., Beaudenon, S., and Howley, P. M., 1995, A family of proteins structurally and functionally related to the E6-AP ubiquitin-protein ligase, *Proc. Natl. Acad. Sci. USA* **92:**2563–2567.

Jentsch, S., McGrath, J. B., and Varshavsky, A., 1987, The yeast DNA repair gene *RAD6* encodes a ubiquitin-conjugating enzyme, *Nature* **329:**131–134.

King, R. W., Peters, J.-M., Tugendreich, S., Rolfe, M., Hieter, P., and Kirschner, M. W., 1995, A 20S complex containing *CDC27* and *CDC16* catalyzes the mitosis-specific conjugation of ubiquitin to cyclin B, *Cell* **81:**279–288.

Lahav-Baratz, S., Sudakin, V., Ruderman, J. V., and Hershko, A., 1995, Reversible phosphorylation controls the activity of cyclosome-associated cyclin-ubiquitin ligase, *Proc. Natl. Acad. Sci. USA* **92:**9303–9307.

Lanker, S., Valdiviseo, M. H., and Wittenberg, C., 1996, Rapid degradation of the G1 cyclin Cln2 induced by CDK-dependent phosphorylation, *Science* **271:**1597–1601.

Lowe, J., Stock, D., Jap, B., Zwickl, P., Baumeister, W., and Huber, R., 1995, Crystal structure of the 20S proteasome from the archaeon *T. acidophyllum* at 3.4A resolution, *Science* **268:**533–539.

Luca, F. C., and Ruderman, J. V., 1989, Control of programmed cyclin destruction in a cell-free system, *J. Cell Biol.* **109:**1895–1909.

Mansfield, E., Hersperger, E., Biggs, J., and Shearn, A., 1994, Genetic and molecular analysis of *hyperplastic disks*, a gene whose product is required for the regulation of cell proliferation in *Drosophila melanogaster* imaginal disks and germ cells, *Dev. Biol.* **165:**507–526.

McKinney, J. D., and Cross, F. R., 1995, *FAR1* and the G1 phase specificity of cell cycle arrest by mating factor in *Saccharomyces cerevisiae*, *Mol. Cell. Biol.* **15:**2509–2516.

Moazed, D., and Johnson, A. D., 1996, A deubiquitinating enzyme interacts with SIR4 and regulates silencing in *S. cerevisiae*, *Cell* **86:**667–677.

Murakami, Y., Matsufuji, S., Kameji, T., Hayashi, S.-I., Igarashi, K., Tamura, T., Tanaka, K., and Ichihara, A., 1992, Ornithine decarboxylase is degraded by the 26S proteasome without ubiquitination, *Nature* **360:**597–599.

Muralidhar, M. G., and Thomas, J. B., 1993, The Drosophila *bendless* gene encodes a neural protein related to ubiquitin-conjugating enzymes, *Neuron* **11:**253–266.

Nefsky, B., and Beach, D., 1996, Pub1 acts as an E6-AP-like protein ubiquitin ligase in the degradation of cdc25, *EMBO J.* **15:**1301–1312.

Nishisawa, M., Furuno, M., Okazaki, K., Tanaka, H., Ogawa, Y., and Sagata, N., 1993, Degradation of Mos by the N-terminal prolife (Pro^2)-dependent ubiquitin pathway on fertilization of Xenopus eggs: Possible significance of natural selection for Pro^2 in Mos, *EMBO J.* **12:**4021–4027.

Papa, F. R., and Hochstrasser, M., 1993, The yeast *DOA4* gene encodes a deubiquitinating enzyme related to the product of the human *tre-2* oncogene, *Nature* **366:**313–319.

Peters, J.-M., 1994, Proteasomes: Protein degradation machines of the cell, *Trends Biochem. Sci.* **19:** 377–382.

Reiss, Y., and Hershko, A., 1990, Affinity purification of ubiquitin-protein ligase on immobilized protein substrates. Evidence for the existence of separate NH_2-terminal binding sites on a single enzyme, *J. Biol. Chem.* **265:**3685–3690.

Reiss, Y., Kaim, D., and Hershko, A., 1988, Specificity of binding of NH_2-terminal residue of proteins to ubiquitin-protein ligase. Use of amino acid derivatives to characterize specific binding sites, *J. Biol. Chem.* **263:**2693–2698.

Reiss, Y., Heller, H., and Hershko, A., 1989, Binding sites of ubiquitin-protein ligase. Binding of ubiquitin-protein conjugates and of ubiquitin-carrier protein, *J. Biol. Chem.* **264:**10378–10383.

Rogers, S., Wells, R., and Rechsteiner, M., 1986, Amino acid sequences common to rapidly degraded proteins: The PEST hypothesis, *Science* **234:**364–368.

Sadoul, R., Fernandez, P.-A., Quiquerez, A.-L., Martinou, I., Maki, M., Schroter, M., Becherer, J. D., Irmler, M., Tschopp, J., and Martinou, J.-C., 1996, Involvement of the proteasome in the programmed cell death of NGF-deprived sympathetic neurons, *EMBO J.* **15:**3845–3852.

Scheffner, M., Huibregtse, J. M., Vierstra, R. D., and Howley, P. M., 1993, The HPV-16 E6 and E6-AP complex functions as a ubiquitin-protein ligase in the ubiquitination of p53, *Cell* **75:**495–505.

Schimke, R. T., and Doyle, D., 1970, Control of enzyme levels in animal tissues, *Annu. Rev. Biochem.* **39:**929–976.

Schneider, B. L., Yang, Q-H., and Futcher, A. B., 1996, Linkage of replication to Start by the cdk inhibitor Sic1, *Science* **272:**560–562.

Schoenheimer, R., 1942, *The Dynamic State of Body Constituents*, Harvard University Press, Cambridge, MA.

Schwob, E., Bohm, T., Mendenhall, M. D., and Nasmyth, K., 1994, The B-type cyclin kinase inhibitor $p40^{SIC1}$ controls the G1 to S transition in S. cerevisiae, *Cell* **79:**233–244.

Simpson, M. V., 1953, The release of labeled amino acids from the proteins of rat liver slices, *J. Biol. Chem.* **201:**143–154.

Sudakin, V., Ganoth, D., Dahan, A., Heller, H., Hershko, J., Luca, F. C., Ruderman, J. V., and Hershko, A., 1995, The cyclosome, a large complex containing cyclin-selective ubiquitin ligase activity, targets cyclins for destruction at the end of mitosis, *Mol. Biol. Cell* **6:**185–198.

Van Nocker, S., Sadis, S., Rubin, D. M., Glickman, M., Fu, H., Coux, O., Wefes, J., Finley, D., and Vierstra, R. D., 1996, The multiubiquitin-binding protein McB1 is a component of the 26S proteasome in *Saccharomyces cerevisiae* and plays a nonessential, substrate-specific role in protein turnover, *Mol. Cell. Biol.* **16:**6020–6028.

Wilkinson, K. D., Urban, M. K., and Haas, A. L., 1980, Ubiquitin is the ATP-dependent proteolysis factor 1 of rabbit reticulocytes, *J. Biol. Chem.* **255:**7529–7532.

Willems, A. R., Lanker, S., Patton, E. E., Craig, K. L., Nason, T. T., Mathias, N., Kobayashi, R., Wittenberg, C., and Tyers, M., 1996, Cdc53 targets phosphorylated Gl cyclins for degradation by the ubiquitin pathway, *Cell* **86:**453–463.

Won, K.-A., and Reed, S. I., 1996, Activation of cyclin E/CDK2 is coupled to site-specific autophosphorylation and ubiquitin-dependent degradation, *EMBO J.* **15:**4182–4193.

Yaglom, J., Liskens, M. H. K., Sadis, S., Rubin, D. M., Futcher, B., and Finley, D., 1995, $p34^{cdc28}$-mediated control of Cln3 cyclin degradation, *Mol. Cell. Biol.* **15:**731–741.

Yu, H., King, R. W., Peters, J.-M., and Kirschner, M. W., 1996, Identification of a novel ubiquitin-conjugating enzyme involved in mitotic cyclin degradation, *Curr. Biol.* **6:**455–466.

Zhu, Y., Carroll, M., Papa, F. R., Hochstrasser, M., and D'Andrea, A.D., 1996, Dub-1, a deubiquitinating enzyme with growth-suppressing activity, *Proc. Natl. Acad. Sci. USA* **93:**3275–3279.

CHAPTER 2

Polyubiquitin Chains

Cecile M. Pickart

1. INTRODUCTION

It is now over 15 years since Hershko, Rose, and co-workers proposed that ubiquitin functions as a covalent signal for proteolysis (Hershko *et al.*, 1980). In the intervening period, with the recognition that the step of ubiquitin attachment makes a major contribution to selectivity in degradation, much effort has been directed toward characterizing substrate-based uniquitination signals and the conjugating enzymes that recognize them. However, it has recently become clear that much remains to be understood about the covalent proteolytic signal itself. We now know that the targeting signal most often comprises a polymeric ubiquitin chain, and that the 26 S proteasome harbors specific components that recognize this chain with high affinity. However, the molecular aspects of chain recognition are only beginning to be understood, and the proteasomal components that bind polyubiquitin chains remain largely uncharacterized.

This chapter will review the structure, function, recognition, and metabolism of polymeric ubiquitin chains. It begins, in Section 2, with a review of the genetics and biochemistry of ubiquitin itself. This is followed in Section 3 by a detailed review of the function, properties, and metabolism of chains linked by K48–G76 isopeptide bonds. Much less is known about these issues for chains assembled

Cecile M. Pickart • Department of Biochemistry, Johns Hopkins University, Baltimore, Maryland 21205.

Ubiquitin and the Biology of the Cell, edited by Peters *et al.* Plenum Press, New York, 1998.

through other Lys residues; the current state of knowledge is summarized in Section 4. Section 5 presents major conclusions and outstanding questions.

2. UBIQUITIN

A meaningful discussion of polyubiquitin chains requires a brief review of some of the structural, biological, and chemical properties of ubiquitin (Section 2.1; see also Wilkinson, 1988). In addition, ubiquitin genes and ubiquitin-related proteins are briefly discussed (Sections 2.2 and 2.3). The properties of mutant (Section 2.4) and chemically modified ubiquitins (Section 2.5) are then summarized, followed by a brief discussion of methods for detection of ubiquitination (Section 2.6).

2.1. Ubiquitin Conservation and Structure

There are only three amino acid substitutions, all conservative, in mammalian versus yeast ubiquitin (P19S, E24D, A28S). Ubiquitin is thus the most conserved protein known in eukaryotes (Özkaynak *et al.*, 1984). Although it is apparently absent from most prokaryotes, including *E. coli*, ubiquitin is present in the cyanobacterium *Anabaena variabilis* (Durner and Boger, 1995), and may be present in archaebacteria (Wolf *et al.*, 1993). In *Anabaena*, ubiquitin can conjugate to endogenous proteins, including dinitrogenase reductase, in extracts prepared from differentiated nitrogen-fixing heterocysts (Durner and Boger, 1995). It is thus likely ubiquitin functions in proteolysis in blue-green algae.

Ubiquitin's extraordinary conservation suggests that the entire molecule has been under intense selective pressure. The molecular details of ubiquitin's proteolytic signaling function, including the sequences of many of the enzymes that carry it out, are highly conserved from yeast to mammals. However, ubiquitin also has nonproteolytic functions: At the least, it is a chaperone in ribosome biogenesis (Finley *et al.*, 1989), a recognition element at the cell surface (Siegelman *et al.*, 1986), and a targeting element within the cell (Hicke and Riezman, 1996). Ubiquitin folds rapidly, a property that may be relevant to its chaperone function; indeed, ubiquitin has become a model for the study of protein folding and dynamics (e.g., Khorasanizadeh *et al.*, 1993; Wand *et al.*, 1996). It seems likely that simultaneous selection for these multiple functions, and perhaps other, unknown functions, has brought about ubiquitin's extreme sequence conservation.

The 1.8-Å X-ray structure of ubiquitin shows a compact, globular protein with a protruding C-terminus (Fig. 1; Vijay-Kumar *et al.*, 1987a). All known ubiquitin functions involve covalent bond formation with the C-terminal carboxyl group. The molecular structure is thus in accord with the requirement that G76 be available for covalent interaction with Lys ε-amino groups of other proteins. The

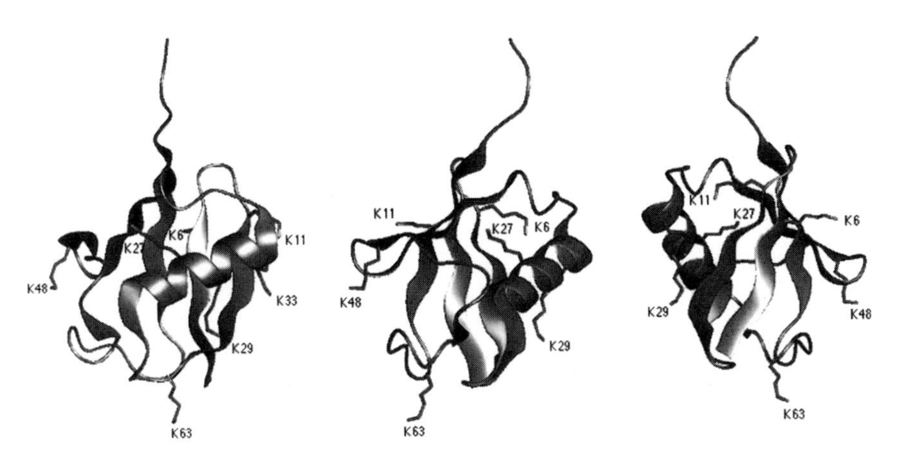

Figure 1. Ubiquitin structure. Helices and β-strands are shown as ribbons. The side chains of the seven Lys residues are indicated. The center and right images were obtained by two successive rightward rotations of the left image about its *y*-axis.

solution structure of ubiquitin confirms that G75 and G76 are conformationally flexible (Di Stefano and Wand, 1987; Weber *et al.*, 1987).

Ubiquitin has a mixed α/β structure, with five β-strands and a 3.5-turn α-helix that lies across the β-sheet (Fig. 1; Vijay-Kumar *et al.*, 1987a). The molecule has a pronounced hydrophobic core. Nearly 90% of the polypeptide chain is involved in hydrogen-bonding interactions that maintain secondary structure. Many turns of the polypeptide chain are needed for a small protein to achieve a globular conformation, and ubiquitin's seven reverse turns and a short 3_{10} helix represent about a third of the total secondary structure. The hydrophobic core, and the extensive hydrogen bonding, provide a structural basis for ubiquitin's unusually high chemical and thermal stability.

All of the acidic and basic side chains are on the surface of the ubiquitin molecule. The seven Lys residues, at least five of which can serve as sites of ubiquitin–ubiquitin conjugation (Section 2.4), are highlighted in Fig. 1. The side chain of K63 is near M1, at the opposite end of the molecule from G76. K6 and K11 are located at the distal and proximal ends of the first and second β-strands, respectively; K27, K29, and K33 project from the α-helix, on the opposite side of the molecule from K6 and K11. K48, which is the major site of chain formation (Section 3), lies in a solvent-exposed loop, with its side chain near the side chain of Y59.

In agreement with the results of early chemical acylation studies (Jabusch and Deutsch, 1985), the crystal structure shows that the side chains of K6, K33, and K63 are solvent-exposed, while the side chains of K11, K27, K29, and K48 participate in hydrogen bonds or salt bridges (Vijay-Kumar *et al.*, 1987a). However, as K11, K29, and K48 are sites of ubiquitin–ubiquitin conjugation, interactions involving these side chains cannot be strong enough to block their utilization by conjugating enzymes. Nor do interactions involving K48 contribute strongly to stability, based on the unchanged folding of the ubiquitin units in K48-linked di- and tetraubiquitin (Cook *et al.*, 1992, 1994). K27 is the least exposed Lys residue, and it has not been implicated in ubiquitin–ubiquitin conjugation. The side chain of K27 forms a salt bridge with D52. Both of these residues are strictly conserved in all known ubiquitin-related proteins (see below).

Only three hydrophobic side chains—L8, I44, and V70—are exposed on the surface of the ubiquitin molecule. These surface residues are juxtaposed, forming a patch that is functionally significant for proteolytic targeting (Sections 2.4 and 3.2.1). The acidic and basic side chains on the ubiquitin surface are largely confined to distinct faces of the molecule (Vijay Kumar *et al.*, 1987a; Wilkinson, 1988). The functional significance of this partitioning remains unclear for the acidic residues, but a subset of the basic residues is important for ubiquitin recognition by the activating enzyme, E1 (Section 2.4).

2.2. Ubiquitin Genes

Ubiquitin genes are unique, in that all of them specify fusions between ubiquitin and other polypeptides, including other ubiquitins. Thus, proteolytic processing is necessary to produce functional ubiquitin, a property that is shared by a subset of ubiquitin-related proteins (Section 2.3). There are three types of ubiquitin genes (Fig. 2). Their structures, like the sequences of the encoded proteins, are highly conserved.

2.2.1. Polyubiquitin Genes

The first ubiquitin genes to be characterized encoded tandemly repeated copies of the ubiquitin coding sequence (Dworkin-Rastl *et al.*, 1984; Özkaynak *et*

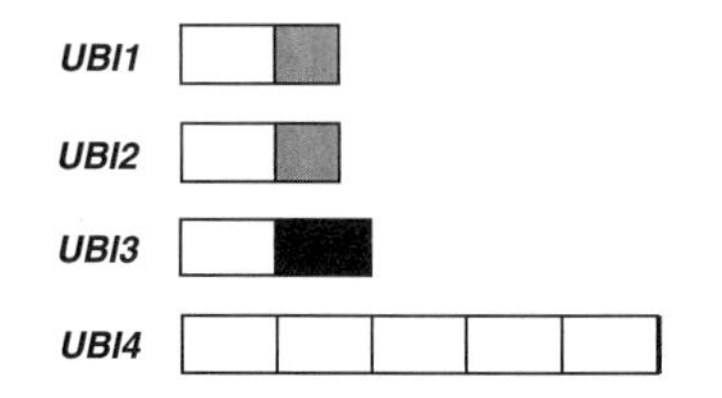

Figure 2. Ubiquitin genes. The specific structures shown are from *S. cerevisiae* (Özkaynak *et al.*, 1987). The white rectangles are ubiquitin coding units; the shaded rectangles encode the respective tails. The ubiquitin-coding portions of *UBI1* and *UBI2* are interrupted by different exons.

al., 1984). In *Saccharomoyces cerevisiae*, the polyubiquitin gene *UBI4* encodes five spacerless copies of the coding sequence; the C-terminal copy is extended by one amino acid (Özkaynak *et al.*, 1987). Polyubiquitin genes are universally distributed, although the number of repeats and the identity of the extra amino acid vary widely. In general, these genes provide ubiquitin under conditions of stress. Induced expression of the *UBI4* gene is essential for the viability of stressed yeast cells; stress conditions result in a strong increase in fractional ubiquitin conjugation, which in the absence of *UBI4* induction creates a deficiency in the level of monomeric ubiquitin (Finley *et al.*, 1987). Synthesis as a polyprotein increases the efficiency of ubiquitin production, as processing of the linear polyubiquitin protein, and removal of the extra residue from the C-terminal ubiquitin unit, are essentially cotranslational. Because ubiquitin activation requires G76 as the C-terminal residue (Wilkinson and Audhya, 1981), the extra residue would prevent activation and conjugation of linear polyubiquitin in the unlikely event of a failure in ubiquitin processing mechanisms (Özkaynak *et al.*, 1984, 1987).

2.2.2. Ubiquitin–CEP Genes

The yeast *UBI4* gene is dispensable in nonstressed cells (Finley *et al.*, 1987). Other genes, known as *UBI1* through *UBI3*, supply most of the ubiquitin for vegetative growth. These genes specify an alternative class of fusions between ubiquitin (at the N-terminus) and small carboxyl extension proteins (CEPs) (Fig. 2). The yeast *UBI1* and *UBI2* genes encode fusions between ubiquitin and the same 52-residue polypeptide, whereas the *UBI3*-encoded protein has a different, 76-residue extension (Özkaynak *et al.*, 1987). Both extensions are basic, and each contains a zinc-binding motif. The ubiquitin–CEP fusions are processed, as suggested initially by the finding that the *UBI1* through *UBI3* genes provide adequate ubiquitin for vegetative growth. The mature extension proteins are localized in ribosomes (Finley *et al.*, 1989; Redman and Rechsteiner, 1989). The longer extension is protein S27a of the 40 S subunit and the shorter one is protein L40 of the 80 S subunit (Finley *et al.*, 1989; Redman and Rechsteiner, 1989; Chan *et al.*, 1995).

Despite the short lifetime of the primary ubiquitin–CEP translation products, slow growth and ribosomal derangement result when the extension peptides are expressed from single-copy genes independently of fused ubiquitin (Finley *et al.*, 1989). Thus, expression of these ribosomal proteins as ubiquitin fusions has functional relevance, as expected from the extreme conservation of the gene structure. The fused ubiquitin appears to function as a chaperone, leading to increased levels of the extension peptides; it is possible that the rapid folding of the initial ubiquitin module prevents degradation of the extension peptide that might otherwise begin while it was still being synthesized. Consistent with this model, an N-terminal ubiquitin module can facilitate the expression of diverse heterologous proteins, in both prokaryotic and eukaryotic cells (see Baker *et al.*, 1994, and references therein).

2.3. Ubiquitin-Related Genes/Proteins

The number of genes known to encode ubiquitin-related proteins continues to grow. The ubiquitin homology (UbH) family is structurally and functionally diverse (Table IA). Known family members can be divided into two classes, processed and unprocessed, based on the sequence at the C-terminus of the UbH domain (Table I). The presence of an -LRGG sequence, followed by any residue except Pro, indicates a high likelihood that cleavage by ubiquitin processing proteases will produce a molecule with a ubiquitin-like C. terminus, and with potential to undergo conjugation. In the absence of these features, processing is unlikely (Bachmair *et al.*, 1986; Butt *et al.*, 1988; Johnson *et al.*, 1992, 1995). Although all of the UbH proteins are homologous to ubiquitin, the presence of substantial sequence divergence suggests that they have evolved either to retain only a subset of ubiquitin functions, or in some cases to perform species-specific functions (D'Cunha *et al.*, 1996; Nakamura *et al.*, 1996). The brief discussion below is limited to UbH proteins for which functional information is available. There are many other UbH proteins in databases; a few of these are included in Table I.

2.3.1. Processed UbH Proteins

The presence in these proteins of a C-terminal GG sequence, and the existence of a family of E1 proteins (McGrath *et al.*, 1991; Dohmen *et al.*, 1995), raise the possibility that there may be novel pathways for activation and ligation of some processed UbH proteins (Section 2.3.1b).

2.3.1a. Viral Ubiquitin. Although the baculovirus UbH protein, known as vUb, is 79% identical to ubiquitin (Guarino, 1990; Table I), the two known baculovirus ubiquitins are more similar to each other than to animal ubiquitin (see Reilly and Guarino, 1996). This suggests that vUb has evolved to perform a specific function(s), but the nature of this function is unclear. Monomeric vUb is localized on the inner face of the viral envelope by virtue of covalent linkage to a phospholipid moiety (Guarino *et al.*, 1995). The vUb gene is expressed late in infection; its disruption causes a 5- to 10-fold reduction in the yield of viral particles and the titer of infectious virus (Reilly and Guarino, 1996). These results, and the localization of vUb in the viral envelope, are consistent with a role for vUb in virus assembly or budding.

However, it is unlikely that all of the functions of vUb depend on its presence in the viral envelope, as 80% of the modified ubiquitin in this location is host cell-derived (Guarino *et al.*, 1995), and there is a normal level of lipidated host ubiquitin in vUb-deleted virus (Reilly and Guarino, 1996). One possibility is that vUb also functions in the ubiquitination of specific viral or host cell proteins. vUb can partially inhibit proteolysis in reticulocyte lysate (Haas *et al.*, 1996). This effect might help to protect late viral proteins against degradation by the host cell ubiquitin pathway.

2.3.1b. UCRP. In mammals, the synthesis of a 15-kDa ubiquitin cross-reactive protein (UCRP), the product of the *ISG15* gene, is strongly induced by β-interferon (Haas *et al.*, 1987). UCRP was first identified by its cross-reaction with antiubiquitin antibodies (Haas *et al.*, 1987; Loeb and Haas, 1992). It consists of two UbH domains, which are 40% (N-terminal domain) and 53% similar (C-terminal domain) to ubiquitin. The ubiquitin C-terminus is conserved in the second, but not the first, UbH domain (Haas *et al.*, 1987), and *in vivo* processing at the second UbH C-terminus generates a linear diubiquitin analogue (Knight *et al.*, 1988). A fraction of the mature protein is secreted from monocytes and lymphocytes, where it can act as a species-specific immunoregulatory cytokine (D'Cunha *et al.*, 1996). Intracellularly, UCRP covalently conjugates to a subset of proteins that cofractionate with intermediate filaments (Loeb and Haas, 1992, 1994). The specific targets and consequences of UCRP conjugation are unknown. This reaction requires ATP and the presence of the C-terminal GG dipeptide, but does not seem to involve the El enzyme that mediates ubiquitin activation (Narasimhan *et al.*, 1996). It is thus likely that cells possess a distinct enzymology dedicated to UCRP–protein ligation.

2.3.1c. Fau/S30. The gene encoding protein S30 of the mammalian 40 S ribosomal subunit, also known as the *fau* gene, encodes a 133-residue protein with an N-terminal UbH domain that is 58% similar to ubiquitin (Kas *et al.*, 1992; Olvera and Wool, 1993; Michiels *et al.*, 1993). In yeast and plants, S30 is expressed in an unfused format, indicating that the linkage of S30 to a UbH domain occurred late in evolution (Baker *et al.*, 1996). It has yet to be determined whether the UbH domain serves a chaperone function in fau biosynthesis (Section 2.2.2). Mature S30 consists of residues 75 through 133 of the primary translation product, indicating that the UbH domain is released by processing, which can be seen following *in vitro* translation of *fau* mRNA in reticulocyte lysate (Olvera and Wool, 1993). The processed UbH domain is associated, in a structurally undefined manner, with an extracellular lymphokine known as monoclonal nonspecific suppressor factor (Nakamura *et al.*, 1995, 1996). The UbH domain of this factor is sufficient for its species-specific activity in inhibition of B-cell proliferation (Nakamura *et al.*, 1996). Interestingly, although ubiquitin does not inhibit proliferation, it blocks inhibition mediated by the fau ubH domain. Therefore, at least some of the divergent residues in the fau UbH domain are significant for its

Table I
UbH Domain Proteins

A. Properties of UbH domain proteins

Name	Access. No.	Proc.[a]	Homology[b] Iden.	Homology[b] Sim.	UbH[c] pI	Function	References
vUb	M30305	+	79	89	6.6	Baculovirus budding/infection	Guarino *et al.* (1995)
Nedd8	D10918	(+)	59	70	6.6	Unknown	Kumar *et al.* (1993)
fau	X62671	+	38	58	4.3	Extension peptide: S30 UbH domain: extracellular lymphokine	Olvera and Wool (1993), Nakamura *et al.* (1995)
UCRP		+	31	53	5.2	15 kDa; intracellular conjugation; extracellular cytokine	Haas *et al.* (1987), D'Cunha *et al.* (1996)
An1a	L08474	−	51	63	5.0	77 kDa; function unknown; An1b closely related	Linnen *et al.* (1993)
GDX	J03589	(−)	41	63	9.2	16 kDa; function unknown	Toniolo *et al.* (1988)
DSK2	L40587	−	34	57	6.9	40 kDa; spindle pole body duplication	Biggins *et al.* (1996)
BAT3	M31293	(−)	33	50	9.1	228 kDa; selectively expressed in testis	Banerji *et al.* (1990), Wang and Liew (1994)
Elongin B	L42855	−	28	58	5.8	15 kDa; elongation factor assembly	Garrett *et al.* (1995)
RAD23	L25428	−	26	43	8.6	42 kDa; excision repair	Watkins *et al.* (1993)

B. Sequences of UbH domain proteins

1. Processed proteins

```
                 10         20         30         40         50         60         70
              + ○    +                  +             ○    +                  +      ○ × #
Ub     MQIFVKTLTG KTITLEVEPSDTIENVKAKI QDKEGIPPDQ QRLIFAGKQL EDGRTLSDYN IQKESTLHLV LRLRGG X
vUb    ----I----- ----a-t--aE-VadL---- a----V-v-- ---------- --sK-M---- --------M- ------ y*
Nedd8  -l-k------ -e-eIDi--T-kV-rI-eRV eE------q- ----ys---M n-eK-a---k -lgg-v---- -a---- gglgq*
fau    -QL--R..aq elh----tgqE-VaQI--hv asl---a-Ed -vVLL--sp- --ea--gqcg VealT--eVa g-Ml-- kvhgs...
UCRP   Ls-L-Rnnk- RSs-Y--rlTq-VahL-eeV sgl--Vqd-l fw-t-e--p- --qlp-gE-g Lkpl--VfMn ------ gtepggrs*
```

2. Unprocessed proteins

```
              + ○    +                   +              ○    +                   +      ○ × #
Ub       MQIFVKTLTG KTITL.EVEPS DTIENVKAKI QDKEGIPPDQ QRLIFAGKQL EDGRTLSDYN. IQKESTLHLV LRLRGGX*
GDX      --Lt--a-q- RecS-.q-ped ElVstL--lV sE-lnV-vr- ---L-k--a- A--Kr----s. -gpn-k-n-- Vkplekvlle...
DSK2   msLN-hI-Sggd -.weV.n-a-e s-VlQF-ea- Nkan---vAN ----YS--i- k-Dq-VeS-h. --dghSV--- ksqpkpqtas
BAT3      LeVL---ds q-r-F.i-gAq mnVkeF-eh- rasvs--sEk ----Yq-Rv- q-dKk-qE--. Vg.gkvI--- e-appqthlp...
Anla   ....-EL-Ie---- tcFe-.r-s-y E-Vts----- -rl----va- -h--rnnme- --ecS--g--. -segc--kM- -aM---pint
E1.b     -dI-Lmirrh --tiFtdake- s-VfeL-riV egilkr--Ee ---Ykddql- D--K--gecgF@p-apa-Vg-a F-addtfeal...
RAD23  VsLt.-knfkke -vpld..L--- n--let---L aqsiscees- iK--YS--v- q-sR-V-EcgL kdgDqvVfM- sqkKstktkv
```

[a]+, processed; −, unprocessed; without parentheses, demonstrated processing status; in parentheses, assumed processing status.
[b]Where relevant, calculated using all naturally occurring substitutions at a given position as "wild type."
[c]Calculated based on UbH domain only (sequences in part B of this table).
Notes. *Ub*: sequence of human ubiquitin; in comparing other sequences, naturally occurring substitutions at a given position were taken as equivalent to the residue shown (T14A; P19SAQG; E24D, A28STQ; E51G; D52E; S57A; see Haas *et al*., 1996). The identity of residue 77 in Ub varies.

Hyphens, identities; *uppercase letters*, conservative substitutions; *dots*, gaps; *crosses*, known sites of Ub chain initiation; *circles*, residues implicated in proteolytic targeting by K48-linked chains; × Arg residue most critical for interaction with E1; #, arg residue important in conjugate recognition; *asterisks*, stop codons.

Only the C-terminal UbH domain of UCRP is shown.

@, position of a six-residue insert (TSQTAR) in elongin B.

extracellular function. The recombinant UbH domain is reported to inhibit *in vitro* histone ubiquitination (Nakamura *et al.*, 1996), suggesting that this domain may interact with ubiquitin activating and conjugating enzymes, despite its acidic pI (Table IB). The Finkel–Biskis–Reilly murine sarcoma virus expresses an antisense version of *fau*, but the functional significance of this phenomenon has yet to be determined (Michiels *et al.*, 1993).

2.3.1d. GMP-1/Sumo-1. The 101-residue mammalian protein known as GMP-1 (Matunis *et al.*, 1996) or Sumo-1 (Mahajan *et al.*, 1997) was identified as a covalent modifier of RanGAP1, a GTPase activating protein that functions in nuclear import. (Sumo-1 also has several other names; see Johnson and Hochstrasser, 1997, and references therein.) Sumo-1 has a 14-residue N-terminal extension, followed by a UbH domain that is 18% identical (51% similar) to ubiquitin. Modification by Sumo-1 serves a targeting function, resulting in the localization of RanGAP1 to the nuclear pore complex (Matunis *et al.*, 1996). This occurs through the Sumo-dependent interaction of RanGAP1 with the nuclear pore protein known as RanBP2 (Mahajan *et al.*, 1997). Even though the molecular basis of this targeting is not yet fully understood, it is evident that the modification of RanGAP1 by Sumo-1 is vital for productive translocation (Mahajan *et al.*, 1997). These exciting results provide a model for the function of at least some UbH proteins, including UCRP and the intra- and interspecies homologues of Sumo-1 itself (see Johnson and Hochstrasser, 1997). As with UCRP, it is likely that Sumo-1 and its homologues are processed, activated, conjugated, and released from conjugates by enzymatic mechanisms that are related to, but distinct from, the corresponding mechanisms for ubiquitin.

2.3.2. Unprocessed UbH Proteins

These proteins have N-terminal UbH domains that exhibit 40 to 60% similarity to ubiquitin, with marked divergence in the C-terminal portion of the UbH domain (Table IB). They are either known to be unprocessed (RAD23, elongin B, Anla, DSK2), or are likely to be (GDX, BAT3). The high sequence divergence seen in the unprocessed UbH domains may reflect the removal of selective pressures associated with a subset of ubiquitin functions (see above), probably including proteolytic targeting (see below). In two cases, there is evidence that the UbH domain plays a chaperonelike role, as seen for ubiquitin in ribosome biogenesis (Section 2.2.2). Structure–function studies on these proteins may thus provide insight into this function of ubiquitin.

2.3.2a. Elongin B. Elongin is a heterotrimeric transcription elongation factor whose 118-residue B subunit is an unprocessed ubiquitin-related protein (Garrett *et al.*, 1995). The first 84 residues of elongin B are 58% similar to ubiquitin, and homology modeling suggests that there is retention of ubiquitinlike folding, with the acquisition of new surface hydrophobic residues and a surface loop (Garrett *et al.*, 1995). *In vitro*, elongin B promotes assembly of the transcriptionally active elongin heterotrimer, and dramatically enhances the thermal stability of transcription elongation activity (Aso *et al.*, 1995). This chaperonelike function seems likely to reside in the UbH domain (see above). Binding of the von Hippel–Lindau (VHL) tumor suppressor protein to elongin, which is mediated by the B and C subunits, inhibits the A subunit (Duan *et al.*, 1995). This interaction is abrogated by VHL mutations that abolish tumor suppression (Kibel *et al.*, 1995).

2.3.2b. RAD23. The yeast nucleotide excision repair gene *RAD23*, and its human homologues *HHR23A* and *HHR23B*, encode approximately 400-residue proteins with N-terminal UbH domains that are predicted to have ubiquitinlike folding (Watkins *et al.*, 1993; Masutani *et al.*, 1994). Deletion of the UbH domain of RAD23 makes yeast cells UV-sensitive, even when the mutant protein is overexpressed; this phenotype can be rescued by replacing the UbH domain with ubiquitin (Watkins *et al.*, 1993), suggesting that the divergent residues in this domain are not significant for its function. In the context of an uncleaved linear fusion protein, ubiquitin can provide a site for polyubiquitin chain initiation, thus acting as a degradation signal (Johnson *et al.*, 1992, 1995; Section 3.1), but the RAD23 UbH domain does not appear to function in this manner. In fact, the intracellular level of a mutant RAD23 protein lacking the UbH domain is much lower than the level of the wild-type protein (Watkins *et al.*, 1993). It is possible that the UbH domain acts as a chaperone in RAD23 synthesis, analogous to the role of ubiquitin in ribosomal protein biosynthesis (see above). *In vitro*, yeast RAD23 binds to two other excision repair proteins, TFIIH and RAD14, promoting the formation of a RAD14–TFIIH complex (Guzder *et al.*, 1995). This activity is reminiscent of the chaperone activity of elongin B (see above). Recent studies suggest that yeast RAD23 also plays a role in spindle pole body duplication (Biggins *et al.*, 1996; see below).

2.3.2c. DSK2. The yeast *DSK2* gene was identified by the ability of mutant *dsk* alleles to suppress a mutation in a gene, *KAR1*, that encodes a component of the spindle pole body (analogous to the centrosome of higher cells; Biggins *et al.*, 1996). DSK2 is a stable 373-residue nuclear protein with an unprocessed N-terminal UbH domain. Although DSK2 disruption elicits no detectable phenotype, *rad23Δdsk2Δ* cells are temperature sensitive for growth as a result of a block in spindle pole body duplication. Thus, RAD23 and DSK2 appear to function together in spindle pole body duplication, while RAD23 has a second, unique function in DNA repair (see above). Two suppressor alleles of *DSK2* specify the same H68Y mutation in the UbH domain, indicating that this region is important for DSK2 function (Biggins *et al.*, 1996). Although DSK2 is not itself a component of the spindle pole body, the suppressor alleles restore proper localization of another spindle pole body component, CDC31. The precise functions of DSK2 and RAD23 in spindle pole body duplication remain to be determined.

2.4. Ubiquitin Mutants

An ever-growing number of mutant ubiquitins have been tested for degradative activity in ubiquitin-depleted reticulocyte lysate (fraction II). This assay usually monitors the N-end rule pathway, in which substrates are recognized and ubiquitinated by E3α and E2-14K, based on the properties of their N-terminal residues (Bachmair *et al.*, 1986; Chapter 8). Purified conjugating enzymes that function in this and other ubiquitination pathways have also been studied. Because the presence of ubiquitin is essential for viability in yeast (Finley *et al.*, 1994), studies in this system have tended to focus on dominant mutations that produce phenotypes when mutant and wild-type proteins are coexpressed. Ubiquitin mutations that generate phenotypes in one or more of these systems are compiled in Table II, and discussed briefly below. Certain mutations that provide insight into

Table II
Site-Specific Mutations in Ubiquitin

Mutation	System/enzyme[a]	Phenotype/conclusion[b]	References[c]
G76A	Yeast	In context of linear Ub fusion, inhibits processing proteases	1
	Overexpression/yeast	Slow growth	1
	Overexpression/yeast	Stress sensitivity; impaired proteolysis; isopeptidase inhibition	2
	Sole expression/yeast	Lethal	10
	Retic. fr. II	Supports degradation at 10–25% wt rate	3, 4
	E1	Ternary complex destabilized: thiol ester but no adenylate	4
	E3α, E2-25K	Supports conjugation at ~25% wt rate	4
	Isopeptidase T	10-fold inhibition for G76A as scissile bond; 100-fold inhibition for G76A at proximal chain terminus	5
L73Δ	Retic. fr. II, E1	No activity	3
G75,76Δ	Overexpression/yeast	Increased transcription of *UB14* polyubiquitin gene	cited in 6
Y59F	Retic. fr. II	Supports degradation at 70–100% wt rate	3
	E1	Properties in PP_i–ATP exchange similar to wt	3
	E2-25K	Supports chain synthesis at 30% wt rate (V/K)	7

Table II (*Continued*)

Mutation	System/enzyme[a]	Phenotype/conclusion[b]	References[c]
H68K	Overexpression/yeast	Slow growth; does not complement *ubi4*Δ	1, 10
	Retic. fr. II	Supports degradation at 30% wt rate	3
	E1	PP_i–ATP exchange: ~10% wt (V/K)	3
Y59F,H68K	Retic. fr. II	Supports degradation at ~30% wt rate	3
	E1	PP_i–ATP exchange: 5% wt (V/K)	3
K48R	Retic. fr. II	No activity in degradation (X-βgal substrate); chain formation blocked	9
	Overexpression/yeast	Weak inhibition of protein turnover	10
	Sole expression/yeast	Lethal (G2/M phase arrest); inhibition of proteolysis	10
	Yeast	In context of uncleaved Ub-βgal fusion substrate, inhibits degradation	11, 16
	CDC34/E2-35K	Chain formation blocked (autoubiquitination)	12, 13
	RAD6/E2-14K	Chain formation/degradation (with E3α) blocked	13
K63R	Overexpression/yeast	Does not complement *ubi4*Δ; K63 site of Ub–Ub conjugation	6
	Sole expression/yeast	Sensitivity to DNA damage and UV; elimination of novel chain-linked conjugate	8
K29R	Overexpression/yeast	K29 site of Ub–Ub conjugation	6
	Sole expression/yeast	Growth defect	8
	Yeast	In context of uncleaved Ub-βgal and Ub-DHFR fusions, inhibits degradation	16
K27R	Sole expression/yeast	Growth defect	8
K33R	Sole expression/yeast	Growth defect	8
K6R	RAD6	E3-independent histone multiubiquitination blocked	13
K11R	E2-EPF/fr. II	Autoubiquitination and (E3-dependent) degradation blocked	13
R54L	E1	10-fold affinity decrease	14
R42L	E1	Weakened binding of adenylate	14
R72L	E1	100-fold affinity decrease; weakened binding of adenylate	14
L8A	Sole expression/yeast	Lethal	cited in 15
L8A,I44A	Retic. fr. II	Impaired recognition of K48-linked chains; similar effects for other double mutations at residues 8, 44, 70	15
L8A,I44A	S5a, proteasome	Binding of multiubiquitin chains abolished	15

[a] "Overexpression": expression from a high-copy plasmid in the presence of at least one gene producing the wild-type protein. "Sole expression": mutant is the only form of ubiquitin present.

[b] For *in vitro* assays, effects are on V_{max} unless otherwise stated.

[c] References: (1) Butt *et al.* (1988); (2) Hodgins *et al.* (1992); (3) Ecker *et al.* (1987); (4) Pickart *et al.* (1994); (5) Wilkinson *et al.* (1995); (6) Arnason and Ellison (1994); (7) Pickart *et al.* (1992); (8) Spence *et al.* (1995); (9) Chau *et al.* (1989); (10) Finley *et al.* (1994); (11) Johnson *et al.* (1992); (12) Chau *et al.* (1989); (13) Baboshina and Haas (1996); (14) Burch and Haas (1994); (15) Beal *et al.* (1996); (16) Johnson *et al.* (1995).

the structure and function of polyubiquitin chains are discussed further in Sections 3 and 4.

A general insight derived from *in vitro* studies is that one of the simplest assays for ubiquitin—activity in PP_1–ATP exchange catalyzed by E1—is not a good predictor of overall proteolytic competence, at least in fraction II. There are two reasons for this (Pickart *et al.*, 1994; Burch and Haas, 1994). First, the exchange reaction is kinetically complex, and may be strongly influenced by microscopic steps that are not part of the forward reaction. Second, in fraction II the E1 reaction is much faster than downstream conjugation steps; thus, a substantial decrease in the E1 rate may fail to inhibit proteolysis.

2.4.1. Naturally Occurring Substitutions

Relative to animal ubiquitin, the three conservative substitutions seen in yeast (P19S, E24D, A28S) and oat (P19S, E24D, S57A) ubiquitins are fully permissive for protein degradation in reticulocyte fraction II (Vierstra *et al.*, 1986; Wilkinson *et al.*, 1986; Ecker *et al.*, 1987). Apart from these substitutions, yeast and oat ubiquitins are structurally identical to animal ubiquitin (Vijay-Kumar *et al.*, 1987b).

2.4.2. Mutations at the C-Terminus

Given the central role of the carboxyl group of G76 in the proteolytic function of ubiquitin, alteration of the ubiquitin C-terminus is expected to abrogate function. Deletion of the C-terminal GG dipeptide has long been known to abolish degradative activity (Wilkinson and Audhya, 1981; see also Ecker *et al.*, 1987), and even modest alterations at the C-terminus have severe effects. For example, yeast cells are inviable when G76A-ubiquitin is the only form of ubiquitin present (Finley *et al.*, 1994). Cells expressing high levels of G76A-ubiquitin in the presence of wild-type ubiquitin grow slowly (Butt *et al.*, 1988), and exhibit several phenotypes characteristic of ubiquitin deficiency, including defective proteolysis in combination with elevated conjugate levels (Hodgins *et al.*, 1992). These cells fail to disassemble polyubiquitin chains assembled from the mutant ubiquitin; consequently, the chains accumulate to high levels (Hodgins *et al.*, 1992). It is now known that isopeptidase T, the major enzyme responsible for disassembly of unanchored chains (Section 3.3.2), is strongly sensitive to the G76A mutation, especially when it is present at the free chain terminus (Wilkinson *et al.*, 1995). Isopeptidase T is even more strongly inhibited by removal of the GG dipeptide at this site (Wilkinson *et al.*, 1995), providing an explanation for accumulation of chains (M. Hochstrasser, personal communication), and induction of the *UBI4* gene (Arnason and Ellison, 1994), in cells overexpressing G77,76Δ-ubiquitin. Slow chain disassembly appears to be the principal basis of the cellular phenotypes associated with the G76A mutation, but whether these deleterious effects follow from the high level of chains (causing blockade of the 26 S proteasome), or from the low level of free ubiquitin (causing inhibition of substrate conjugation), or both, is not clear. The G76A mutation also impairs conjugation, reducing the rate of the E3α reaction by fourfold (Pickart *et al.*, 1994). Finally, in the context of linear ubiquitin fusions, mutating G76 strongly inhibits ubiquitin processing proteases (Butt *et al.*, 1988; Johnson *et al.*, 1992).

2.4.3. Tyr and His Mutations

Simultaneous diiodination of Y59 and H68 is permissive for proteolysis *in vitro* (Cox *et al.*, 1986; Section 2.5.3), suggesting that mutations at these positions might be tolerated. The hydroxyl group of Y59 is involved in a hydrogen bond that may be important for ubiquitin stability (Vijay-Kumar *et al.*, 1987a); Y59F-ubiquitin exhibits poor solubility (Ecker *et al.*, 1987). However, Y59F-ubiquitin is only slightly defective in degradation *in vitro* (Ecker *et al.*, 1987). An H68K mutation causes a stronger degradative defect (Ecker *et al.*, 1987), and confers a slow growth phenotype when expressed in yeast in the presence of the wild-type protein (Butt *et al.*, 1988). The molecular basis of these H68K phenotypes is unknown.

2.4.4. Lys Mutations

Following the demonstration by Chau *et al.* (1989) that K48 is a site of functionally important ubiquitin–ubiquitin conjugation (Section 3.1), the possibility of a similar function for other Lys residues has been systematically addressed by mutagenesis. The results show that at least four other Lys residues can be sites of ubiquitin–ubiquitin conjugation. It is a point not often appreciated that the results of these experiments generally implicate a given Lys residue only in chain *initiation* (i.e., in the bond that links ubiquitins *a* and *b* in Fig. 3A). To prove that the same linkage is propagated throughout the chain, chemical mapping must be done, or the chain must be assembled from a single-Lys ubiquitin derivative in the absence of any other source of ubiquitin. Nonetheless, the mutagenesis results have been highly informative, with two of the Lys-to-Arg mutations generating selective phenotypes in yeast.

The various Lys-to-Arg mutations impact only minimally on the ubiquitin thiol ester-forming activities of E1 and the several E2s that have so far been examined in this regard (Baboshina and Haas, 1996). Therefore, mutant phenotypes, where observed, have been interpreted to derive from inappropriate chain termination. This interpretation should be applied with caution, however, in view of the potential for unexpected structure–function requirements in chain synthesis, recognition, or disassembly.

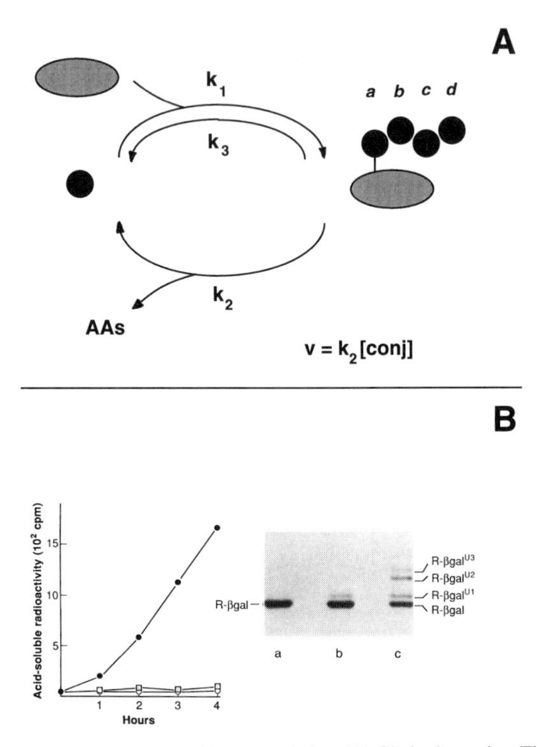

Figure 3. Dynamics and function of K48-linked chains. (A) Chain dynamics. The oval represents a target protein; the circle represents ubiquitin; AAs are amino acids. The constant k_1 is an aggregate *V/K* for substrate conjugation. The constant k_2 corresponds to *V/K* for degradation of the conjugate by the 26 S proteasome. The constant k_3 is an aggregate *V/K* for isopeptidase-dependent conjugate disassembly. Four ubiquitins are shown in the chain. Ubiquitin *a* is the proximal ubiquitin in the chain; G76 of this ubiquitin forms an isopeptide bond with a Lys residue of the substrate. Ubiquitin *d* is the distal ubiquitin in the chain; K48 of this ubiquitin is available for extending the chain by an additional ubiquitin. The isopeptide bond between ubiquitins *a* and *b* is the initiating linkage in the chain.

(B) Uniform K48-linked chain is competent in proteolytic targeting. Left: K48R-ubiquitin inhibits *in vitro* degradation. Degradation of radiolabeled R-βgal in reticulocyte fraction II, either without added ubiquitin (open circles), in the presence of wild-type ubiquitin (filled circles), or in the presence of K48R-ubiquitin (stippled squares). Right: K48R-ubiquitin blocks polyubiquitination. Aliquots were removed from the incubations (left panel) at 30 min and analyzed by SDS–PAGE and autoradiography: Lane a, without ubiquitin; lane b, K48R-ubiquitin; lane c, wild-type ubiquitin. Individual ubiquitinated species of R-βgal are indicated on the right (structures based on the chemical mapping described in the text). Figure reprinted with permission from Chau *et al.* (1989), *Science* **243:**1576–1583. Copyright 1989, American Association for the Advancement of Science.

K48R-ubiquitin. In vitro, K48R-ubiquitin supports neither degradation, nor the formation of high-molecular-weight conjugates. Direct chemical mapping shows that substrate-borne chains assembled from wild-type ubiquitin are linked predominantly or exclusively through K48 (Chau *et al.*, 1989; Section 3.1). When yeast cells are induced to express K48R-ubiquitin selectively, they arrest in late G2 or M phase of the cell cycle, and concomitantly exhibit a strong defect in the turnover of short-lived and canavanyl proteins (Finley *et al.*, 1994). These and other data indicate that the most commonly utilized degradation signal in the ubiquitin pathway is a uniform polyubiquitin chain linked by K48–G76 isopeptide bonds (Section 3.1). Because incorporation of K48R-ubiquitin into such a chain should terminate its growth, this derivative might be expected to exert a strong dominant negative effect. However, overexpression of K48R-ubiquitin causes only weak inhibition of ubiquitination and protein turnover (Arnason and Ellison, 1994; Finley *et al.*, 1994). The failure of K48R-ubiquitin to act as a dominant negative apparently reflects rapid disassembly and reassembly of chains within cells (Section 3.3.2), as suggested by the stronger inhibition of proteolysis seen on overexpression of a K48R,G76A double mutant (Finley *et al.*, 1994; Yaglom *et al.*, 1995). (The G76A mutation impairs chain disassembly; see above.)

K63R-ubiquitin. Overexpression of a tagged, C-terminally blocked ubiquitin derivative, in which K63 is the only Lys residue, results in elevated levels of tagged diubiquitin. This observation is most simply explained if K63 is a site of chain initiation (Arnason and Ellison, 1994). The K63R mutation alters the spectrum of intracellular ubiquitin conjugates in a manner suggesting that chains harboring K63 linkages are relatively abundant (Spence *et al.*, 1995). Yeast cells expressing K63R-ubiquitin as the only form of ubiquitin grow normally, and are competent in general proteolysis (Spence *et al.*, 1995). However, two selective phenotypes—a defect in error-prone DNA repair and enhanced sensitivity to stress—are correlated with expression of K63R-ubiquitin (Arnason and Ellison, 1994; Spence *et al.*, 1995; Section 4).

Other Lys-to-Arg mutants. Overexpression of a tagged ubiquitin derivative in which K29 is the only Lys residue results in an elevated level of tagged diubiquitin, indicating that K29 can be a chain initiation site (Arnason and Ellison, 1994). Chains initiating at K29 can target proteins for degradation (Johnson *et al.*, 1995; Section 4), although such chains do not appear to be very abundant (Spence *et al.*, 1995). The basis of the slow growth caused by sole expression of K29R-ubiquitin remains to be determined (Spence *et al.*, 1995).

Expression of K27R- or K33R-ubiquitin as the sole form of ubiquitin in yeast causes slow growth (Spence *et al.*, 1995), but as yet there is no evidence for chain initiation at K27 or K33 (Arnason and Ellison, 1994; Spence *et al.*, 1995). Overexpression of single-Lys ubiquitin mutants in yeast cells provided no evidence for chain initiation at K6 or K11 (Arnason and Ellison, 1994), and cells expressing either mutant as the sole form of ubiquitin grow normally (Spence *et al.*, 1995). However, *in vitro* studies with purified ubiquitin conjugating enzymes indicate that K6 and K11 can each be used as a site of chain initiation (Baboshina and Haas, 1996; Section 4).

2.4.5. Arg Modifications/Mutations

Modification of R42 and/or R72 with the bulky, negatively charged reagent 4-(oxyacetyl)phenoxyacetic acid strongly inhibits PP_i–ATP exchange catalyzed by E1 (Duerksen-Hughes *et al.*, 1987). Site-specific mutagenesis indicates that R42, R54, and especially R72 are all important for binding of ubiquitin at various stages in the E1 reaction (Burch and Haas, 1994). Mutation or modification of R74 is permissive for E1 and for downstream conjugation steps (Duerksen-Hughes *et al.*, 1987; Burch and Haas, 1994), but conjugates assembled from R74-modified ubiquitin are not degraded (Duerksen-Hughes *et al.*, 1987; Section 3.2.1).

2.4.6. Mutations of Surface Hydrophobic Residues

The side chains of L8, I44, and V70 form a hydrophobic patch on one side of ubiquitin, fairly close to the C-terminus. As determined in fraction II, mutation of pairs of these residues to Ala strongly inhibits degradation, but has little effect on steady-state conjugate level (Beal *et al.*, 1996). The implication of these and other results is that mutation of these residues abrogates the recognition of K48-linked polyubiquitin chains by the 26 S proteasome (Beal *et al.*, 1996; Section 3.2.1). L8A-ubiquitin cannot serve as the sole source of ubiquitin in yeast cells (see Beal *et al.*, 1996), although more detailed studies are needed to determine whether impaired conjugate recognition is the principal basis of this effect. Single and double mutations at L8, I44, and V70 modestly inhibit E3α (Beal *et al.*, 1996). These mutations also affect ubiquitin recognition by E1: L8A-ubiquitin binds to E1 about 4-fold more weakly, and L8A,I44A-ubiquitin about 20-fold more weakly, than wild-type ubiquitin (R. Beal and C. Pickart, unpublished results of competition in E1-ubiquitin thiol ester formation). These effects on conjugating enzymes are not strongly manifested as a decreased steady-stage conjugate level in fraction II (Beal *et al.*, 1996), presumably because the steps mediated by E1, E2-14K, and E3α all remain rapid relative to conjugate degradation.

2.5. Chemically Modified Ubiquitin

Besides mutant ubiquitins, several chemically modified forms of ubiquitin have figured importantly in dissection of the proteolytic pathway, including the function of polyubiquitin chains.

2.5.1. Ubiquitin Aldehyde (Ubal)

Originally generated in small quantities by reduction of a ubiquitin thiol ester adduct of ubiquitin carboxyl-terminal hydrolase (Pickart and Rose, 1986), Ubal is a derivative in which the carboxyl group of G76 has been converted to an aldehyde. It binds tightly to the active-site thiols of isopeptidases and carboxyl-terminal hydrolases ($K_d \sim 1$ pM), by mimicking the tetrahedral intermediate in catalysis (Pickart and Rose, 1986; Hershko and Rose, 1987). Ubal can now be generated on a large scale (Mayer and Wilkinson, 1989); it has found wide application in mechanistic analysis of isopeptidases (e.g., Hadari *et al.*, 1992), and in enhancing the sensitivity of detection of ubiquitin conjugates *in vitro* (e.g., Chen *et al.*, 1995).

2.5.2. Methyl-Ubiquitin (MeUb)

Exhaustive treatment of ubiquitin with cyanoborohydride and formaldehyde methylates all seven Lys residues, as well as the α-amino group (Hershko and Heller, 1985). This set of modifications appears to be functionally equivalent to mutating all seven Lys residues to Arg. MeUb conjugates to target proteins, but lacks sites for ubiquitin–ubiquitin conjugation; MeUb conjugates are poorly recognized by the 26 S proteasome (Hershko and Heller, 1985; Section 3.1). Inhibition of degradation by MeUb has been used to substantiate the ubiquitin dependence of substrate turnover in crude lysates (e.g., Hershko *et al.*, 1991). Incorporation of MeUb terminates the elongation of polyubiquitin chains; a loss of electrophoretically retarded conjugation products on substitution of MeUb for the native protein can thus suggest that a substrate of interest bears chains (e.g., Chen *et al.*, 1995).

2.5.3. Iodinated Ubiquitin

The availability of radioiodinated ubiquitin allowed the initial demonstration of ubiquitin–substrate ligation (Hershko *et al.*, 1980), and this derivative remains a critical tool in mechanistic analyses of conjugating enzymes. As typically prepared using chloramine-T (Hershko *et al.*, 1980), radioiodinated (I_4)-ubiquitin is diiodinated on both H68 and Y59, and oxidized at M1 (Cox *et al.*, 1986). In general, the substrate properties of I_4-ubiquitin closely mimic those of native ubiquitin. Differences include a twofold higher V_{max} for I_4-ubiquitin in proteolysis *in vitro*, an effect attributable to H68 modification (Cox *et al.*, 1986), and failure of I_4-ubiquitin to be utilized as an acceptor by the K48-chain-synthesizing enzyme E2-25K (Pickart *et al.*, 1992). The latter effect is attributable to Y59 modification.

2.6. Detection of Ubiquitination

A question frequently asked by newcomers to the field is how best to detect the ubiquitination of a potential substrate. There are two approaches in current use.

2.6.1. Antiubiquitin Antibodies

Antibodies elicited by immunization with full-length ubiquitin can have several different specificities. Injection of cross-linked, SDS-denatured ubiquitin yields antibodies that recognize both free and conjugated ubiquitin on immunoblots, but that show specificity for conjugated ubiquitin in solution (Haas and Bright, 1985; see also Takada *et al.*, 1995). An undenatured immunogen yields antibodies with high specificity for the ubiquitin C-terminus, which exhibit preferential reactivity with monoubiquitin (and unanchored chains) in solution (Haas *et al.*, 1985; Takada *et al.*, 1996). A third type of antibody was produced by Fujimoro *et al.* (1994), who immunized mice with a crude preparation of polyubiquitinated proteins; one of the resulting monoclonal antibodies exhibited apparent specificity for polyubiquitin chains, as suggested by its failure to recognize either monoubiquitin, or conjugates produced using MeUb.

Affinity-purified polyclonal antibodies specific for conjugates have been widely used to detect ubiquitination in cell extracts and fixed cells (Haas and Bright, 1985; Haas, 1988; Chapter 15). When addressing the ubiquitination of a particular substrate, one typically immunoprecipitates with antisubstrate antibodies, then counterscreens the immunoprecipitated proteins by immunoblot analysis with antiubiquitin antibodies (e.g., Watkins *et al.*, 1993; Chen *et al.*, 1995; Nefsky and Beach, 1996). Alternatively, one can immunoprecipitate with ubiquitin antibodies and blot with substrate antibodies (e.g., Kalchman *et al.*, 1996). Bona fide ubiquitinated forms of the substrate should be reactive with both antibodies, and should migrate at a higher molecular mass than the unmodified substrate, often in a 6- to 8-kDa ladder. Conjugates can be stabilized during lysis and immunoprecipitation by including *N*-ethyl maleimide or iodoacetamide in the extraction buffer at about 0.5 mM (e.g., Kalchman *et al.*, 1996). These reagents inactivate endogenous isopeptidases, which otherwise rapidly disassemble conjugates as ATP is depleted postlysis (Riley *et al.*, 1988).

One disadvantage of ubiquitin-directed immunochemical approaches is that commercially available antiubiquitin antibodies are of rather mixed quality—somewhat inexplicably, as excellent antibodies are easily prepared and purified (Haas and Bright, 1985). A more serious disadvantage is that the antibodies may recognize ubiquitin-related proteins. Unconjugated UCRP cross-reacts strongly with affinity-purified (polyclonal) antiubiquitin antibodies (Haas *et al.*, 1987; Loeb and Haas, 1992; Lowe *et al.*, 1995); yeast DSK2 and the fau UbH domain react weakly with antiubiquitin antibodies (Biggins *et al.*, 1996; Nakamura *et al.*,

1996). Antibodies directed against the fau UbH domain precipitate a large number of intracellular proteins (Nakamura *et al.*, 1995); it has not been excluded that this is related in part to cross-reaction with ubiquitin conjugates.

2.6.2. Tagged Ubiquitins

Ubiquitin is currently available with four N-terminal tags: glutathione S-transferase (GST), myc, hemagglutinin (HA), and hexahistidine (His6). GST-ubiquitin is useful for studying ubiquitination *in vitro*, where it represents an alternative to radioiodinated ubiquitin by virtue of a protein kinase site in the GST moiety (Scheffner *et al.*, 1993). Despite the presence of the bulky GST moiety, and the probability that GST-ubiquitin is dimeric (e.g., McTigue *et al.*, 1995), GST-ubiquitin is activated by E1 and is a competent substrate for the E2 and E3 enzymes so far examined in this regard. The GST moiety also provides a potential route to affinity purification of conjugated substrates (or thiol ester-linked enzymes) on glutathione-agarose.

For the detection of ubiquitin conjugates in cells and cell extracts, myc-, HA-, and His6-tagged ubiquitins increasingly represent the method of choice. For intact cells, this approach requires prior transfection or transformation with DNA encoding the tagged ubiquitin derivative. This is slightly disadvantageous in that the tagged ubiquitin must compete with the highly expressed endogenous gene product. However, there are powerful advantages. First and most important, using a "ectopic" tag allows for the most rigorous negative control. Thus, comparing the *specific* reactive species in transfected versus untransfected cells avoids ambiguities that can be associated with using a distinct reagent (nonspecific preimmune serum) for the negative control. Second, excellent antibodies against the myc and HA tags are commercially available. Third, the tag can be used to affinity-purify the conjugates, either immunochemically (myc and HA tags) or on nickel resin (His6 tag; see below). Fourth, the presence of the tag alters the size of ubiquitin, leading to a predictable increase in the sizes of bona fide conjugates. Fifth, this approach can easily be adapted to include useful ubiquitin mutants, e.g., the chain-terminating K48R,G76A derivative (Finley *et al.*, 1994). Finally, where a substrate is introduced by transient transfection, cotransfection of a tagged ubiquitin gene is particularly useful, because the detected conjugates then originate solely from the (transfected) cells expressing the substrate of interest (e.g., Treier *et al.*, 1994).

Some of the tagged ubiquitin derivatives have been characterized with respect to proteolytic functionality. Myc-ubiquitin conjugates normally to substrates in yeast cells and in reticulocyte fraction II (Ellison and Hochstrasser, 1991; Hochstrasser *et al.*, 1991; Johnson *et al.*, 1992; Hodgins *et al.*, 1992). Myc-ubiquitin supports degradation in fraction II at a fourfold reduced rate, apparently related to impaired interaction of myc-tagged polyubiquitin chains with the 26 S proteasome (Ellison and Hochstrasser, 1991). His6-ubiquitin (Beers and Callis, 1993) supports degradation in reticulocyte fraction II with a V_{max} similar to that of wild-type ubiquitin (G. Xia and C. Pickart, unpublished data). The use of His6-ubiquitin permits single-step purification of conjugates on nickel affinity resin, avoiding the low yields and high costs associated with immunopurification (Beers and Callis, 1993; Treier *et al.*, 1994).

3. PROTEOLYTIC TARGETING MEDIATED BY K48-LINKED POLYUBIQUITIN CHAINS

It is now clear that polyubiquitin chains linked through K48–G76 isopeptide bonds represent the predominant *in vivo* targeting signal in the ubiquitin pathway. Biochemical and genetic evidence supporting this conclusion is reviewed in Section 3.1 (see also Section 2.4.4), and data addressing recognition determinants at the level of chains and the proteasome are discussed in Section 3.2. Chain assembly and disassembly is reviewed in Section 3.3.

3.1. Existence and Significance of Targeting by K48-Linked Chains

3.1.1. Early Studies

An isopeptide structure for the ubiquitin–substrate bond was first inferred from its chemical properties (Hershko *et al.*, 1980). The presence in the bond of G76 of ubiquitin (Hershko *et al.*, 1981) implied that activated ubiquitin reacted with a nucleophilic group of the substrate. Multiple molecules of ubiquitin were conjugated to a single molecule of substrate, so the simplest model invoked Lys ε-amino groups of the substrate as the sites of ubiquitin ligation. In these early studies, a near congruence between the number of conjugated forms of a given substrate, and the number of substrate Lys residues, appeared to support this hypothesis (Hershko *et al.*, 1980).

However, it soon became apparent that conjugates were often larger than predicted by a model in which the maximum number of conjugated ubiquitins was equal to the number of substrate Lys residues (Hershko *et al.*, 1984; Hershko and Heller, 1985; Hough *et al.*, 1986). Interest in the structural basis of this apparent anomaly was increased by the observation that the largest conjugates were preferentially degraded by the 26 S proteasome (Hershko *et al.*, 1984; Hough *et al.*, 1986). Substituting MeUb for ubiquitin blocked the formation of excessively large conjugates, and made the conjugate profile conform more closely to that expected for the one-ubiquitin-per-Lys model (Hershko and Heller, 1985). Because MeUb lacks sites for ubiquitin–ubiquitin conjugation (Section 2.5.2), this result suggested that ubiquitin "chains" were present in aberrantly large conjugates. An isopeptide structure for such chains was suggested by the stability of large conjugates to treatment with cyanogen bromide. This result ruled out a linear chain structure, which was already unlikely given the metabolic instability of α-linked ubiquitin fusions, and the lack of a posttranslational route to linear polyubiquitin (Section 2.2).

3.1.2. Chain Linkage and Targeting Function

Compelling evidence for the existence and significance of isopeptide-linked ubiquitin chains was provided by Chau and co-workers, who characterized the structures of purified conjugates of the well-studied Ub-X-βgal substrates of the N-end rule pathway (Chau *et al.*, 1989). These artificial substrates are rapidly deubiquitinated by ubiquitin processing proteases in fraction II; depending on the identity of residue X, the βgal moiety may then be ubiquitinated by E3α in cooperation with E2-14K (Bachmair *et al.*, 1986; Gonda *et al.*, 1989; Chapter 8). An elegant chemical mapping approach was applied to show that the multiple ubiquitins conjugated to βgal (up to four in these mapping studies) were organized as a chain linked exclusively through isopeptide bonds involving K48 of one ubiquitin and G76 of the next (Fig. 3A). A uniform linkage within the chain was subsequently confirmed using K48C-ubiquitin that was first reductively methylated to block the six remaining Lys residues, then alkylated to restore an ε-amino group at residue 48. The resulting molecule, containing only the artificial "K48," had the same V_{max} as wild-type ubiquitin in βgal degradation (Gregori *et al.*, 1990).

Each molecule of conjugated βgal was found to bear a single ubiquitin chain, which was linked to one of a pair of Lys residues near the βgal N-terminus (Chau *et al.*, 1989). When K48R-ubiquitin was used, only monoubiquitinated βgal was formed, and there was negligible turnover, indicating that the monoubiquitinated conjugate was not recognized by the 26 S proteasome in fraction II (Chau *et al.*, 1989; Gregori *et al.*, 1990; Fig. 3B). These results indicate that βgal is efficiently targeted for degradation by the ligation of a single K48-linked ubiquitin chain (Fig. 3). As discussed above (Section 2.4), the strong inhibition of proteolysis attendant on selective expression of K48R-ubiquitin in yeast cells indicates that proteolytic targeting is primarily achieved through ligation of polyubiquitin chains assembled through K48 linkages (Finley *et al.*, 1994).

Uniquely among the Ub-X-βgal substrates, Ub-P-βgal is very slowly deubiquitinated (Bachmair *et al.*, 1986). However, it is rapidly degraded. Degradation of this substrate involves extension of a chain from K48 of the fused ubiquitin moiety, as shown by the finding that a K48R mutation in this moiety increases the half-life of βgal by a factor of 10 (Johnson *et al.*, 1992). It is not known whether the K48 linkage is propagated throughout the chain.

Despite the apparent predominance *in vivo* of a targeting mechanism involving K48-linked chains, *in vitro* results show that chains are not absolutely required for recognition by the 26 S proteasome. For example, a subsaturating concentration of monoubiquitinated α-globin is degraded by the purified proteasome at a significant rate, whereas unconjugated α-globin is inert (Shaeffer and Kania, 1995). Thus, a conjugate bearing only a single ubiquitin is a competent substrate for the proteasome (see also Haas *et al.*, 1990), although adding a second ubiquitin to α-globin increases the degradation rate severalfold (Shaeffer and Kania, 1995). Here the diubiquitinated conjugates were probably a mixture of globin linked to two single ubiquitins and globin linked to a diubiquitin chain (Shaeffer, 1994). Within cells, conjugates bearing one or a few ubiquitins must compete both with abundant chain-bearing conjugates and with unanchored chains (see below). Although monoubiquitinated α-globin appears to compete successfully (Shaeffer, 1994), it is not clear that this will generally be the case. For example, monoubiquitinated βgal is not degraded in fraction II (Chau *et al.*, 1989), and MeUb, which prevents chain formation, is generally an inhibitor of degradation (see above).

3.1.3. Rationale for the Chain Targeting Signal: Principles and Evidence

The findings discussed above indicate that a K48-linked polyubiquitin chain is an effective signal for substrate degradation by the 26 S proteasome (Fig. 3), but leave open the question of exactly why this is so. Five possibilities are outlined below, together with some specific predictions of each model (see also Ellison and Hochstrasser, 1991). These models are considered in the context of the simple scheme shown in Fig. 3A. In this model, k_1 is an aggregate rate constant for conjugate formation, k_2 corresponds to *V/K* for conjugate degradation by the 26 S proteasome, and k_3 is an aggregate rate constant for conjugate disassembly by isopeptidases. The rate of degradation is given by $v = k_2$[conjugate].

A first model postulates that a chain has a mechanistic advantage at the stage of conjugate turnover because a chain is intrinsically better recognized by the proteasome. That is, k_2 is higher (Fig. 3A) for a substrate bearing an *n*-ubiquitin chain, relative to the same substrate bearing *n* single ubiquitins. This effect could arise in two fundamentally different ways. The first is a simple increase in probability related to a higher local concentration of ubiquitin. (Implicit in this model is an assumption that separated single ubiquitins, for example on opposite faces of the substrate protein, would not be similarly accessible to a chain receptor, whereas successive ubiquitins in a chain would be.) Alternatively, a chain might be better recognized as a result of the presence in the chain of structural features that are absent, or muted, in monomeric ubiquitin. The latter idea is especially attractive, as it could allow for differential recognition of chains assembled through different Lys residues, and provide an explanation for the selective phenotypes elicited by certain Lys-to-Arg mutations (Sections 2.4 and 4).

Either version of the enhanced recognition model predicts that a substrate bearing a single *n*-ubiquitin chain is recognized better by the proteasome than a substrate bearing *n* single ubiquitins. Unfortunately, this simple test is not (yet) experimentally accessible, because of inability to control the extent of either type of ubiquitination as catalyzed by current available conjugating enzymes. If enhanced recognition derives solely from increased probability, the dependence of degradation rate on chain length should be linear, and recognition should be independent of the linkage within the chain. If enhanced recognition derives from a specialized chain structure, the relationship between degradation and chain length could be nonlinear, and different chains could be recognized differently.

A second model postulates that chains have a mechanistic advantage at the stage of conjugate synthesis. That is, recognition by the proteasome could be relatively independent of the structural organization of the conjugated ubiquitins, but mechanistic imperatives at the level of conjugation could dictate that a chain is the predominant product of this first phase of the degradation process. If chains have an advantage in conjugate synthesis, this will facilitate degradation by increasing conjugate concentration. In the extreme case, this second model could accommodate similar k_2 values for the two hypothetical conjugates described above.

A third model postulates that chains have an advantage in targeting because they are less readily disassembled by isopeptidases (Ellison and Hochstrasser, 1991). In the context of Fig. 3A, this corresponds to lowering the value of k_3. This will lead to a faster degradation by increasing the concentration of conjugates.

A fourth model postulates that chains confer a specific advantage at a postrecognition phase of degradation (Rubin and Finley, 1995). Because the active sites of the 20 S catalytic complex are localized within a narrow pore that cannot accommodate a structured polypeptide chain (Löwe *et al.*, 1995; Seemuller *et al.*, 1995), substrate unfolding and translocation must precede peptide bond hydrolysis. The presence in the conjugate of multiple single ubiquitins could impede translocation by "tethering" the substrate to recognition components at multiple positions within the linear substrate sequence. Using a chain for recognition would minimize this problem. This model could accommodate similar binding of the two hypothetical conjugates described above, but it predicts that the chain-bearing substrate will be degraded faster.

Finally, it has been suggested that the role of ubiquitin in proteolysis could involve more than just targeting (e.g., Wilkinson, 1988; Johnston *et al.*, 1995). Ubiquitin promotes the folding of α-linked polypeptides (Sections 2.2 and 2.3); it is conceivable that ubiquitin could promote the *un*folding of an ε-linked polypeptide. As noted above, unfolding is apparently a prerequisite for degradation. This requirement presumably explains why formation of a high-affinity ligand–substrate complex inhibits the turnover of a (polyubiquitinated) substrate (Johnson *et al.*, 1995). Thus, a final model is that in addition to targeting the substrate to the proteasome, the chain facilitates substrate unfolding. This model predicts that the chain will detectably destabilize the structure of the substrate.

These five models are not mutually exclusive. It would make the most sense for the proteasome to recognize efficiently the predominant structure generated by conjugating enzymes, and there is some evidence that attachment of a chain is the principal outcome of substrate conjugation (Section 3.3.1). There is also some evidence that the presence of a K48-linked chain enhances conjugate recognition. However, as outlined below, the available data do not yet indicate what quantitative advantage in rate or affinity derives from using chains at the level of conjugate degradation or conjugate synthesis. As discussed in Section 3.3.2, the third model appears to hold, in that a substrate-linked chain is partially protected from disassembly. The translocation barrier and unfolding models are not addressed by currently available data.

As noted above, degradation is severely impaired when chain assembly is blocked by the imposition of mutant or modified forms of ubiquitin. In fraction II, MeUb and K48R-ubiquitin support the degradation of several N-end rule substrates at rates that are $\leq 20\%$ of the rate for wild-type ubiquitin (Chau *et al.*, 1989; Gregori *et al.*, 1990; Baboshina and Haas, 1996; Beal *et al.*, 1996). Overall, these results leave no doubt that chains are used in targeting, but do not prove that chains provide a mechanistic advantage at the level of conjugate recognition, because it cannot be excluded that the slow rates obtained with the mutants reflect the low level of overall ubiquitination versus the specific absence of chains. In addition, although these rates were obtained at saturating ubiquitin concentrations, the complexity of fraction II makes it impossible to assign the observed V_{max} values to the actual proteasome step. For example, MeUb may inhibit lysozyme degradation modestly rather than severely (Hershko and Heller, 1985) because conjugate formation, not conjugate turnover, is rate-limiting for the degradation of this substrate (Dunten *et al.*, 1991).

In assays of the purified 26 S proteasome, lysozyme conjugates bearing three to six ubiquitins are degraded about 10 times more slowly than lysozyme conjugates bearing 10 to 15 ubiquitins (Hershko *et al.*, 1984; Hough *et al.*, 1986). Even though the conjugates used in this work were structurally heterogeneous, the apparently nonlinear dependence of degradation rate on the number of conjugated ubiquitins is most consistent with a model in which chains are inherently better recognized by the proteasome (see above).

3.2. Chain Recognition Determinants

3.2.1. Binding Determinants within the Chain

In principle, the question of whether assembly of a K48-linked chain creates or amplifies a special recognition element could be addressed by determining the molecular structure of a chain. We used ubiquitin conjugating enzyme E2-25K (Chen and Pickart, 1990; Chen *et al.*, 1991) to synthesize uniformly K48-linked tetraubiquitin for crystallographic analysis (Cook *et al.*, 1994). The structure shows a dimer of dimers that can be satisfyingly extended to accommodate additional ubiquitins (Fig. 4A). The ubiquitins in the dimeric unit of the tetramer are oriented quite differently from those in isolated diubiquitin, whose structure is also known (Cook *et al.*, 1992). This had been expected based on the specific symmetry of the diubiquitin structure, which could not readily accommodate additional ubiquitins (see below). The folding of each of the four ubiquitins in the tetramer (and dimer) is nearly identical to the folding of monomeric ubiquitin, as expected (Section 2.1). The noncovalent ubiquitin–ubiquitin interactions in the tetramer are entirely electrostatic. Residues whose side chains are prominently involved in stabilizing salt bridges and hydrogen bonds between two or more ubiquitin units include E23, D39, R42, E51, Y59, R72, and R74.

If recognition by the proteasome of a K48-linked chain is indeed distinct from recognition of monoubiquitin, two kinds of point mutations would be expected to affect targeting (Cook *et al.*, 1994). One type would involve residues whose side chains interact directly with recognition components of the 26 S proteasome. A second class of mutations would involve residues that participate in ubiquitin–ubiquitin interactions within the chain. Mutations in these latter residues would destabilize the chain structure, so that side chains of the first type could not be optimally presented to recognition components.

In examining the tetramer structure for candidate residues of the first type, it was noted that the three residues whose side chains comprise the "hydrophobic patch" in monoubiquitin—L8, I44, and V70—remained on the surface in tetraubiquitin (Fig. 4B). Directed mutagenesis was used to test the significance of this structural feature. It was found that double mutations (to Ala) at residues 8, 44, and 70 inhibit degradation severely, without strongly affecting steady-state conjugate level (as determined in fraction II). The simplest interpretation of these results is that the side chains of L8, I44, and V70 are important for chain recognition by the 26 S proteasome. This inference was confirmed by showing that unanchored chains assembled from wild-type ubiquitin, but not L8A,I44A-ubiquitin, inhibit degradation (Beal *et al.*, 1996).

The side chains of L8,I44 and V70 form a repeating hydrophobic patch on two of the four extended surfaces of the chain (Fig. 4B). Because these surface patches are remote from ubiquitin–ubiquitin contact regions, they should not contribute to chain stability. Thus, the negative impact of the Ala mutations is most consistent with a direct interaction of L8, I44, and V70 with recognition components. That recognition is abrogated on decreasing side chain size (the predominant effect of mutation to Ala) indicates that recognition of chains by the proteasome involves close hydrophobic contacts with the surface patches. Consistent with this model, subsequent work has shown that the *in vitro* degradative competence of ubiquitin is proportional to the size of the side chain at residue 8 (Beal *et al.*, 1998).

These results identify ubiquitin's surface hydrophobic patch as an element

Figure 4. Structure of K48-linked chain. (A) Ribbon. The relative orientations of the four ubiquitins in K48-linked tetraubiquitin are shown. The proximal ubiquitin is at the lower right. The distal ubiquitin is at the upper left. (B) Side chains involved in targeting. The side chains of L8, I44, and V70 are indicated.

that interacts with the proteasome. However, to prove that the structure of the chain is required to present this element, it is necessary to show that altering the structure of the (K48-linked) chain abrogates recognition of the patch. Demonstrating this by directed mutagenesis of residues involved in ubiquitin–ubiquitin interactions within the chain will be difficult. Several interactions may need to be eliminated before the structure is significantly destabilized; also, it may be hard to prove that the structure has been altered. Nonetheless, the properties of certain existing mutants are suggestive (Cook *et al.*, 1994). For example, the E24A substitution in vUb (Section 2.3.1a) will destabilize hydrogen bond contacts that occur among all four ubiquitins in tetraubiquitin. This and other substitutions in vUb (Cook *et al.*, 1994) may contribute to the reduced proteolytic activity of vUb *in vitro* (Haas *et al.*, 1996). R72 and R74 participate in several stabilizing interactions in tetraubiquitin. Mutation of R72 to Leu, and chemical modification of R74, eliminate the activity of ubiquitin in degradation (A. Haas, personal communication; Duerksen-Hughes *et al.*, 1987). In the latter case the effect is definitively the result of poor conjugate recognition (Section 2.4). Another way to show that the structure of the K48-linked chain is important for recognition of the surface patch is to show that chains linked through a different Lys residue are recognized differently from K48-linked chains. So far it has not been possible to show differential recognition of chains based on the presence of alternative linkages (see Section 4). However, only one recognition component is currently available for such determinations (Section 3.3.2).

In the crystal structure of K48-linked diubiquitin, the side chains of L8, I44, and V70 are packed into a hydrophobic pocket between the two ubiquitin units (Cook *et al.*, 1992). This close-packing interaction, and several hydrogen bonds, apparently stabilize the structure. The biological relevance of the diubiquitin structure remains uncertain. The symmetry of diubiquitin requires that the next ubiquitin in the chain must participate in different interactions than the first two. Also, the hydrophobic patches implicated in the interaction of chains with the proteasome (Fig. 4B) are not exposed in diubiquitin. It is probable that the crystal structure of diubiquitin does not represent the predominant solution conformation, as suggested by the results of preliminary NMR studies (R. Cohen, personal communication). Overall, the structure of tetraubiquitin, which predicts that the same ubiquitin–ubiquitin interactions will be replicated throughout the chain, provides the best current model for the structure of higher-order chains.

3.2.2. Binding Determinants within the 26 S Proteasome

The 26 S proteasome is assembled in an ATP-dependent reaction from catalytic (20 S) and regulatory (19 S) subcomplexes (Chapter 6). Components of the 26 S proteasome concerned with chain recognition are expected to reside in the 700-kDa 19 S complex, which confers on the proteasome the ability to recognize ubiquitin conjugates. The majority of the 20 or so subunits of the 19 S complex are functionally uncharacterized (Coux *et al.*, 1996). In view of the centrality of ubiquitin chain recognition in degradation, it would not be surprising if this function were shared by several functionally redundant subunits.

A blot overlay assay was used to implicate a 50-kDa subunit of the 19 S complex in the recognition of K48-linked chains (Deveraux *et al.*, 1994; Van Nocker *et al.*, 1996a,b). S5a, also known as MBP1 or MCB1, binds substrate-bound and unanchored polyubiquitin chains in a manner that depends strongly on chain length. In the blot overlay assay, the K_d values for unanchored diubiquitin and triubiquitin are ~20 and ~9 μM, respectively (Deveraux *et al.*, 1994; Baboshina and Haas, 1996). These values are approximately as expected if the chain increases binding by a simple probability effect (Section 3.1). However, tetraubiquitin binds much more tightly than di- or triubiquitin (its actual K_d is unknown; Deveraux *et al.*, 1994), and a mixture of chains with an average length of seven ubiquitins binds with K_d ~20 nM (Baboshina and Haas, 1996). These data, although as yet fragmentary, show that there is a strong and nonlinear dependence of binding on chain length above $n = 4$. The apparent break point at $n = 4$ (Deveraux *et al.*, 1994) may be related to a change in chain structure, as tetraubiqutin may be the smallest K48-linked chain whose predominant conformation exposes the hydrophobic patches (Cook *et al.*, 1994; Cummings *et al.*, 1995).

Despite its ability to bind polyubiquitin chains with high affinity (Deveraux *et al.*, 1994), and despite the sensitivity of this binding to mutations (in ubiquitin) that abrogate chain recognition by the 26 S protease (Beal *et al.*, 1996), S5a does not play a generalized role in conjugate recognition by the 26 S proteasome. This follows from the finding that disruption of the yeast S5a gene elicits no pronounced defect in general proteolysis, although one (ubiquitin fusion) test substrate is strongly stabilized (Van Nocker *et al.*, 1996b). Nonetheless, the strong correlation between chain binding to S5a and functionality in degradation suggests that key features of chain binding by S5a may be replicated in other recognition components.

These results require that there is at least one other receptor for K48-linked chains in the 19 S complex. As yet, neither biochemistry nor genetics has provided insight into the identity of this component(s). The original method used to characterize S5a (Deveraux *et al.*, 1994), and the method used to clone its cDNA (Van Nocker *et al.*, 1996a), require that the isolated subunit binds chains—a property confirmed by the ability of purified S5a to inhibit proteolysis *in vitro* (Deveraux *et al.*, 1995). The blot overlay assay also requires that isolated S5a is able to refold following denaturation in SDS. In view of these considerations, it is not surprising that the blot overlay assay fails to detect other recognition components.

Mutation of a recognition component is expected to cause a general defect in ubiquitin-mediated proteolysis, probably accompanied by hyperaccumulation of conjugates. A chain-recognizing component is not expected to discriminate among different substrates as long as they bear chains, i.e., most substrates of the pathway should be similarly stabilized in the mutant cells. Deletion of the gene encoding an important recognition component is expected to be lethal. Unfortunately, these properties are probably shared by many components of the 19 S, including those that do not function explicitly in chain recognition (e.g., DeMarini *et al.*, 1995).

3.3. Metabolism of K48-Linked Chains

3.1.1. Chain Assembly

There are two limiting mechanisms for ligation of K48-linked chains to substrates (Fig. 5), and both of them appear to operate. First, unanchored chains may be assembled, using monoubiquitin as the chain initiator, by ubiquitin-specific conjugating enzymes, and then utilized in this preassembled form by substrate-specific conjugating enzymes (Fig. 5A). Alternatively, the chain initiator may be the first ubiquitin ligated to the target by a substrate-specific conjugating enzyme (Fig. 5B). In this second mechanism, chain elongation could be

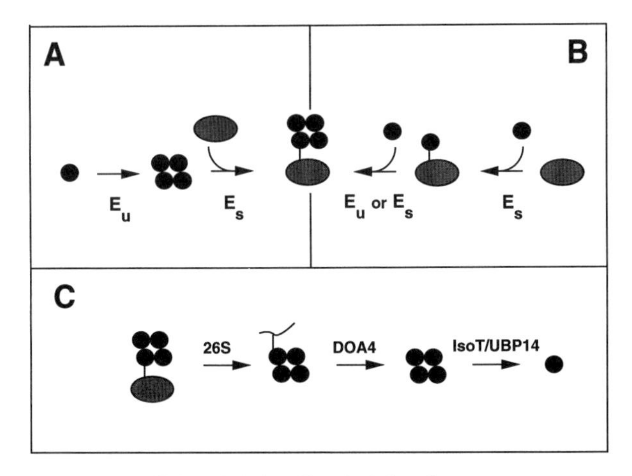

Figure 5. Metabolism of K48-linked chains. The shaded oval represents a target protein; the black circles are ubiquitins. (A,B) Models for chain assembly. Panel A depicts a model in which ubiquitin-specific conjugating enzymes (E_u) preassemble unanchored K48-linked chains, which are then utilized by substrate-specific conjugating enzymes (E_S). "Enzymes" is a generic term that encompasses E2s, E3s, and any other factors required for the respective conjugative processes. Panel B depicts a model in which substrate-specific conjugating enzymes ligate the initial ubiquitin; a chain is then extended from this ubiquitin by either type of conjugating enzyme. (C) Model for chain disassembly. See text.

catalyzed by these same substrate-specific conjugating enzymes, by distinct ubiquitin-specific conjugating enzymes, or by both.

The discovery of E2 enzymes that assemble K48-linked chains from isolated ubiquitin led to the idea that chain assembly might be mechanistically separated from substrate conjugation. Such enzymes have been characterized in mammals (E2-25K; Chen and Pickart, 1990; Chen *et al.*, 1991) and in plants (UBC7; Van Nocker and Vierstra, 1991; Van Nocker *et al.*, 1996c). The two known chain-synthesizing E2 proteins are not highly related at the primary sequence level, despite their biochemical similarities, and it remains uncertain whether these enzymes function in chain assembly within cells. For example, sevenfold over-expression of *Arabidopsis* UBC7 does not augment the cellular level of un-anchored chains or polyubiquitinated conjugates (Van Nocker *et al.*, 1996c), and the *V/K* value of E2-25K (Chen and Pickart, 1990) predicts that chain synthesis catalyzed by this enzyme will be negligible at cellular concentrations of ubiquitin. It remains possible that these enzymes function in chain assembly in cooperation with uncharacterized E3 enzymes.

Regardless of whether they originate in a dedicated conjugation pathway (above) or in conjugate degradation (Section 3.3.2), unanchored chains definitely exist within cells. Ubiquitin immunoblots of extracts from plant and animal tissues reveal significant levels of unanchored chains (Van Nocker and Vierstra, 1993; Haldeman *et al.*, 1995; Spence *et al.*, 1995), and only the K48–G76 linkage is detected when plant tissue-derived diubiquitin is analyzed by tryptic peptide mapping (Van Nocker and Vierstra, 1993). In yeast, mutation of Lys residues other than K48 does not decrease the level of unanchored chains, providing additional evidence that these species harbor primarily K48–G76 linkages (Spence *et al.*, 1995).

Diubiquitin and higher chains are activated by El, and transferred to E2s, with the same kinetics as monoubiquitin (Chen and Pickart, 1990). The resulting chain-bearing E2 thiol ester adducts are competent in E3–independent substrate conjugation (Van Nocker and Vierstra, 1993). Barring the unlikely possibility that E3 enzymes discriminate strongly against them, preassembled chains should thus make a significant contribution to substrate conjugation whenever the level of chains is significant relative to the level of monoubiquitin. This condition does not appear to be met in differentiating erythroid cells (Haldeman *et al.*, 1995), but may well be achieved in other tissues (Van Nocker and Vierstra, 1993). On the other hand, too massive an accumulation of unanchored chains is deleterious: *In vivo* and *in vitro*, unanchored chains inhibit degradation by competing with polyubi-quitinated substrates for binding to the 26 S proteasome (Papa and Hochstrasser, 1993; Beal *et al.*, 1996; Section 3.3.2).

In view of this last consideration, it is not surprising that assembly of the chain can be delayed until the first ubiquitin is conjugated to the substrate (Fig. 5B). In the best-understood example of this type of assembly, E3α and E2-14K utilize monoubiquitin as the substrate in assembling K48-linking chains on N-end rule substrates (Hough *et al.*, 1986; Chau *et al.*, 1989). This pair of enzymes forms a stable complex (Reiss *et al.*, 1989; Dohmen *et al.*, 1991; Berleth *et al.*, 1992). Recent kinetic studies indicate that chain elongation catalyzed by the complex proceeds in a highly processive fashion once the first ubiquitin has been ligated to the substrate (A. Haas, personal communication). Processive chain elongation may rely on a ubiquitin (or chain) binding site in E3α (Reiss *et al.*, 1989). It seems likely that such a processive mechanism will be the rule rather than the exception. To the extent that it is partly based in ubiquitin recognition, such a mechanism can have a substantial kinetic advantage, because an identical conjugation "site" is regenerated in each round of ubiquitination.

There may be ubiquitin-specific conjugating enzymes in addition to the two E2 enzymes discussed above. Among the yeast genes required for degradation of ubiquitin fusion substrates are *UFD2* and *UFD4*. The respective gene products are apparently E3 enzymes that function in the assembly of ubiquitin chains. The 110-kDa UFD2 protein cooperates with the UBC4/UBC5 E2 enzymes in elongating a ubiquitin chain from K48 of the ubiquitin moiety in a Ub-βgal test substrate (Johnson *et al.*, 1992, 1995). UFD2 has no homology to known E3 enzymes. UFD4 functions in the extension of a ubiquitin chain from K29 to the ubiquitin moiety in Ub-βgal and Ub-DHFR test substrates; one region of UFD4 is homologous to the active-site hect domain of the E6-AP E3 enzyme (Johnson *et al.*, 1995; Section 4). The exact biochemical activities of UFD2 and UFD4 remain undefined; both proteins exhibit specificity with regard to the site of chain initiation, but subsequent linkages within the respective chains have not yet been characterized. Chain propagation could therefore involve a distinct linkage. It is attractive to speculate that these enzymes are "ubiquitin-specific" E3s that function in processive elongation of chains from the mono- or oligo-ubiquitinated products of substrate-specific conjugating enzymes, or even E3s that function in assembly of unanchored chains (Fig. 5A,B). However, such a function is not easily reconciled with the finding that UFD2 discriminates among ubiquitin fusion test substrates based on the identity of the nonubiquitin moiety (Johnson *et al.*, 1995).

3.3.2. Chain Disassembly

Ellison and Hochstrasser (1991) first pointed out that a polyubiquitin chain has a potential advantage as a targeting signal by virtue of persistence: Removal of a single ubiquitin from the distal end of the chain will leave many ubiquitins in close proximity to recognition components. It is now evident that chains are chiefly disassembled by a mechanism that maximizes persistence while minimizing deleterious effects of the chain once substrate degradation is complete. Because this process is reviewed in detail elsewhere (Chapter 4), only a few salient points are discussed here.

Mutations in the yeast *DOA4* isopeptidase gene inhibit ubiquitin-mediated proteolysis, and cause pronounced accumulation of small peptides conjugated to polyubiquitin chains (Papa and Hochstrasser, 1993). Disruption of the *UBP14* gene, which encodes a different isopeptidase, also inhibits degradation, but in this case, unanchored chains, rather than peptide-linked chains, accumulate (M. Hochstrasser, personal communication). The two types of chains can be distinguished by two-dimensional IEF/SDS–PAGE (e.g., Spence *et al.*, 1995; Haldeman *et al.*, 1995). In each case, inhibition of proteolysis apparently reflects the accumulation of chain species that bind to the 26 S proteasome in competition with conjugated substrates. The structure of the accumulated chains in *doa4* cells suggests that DOA4's normal role is to remove the largely intact chain from the limit degradation product at a terminal stage of breakdown. On the other hand, biochemical data indicate that the mammalian *UBP14* homologue, known as isopeptidase T (Hadari *et al.*, 1992), preferentially disassembles unanchored chains by sequentially removing ubiquitin units from the (free) proximal chain terminus (Wilkinson *et al.*, 1995). The accumulation of unanchored chains in *ubp14* disruptants is consistent with isopeptidase T representing the major *in vivo* route to disassembly of these species.

These data support a two-step mechanism of chain disassembly (Fig. 5C): When degradation of the substrate is nearly complete, DOA4 releases the chain by cleaving the isopeptide bond that links the proximal ubiquitin to the Lys ε-amino group of the substrate; this cleavage exposes G76 of the proximal ubiquitin, leading to disassembly of the chain by isopeptidase T. Although it is clear that there are enzymes that can disassemble a chain from its distal end (see below), the phenotypes of *doa4* and *ubp14* mutants strongly suggest that the sequential actions of these enzymes represent the predominant mechanism of disassembly of K48-linked chains. This mechanism maximizes the persistence of the chain targeting signal, but minimizes inhibition by the chain once degradation is complete.

Besides the DOA4–isopeptidase T route for chain disassembly, which is centered on the proximal terminus of the chain, there exist other chain-disassembling enzymes with different specificities (this is apparently why K48R-ubiquitin does not act as a dominant negative; Section 2.4.4). One such enzyme, localized in the 19 S complex, removes ubiquitins sequentially from the distal chain end (Lam *et al.*, 1997). Long chains, will be least affected by such distal-end "nibbling." Thus, the properties of this enzyme will tend to give long chains an additional advantage in targeting.

4. OTHER POLYUBIQUITIN CHAINS

At least four Lys residues besides K48 are known to be sites of chain initiation (Section 2.4.4). The significance of this phenomenon has yet to be fully elucidated. It remains possible that all chains function in proteolytic targeting in an essentially equivalent manner. However, considerable circumstantial evidence argues against this model, including the defined structure of K48-linked tetraubiquitin (Section 3.2.1), the predominant utilization of the K48 linkage (versus other linkages) in proteolysis (Section 3.1), and the specific DNA repair defect observed in cells expressing K63R-ubiquitin (Section 2.4.4). It is attractive to hypothesize that alternatively linked chains are selectively recognized, leading to distinct functions for different chains. This would be possible if different chains have different structures, and thus present distinct recognition elements on the chain surface. This model remains speculative; indeed, certain data argue against it.

Several major unresolved issues provide a useful framework for reviewing the relevant data. The first such issue is the "primary" structure of alternatively linked chains: When a novel linkage is present, is it uniformly propagated throughout the chain? As discussed above (Section 2.4), strict proof of uniform linkage requires chemical mapping, or assembly of the chain from a ubiquitin derivative containing only one Lys residue. So far this criterion has been satisfied only for K48-linked chains (Section 3.1).

A second issue is the "quaternary" structure of alternatively linked chains: Do such chains adopt a conformation distinct from the K48 chain conformation? It is unlikely *a priori* that the K48 chain conformation is broadly replicated in other uniformly linked chains. For example, the ε-amino group of K63 is more than 20 Å distant from that of K48, and it would appear impossible to make a K63-linked tetramer adopt the conformation observed in K48-linked tetraubiquitin (Spence *et al.*, 1995). However, given the high solvent exposure of most of the Lys residues and the C-terminus, it remains possible that chains harboring certain linkages can adopt a "beads on a string" conformation in which there are minimal noncovalent ubiquitin–ubiquitin interactions (Cook *et al.*, 1992).

A third issue is whether there exist recognition components specific for alternatively linked chains. Selective recognition, presumably dependent on unique chain structures, is the simplest explanation for the specific functions observed for K48- and K63-linked chains. Progress in this area has been impeded by the lack of information about chain recognition components. The only characterized component, S5a, does not discriminate strongly among chains harboring K48, K6, and K11 linkages (Baboshina and Haas, 1996; see below). The significance of this apparent promiscuity is difficult to evaluate without a fuller knowledge of the structures of these chains, and the mode(s) of their binding by S5a.

A final issue is the nature of the specificity determinants in the formation of novel chains. It appears that linkage-specific E2/E3 pairs may exist (Section 3.3.1; see below), and regulated expression of such enzymes could provide a mechanism to alter the spectrum of chain linkages in the cell. However, the structural features of ubiquitin that influence the specificities of such enzymes remain unknown.

Chains initiating at K29. K29-linked diubiquitin synthesis dependent on the presence of the UBC4/UBC5 E2 enzymes has been observed in yeast cells (Arnason and Ellison, 1994; Section 2.4.4), and K29 can initiate a targeting-competent chain at the ubiquitin moiety of a Ub-DHFR test substrate (Johnson *et al.*, 1995). In addition to UBC4 or UBC5, this assembly requires the 140-kDa E3 encoded by the *UFD4* gene (Section 3.3.1). Unexpectedly, a Ub-βgal test substrate is strongly stabilized in yeast by making *either* a K48R or a K29R mutation in the fused ubiquitin moiety (Johnson *et al.*, 1995). The basis of this effect is unclear. It could result from a requirement for ligation of two chains to the initiating ubiquitin, or from unexpected structure–function requirements of UFD4 or UFD2 (UFD2 is involved in extending a chain from K48 of Ub-βgal; Section 3.3.1).

It is not known whether chains that initiate at K29 are propagated exclusively through K29–G76 linkages. In the affirmative, K29-linked chains must interact productively with the proteasome. However, uniform K29-linked chains cannot be broadly utilized in targeting. This follows from the general proteolytic competence and unaltered conjugate profile of yeast cells expressing the K29R protein as the sole form of ubiquitin (Spence *et al.*, 1995).

Chains initiating at K6. The yeast RAD6 E2 enzyme elaborates chains on histones *in vitro* (Jentsch *et al.*, 1987; Berleth and Pickart, 1989; Haas *et al.*, 1991). Only monoubiquitinated histone is formed when K6R-ubiquitin is substituted for wild-type ubiquitin (Baboshina and Haas, 1996). Thus, K6 is a site of chain initiation in this reaction, and it is possible that the K6–G76 linkage is propagated throughout the chain. The ubiquitinated histone product assembled from wild-type ubiquitin binds to immobilized S5a with the same affinity as does autoubiquitinated CDC34 (which bears K48-linked chains; Banerjee *et al.*, 1993). This result is difficult to reconcile with a critical role for the K48 chain structure in "presenting" the hydrophobic patches (Section 3.2.1), as the side chain of K6 is ~18 Å from that of K48. The resolution of this dilemma awaits the structural characterization of a K6-linked chain.

When RAD6 acts autonomously, it assembles chains through K6 (see above). However, when RAD6 acts on N-end rule substrates in concert with mammalian E3α, chains are assembled through K48. The linkage specificity of RAD6 thus depends on the enzymatic mechanism of chain synthesis. This is just one example of such linkage switching. Remarkably, the UBC4/UBC5 E2 enzymes can initiate chains at K29 (Arnason and Ellison, 1994; Johnson *et al.*, 1995), K48 (Johnson *et al.*, 1992), or K63 (Arnason and Ellison, 1994; Spence *et al.*, 1995). Whether the chain initiates at K29 or K48 may well depend on the identity of the cooperating E3 (see above); nothing is yet known about additional enzymatic requirements in the synthesis of K63–G76 bonds. If different chains are selectively recognized, this kind of linkage switching could provide a mechanism for achieving selectivity in the fates of ubiquitinated proteins.

Chains initiating at K11. The keratinocyte-specific enzyme E2-EPF supports degradation of an N-end rule substrate in E2-depleted fraction II (Liu *et al.*, 1996; Baboshina and Haas, 1996). This activity involves a chain that initiates at K11, as substitution of K11R-ubiquitin for wild-type ubiquitin severely inhibits degradation; however, these chains may contain K48 linkages as well, as there is partial inhibition by K48R-ubiquitin (Liu *et al.*, 1996; Baboshina and Haas, 1996). E2-EPF catalyzes an autoubiquitination reaction that is strongly suppressed when it is carried out with K11R-ubiquitin; the autoubiquitinated E2 binds to immobilized S5a with an affinity similar to autoubiquitinated CDC34 bearing K48-linked chains (Baboshina and Haas, 1996). If the E2-EPF-linked chains are indeed uniformly linked, then K11- and K48-linked chains must bind similarly to S5a. Consistent with this idea, the binding of autoubiquitinated E2-EPF to S5a is strongly inhibited when L8A,I44A-ubiquitin is used in chain assembly (A. Haas, personal communication). The same questions apply as with K6-linked chains (above).

Chains initiating at K63. Yeast cells expressing solely K63R-ubiquitin are proficient in general proteolysis and contain normal levels of unanchored chains and most conjugates (Spence *et al.*, 1995). However, polybuiquitinated forms of a small, basic protein are strikingly absent, indicating that this *in vivo* substrate bears a chain that initiates and/or propagates through K63. Ubiquitinated forms of this substrate comprise a significant fraction of the total conjugate pool in *S. cerevisiae*.

Yeast cells expressing solely K63R-ubiquitin are deficient in error-prone DNA repair carried out by the RAD6 pathway, indicating that such DNA repair depends in some fashion on the assembly of a K63-linked chain (Spence *et al.*, 1995). The DNA repair defect is masked when K63R-ubiquitin is overexpressed in a wild-type ubiquitin background (Arnason and Ellison, 1994). On the other hand, cells coexpressing K63R-ubiquitin and wild-type ubiquitin in a *ubi4*Δ background exhibit a stress-sensitive phenotype despite general proteolytic competence, suggesting that protection against stress depends on assembly of K63-linked chain (Arnason and Ellison, 1994). The finding that thermal sensitivity is exaggerated when K63C-ubiquitin is overexpressed (Spence *et al.*, 1995) raises the possibility that the stress phenotype elicited by overexpression of K63 mutants reflects structure–function requirements rather than inappropriate chain termination, as K63R- and K63C-ubiquitin should function similarly as chain terminators (as seen for the DNA repair phenotype; Spence *et al.*, 1995). However, it is also possible that the exaggerated K63C phenotype reflects selective destabilization of this derivative in stressed cells, reflecting disulfide bond formation or other oxidative modifications.

It is not clear whether the two implied functions of K63-linked chains—in DNA repair and stress resistance—are related or distinct. Nor is it proven that the chains mediating these functions harbor a uniform K63–G76 linkage, although such a structure is consistent with the finding that among the Lys-to-Arg mutations, only K63R abrogates ubiquitination of the basic substrate (Spence *et al.*, 1995). Nor is the fate of a conjugate bearing a K63-linked chain known. Nonetheless, these results provide the strongest evidence to date that an alternatively linked chain can be selectively recognized. The formation of K63-linked chains, unanchored or substrate-conjugated, depends on UBC4 and UBC5 (Arnason and Ellison, 1994; Spence *et al.*, 1995). Expression of these E2 enzymes is induced in stress conditions (Seufert and Jentsch, 1990); this may facilitate assembly of K63-linked chains in stressed cells.

5. CONCLUDING REMARKS

The role of ubiquitin as a covalent proteolytic signal, and the dependence of proteolytic targeting on the formation of a ubiquitin–substrate isopeptide bond, were well established by 1985. A major conclusion of the last decade is that ubiquitin–ubiquitin isopeptide bonds play an equally important role in proteolytic targeting. This surprising feature has piqued interest from unexpected quarters (see Barrett, 1996). Even though important aspects of chain function remain to be elucidated, it seems clear that K48-linked chains provide a mechanistic advantage at the levels of synthesis, recognition, and persistence of the targeting signal. Unresolved questions of particular interest include the identities and properties of the proteolytic components mediating recognition of K48-linked chains, and the enzymatic mechanisms leading to the assembly of chains on proteolytic substrates.

The ubiquitin pathway is distinguished by an unusual combination of high volume and high specificity. Nearly all short-lived proteins are substrates, but the half-lives of individual proteins can be acutely and independently regulated. Selectivity in degradation arises principally at the level of ubiquitination. The existence of multiple E2 and E3 enzymes provides both intrinsic specificity, through the recognition of substrate-based ubiquitination signals by E3 enzymes, and combinatorial specificity enhancement, through the formation of E2–E3 pairs with distinct conjugative properties. The existence of ubiquitin chains harboring distinct isopeptide linkages, and the specific functions associated with two such linkages, suggest that still another specificity mechanism may exist, operating at the level of the fates of ubiquitinated proteins, in which conjugates can be "sorted" based on the linkage in a polyubiquitin chain. Proof of this novel mechanism awaits a more complete characterization of the structure, function, and recognition of alternatively linked chains.

Acknowledgments. Our work on polyubiquitin chains is supported by grant DK46984 and a Research Career Development Award, both from the NIH. I am indebted to Rick Beal, Julie Piotrowski, Ann Hofmann, and Glen Lawson for valuable comments on the manuscript, and to David Noll for assistance in preparing the structure figures. I thank Rick Beal, Bob Cohen, Art Haas, Mark Hochstrasser, Amy Lam, and Gang Xia for permission to cite unpublished data.

6. REFERENCES

Arnason, T., and Ellison, M. J., 1994, Stress resistance in *Saccharomyces cerevisiae* is strongly correlated with assembly of a novel type of multiubiquitin chain, *Mol. Cell. Biol.* **14:**7876–7883.

Aso, T., Lane, W. S., Conaway, J. W., and Conaway, R. C., 1995, Elongin (SIII): A multisubunit regulator of elongation by RNA polymerase II, *Science* **269:**1439–1443

Baboshina, O. V., and Haas, A. L., 1996, Novel multiubiquitin chain linkages catalyzed by the conjugating enzymes E2-EPF and RAD6 are recognized by 26S proteasome subunit 5, *J. Biol. Chem.* **271:**2823–2831.

Bachmair, A., Finley, D., and Varshavsky, A., 1986, In vivo half-life of a protein is a function of its amino-terminal residue, *Science* **234:**179–186.

Baker, R. T., Smith, S. A., Marano, R., McKee, J., and Board, P. G., 1994, Protein expression using cotranslational fusion and cleavage of ubiquitin, *J. Biol. Chem.* **269:**25381–25386.

Baker, R. T., Williamson, N. A., and Wettenhall, R. E. H., 1996, The yeast homolog of mammalian ribosomal protein S30 is expressed from a duplicated gene without a ubiquitin-like protein fusion sequence, *J. Biol. Chem.* **271:**13549–13555.

Banerjee, A., Gregori, L., Xu, Y., and Chau, V., 1993, The bacterially expressed yeast CDC34 gene product can undergo autoubiquitination to form a multiubiquitin chain-linked protein, *J. Biol. Chem.* **268:**5668–5675.

Banerjee, J., Sands, J., Strominger, J. L., and Spies, T., 1990, A gene pair from the human major histocompatibility complex encodes large proline-rich proteins with multiple repeated motifs and a single ubiquitin-like domain, *Proc. Natl. Acad. Sci. USA* **87:**2374–2378.

Barrett, A., 1996, The mysteries of ubiquitin, *Story* **Summer 1996:**69–82.

Beal, R., Deveraux, Q., Xia, G., Rechsteiner, M., and Pickart, C., 1996, Surface hydrophobic residues of multiubiquitin chains essential for proteolytic targeting, *Proc. Natl. Acad. Sci. USA* **93:** 861–866.

Beal, R., Toscano-Cantatfa, D., Young, P., Rechsteiner, M., and Pickart, C., 1998, The hydrophobic effect contributes to polyubiquitin chain recognition, *Biochemistry* **37**: in press.

Beers, E. P., and Callis, J., 1993, Utility of polyhistidine-tagged ubiquitin in the purification of ubiquitin–protein conjugates and as an affinity ligand for the purification of ubiquitin-specific hydrolases, *J. Biol. Chem.* **268:**21645–21649.

Berleth, E. S., and Pickart, C. M., 1989, Several mammalian ubiquitin carrier proteins, but not E2-20K, are related to the 20-kDa yeast E2, RAD6, *Biochem. Biophys. Res. Commun.* **171:**705–710.

Berleth, E. S., Kasperek, E. J., Grill, S. P., Braunscheidel, J. A., Graziani, L. A., and Pickart, C. M., 1992, Inhibition of ubiquitin-protein ligase (E3) by mono- and bifunctional phenylarsenoxides, *J. Biol. Chem.* **267:**16403–16411.

Biggins, S., Ivanovska, I., and Rose, M. D., 1996, Yeast ubiquitin-like genes are involved in duplication of the microtubule organizing center, *J. Cell Biol.* **133:**1331–1346.

Burch, T. J., and Haas, A. L., 1994, Site-directed mutagenesis of ubiquitin. Differential roles for arginine in the interaction with ubiquitin-activating enzyme, *Biochemistry* **33:**7300–7308.

Butt, T. R., Khan, M. I., Marsh, J., Ecker, D. J., and Crooke, S. T., 1988, Ubiquitin-metallothionein fusion protein expression in yeast: A genetic approach for analysis of ubiquitin functions, *J. Biol. Chem.* **263:**16364–16371.

Chan, Y.L., Suzuki, K., and Wool, I. G., 1995, The carboxyl extensions of two rat ubiquitin fusion proteins are ribosomal proteins S27a and L40, *Biochem. Biophys. Res. Commun.* **215:**682–690.

Chau, V., Tobias, J. W., Bachmair, A., Marriott, D., Ecker, D. J., Gonda, D. K., and Varshavsky, A., 1989, A multiubiquitin chain is confined to specific lysine in a targeted short-lived protein, *Science* **243:**1576–1583.

Chen, Z., and Pickart, C. M., 1990, A 25-kilodalton ubiquitin carrier protein (E2) catalyzes multiubiquitin chain synthesis via lysine 48 of ubiquitin, *J. Biol. Chem.* **265:**21835–21842.

Chen, Z., Niles, E. G., and Pickart, C. M., 1991, Isolation of a cDNA encoding a mammalian multiubiquitinating enzyme (E2-25K) and overexpression of the functional enzyme in *E. coli*, *J. Biol. Chem.* **266:**15698–15704.

Chen, Z., Hagler, J., Palombella, V. J., Melandri, F., Scherer, D., Ballard, D., and Maniatis, T., 1995, Signal-induced site-specific phosphorylation targets IκBα to the ubiquitin-proteasome pathway, *Genes Dev.* **9:**1586–1597.

Cook, W. J., Jeffrey, L. C., Carson, M., Chen, Z., and Pickart, C. M., 1992, Structure of a diubiquitin conjugate and a model for interaction with ubiquitin conjugating enzyme (E2), *J. Biol. Chem.* **267:**16467–16471.

Cook, W. J., Jeffrey, L. C., Kasperek, E., and Pickart, C. M., 1994, Structure of tetraubiquitin shows how multiubiquitin chains can be formed, *J. Mol. Biol.* **236:**601–609.

Coux, O., Tanaka, K., and Goldberg, A. L., 1996, Structure and functions of the 20S and 26S proteasomes, *Annu. Rev. Biochem.* **65:**801–847.

Cox, M. J., Haas, A. L., and Wilkinson, K. D., 1986, Role of ubiquitin conformations in the specificity of protein degradation: Iodinated derivatives with altered conformations and activities, *Arch. Biochem. Biophys.* **250:**400–409.

Cummings, M. D., Hart, T. N., and Read, R. J., 1995, Monte Carlo docking with ubiquitin, *Prot. Sci.* **4:**885–899.

D'Cunha, J., Knight, E., Haas, A. L., Truitt, R. L., and Borden, E. C., 1996, Immunoregulatory properties of ISG15, an interferon-induced cytokine, *Proc. Natl. Acad. Sci. USA* **93:**211–215.

DeMarini, D. J., Papa, F. R., Swaminathan, S., Ursic, D., Rasmussen, T. P., Culbertson, M. R., and Hochstrasser, M., 1995, The yeast *SEN3* gene encodes a regulatory subunit of the 26S proteasome complex required for ubiquitin-dependent protein degradation in vivo, *Mol. Cell. Biol.* **15:**6311–6321.

Deveraux, Q., Ustrell, V., Pickart, C., and Rechsteiner, M., 1994, A 26S protease subunit that binds ubiquitin conjugates, *J. Biol. Chem.* **269:**7059–7061.

Deveraux, Q., Van Nocker, S., Mahaffey, D., Vierstra, R., and Rechsteiner, M., 1995, Inhibition of ubiquitin-mediated proteolysis by the *Arabidopsis* 26S protease subunit S5a, *J. Biol. Chem.* **270:**29660–29663.

Di Stefano, D. L., and Wand, A. J., 1987, Two-dimensional ^{1}H NMR study of human ubiquitin: A main chain directed assignment and structure analysis, *Biochemistry* **26:**7272–7281.

Dohmen, R. J., Madura, K., Bartel, B., and Varshavsky, A., 1991, The N-end rule is mediated by the UC2(RAD6) ubiquitin-conjugating enzyme, *Proc. Natl. Acad. Sci. USA* **88:**7351–7355.

Dohmen, R. J., Stappen, R., McGrath, M. P., Forrova, H., Kolarov, J., Goffeau, A., and Varshavsky, A., 1995, An essential yeast gene encoding a homolog of ubiquitin-activating enzyme, *J. Biol. Chem.* **270:**18099–18109.

Duan, D. R., Pause, A., Burgess, W. H., Aso, T., Chen, D. Y. T., Garrett, K. P., Conaway, R. C., Conaway, J. W., Linehan, W. M., and Klausner, R. D., 1995, Inhibition of transcription elongation by the VHL tumor suppressor protein, *Science* **269:**1402–1406.

Duerksen-Hughes, P. J., Xu, X., and Wilkinson, K. D., 1987, Structure and function of ubiquitin: Evidence for differential interactions of arginine-74 with the activating enzyme and the proteases of ATP-dependent proteolysis, *Biochemistry* **26:**6980–6987.

Dunten, R. L., Cohen, R. E., Gregori, L., and Chau, V., 1991, Specific disulfide cleavage is required for ubiquitin conjugation and degradation of lysozyme, *J. Biol. Chem.* **266:**3260–3267.

Durner, J., and Boger, P., 1995, Ubiquitin in the prokaryote *Anabena variabilis*, *J. Biol. Chem.* **270:** 3720–3725.

Dworkin-Rastl, E., Shrutkowski, A., and Dworkin, M. B., 1984, Multiple ubiquitin mRNAs during *Xenopus laevis* development contain tandem repeats of the 76 amino acid coding sequence, *Cell* **39:**321–325.

Ecker, D. J., Butt, T. R., Marsh, J., Sternberg, E. J., Margolis, N., Monia, B. P., Jonnalagadda, S., Khan, M. I., Weber, P. L., Mueller, L., and Crooke, S. T., 1987, Gene synthesis, expression, structures, and functional activities of site-specific mutants of ubiquitin, *J. Biol. Chem.* **262:**14213–14221.

Ellison, M. J., and Hochstrasser, M., 1991, Epitope-tagged ubiquitin: A new probe for analyzing ubiquitin function, *J. Biol. Chem.* **266:**21150–21157.

Finley, D., Özkaynak, E., and Varshavsky, A., 1987, The yeast polyubiquitin gene is essential for resistance to high temperatures, starvation, and other stresses, *Cell* **48:**1035–1046.

Finley, D., Bartel, B., and Varshavsky, A., 1989, The tails of ubiquitin precursors are ribosomal proteins whose fusion to ubiquitin facilitates ribosome biogenesis, *Nature* **338:**394–401.

Finley, D., Sadis, S., Monia, B. P., Boucher, P., Ecker, D. J., Crooke, S. T., and Chau, V., 1994, Inhibition of proteolysis and cell cycle progression in a multiubiquitination-deficient yeast mutant, *Mol. Cell. Biol.* **14:**5501–5509.

Fujimuro, M., Sawada, H., and Yokosawa, H., 1994, Production and characterization of monoclonal antibodies specific to multi-ubiquitin chains of polyubiquitinated proteins, *FEBS Lett.* **349:** 173–180.

Garrett, K. P., Aso, T., Bradsher, H. N., Foundling, S. I., Lane, W. S., Conaway, R. C., and Conaway, J. W., 1995, Positive regulation of general transcription factor SIII by a tailed ubiquitin homolog, *Proc. Natl. Acad. Sci. USA* **92:**7172–7176.

Goebl, M. G., Yochem, J., Jentsch, S., McGrath, J. P., Varshavsky, A., and Byers, B., 1988, The yeast cell cycle gene *CDC34* encodes a ubiquitin-conjugating enzyme, *Science* **241:**1331–1335.

Gonda, D. K., Bachmair, A., Wunning, I., Tobias, J. W., Lane, W. S., and Varshavsky, A., 1989, Universality and structure of the N-end rule, *J. Biol. Chem.* **264:**16700–16712.

Gregori, L., Poosch, M. S., Cousins, G., and Chau, V., 1990, A uniform isopeptide-linked multiubiquitin chain is sufficient to target substrate for degradation in ubiquitin-mediated proteolysis, *J. Biol. Chem.* **265:**8354–8357.

Guarino, L. A., 1990, Identification of a viral gene encoding a ubiquitin-like protein, *Proc. Natl. Acad. Sci. USA* **87:**409–413.

Guarino, L. A., Smith, G., and Dong, W., 1995, Ubiquitin is attached to membranes of baculovirus particles by a novel type of phospholipid anchor, *Cell* **80:**301–309.

Guzder, S. N., Bailly, V., Sung, P., Prakash, L., and Prakash, S., 1995, Yeast DNA repair protein RAD23 promotes complex formation between transcription fact TFIIH and DNA damage recognition factor RAD14, *J. Biol. Chem.* **270:**8385–8388.

Haas, A. L., 1988, Immunochemical probes of ubiquitin pool dynamics, in *Ubiquitin* (M. Rechsteiner, ed.), pp. 173–206, Plenum Press, New York.

Haas, A. L., and Bright, P. M., 1985, The immunochemical detection and quantitation of intracellular ubiquitin–protein conjugates, *J. Biol. Chem.* **260:**12464–12473.

Haas, A. L., Murphy, K. E., and Bright, P. M., 1985, The inactivation of ubiquitin accounts for the inability to demonstrate ATP, ubiquitin-dependent proteolysis in liver extracts, *J. Biol. Chem.* **260:**4694–4703.

Haas, A. L., Ahrens, P., Bright, P. M., and Ankel, H., 1987, Interferon induces a 15-kilodalton protein exhibiting marked homology to ubiquitin, *J. Biol. Chem.* **262:**11315–11323.

Haas, A., Reback, P. M., Pratt, G., and Rechsteiner, M., 1990, Ubiquitin-mediated degradation of histone H3 does not require the substrate-binding ubiquitin protein ligase, *E3*, or attachment of polyubiquitin chains, *J. Biol. Chem.* **265:**21664–21669.

Haas, A. L., Reback, P. B., and Chau, V., 1991, Ubiquitin conjugation by the yeast RAD6 and CDC34 gene products: Comparison to their putative rabbit homologs, E2-20K and E2-32K, *J. Biol. Chem.* **266:**5104–5112.

Haas, A. L., Katzung, D. R., Reback, P. M., and Guarino, L. A., 1996, Functional characterization of the ubiquitin variant encoded by the baculovirus *Autographa californica*, *Biochemistry* **35:**5385–5394.

Hadari, T., Warms, J. V. B., Rose, I. A., and Hershko, A., 1992, A ubiquitin C-terminal isopeptidase that acts on polyubiquitin chains, *J. Biol. Chem.* **267:**719–727.

Haldeman, M. T., Finley, D., and Pickart, C. M., 1995, Dynamics of ubiquitin conjugation during erythroid differentiation *in vitro*, *J. Biol. Chem.* **270:**9507–9516.

Hershko, A., and Heller, H., 1985, Occurrence of a polyubiquitin structure in ubiquitin–protein conjugates, *Biochem. Biophys. Res. Commun.* **128:**1079–1086.

Hershko, A., and Rose, I. A., 1987, Ubiquitin aldehyde: A general inhibitor of ubiquitin-recycling processes, *Proc. Natl. Acad. Sci. USA* **84:**1829–1833.

Hershko, A., Ciechanover, A., Heller, H., Haas, A. L., and Rose, I. A., 1980, Proposed role of ATP in protein breakdown: Conjugation of proteins with multiple chains of the polypeptide of ATP-dependent proteolysis, *Proc. Natl. Acad. Sci. USA* **77:**1783–1786.

Hershko, A., Ciechanover, A., and Rose, I. A., 1981, Identification of the active amino acid residue of the polypeptide of ATP-dependent protein breakdown, *J. Biol. Chem.* **256:**1525–1528.

Hershko, A., Leshinsky, E., Ganoth, D., and Heller, H., 1984, ATP-dependent degradation of ubiquitin–protein conjugates, *Proc. Natl. Acad. Sci. USA* **81:**1619–1623.

Hershko, A., Ganoth, D., Pehrson, J., Palazzo, R. E., and Cohen, L. H., 1991, Methylated ubiquitin inhibits cyclin degradation in clam embryo extracts, *J. Biol. Chem.* **266:**16376–16379.

Hicke, L., and Riezman, H., 1996, Ubiquitination of a yeast plasma membrane receptor signals its ligand-stimulated endocytosis, *Cell* **84:**277–287.

Hochstrasser, M., Ellison, M. J., Chau, V., and Varshavsky, A., 1991, The short-lived MATα2 transcriptional regulator is ubiquitinated in vivo, *Proc. Natl. Acad. Sci. USA* **88:**4604–4610.

Hodgins, R. R. W., Ellison, K. S., and Ellison, M. J., 1992, Expression of a ubiquitin derivative that conjugates to protein irreversibly produces phenotypes consistent with a ubiquitin deficiency, *J. Biol. Chem.* **267:**8807–8812.

Hough, R., Pratt, G., and Rechsteiner, M., 1986, Ubiquitin–lysozyme conjugates: Identification and characterization of an ATP-dependent protease from rabbit reticulocyte lysates, *J. Biol. Chem.* **261:**2400–2408.

Jabusch, J. R., and Deutsch, H. F., 1985, Localization of lysine acetylated in ubiquitin reacted with *p*-nitrophenyl acetate, *Arch. Biochem. Biophys.* **238:**170–177.

Johnson, E. S., Bartel, B., Seufert, W., and Varshavsky, A., 1992, Ubiquitin as a degradation signal, *EMBO J.* **11:**497–505.

Johnson, E. S., Ma, P. C. M., Ota, I. M., and Varshavsky, A., 1995, A proteolytic pathway that recognizes ubiquitin as a degradation signal, *J. Biol. Chem.* **270:**17442–17456.

Johnson, P. R., and Hochstrasser, M., 1997, Sumo-1: ubiquitin gains weight, *Trends Cell Biol.* **7:**408–413.

Johnston, J. A., Johnson, E. S., Waller, P. R. H., and Varshavsky, A., 1995, Methotrexate inhibits proteolysis of dihydrofolate reductase by the N-end rule pathway, *J. Biol. Chem.* **270:**8172–8178.

Kalchman, M. A., Graham, R. K., Koide, H. B., Xia, G., Hodgson, J. G., Graham, K. C., Goldberg, Y. P., Gietz, R. D., Pickart, C. M., and Hayden, M. R., 1996, Huntingtin is ubiquitinated and interacts with a specific ubiquitin-conjugating enzyme, *J. Biol. Chem.* **271:**19385–19394.

Kas, K., Michiels, L., and Merregaert, J., 1992, Genomic structure and expression of the human fau gene: Encoding the ribosomal protein S30 fused to a ubiquitin-like protein, *Biochem. Biophys. Res. Commun.* **187:**927–933.

Khorasanizadeh, S., Peters, I. D., Butt, T. R., and Roder, H., 1993, Folding and stability of a tryptophan-containing mutant of ubiquitin, *Biochemistry* **32:**7054–7063.

Kibel, A., Iliopoulos, O., DeCaprio, J. A., and Kaelin, W. G., 1995, Binding of the von Hippel-Lindau tumor suppressor protein to elongin B and C, *Science* **269:**1444–1446.

Knight, E., Fahey, D., Cordova, B., Hillman, M., Kutny, R., Reich, N., and Blomstrom, D., 1988, A 15-kDa interferon-induced protein is derived by COOH-terminal processing of a 17-kDa protein, *J. Biol. Chem.* **263:**4520–4522.

Kumar, S., Yoshida, Y., and Noda, M., 1993, Cloning of a cDNA which encodes a novel ubiquitin-like protein, *Biochem. Biophys. Res. Commun.* **195:**393–399.

Lam, Y. A., Xu, W., DeMartino, G. N., and Cohen, R. E., 1997, Editing of ubiquitin conjugates by an isopeptidase in the 26S proteasome, *Nature* **385:**737–740.

Linnen, J. M., Bailey, C. P., and Weeks, D. L., 1993, Two related localized mRNAs from *Xenopus laevis* encode ubiquitin-like fusion proteins, *Gene* **128:**181–188.

Liu, Z., Haas, A. L., Diaz, L. A., Conrad, C. A., and Giudice, G. J., 1996, Characterization of a novel keratinocyte ubiquitin carrier protein, *J. Biol. Chem.* **271:**2817–2822.

Loeb, K. R., and Haas, A. L., 1992, The interferon-inducible 15-kDa ubiquitin homolog conjugates to intracellular proteins, *J. Biol. Chem.* **267:**7806–7813.

Loeb, K. R., and Haas, A. L., 1994, Conjugates of a ubiquitin cross-reactive protein distribute in a cytoskeletal pattern, *Mol. Cell. Biol.* **14:**8408–8419.

Lowe, J., McDermott, H., Loeb, K., Landon, M., Haas, A. L., and Mayer, R. J., 1995, Immunohistochemical localization of ubiquitin cross-reactive protein in human tissues, *J. Pathol.* **177:** 163–179.

Löwe, J., Stock, D., Jap, B., Zwickl, P., Baumeister, W., and Huber, R., 1995, Crystal structure of the 20s proteasome from the archaeon *T. acidophilum* at 3.4 Å resolution, *Science* **268:**533–539.

Mahajan, R., Delpin, C., Guan, T., Gerace, L., and Melchior, F., 1997, A small ubiquitin-related polypeptide involved in targeting RanGAP1 to nuclear pore complex protein RanBP2, *Cell* **88:**97–107.

Masutani, C., Sugasawa, K., Yanagisawa, J., Sonoyama, T., Ui, M., Enomoto, T., Takio, K., Tanaka, K., van der Spek, P. J., Bootsma, D., Hoeijmakers, H. J. H., and Hanaoka, F., 1994, Purification and cloning of a nucleotide excision repair complex involving the xeroderma pigmentosum group C protein and a human homologue of yeast RAD23, *EMBO J.* **13:**1831–1843.

Matunis, M. J., Coutavas, E., and Blobel, G., 1996, A novel ubiquitin-like modification modulates the partitioning of the Ran-GTPase-activating protein RanGAP1 between the cytosol and the nuclear pore complex, *J. Cell Biol.* **135:**1457–1470.

Mayer, A. N., and Wilkinson, K. D., 1989, Detection, resolution, and nomenclature of multiple ubiquitin carboxyl-terminal esterases from bovine calf thymus, *Biochemistry* **28:**166–172.

McGrath, J. P., Jentsch, S., and Varshavsky, A., 1991, *UBA1:* An essential yeast gene encoding ubiquitin-activating enzyme, *EMBO J.* **10:**227–236.

McTigue, M. A., Williams, D. R., and Tainer, J. A., 1995, Crystal structures of a schistosomal drug and vaccine target: Glutathione S-transferase from *S. japonica* and its complex with the leading antischistosomal drug praziquantel, *J. Mol. Biol.* **246:**21–27.

Michiels, L., Van der Rauwelaert, E., Van Hasselt, F., Kas, K., and Merregaert, J., 1993, *fau* cDNA encodes a ubiquitin-like-S30-fusion protein and is expressed as an antisense sequence in the Finkel-Biskis-Reilly murine sarcoma virus, *Oncogene* **8:**2537–2546.

Nakamura, M., Xavier, R. M., Tsunematsu, T., and Tanigawa, Y., 1995, Molecular cloning and characterization of a cDNA encoding monoclonal nonspecific suppressor factor, *Proc. Natl. Acad. Sci. USA* **92:**3463–3467.

Nakamura, M., Xavier, R. M., and Tanigawa, Y., 1996, Ubiquitin-like moiety of the monoclonal nonspecific suppressor factor β is responsible for its activity, *J. Immunol.* **156:**533–538.

Narasimhan, J., Potter, J. L., and Haas, A. L., 1996, Conjugation of the 15-kDa interferon-induced ubiquitin homolog is distinct from that of ubiquitin, *J. Biol. Chem.* **271:**324–330.

Nefsky, B., and Beach, D., 1996, Pub1 acts as an E6-AP-like protein ubiquitin ligase in the degradation of cdc25, *EMBO J.* **15:**1301–1312.

Olvera, J., and Wool, I. G., 1993, The carboxyl extension of a ubiquitin-like protein is rat ribosomal protein S30, *J. Biol. Chem.* **268:**17967–17974.

Özkaynak, E., Finley, D., and Varshavsky, A., 1984, The yeast ubiquitin gene: Head-to-tail repeats encoding a polyubiquitin precursor, *Nature* **312:**663–666.

Özkaynak, E., Finley, D., Solomon, M. J., and Varshavsky, A., 1987, The yeast ubiquitin genes: A family of natural gene fusions, *EMBO J.* **6:**1429–1439.

Papa, F. R., and Hochstrasser, M., 1993, The yeast *DOA4* gene encodes a deubiquitinating enzyme related to a product of the human *tre-2* oncogene, *Nature* **366:**313–319.

Pickart, C. M., and Rose, I. A., 1986, Mechanism of ubiquitin carboxyl-terminal hydrolase: Borohydride and ubiquitin inactivate in the presence of ubiquitin, *J. Biol. Chem.* **261:**10210–10217.

Pickart, C. M., Haldeman, M. T., Kasperek, E. M., and Chen, Z., 1992, Iodination of tyrosine 59 of ubiquitin selectively blocks ubiquitin's acceptor activity in diubiquitin synthesis catalyzed by E2-25K, *J. Biol. Chem.* **267:**14418–14423.

Pickart, C. M., Kasperek, E. M., Beal, R., and Kim, A., 1994, Substrate properties of site-specific mutant ubiquitin protein (G76A) reveal unexpected mechanistic features of ubiquitin-activating enzyme (E1), *J. Biol. Chem.* **269:**7115–7123.

Redman, K. L., and Rechsteiner, M., 1989, Identification of the long ubiquitin extension as ribosomal protein S27a, *Nature* **338:**438–440.

Reilly, L. M., and Guarino, L. A., 1996, The viral ubiquitin gene of *Autographa californica* nuclear polyhedrosis virus is not essential for viral replication, *Virology* **218:**243–247.

Reiss, Y., Heller, H., and Hershko, A., 1989, Binding sites of ubiquitin-protein ligase, *J. Biol. Chem.* **264:**10378–10383.

Riley, D. A., Bain, J. L. W., Ellis, S., and Haas, A. L., 1988, Quantitation and immunocytochemical localization of ubiquitin conjugates within rat red and white skeletal muscles, *J. Histochem. Cytochem.* **36:**621–632.

Rubin, D. M., and Finley, D., 1995, The proteasome: A protein-degrading organelle? *Curr. Biol.* **5:** 854–858.

Scheffner, M., Huibregtse, J. M., Vierstra, R. D., and Howley, P. M., 1993, The HPV-16 E6 and E6-AP complex functions as a ubiquitin-protein ligase in the ubiquitination of p53, *Cell* **75:**495–505.

Seemüller, E., Lupas, A., Stock, D., Löwe, J., Huber, R., and Baumeister, W., 1995, Proteasome from *Thermoplasma acidophilum*: A threonine protease, *Science* **268:**579–582.

Seufert, W., and Jentsch, S., 1990, Ubiquitin-conjugating enzymes UBC4 and UBC5 mediate selective degradation of short-lived and abnormal proteins, *EMBO J.* **9:**543–550.

Shaeffer, J. R., 1994, Heterogeneity in the structure of the ubiquitin conjugates of human α globin, *J. Biol. Chem.* **269:**29530–29536.

Shaeffer, J. R., and Kania, M. A., 1995, Degradation of monoubiquitinated α-globin by 26S proteasomes, *Biochemistry* **34:**4015–4021.

Siegelman, M., Bond, M. W., Gallatin, W. M., St. John, T., Smith, H. T., Fried, V. A., and Weissman, I. L., 1986, Cell surface molecule associated with lymphocyte homing is a ubiquitinated branched-chain glycoprotein, *Science* **231:**823–829.

Spence, J., Sadis, S., Haas, A. L., and Finley, D., 1995, A ubiquitin mutant with specific defects in DNA repair and multiubiquitination, *Mol. Cell. Biol.* **15:**1265–1273.

Takada, K., Nasu, H., Hibi, N., Tsukada, Y., Ohkawa, K., Fujimuro, M., Sawada, H., and Yokosawa, H., 1995, Immunoassay for the quantification of intracellular multi-ubiquitin chains, *Eur. J. Biochem.* **233:**42–47.

Takada, K., Hibi, N., Tsukada, Y., Shibasaki, T., and Ohkawa, K., 1996, Ability of ubiquitin radioimmunoassay to discriminate between monoubiquitin and multi-ubiquitin chains, *Biochim. Biophys. Acta* **1290:**282–288.

Toniolo, D., Persico, M., and Alcalay, M., 1988, A "housekeeping" gene on the X chromosome encodes a protein similar to ubiquitin, *Proc. Natl. Acad. Sci. USA* **85:**851–855.

Treier, M., Staszewski, L. M., and Bohmann, D., 1994, Ubiquitin-dependent c-Jun degradation *in vivo* is mediated by the δ domain, *Cell* **78:**787–798.

Van Nocker, S., and Vierstra, R. D., 1991, Cloning and characterization of a 20-kDa ubiquitin carrier protein from wheat that catalyzes multiubiquitin chain formation *in vitro*, *Proc. Natl. Acad. Sci. USA* **88:**10297–10301.

Van Nocker, S., and Vierstra, R. D., 1993, Multiubiquitin chains linked through lysine 48 are abundant *in vivo* and are competent intermediates in the ubiquitin proteolytic pathway, *J. Biol. Chem.* **268:** 24766–24773.

Van Nocker, S., Deveraux, Q., Rechsteiner, M., and Vierstra, R.D., 1996a, *Arabidopsis MBP1* gene encodes a conserved ubiquitin recognition component of the 26S proteasome, *Proc. Natl. Acad. Sci. USA* **93:**856–860.

Van Nocker, S., Sadis, S., Rubin, D. M., Glickman, M., Fu, H., Coux, O., Wefes, I., Finley, D., and Vierstra, R. D., 1996b, The multiubiquitin-chain-binding protein Mcb1 is a component of the 26S proteasome in *Saccharomyces cerevisiae* and plays a nonessential, substrate-specific role in protein turnover, *Mol. Cell. Biol.* **16:**6020–6028.

Van Nocker, S., Walker, J. M., and Vierstra, R. D., 1996c, The *Arabidopsis thaliana UBC7/13/14* genes encode a family of multiubiquitin chain-forming E2 enzymes, *J. Biol. Chem.* **271:**12150–12158.

Vierstra, R. D., Langan, S. M., and Schaller, G. E., 1986, Complete amino acid sequence of ubiquitin from the higher plant *Avena sativa*, *Biochemistry* **25:**3105–3108.

Vijay-Kumar, S., Bugg, C. E., and Cook, W. J., 1987a, Structure of ubiquitin refined to 1.8 Å resolution, *J. Mol. Biol.* **194:**531–544.

Vijay-Kumar, S., Bugg, C. E., Wilkinson, K. D., Vierstra, R. D., Hatfield, P. M., and Cook, W. J., 1987b, Comparison of the three-dimensional structures of human, yeast, and oat ubiquitin, *J. Biol. Chem.* **262:**6396–6399.

Wand, A. J., Urbauer, J. L., McEvoy, R. P., and Bieber, R. J., 1996, Internal dynamics of human ubiquitin revealed by ^{13}C-relaxation studies on randomly fractionally labeled protein, *Biochemistry* **35:**6116–6125.

Wang, R., and Liew, C. C., 1994, The human BAT3 ortholog in rodents is predominantly and developmentally expressed in testis, *Mol. Cell. Biochem.* **136:**49–57.

Watkins, J. F., Sung, P., Prakash, L., and Prakash, S., 1993, The *Saccharomyces cerevisiae* DNA repair gene *RAD23* encodes a nuclear protein containing ubiquitin-like domain required for biological function, *Mol. Cell.Biol.* **13:**7757–7765.

Weber, P. L., Brown, S. C., and Mueller, L., 1987, Sequential ^{1}H NMR assignments and secondary structure identification of human ubiquitin, *Biochemistry* **26:**7272–7290.

Wilkinson, K. D., 1988, Purification and structural properties of ubiquitin, in *Ubiquitin* (M. Rechsteiner, ed.), pp. 5–38, Plenum Press, New York.

Wilkinson, K. D., and Audhya, T. K., 1981, Stimulation of ATP-dependent proteolysis requires ubiquitin with the COOH-terminal sequence Arg-Gly-Gly, *J. Biol. Chem.* **256:**9235–9241.

Wilkinson, K. D., Cox, M. J., O'Connor, L. B., and Shapira, R., 1986, Structure and activities of a variant ubiquitin from bakers' yeast, *Biochemistry* **25:**4999–5004.

Wilkinson, K. D., Tashayev, V. L., O'Connor, L. B., Larsen, C. N., Kasperek, E. M., and Pickart, C. M., 1995, Metabolism of the polyubiquitin degradation signal: Structure, mechanism, and role of isopeptidse T, *Biochemistry* **34:**14535–14546.

Wolf, S., Lottspeich, F., and Baumeister, W., 1993, Ubiquitin found in the archaebacterium *Thermoplasma acidophilum*, *FEBS Lett.* **326:**42–44.

Yaglom, J., Linskens, M. H., Sadis, S., Rubin, D. M., Futcher, B., and Finley, D., 1995, p34^{Cdc28}-mediated control of Cln3 cyclin degradation, *Mol. Cell. Biol.* **15:**731–741.

CHAPTER 3

The Ubiquitin-Conjugation System

Martin Scheffner, Susan Smith,
and Stefan Jentsch

1. INTRODUCTION

The ubiquitin/proteasome system is believed to be the major nonlysosomal proteolytic system of eukaryotic cells. It is present in the cytosol and the nucleus but apparently absent from the lumen of membrane-enclosed organelles, i.e., the endoplasmic reticulum, the Golgi and vesicular system, mitochondria, chloroplasts, and peroxisomes (for reviews see Finley and Chau, 1991; Jentsch, 1992a,b; Hershko and Ciechanover, 1992; Ciechanover, 1994; Hochstrasser, 1995; Smith *et al.*, 1996). Substrates of this pathway include soluble proteins, subunits of oligomeric protein complexes, and integral membrane proteins. An important function of this pathway is the elimination of abnormal proteins (e.g., misfolded, misassembled) generated under normal and, in particular, stress conditions. Moreover, it is assumed that most naturally short-lived proteins of the cytosol and the nucleus are degraded by this pathway. Known substrates include proteins with important regulatory roles such as transcription factors, cell cycle regulators, and signal transducers.

Martin Scheffner • Deutsches Krebsforschungszentrum, Angewandte Tumorvirologie, 69120 Heidelberg, Germany. **Susan Smith and Stefan Jentsch** • Zentrum für Molekulare Biologie der Universität Heidelberg, 69120 Heidelberg, Germany. *Present address of S.S.:* Mitotix, Inc., One Kendall Square, Building 600, Cambridge, Massachusetts 02139.

Ubiquitin and the Biology of the Cell, edited by Peters *et al.* Plenum Press, New York, 1998.

Despite the large number of structurally unrelated substrates, the ubiquitin/proteasome system is remarkably selective. Substrate proteins are thought to bear specific degradation signals that serve as recognition sites for components of this pathway. These signals are expected to be rather complex in its organization because they should indicate not only whether the protein has to be degraded, but also at what rate and when. Cryptic degradation signals may be present in almost any protein and may be recognized when a protein is in a nonnative conformation or when the protein fails to find it is natural partners. To decode each of these putative signals, nature has evolved a specialized recognition apparatus, the ubiquitin-conjugation system, which operates functionally and possibly specially detached from the ubiquitin-conjugate-specific protease, the proteasome (reviewed in Jentsch and Schlenker, 1995; see Chapters 5 and 6). Enzymes and assessory factors of the ubiquitin-conjugation system recognize the substrates destined for degradation and earmark the proteins by the covalent attachment of ubiquitin to internal lysine residues of substrate proteins. Proteins tagged with ubiquitin molecules may have different fates (see Chapter 13). It appears, however, that most ubiquitin–protein conjugates are targeted for degradation by the proteasome.

In this chapter we discuss the enzymatic components of the ubiquitin-conjugation system and review the functions mediated by these enzymes. Emphasis is given to a conceptional discussion of the possible organization of this system and a speculative view as to how this pathway might be regulated.

2. STRUCTURE AND SIGNIFICANCE OF UBIQUITIN–PROTEIN CONJUGATES

Early studies on chromatin proteins identified nonhistone protein A24 as a covalent adduct of histone H2A and ubiquitin (termed uH2A). H2A was found to be modified at Lys-114 by the covalent attachment of a ubiquitin molecule (Goldknopf and Busch, 1977; Hunt and Dayhoff, 1977). The C-terminus of ubiquitin was shown to be linked to the ε-amino group of the lysine residue of the histone leading to a branched protein–protein conjugate. This type of ubiquitin–substrate linkage appears to be characteristic for all ubiquitin-conjugation reactions. No other amino group, including the α-amino group at the amino terminus of a protein, has been unambiguously shown to be modified by ubiquitin conjugation *in vivo*. Histones are largely modified by the attachment of a single ubiquitin molecule (termed *monoubiquitination*) and these conjugates are thought to be metabolically stable. Some additional proteins also appear to be modified by monoubiquitination (Ball *et al.*, 1987) but the biological significance of this type of modification remains largely enigmatic. A discussion of possible functions of ubiquitin conjugation other than the degradation of the conjugate by the proteasome can be found in Chapter 13.

Modification of proteins by the attachment of a single ubiquitin molecule may be an exception rather than the rule. Most cellular proteins that are recognized by the ubiquitin-conjugation system are thought to be modified by the attachment of so-called "multiubiquitin chains" (the reaction is termed *multiubiquitination*) (Chau *et al.*, 1989). Such chains are formed when a ubiquitin molecule is a substrate of its own enzymatic machinery. Ubiquitin has seven lysine residues and several of them may be used as a target for further ubiquitination *in vivo*. Depending on the lysine used for ubiquitin–ubiquitin linkage, structurally different multiubiquitin chains are formed (e.g., lysine 23, 48, 63 chains). Yeast cells expressing ubiquitin variants lacking specific lysine residues confer characteristic mutant phenotypes (Arnason and Ellison, 1994; Finley *et al.*, 1994; Spence *et al.*, 1995), but it is not known whether distinct multiubiquitin chains have intrinsically different, functionally relevant properties. It is becoming apparent that the type of ubiquitin–ubiquitin linkage is dictated by both the substrate and the enzymes of the ubiquitin-conjugation system (Johnson *et al.*, 1995). Multiubiquitination is thought to be a processive event and may proceed after the first ubiquitin moiety has been attached to the substrate. This reaction may yield very long chains with up to 10 or more ubiquitin molecules which may be linked to several lysine residues of the same target protein (Johnson *et al.*, 1995). How the proteasome selects multiubiquitinated proteins is not known but ubiquitin chain-specific receptor proteins are thought to be involved (see Chapter 6).

3. BASIC ENZYMES OF THE UBIQUITIN-CONJUGATION SYSTEM

3.1. Organization of the Ubiquitin-Conjugation System

All known genes for ubiquitin encode precursor proteins that are rapidly processed after translation by the action of ubiquitin-specific hydrolases (Özkaynak *et al.*, 1987; Finley *et al.*, 1989; Finley and Chau, 1991). This reaction yields free ubiquitin with a C-terminal glycine residue. Because a C-terminal glycine residue is essential for conjugation, unprocessed ubiquitin precursors (including the products of "polyubiquitin" genes that have additional amino acids at their C-termini) are inert for conjugation (Finley and Chau, 1991). Prior to its conjugation to a target protein, ubiquitin must undergo a series of reactions (Fig. 1). The E1 or ubiquitin-activating enzyme catalyzes the "activation" of ubiquitin (Ciechanover *et al.*, 1982; Haas and Rose, 1982; Haas *et al.*, 1982; Hershko *et al.*, 1983). This reaction, which requires ATP hydrolysis and proceeds via an adenylate intermediate, results in the formation of a high-energy thioester bond between the C-terminal glycine residue of ubiquitin and a specific cysteine residue in the E1

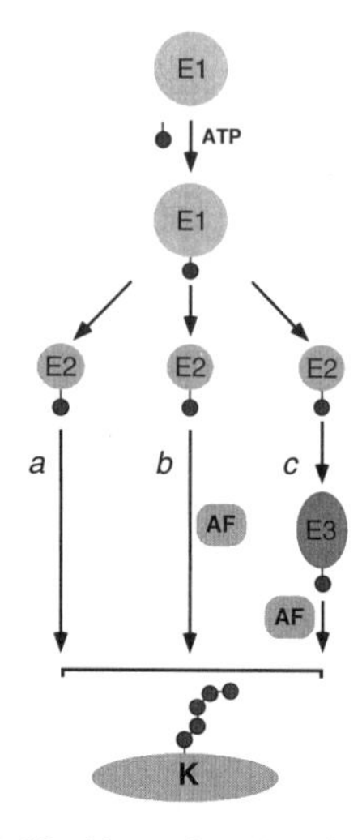

Figure 1. Suggested pathways of ubiquitin conjugation (from top to bottom). Ubiquitin (lollipop symbol) is activated in an ATP-dependent reaction and is covalently linked to ubiquitin-activating (E1) enzyme. Ubiquitin can then be transferred to any of a number of ubiquitin-conjugating (E2) enzymes. Finally, the ubiquitin moiety is transferred to a substrate by one of three possibilities. In path *a* ubiquitin may be transferred from the E2 directly to the substrate. In path *b*, an ancillary factor (AF) is required to facilitate substrate selection and subsequent ubiquitination. Alternatively, in path *c* ubiquitin is accepted as a thioester by the E3 before being conjugated to the substrate with (or without) the aid of an ancillary factor (AF). K, lysine residue in substrate protein.

enzyme. In this form, ubiquitin may now be transferred to an E2 or ubiquitin-conjugating enzyme (also termed *ubiquitin carrier protein*), which can, in a similar fashion to E1, accept ubiquitin as a thioester (Hershko *et al.*, 1983; Pickart and Rose, 1985). *In vitro*, E2 enzymes may transfer ubiquitin directly to substrates (Fig. 1, path *a*) (Pickart and Rose, 1985; Jentsch *et al.*, 1987). It is assumed, however, that ubiquitin–protein conjugation *in vivo* may require accessory factors for substrate recognition. Interestingly, it appears that such factors can perform this function in different ways. One class, which we term *ancillary factors*, appears to have a role in the localization of E2 enzymes to putative substrates or to modulate the specificity of the enzymes (Bailly *et al.*, 1994; Fig. 1, path *b*). Yet there is a second group that in addition to playing a role in substrate selection also shares with E1 and E2s the ability to accept ubiquitin as a thioester. The accessory proteins that can form a thioester with ubiquitin and participate in substrate recognition will be referred to as *E3 enzymes* herein (a term often used in a broader sense; see below). Hence, ubiquitin may be transferred by a thioester cascade (Fig. 1, path *c*) from the E1 to an E2 and finally, via an E3, to the substrate (Scheffner *et al.*, 1995). Interestingly it appears that E3 enzymes sometimes also require ancillary factors (specificity factors) for substrate recognition (Huibregtse *et al.*, 1991; Scheffner *et al.*, 1995).

The existence of at least three different classes of proteins that are involved in protein ubiquitination, i.e., El, E2, E3, supposedly contributes to ensure the required specificity and selectivity of ubiquitin conjugation. However, what is the logic behind the proposed ubiquitin thioester cascade? A possibility is that E3 enzymes act as catalytically active (i.e., ubiquitin-enzyme thioester-forming) partner proteins of a limited set of substrates. In a complex with potential substrates, E3 proteins may even have other functions in addition to ubiquitin conjugation (Imhof and McDonnell, 1996). Further, E3s might be particularly sensitive to regulation, as certain E3s possess binding sites for specificity factors (e.g., E6; see below) or other regulatory domains. In this scheme E3s bound to substrates are conjugation incompetent until they receive ubiquitin from a ubiquitin-loaded E2. The advantage of such a system would be that a given E2 protein could feed into several different E3s thereby allowing the simultaneous ubiquitination of different substrates. The function of an El enzyme is expected to be more general as it presumably will not discriminate much between different E2 enzymes. Such a scenario may indicate that some E3s, and maybe E2s and El as well, are integrated into a common structure or complex to ensure efficient and processive ubiquitination of target proteins.

Initial studies of the enzymes of the ubiquitin-conjugation system were assisted by "covalent affinity" chromatography (Ciechanover *et al.*, 1982). This method takes advantage of the fact that the El enzyme is capable of forming a thioester with immobilized ubiquitin (ubiquitin Sepharose) in the presence of ATP. Moreover, El enzymes bound to the resin can be exchanged with E2 (Hershko *et al.*, 1983; Pickart and Rose, 1985) and, to a lesser extent, E3 enzymes. These enzymes can be specifically recovered from such columns by elution with dithiothreitol (DTT). This method was instrumental for the cloning of the first genes encoding components of the ubiquitin-conjugation system (Jentsch *et al.*, 1987).

The yeast *Saccharomyces cerevisiae* has proved a powerful model system to study the enzymes of this pathway. Biochemical and genetic studies complemented by the sequence analysis of the entire yeast genome led to the identification of one functional El enzyme, 13 E2 enzymes, and 6 putative E3s in yeast (Table I). Thus, it appears that an El enzyme can activate many E2s, which may then function together with different E3s and ancillary factors in substrate recognition and ubiquitination. Below, we will briefly introduce the known components of the ubiquitin system and describe their role in specific processes.

3.2. E1, Ubiquitin-Activating Enzymes

Ubiquitin-activating enzymes are abundant proteins of the cytosol and the nucleus. Genes encoding El enzymes have been cloned from various organisms

Table I
Enzyme Factors, Functions, and Substrates of the Yeast Ubiquitin-Conjugation System[a]

Factors	Functions	Substrates
El enzymes		
UBA1	Required to activate the E2s listed below, essential	
UBA2	Unknown, essential	
E2 enzymes		
UBC1	Bulk degradation of short-lived and abnormal proteins	
UBC2/RAD6	DNA repair, N-end rule, sporulation	GCN4, GPA1
UBC3/CDC34	Required for G1/S transition of the cell cycle, essential	FAR1, SIC1, CLN3, GCN4
UBC4, UBC5	Bulk degradation of short-lived and abnormal proteins	MATα2, CLN3, STE2
UBC6	ER-membrane protein degradation	MATα2, mutant SEC61
UBC7	Cadmium tolerance	MATα2, mutant SEC61
UBC8	Unknown, nonessential	
UBC9	Required for G2/M transition of the cell cycle, essential	CLB2, CLB5
UBC10/PAS2	Peroxisome biogenesis	
UBC11	Unknown, nonessential	
UBC12	Unknown, nonessential	
UBC13	Unknown, nonessential	
E3 enzymes		
UBR1	N-end rule, interacts with UBC2/RAD6	GPA1
UFD4	Unknown, nonessential	
RSP5	Essential	GAP1, FUR4
TOM1 (U33050)	Unknown	
ORF (Z49536)	Unknown, nonessential	
ORF (Chr. VII)	Unknown	
Ancillary factors		
RAD18	Binds single-stranded DNA, interacts with UBC2/RAD6, may localize UBC2/RAD6 to DNA	
CDC16, CDC23	TPR proteins, required for destruction box-mediated degradation of CLB2, essential	

[a]See text for details (references, see text; and our unpublished data).

including yeast (McGrath *et al.*, 1991), wheat (Hatfield *et al.*, 1990), and man (Zachsenhaus and Sheinin, 1990; Handley *et al.*, 1991). In those cases where the complete sequence is available, the sequences predict enzymes of roughly 100 kDa that contain the nucleotide-binding motif Gly-X-Gly-X-X-Gly. A putative "active-site" cysteine residue required for ubiquitin-enzyme thioester formation has been mapped to approximately the center of the amino acid sequence (Hatfield and Vierstra, 1992). Mutant El enzymes lacking this cysteine residue are enzymatically inactive. El seems to act as a homodimer (Ciechanover *et al.*, 1982) and is expected to complex with E2 and E3 enzymes in the multiubiquitination of substrates.

Currently, one functional *ub*iquitin-*a*ctivating enzyme, UBA1, is known in yeast which can charge multiple E2 enzymes with ubiquitin. As expected, the *UBA1* gene is essential for viability (McGrath *et al.*, 1991). The product of *UBA2*, which is also an essential gene, shares significant similarity with UBA1 including the presumed nucleotide binding site and the region flanking the putative active-site cysteine residue (Dohmen *et al.*, 1995). However, the function of UBA2 remains enigmatic as it cannot activate ubiquitin *in vitro* and may not be involved in this pathway (Dohmen *et al.*, 1995). The yeast genome sequencing project has revealed the existence of three additional open reading frames that share weak homology with UBA1 (accession numbers S48571, S52528, and U25842). Interestingly, two of these proteins (S52528 and U25842), like the auxin-resistance protein AXR1 from *Arabidopsis thaliana* (Leyser *et al.*, 1993), lack the presumptive active-site cysteine residue (required for ubiquitin-enzyme thioester formation) found in UBA1 and its homologues from other organisms. The roles of each of these El-related proteins are unknown.

Analogous to yeast, higher eukaryotes appear to encode several proteins homologous to El. The human gene for one major functional El enzyme is located on the short arm of the X chromosome (Xp11.23) and does not escape X chromosome inactivation (Zacksenhaus and Sheinin, 1990). Interestingly, distinct temperature-sensitive mutations in this gene (or homologues from other mammals) arrest respective mutant cell lines in either the S phase (e.g., tsA1S9 cells) or at the S/G2 boundary (e.g., ts85, ts20, tsBN75 cell lines) (Zacksenhaus and Sheinin, 1990; Ayusawa *et al.*, 1992; reviewed in Jentsch *et al.*, 1991). This suggests that the respective mutant proteins may differ in their defects, e.g., in their binding properties toward specific E2s. Moreover, the observed cell cycle defects of these mutant cell lines provided preliminary evidence for a role of the ubiquitin system in controlling cell cycle progression. Another gene for a protein homologous to El is located on the Y chromosome of rodents (Mitchell *et al.*, 1991; Kay *et al.*, 1991) and marsupials (Mitchell *et al.*, 1992). The respective mouse gene, *Sby*, exhibits a testis-specific expression and is a candidate for a spermatogenesis gene required for survival and proliferation of spermatogonia. A homologous gene has not been found on the human Y chromosome, however (Kay *et al.*, 1991). Another gene, termed *D8*, encoding a distinct relative of El has been mapped to 3p21, a region thought to carry a novel tumor suppressor gene (Kok *et al.*, 1993). The *D8* gene shows a reduced expression in certain lung cell carcinomas but it is unclear whether this contributes to tumor formation.

3.3. E2, Ubiquitin-Conjugating Enzymes

3.3.1. Structure and Function

*U*biquitin-*c*onjugating (UBC) enzymes are encoded by a gene family and differ in their properties and intracellular localization (reviewed in Jentsch *et al.*, 1990; Jentsch, 1992a,b). Structurally, E2s are related proteins bearing a (~160 amino acid) highly conserved (35–40% identity) catalytic domain termed the *UBC domain*. Within this domain E2 enzymes possess a specific ("active site") cysteine residue required for ubiquitin-E2 thioester formation. Replacement of this cysteine residue by other amino acids results in inactive E2 enzymes (Sung *et al.*, 1991a; Sommer and Jentsch, 1993; Seufert *et al.*, 1995). Overexpression of inactive E2 enzymes in cells may cause dominant-negative defects, suggesting that E2 enzymes are part of homo- or heterooligomeric complexes (Bailly *et al.*, 1994; R. Heinlein and S. Jentsch, unpublished). Indeed, biochemical and genetic data indicate that certain E2 enzymes function as homodimers (Pickart and Rose, 1985) or heterodimers with other E2s (Chen *et al.*, 1993) and they can physically interact with other proteins (Bailly *et al.*, 1994).

Many E2 enzymes are small proteins of roughly 16 kDa and consist of the UBC domain only (class I E2s). Some E2 enzymes have additional C-terminal extensions (class II E2s). Such sequences can be important for UBC function and may mediate substrate specificity or intracellular localization (Jentsch *et al.*, 1987; Sung *et al.*, 1988; Goebl *et al.*, 1988; Kolman *et al.*, 1992; Silver *et al.*, 1992; Sommer and Jentsch, 1993). Other E2 enzymes lack C-terminal extensions but possess additional N-terminal sequences (class III E2s) (Matuschewski *et al.*, 1996; Nuber *et al.*, 1996). Lastly, there are some E2s that have both N- and C-terminal extensions (class IV E2s) (Jentsch *et al.*, 1991; H.-P. Hauser, M. Bardroff, and S. Jentsch, unpublished).

The three-dimensional structure of some class I E2s has been determined by X-ray crystallography (Cook *et al.*, 1993). The UBC domains of these enzymes fold into a conserved structure with four α-helices and a four-stranded antiparallel β-sheet. The "active site" cysteine is located in a cleft between two loops. The N- or C-terminal extensions of the class II–IV E2s are expected to protrude from the UBC domain and may be accessible to interacting partners.

3.3.2. Yeast E2 Enzyme Family

In the genome of *S. cerevisiae* 13 genes for *ub*iquitin-*c*onjugating enzymes (*UBC* genes) can be identified (Fig. 2). All yeast E2s are small proteins: 8 are class I and 5 are class II enzymes possessing short C-terminal extensions. The disruption of the genes encoding various E2s confers different phenotypes to the mutant yeast strains indicating that the encoded enzymes mediate distinct cellular functions (reviewed in Jentsch, 1992a,b; Smith *et al.*, 1996). Only UBC3/CDC34 and UBC9 are encoded by essential genes (Goebl *et al.*, 1988; Seufert *et al.*, 1995; T. Mayer, K. Matuschewski, and S. Jentsch, unpublished data; D. Finley, personal communication). Both of these E2s play an essential role in cell cycle progression. The remaining 11 E2s are nonessential for vegetative growth under normal conditions. Mutations in *UBC1*, *UBC2/RAD6*, *UBC4*, *UBC5*, and *UBC7* render cells susceptible to environmental stresses (Reynolds *et al.*, 1985; Jentsch *et al.*, 1987; Seufert and Jentsch, 1990; Seufert *et al.*, 1990; Jungmann *et al.*, 1993). Additionally, combinations of particular *ubc* mutants are known to be inviable such as *ubc1 ubc4 ubc5* (Seufert *et al.*, 1990) and *ubc4 ubc5 ubc7* (Chen *et al.*, 1993) triple mutants. This suggests that these enzymes together perform essential functions.

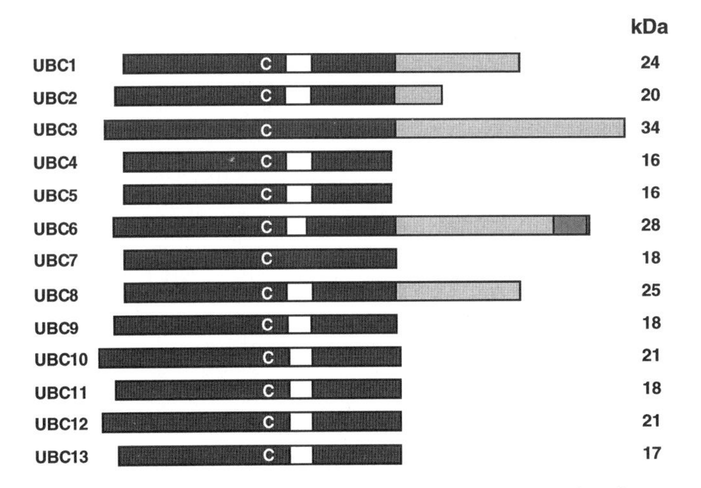

Figure 2. Schematic diagram of protein structures of yeast ubiquitin-conjugating enzymes. The known E2s from yeast are class I and class II enzymes. The sizes of the proteins (long boxes) is given in kDa. The central shaded regions are the conserved domains of the enzymes (UBC domain). Within these regions small domains are missing in some enzymes (white regions). The putative cysteine residues required for enzyme-ubiquitin thioester formation are indicated by the letter C. Protein extensions of class II E2s are indicated as shaded boxes; the signal anchor sequence of UBC6 as a dark shaded box at the C-terminus of the protein.

Some of the cellular functions are described below (for further information see the *SGD* database at http://genome-www.stanford.edu/Saccharomyces/ and the *YPD* database at http://www.proteome.com/YPDhome.html).

3.3.2a. DNA Repair. *UBC2/RAD6* encodes a 172-residue, 20-kDa protein containing the conserved UBC domain and a short acidic C-terminal tail (Reynolds *et al.*, 1985; Jentsch *et al.*, 1987). *In vitro*, this extension is required for the recognition of basic substrates such as histones (Sung *et al.*, 1988), and *in vivo* for sporulation (Morrison *et al.*, 1988). Mutant *UBC2/RAD6* cells exhibit a number of noticeable phenotypes including slow growth, hypersensitivity to DNA-damaging agents, deficiency in UV-induced mutagenesis, increased retrotransposition, and the failure of homozygous *ubc2/rad6* diploids to sporulate (reviewed in Lawrence, 1994; Picologlou *et al.*, 1990; Kang *et al.*, 1992). Although the actual role that UBC2/RAD6 plays in DNA repair and related pathways remains enigmatic, recent work has indicated that UBC2/RAD6 can stably interact with RAD18 *in vivo* and that this complex is required for DNA repair (Bailly *et al.*, 1994). *RAD18* belongs to the *RAD6* epistasis group of DNA repair genes and is also required for protection against DNA-damaging agents (reviewed in Lawrence, 1994). *RAD18* encodes a 66-kDa protein that can bind to single-stranded DNA (Jones *et al.*, 1988). Thus, RAD18 is believed to act as an ancillary factor by bringing UBC2/RAD6 into close proximity with single-stranded DNA at which a stalled DNA polymerase might be found. It is speculated that UBC2/RAD6 may then target a component of the replication machinery for degradation thereby permitting the DNA repair process to proceed (Bailly *et al.*, 1994). Interestingly, UBC2/RAD6 together with UBR1, a 225-kDa E3 enzyme, is also known to effect the turnover of certain proteins by the so-called "N-end rule" pathway (Sung *et al.*, 1991b; Dohmen *et al.*, 1991). The N-end rule relates the *in vivo* half-life of a protein to the identity of its N-terminal residue (Bachmair *et al.*, 1986; Bachmair and Varshavsky, 1989; reviewed in Varshavsky, 1992, 1996; see Chapter 8).

3.3.2b. Stress Response. Normal physiological and particularly stress conditions generate misfolded proteins which must either be repaired by the action of molecular chaperones or destroyed. The elimination of abnormal proteins is thought to be largely mediated by the ubiquitin/proteasome pathway. The presence of the polyubiquitin gene, *UBI4*, is required to increase the tolerance of yeast cells to elevated temperatures or other stress conditions (Finley *et al.*, 1987) and the expression levels of a number of E2s are also increased on exposure of cells to various forms of stress. For example, the abundance of transcripts of *UBC4* and *UBC5* increases on exposure to heat shock (Seufert and Jentsch, 1990) and those for *UBC5* and *UBC7* are induced by cadmium (Jungmann *et al.*, 1993). Consistent with these observations, elimination of *UBC4*, *UBC5*, and *UBC7* renders cells sensitive to heat shock, amino acid analogues such as canavanine, or exposure to heavy metals (Seufert and Jentsch, 1990; Jungmann *et al.*, 1993). UBC4 and UBC5 share 92% sequence identity and are believed to be enzymatically identical (Seufert and Jentsch, 1990). Severe defects in growth are only observed in the *ubc4 ubc5* double mutant, presumably related to the inability of these cells to efficiently mediate the turnover of misfolded and short-lived proteins (Seufert and Jentsch, 1990). As previously mentioned, certain triple combinations of *ubc* mutations such as *ubc1 ubc4 ubc5* (Seufert *et al.*, 1990) and *ubc4 ubc5 ubc7* (Chen *et al.*, 1993) are not recoverable, suggesting that these combinations of enzymes perform vital functions. The enzymes UBC1, UBC4, UBC5, and UBC7 are related in sequence and together with UBC13 (a nonessential enzyme of unknown function) constitute a separate branch of ubiquitin-conjugating enzymes designated the "UBC4 branch" of E2s. Additional members of this branch are present in higher eukaryotes (see below).

In general it is believed that these E2s, together with specific E3s or ancillary factors, can recognize abnormal proteins and prevent their accumulation by targeting them for degradation. It remains unclear exactly what structural features of an abnormal protein are recognized. However, either exposure of an otherwise inaccessible region of a protein or failure of a protein to form a complex with a physiological partner may be a sufficient signal for degradation. It seems attractive to speculate that enzymes of the ubiquitin-conjugating system are assisted by molecular chaperones in their task to identify proteins with nonnative conformations.

3.3.2c. Membrane Protein Turnover. Not only soluble but also integral membrane proteins are often rapidly degraded. Accumulating evidence suggests that such proteins may be targeted for degradation by the ubiquitin pathway (Sommer and Jentsch, 1993; Jensen *et al.*, 1995b; Ward *et al.*, 1995; Biederer *et al.*, 1996). Yeast *UBC6* encodes an E2 enzyme with a C-terminal transmembrane anchor (Sommer and Jentsch, 1993). UBC6 is an integral membrane protein of the endoplasmic reticulum (ER) membrane (and possibly the nuclear envelope); the catalytic domain faces the cytosol. Mutants in *UBC6* are devoid of detectable deleterious phenotypes. Intriguingly, however, *ubc6* mutants suppress the defect of a temperature-sensitive mutant of *SEC61*, a gene encoding a crucial component of the protein translocation apparatus of the ER (Sommer and Jentsch, 1993). This led to the suggestion that UBC6 may be part of an "ER degradation" pathway for integral membrane proteins such as mutant (misfolded) forms of SEC61 (Sommer and Jentsch, 1993). Indeed, recent studies confirmed that the mutant sec61 protein, which is short lived in wild-type cells, is significantly stabilized in *ubc6* mutants (Biederer *et al.*, 1996). How integral ER membrane proteins are recognized and degraded is unknown but specific proteolytic signals on the substrate, E2 and E3 enzymes, and the proteasome are apparently involved (Jensen *et al.*, 1995b; Ward *et al.*, 1995; Biederer *et al.*, 1996; T. Braun, T. Mayer, and S. Jentsch, unpublished data). Importantly, UBC6 seems to form a complex with (otherwise soluble) UBC7 and both enzymes are equally required for the degradation of certain ER membrane proteins (Biederer *et al.*, 1996) and, surprisingly, also the transcription factor MATα2 (Chen *et al.*, 1993; see below).

In addition to UBC6, one other yeast E2 is membrane bound. This E2, UBC10/PAS2, is a peripheral membrane protein of yeast peroxisomes and required for the biogenesis of this organelle (Wiebel and Kunau, 1992). The gene for this enzyme was identified in a yeast screen for mutants (*pas* mutants) that lack functional peroxisomes. The localization of the UBC10/PAS2 enzyme suggests that substrates may include peroxisomal membrane proteins but the precise role of ubiquitination in peroxisome biogenesis remains speculative at present.

3.3.2d. Cell Cycle Progression. Only two yeast E2 enzymes, UBC3/CDC34 and UBC9, are essential for cell viability and both are involved in controlling cell cycle progression. Central to the regulation of the eukaryotic cell cycle are the cyclin-dependent kinases (CDKs), which regulate a number of target molecules by phosphorylation. The activity of CDKs requires their association with cyclins and this complex can be inactivated by specific CDK inhibitors. The major CDK protein in yeast is encoded by *CDC28* (reviewed in Nasmyth, 1993). In yeast, three cyclins, CLN1–CLN3, are required for G1-specific events such as budding (Hadwiger *et al.*, 1989; Richardson *et al.*, 1989), CLB5 and CLB6 are necessary for the initiation of DNA replication (Schwob and Nasmyth, 1993; Schwob *et al.*, 1994), and CLB1–CLB4 mediate aspects of mitosis (Amon *et al.*, 1993). The abundance of all of these cyclins is thought to be controlled by transcriptional mechanisms and regulated protein turnover. The C-terminal region of each of the CLNs contains so-called "PEST motifs" (see below), which seem to act as phosphorylation sites. Phosphorylation of CLN2 and CLN3 proteins at this site triggers cyclin proteolysis (Yaglom *et al.*, 1995; Lanker *et al.*, 1996). UBC3/CDC34 has been shown to play a role in the turnover of CLN2 and CLN3 (Deshaies *et al.*, 1995; Yaglom *et al.*, 1995). Interestingly, CLN3 turnover additionally involves UBC4 and UBC5 (Yaglom *et al.*, 1995). It remains unclear whether UBC3/CDC34 and UBC4 and UBC5 recognize the same or different signals in CLN3.

In contrast to the CLNs, the CLBs (and in fact all B-type cyclins) lack any obvious PEST region but possess a different signal that confers instability known as the "destruction box" (Glotzer *et al.*, 1991). Of the B-type cyclins in yeast, CLB2 plays the central role in mitosis (Surana *et al.*, 1991; Fitch *et al.*, 1992). *CLB2* is normally only expressed early in G2 when the stability of the protein is also high (Amon *et al.*, 1994; Seufert *et al.*, 1995). It is during the metaphase/anaphase transition that CLB2 becomes unstable. This instability, mediated by a functional destruction box, extends into the G1 phase of the following cell cycle (Amon *et al.*, 1994; Seufert *et al.*, 1995). Interestingly, UBC9, an essential E2 enzyme, participates in the turnover, not only of CLB2 but also of CLB5 (Seufert *et al.*, 1995). Recently, UBC9 was found to be an E2 enzyme that conjugates SMT3, a ubiquitin-related protein, to other proteins (M. Scheffner and S. Jentsch, unpublished data). It is currently unknown which proteins are modified by SMT3 conjugation and how this modification affects cell cycle function. Similar to the transcription factor MATα2 (see below), some yeast B-type cyclins appear to be degraded by at least two different pathways (S. Smith and S. Jentsch, unpublished data). In *Xenopus* egg and clam oocyte extracts, several E2s can function in destruction box-dependent cyclin ubiquitination. They include homologues of yeast UBC4 and UBC2 (King *et al.*, 1995) and UbcH10/E2-C/UBCx, an E2 enzyme not found in yeast (Aristarkhov *et al.*, 1996; Yu *et al.*, 1996). Regulated, destruction box-dependent degradation of M-phase cyclins is controlled at the level of cyclin ubiquitination and requires a protein particle with a sedimentation coefficient of 20 S [termed *anaphase promoting complex* (APC) or *cyclosome*; see below). Recent data indicate that APC's function is not restricted to the M phase of the cell cycle but also presents DNA rereplication in S phase (Heichman and Roberts, 1996).

In yeast, two inhibitors of the CDC28 kinase are known. FAR1 specifically inhibits the CLN-CDC28 kinases (Chang and Herskowitz, 1990; Peter *et al.*, 1993; Peter and Herskowitz, 1994). It is stable in pre-START cells but becomes unstable once cells have passed through this point, possibly as a result of phosphorylation (McKinney *et al.*, 1993). What components of the ubiquitin system recognize FAR1 has not been rigorously examined but there is an indication that UBC3/CDC34 plays a role (McKinney *et al.*, 1993). SIC1 is a potent inhibitor of CLB-CDC28 kinases (Nugroho and Mendenhall, 1994) and is degraded at the G1/S boundary. This instability depends on the presence of a functional UBC3/CDC34 enzyme (Schwob *et al.*, 1994) and the CLN-CDC28 kinase (Schneider *et al.*, 1996). In fact, SIC1 represents the crucial substrate of UBC3/CDC34: Elimination of SIC1 will rescue the G1/S arrest of *ubc3/cdc34* cells (Schwob *et al.*, 1994). Interestingly, UBC3/CDC34-dependent SIC1 degradation requires CDC4, a protein with so-called WD motifs (Schwob *et al.*, 1994).

3.3.2e. Transcription. Protein degradation of transcription factors has been exploited by cells to efficiently and rapidly change a gene expression program in response to extracellular signals. Two well-known examples in yeast are the turnover of the MATα2 protein and the transcription factor, GCN4.

MATα2 is a constitutively unstable protein (Hochstrasser and Varshavsky, 1990; see Chapter 9 for details). Genetics studies revealed that at least four different E2 enzymes, UBC4, UBC5, UBC6, and UBC7, are involved in its degradation (Chen *et al.*, 1993). The signal recognized by UBC4 and UBC5 remains uncharacterized but that for UBC6 and UBC7 resides within the 67 amino-terminal amino acids of MATα2 (Chen *et al.*, 1993). This study also revealed that UBC6 and UBC7 can physically associate, a property that may be required for their activity toward this substrate. Additionally, the localization of UBC6 to the ER and possibly the nuclear envelope via sequences in the UBC6 C-terminal tail (Sommer and Jentsch, 1993) is required for the activity of UBC6 toward not only MATα2 (Chen *et al.*, 1993), but also abnormal membrane proteins of the ER (see above).

GCN4, a well-studied yeast transcriptional activator, is required for the expression of genes involved in the biosynthesis of amino acids and purines. The level of GCN4 is tightly controlled within the cell and depends on the abundance of nutrients (reviewed in Hinnebusch, 1988). In favorable conditions, GCN4 is poorly translated (reviewed in Hinnebusch, 1994) and any protein that is actually synthesized is highly unstable (Kornitzer *et al.*, 1994). The degradation signal has been shown to require several regions of GCN4 and it was further established that UBC3/CDC34 and UBC2/RAD6 act synergistically, presumably in different pathways, to promote GCN4 turnover (Kornitzer *et al.*, 1994; for further details see Chapter 9).

3.3.3. E2 Enzymes from Higher Eukaryotes and Viruses

During the evolution of higher eukaryotes, the family of ubiquitin-conjugating enzymes has significantly been expanded. We currently estimate that mammals express at least 20–30 different members. Homologues to most yeast E2s have been identified in plants and animals. In some cases, e.g., for UBC2/RAD6 and UBC4, multiple human genes homologous to the yeast E2 enzymes have evolved that might be distinctly regulated (Schneider *et al.*, 1990; Koken *et al.*, 1991; Scheffner *et al.*, 1994; Jensen *et al.*, 1995a; Nuber *et al.*, 1996). When expressed in yeast, the homologous genes from higher eukaryotes frequently complement the phenotype of the respective yeast mutant (Koken *et al.*, 1991; Treier *et al.*, 1992; Zhen *et al.*, 1993; Hateboer *et al.*, 1996). This can be given as preliminary evidence for an *in vivo* function analogous to those of the yeast enzymes. However, direct *in vivo* support for this assumption is lacking. Clear functional *in vivo* data come from genetic studies in *Drosophila melanogaster* and *Caenorhabditis elegans* (for further information see *Flybase* at http://morgan.harvard.edu/ and *C. elegans* databases at http://eatworms.swmed.edu/genome.shtml).

Drosophila UbcD1 and *C. elegans* ubc-2 proteins are structural homologues of yeast UBC4 and can substitute for UBC4 in yeast (Treier *et al.*, 1992; Zhen *et al.*, 1993). The gene for *Drosophila* UbcD1 is known as *effete* (*eff*). P-element insertions in *eff* affect telomere behavior and chromosome separation during mitosis and male meiosis (information from *Flybase*). This can result in sterile male flies with a reduced viability. The *C. elegans UBC4* homologue *ubc-2* is an essential gene (*let-70*) and required for larval development (Zhen *et al.*, 1996).

Expression of *ubc-2* is ubiquitous in embryos and early larvae but becomes largely restricted to the nervous system at later stages of development and in adult worms. These seemingly specific functions of higher eukaryotic UBC4 enzymes appear to contrast with the pleiotropic defects of yeast *ubc4 ubc5* mutants (see above) and the rather promiscuous activity of the UBC4 enzyme *in vitro*. In higher eukaryotes, however, loss-of-function mutations in genes for E2 enzymes may be compensated by other genes with the same or overlapping functions. Mutant defects are then only expected in those cells and tissues that do not express the other enzymes with redundant functions.

A second example from *Drosophila* is *bendless* (ben) (Muralidhar and Thomas, 1993; Oh *et al.*, 1994). This gene encodes UbcD3, an enzyme of the UBC4 branch of E2s and homologous to the nonessential yeast UBC13 enzyme (Matuschewski *et al.*, 1996; K. Matuschewski, T. Mayer, and S. Jentsch, unpublished data). Mutants in this gene from *Drosophila* exhibit several, largely neuronal defects including lesions affecting the neuronal connectivity of the giant fiber with the "jumping muscle," and the axons of photoreceptor cells R7 and R8 fail to make the proper right-angle turn into the medulla (hence the term *bendless*). In addition, the rhabdomers of the *Drosophila* composite eye are abnormally shaped (Muralidhar and Thomas, 1993; Oh *et al.*, 1994).

Higher eukaryotes also possess E2 enzymes not found in yeast. Examples are UbcH10/E2-C/UBCx implicated in B-type cyclin ubiquitination (Aristarkhov *et al.*, 1996; Yu *et al.*, 1996) and UbcH7/E2-F1, which is able to ubiquitinate p53 *in vitro* (Nuber *et al.*, 1996). Specifically, the UBC4 branch of E2 enzymes produced new members during evolution including a complete family of highly related E2 enzymes. Different from UBC4, however, members of this family are class III E2 enzymes bearing N-terminal extensions (Matuschewski *et al.*, 1996; Nuber *et al.*, 1996). The UBC domains of these E2s are virtually identical (94% identity) but their extensions differ in both size and sequence. At least four different members of this family are thought to be present in each organism (i.e., *Drosophila*, mouse, man). All four extensions are enriched in serine and threonine residues and may function as phosphorylation sites. Members of this subfamily of E2s can partially substitute for UBC4 in yeast, suggesting that they operate in UBC4-related pathways (Matuschewski *et al.*, 1996). The cellular functions of these enzymes is unknown at present but their extensions are likely to provide specificity to the enzymes.

A striking example of a novel E2 specific for higher eukaryotes is UbcM1, a murine enzyme of unknown function, and its human homologue (H.-P. Hauser, M. Bardroff, and S. Jentsch, unpublished). UbcM1 is a class IV E2 with a short C-terminal extension and a very large N-terminal extension. The enzyme, encoded by a 16-kb mRNA, is 530 kDa in size and a peripheral membrane protein facing the cytosol. UbcM1 localizes to endomembranes including the *trans*-Golgi network and vesicles (M. Bardroff, H.-P. Hauser, and S. Jentsch, unpublished data).

The overall organization of the enzyme with a large N-terminal extension and a C-terminal ubiquitin-accepting domain resembles E3 enzymes with hect domains (see below), suggesting that UbcM1 may combine E2 with E3-like properties. Another enzyme with suggested combined E2 and E3-like functions is E2-230K from vertebrate cells (Berleth and Pickart, 1996). This large protein appears to possess at least two distinct "active-site" cysteines, suggesting that ubiquitin conjugation by E2-230K might be mediated by an *intra*molecular thioester cascade.

Intriguingly, African swine fever virus (ASFV), a double-stranded DNA virus resembling pox virus, encodes an E2 enzyme (Hingamp *et al.*, 1992; Rodriguez *et al.*, 1992). No counterpart has yet been identified in the host genome but the sequence of the viral E2 distantly resembles yeast UBC2/RAD6 and UBC3/CDC34 enzymes. The expression of the gene peaks at late stages of infection but it remains unclear whether the enzyme is required for virus propagation or interferes with host functions.

3.3.4. E2-Related Proteins

Two proteins have been reported to be related to E2 enzymes but appear to lack the "active-site" cysteine residues. One of these proteins is encoded by the murine gene *ftl* (accession number Z67963). This gene was identified through an insertion of a *Ha-ras* transgene. The insertion leads to a loss of function of the *ftl* gene but further effects on adjacent genes cannot be excluded at present. Heterozygous mice carrying this insertion have a dominant mutant phenotype termed *fused toes* (*ft*). These mutant mice exhibit defects in apoptosis (fused toes) and thymic hyperplasia (van der Hoeven *et al.*, 1994). Homozygosity of the mutation leads to a malformation of the brain, decontrolled left–right asymmetry, and the mutant mice die at midgestation. The *ftl* gene product, a 33-kDa protein, shows significant sequence similarity over its entire length with E2 enzymes; yet the protein lacks the "active-site" cysteine residue. A 32-kDa protein with weak similarity to known E2 enzymes is encoded by the yeast gene *UBS1* (Prendergast *et al.*, 1996). The *UBS1* gene was identified as a dosage suppressor of the yeast *cdc34-2* mutant with a temperature-sensitive lesion in the UBC3/CDC34 E2 enzyme. Null mutants in *UBS1* do not exhibit deleterious phenotypes but rather an exacerbation of the cell cycle defect of the *cdc34-2* mutant. The biochemical functions of both the *ftl* and the *UBS1* gene products are unknown but a role as regulators of E2s through a binding of E2 partner proteins seems attractive.

3.4. E3, Ubiquitin–Protein Ligases

Although it is widely believed that E3s play a major role in substrate recognition, little is known about how E3s function. This is at least partially related to the fact that, until very recently, only a few proteins with E3 activity had been identified. This lack of knowledge is also reflected in the broad definition of the activity of E3s used in the literature (see below). We would, therefore, like to emphasize that this chapter represents our point of view on the role of E3s in ubiquitin conjugation.

3.4.1. Definition and Mechanism of E3-Dependent Ubiquitination

Based on studies concerning the ubiquitination of N-end rule substrates (for a detailed discussion of the N-end rule pathway, see Chapter 8), as well as of other artificial and natural substrates *in vitro*, it was concluded that in many cases substrate recognition is not mediated by E2s alone but rather in conjunction with additional proteins. These proteins were termed E3 enzymes, ubiquitin-protein ligases, or ubiquitin recognins, and were loosely defined as proteins that participate with El and E2s in the ubiquitination of proteins that are otherwise not recognized by E2s (Hershko and Ciechanover, 1992; Ciechanover, 1994). In *S. cerevisiae*, the recognition of substrates of the N-end rule pathway is mediated by UBR1 (*ub*iquitin *r*ecogin) (Bartel *et al.*, 1990). Biochemical characterization has shown that UBR1 (and its mammalian homologues) interacts specifically with UBC2/RAD6 as well as with target proteins of the N-end rule pathway (Dohmen *et al.*, 1991; Sung *et al.*, 1991b; Madura *et al.*, 1993; reviewed in Varshavsky, 1992, 1996). It has therefore been proposed that E3s in general function by binding specific E2s and substrate proteins, thereby bringing E2s and substrates into close proximity, and that ubiquitin is then transferred directly from E2 to the substrate (Ciechanover, 1994).

Studies on E6-AP, an E3 enzyme involved in the human papillomavirus (HPV) E6 oncoprotein-induced ubiquitination of p53 (for further details, see Chapter 11), however, have revealed that an essential intermediate step in E6-AP-dependent ubiquitination is the E2-dependent formation of a thioester complex of E6-AP with ubiquitin (Scheffner *et al.*, 1995). This cascade of ubiquitin thioester complexes, with the order of ubiquitin transfer being from El to E2 and from E2 to E6-AP, suggests that ubiquitin is then transferred directly from E6-AP to a substrate and not from the E2 as previously assumed (although there is some evidence for this hypothesis, it should be noted that a role of the E2s in the final attachment of ubiquitin to a substrate cannot be excluded at present). This indicates that E3s do not function simply as docking proteins but, rather, that they indeed have ubiquitin-protein ligase activity (Fig. 1, path *c*). In the meantime, it has been reported that UBR1 also has the capacity to form thioester complexes with ubiquitin (cited in Hochstrasser, 1995). However, it still remains to be shown if this property of UBR1 is essential for it to function as an E3. Nevertheless, based on the above, we would like to propose that similar to El and E2s, E3s should be defined as proteins that have the property to form thioester complexes with ubiquitin in the presence of El and specific E2s. All other proteins, such as CDC2/

CDC28 kinase or the HPV E6 oncoprotein, that are involved in the ubiquitination of a particular target protein, but cannot form ubiquitin thioester complexes, should be considered as ancillary factors.

According to the above definition, proteins that have been identified with E3 activity so far include UBR1, E6-AP, and presumably an E6-AP-related family of proteins termed *hect domain proteins*. Despite their functional similarity, there is no apparent similarity between UBR1 and the E6-AP-related family of proteins at the amino acid sequence level. This may indicate that there are additional E3 proteins, or classes of E3 proteins, yet to be discovered.

3.4.2. UBR1

UBR1, a 225-kDa protein, was originally isolated as an E3 that specifically recognizes artificial fusion proteins according to the N-end rule pathway (Bartel *et al.*, 1990). The N-end rule pathway seems to be conserved throughout evolution with minor modifications (Varshavsky, 1992, 1996). Although an E3 activity (termed E3α), functionally similar to UBR1, has been shown to exist in mammalian cells (Hershko and Ciechanover, 1992, and references therein), attempts to identify the protein, and to isolate a corresponding cDNA, have not yet been successful. It is, therefore, still unclear whether the mammalian E3 and UBR1 are conserved only on the functional level or whether they are also conserved at the amino acid sequence level. Identification and characterization of the mammalian homologue of UBR1 will certainly contribute to elucidating the mechanisms by which N-end rule substrates are recognized. UBR1 is involved in the degradation of several physiological substrates in yeast. Yeast lacking UBR1 are unable to import di- and tripeptides (Alagramam *et al.*, 1995; Varshavsky, 1996), presumably because UBR1 is involved in controlling the turnover both of a transcriptional regulator of *PTR2*, the gene for a peptide transporter, and also of the PTR2 protein itself (Varshavsky, 1996). Furthermore, UBR1 is involved in determining the abundance of GPA1 (Madura and Varshavsky, 1994). *GPA1* encodes the Gα subunit of a heterotrimeric G protein that regulates cell differentiation in response to mating pheromones (for a review see Marsh *et al.*, 1991). The above-mentioned UBR1-mediated functions require the activity of UBC2/RAD6 (for PTR2, GPA1) as well as of UBC4 (for PTR2). Because only UBC2, not UBC4, is a component of the N-end rule pathway, this may indicate that UBR1 is also involved in the recognition of non-N-end rule substrates. This hypothesis is supported by the fact that Gα itself does not bear a destabilizing N-terminal amino acid. It cannot be excluded, however, that GPA1 may be targeted by the N-end rule pathway through "*trans*-recognition" (Johnson *et al.*, 1990) as it has been reported that GPA1 is targeted for degradation only in the presence of SST2 (Madura and Varshavsky, 1994).

3.4.3. Hect Domain Proteins

Similar to El and E2s, hect domain proteins are found in all eukaryotic organisms examined including yeast, *Arabidopsis*, *C. elegans*, *Drosophila*, and mammals (Huibregtse *et al.*, 1995; M. Scheffner, unpublished). Hect domain proteins are large proteins ranging from 90 kDa to more than 200 kDa and are characterized by a C-terminal region, termed *hect domain* (*h*omologous to *E*6-AP *C t*erminus), of approximately 350 amino acids that shows significant similarity to the C-terminus of E6-AP. The C-terminus of E6-AP has been shown to contain a cysteine residue that is necessary for thioester formation with ubiquitin (Scheffner *et al.*, 1995). Because the position of the cysteine residue, as well as several surrounding residues, is highly conserved among all of these proteins, it was speculated that they share the ability to form thioester complexes with ubiquitin and are therefore members of an E6-AP-like family of E3s (Scheffner *et al.*, 1995). This hypothesis has been supported by the demonstration that several of these proteins, including yeast RSP5, a rat-100 kDa protein, and several human proteins, can indeed form thioester complexes in the presence of El and distinct E2s (Huibregtse *et al.*, 1995; S. E. Schwarz and M. Scheffner, unpublished). The size of the hect domain coincides with the minimal region of E6-AP and RSP5 necessary for thioester complex formation (S. E. Schwarz and M. Scheffner, unpublished). This suggests that hect domain proteins have a modular structure consisting of a conserved catalytic domain (i.e., the hect domain), the function of which is similar to the UBC domain of E2s (see above), and different N-terminal extensions.

An intriguing hypothesis is that the N-terminal extensions are involved in defining the substrate specificity of each hect domain protein, as has been shown for the interaction of E6-AP with the HPV E6 oncoprotein and p53 (Huibregtse *et al.*, 1993). Similarly, the N-terminal extension of RSP5, as well as of its apparent mouse homologue, the Nedd4 protein, contains a protein structure motif, the so-called "WW domain" (Sudol *et al.*, 1995). This domain has recently been shown to be involved in the interaction of the Nedd4 protein with a membrane protein having the characteristics of a Na^+ channel (Staub *et al.*, 1996). It has not been reported, however, whether binding of the Nedd4 protein results in ubiquitination of the Na^+ channel protein. Furthermore, it is unclear whether the hect domain and the N-terminal extensions represent two completely separate entities on a hect domain protein or whether the hect domain is also involved in substrate recognition.

Recent database searches suggest that the human genome encodes for as many as 20–25 different hect domain proteins (M. Scheffner, unpublished). The cellular functions, as well as the substrates, of these hect domain proteins are unknown at present. The genome of *S. cerevisiae* encodes five hect domain proteins, UFD4, RSP5 (or NPI1), TOM1, and two other open reading frames (see Table I). UFD4 has been isolated in a screen searching for proteins that are involved in the degradation of fusion proteins bearing a nonremovable N-terminal ubiquitin moiety (Johnson *et al.*, 1995). *RSP5* is an essential gene and has recently been reported to play a role in the turnover of several transporter proteins, including uracil permease (Hein *et al.*, 1995; Galan *et al.*, 1996). Similar to the STE2 protein (Hicke and Riezman, 1996), however, ubiquitination of uracil permease does not result in proteasome-mediated degradation but, rather, results in endocytosis of the permease and subsequent vacuolar degradation (Galan *et al.*, 1996). Furthermore, the RSP5 homologue in *S. pombe*, PUB1, has been implicated in the regulation of the half-life of the cdc25 phosphatase (Nefsky and Beach, 1996) and, as indicated above, the mouse homologue interacts with a Na^+ channel protein (Staub *et al.*, 1996). This may indicate that homologous hect domain proteins may be involved in the turnover of different proteins in different organisms. Finally, RSP5, as well as its human homologue, has been reported to influence transcriptional activation mediated by certain steroid receptors (Imhof and McDonnell, 1996). As the authors of this study provide evidence that, in this particular system, the thioester forming capacity of RSP5 does not appear to be important, one has to consider the possibility that at least some of the hect domain proteins are multifunctional, i.e., they are not exclusively involved in processes requiring ubiquitin conjugation. Similar to UBR1, it appears that the hect domain proteins tested so far interact only with a distinct set of E2s, namely, either with yeast UBC4/UBC5 and homologous E2s from higher organisms, such as human UbcH5, or with human UbcH7 (as in the case of some human hect domain proteins) (Huibregtse *et al.*, 1995; Nuber *et al.*, 1996; S. E. Schwarz and M. Scheffner, unpublished). This demonstrates that a single E2 can communicate with several of the E3s, which may allow efficient and processive ubiquitination of several substrates at a given time. It will be interesting to see whether the aforementioned E2s (yeast UBC2, UBC4/5 and their homologues in other organisms, human UbcH7) represent the only E2s that are involved in E3-dependent ubiquitination (with E3 being defined as a ubiquitin thioester forming enzyme) or whether other E2s interact with other E3s, not related to UBR1 or the hect domain proteins.

4. SUBSTRATE RECOGNITION, SPECIFICITY, AND REGULATION

4.1. Basic Concept of Substrate Recognition

For the viability of a cell it is critical that ubiquitin-dependent degradation, and therefore substrate recognition, is a highly specific and regulated process. The specificity of substrate recognition is achieved by the interplay of *cis*- and *trans*-

acting factors, the *cis* factors being the amino acid sequences or structures of a protein that confer the half-life onto this protein (further referred to as ubiquitination signals or signals) and the *trans* factors being the proteins that mediate the recognition of these signals. In addition, some substrates of the ubiquitin-conjugation system are supposed to be targeted only at distinct times during development or differentiation of the cell cycle. Generally, this can be accomplished by regulating the activity of the *trans* factors that are involved in the recognition of a particular substrate, by regulating the accessibility of the signals on the substrate protein or by regulating both. The specificity and selectivity of protein ubiquitination can be achieved by several mechanisms; for instance, by the presence of a variety of different signals, by regulating the accessibility of a limited number of signals, or by a combination of both. The presence of an apparently large number of factors involved in substrate recognition (see below) suggests that there is indeed a large number of amino acid sequences or protein structures that can serve as ubiquitination signals. This remains to be seen, however, as the signals of only a few target proteins have yet been identified. Obviously, different proteins have different metabolic half-lives. One possibility to account for this is that some signals are more efficiently recognized by the ubiquitin-conjugation system than others. In addition, it appears that at least some substrates contain at least two signals (e.g., MATα2, GCN4) that are presumably recognized by different factors (Hochstrasser and Varshavsky, 1990; Chen *et al.*, 1993; Kornitzer *et al.*, 1994). The presence of multiple signals on a protein could further contribute to finely tune its turnover. In this context, it should be noted that signals, or part of a signal, do not necessarily have to be provided by the target protein itself; rather, they can also be provided by factors in complex with the target protein (a process termed *trans-recognition*) (Johnson *et al.*, 1990).

As outlined above, it is thought that the E2 and E3 enzymes play an important role in substrate recognition. Yeast, for instance, has 13 different E2s and at least 6 different putative E3s. Assuming that each of these enzymes is involved in the recognition of different signals, this would allow the ubiquitination machinery to recognize a significant, but limited, number of target proteins. Apparently, the ubiquitin-conjugation system has evolved additional mechanisms to expand the repertoire of signals that can be recognized. At least some of the E2s have the property to oligomerize with heterologous E2s and the hetero-oligomeric complexes are involved in the turnover of different proteins than the monomeric E2s (termed *combinatorial expansion of specificities* by Hochstrasser, 1995). This mechanism has been suggested by studies investigating the turnover of the transcription factor MATα2, which involves a complex consisting of UBC6 and UBC7 (Chen *et al.*, 1993). Mutants lacking either UBC6 or UBC7, however, have distinguishable phenotypes, suggesting that they are at least in part involved in the degradation of different proteins. Similarly, it seems possible that either some of the E3s may form hetero-oligomeric complexes, or that a given E3 may interact with different E2s (as has been shown for E6-AP that can interact with UbcH5 or with UbcH7; Nuber *et al.*, 1996), and that the different combinations are involved in the recognition of different proteins than the respective monomeric enzymes. However, this is a purely speculative hypothesis at present. Finally, and maybe most importantly, specific binding of E2s or E3s to target proteins may need the activity of additional factors, here designated as *specificity factors*, as suggested by the finding that E6-AP requires the HPV E6 oncoprotein to target ubiquitination of p53 in an *in vitro* system (Huibregtse *et al.*, 1991; Scheffner *et al.*, 1993). In the following we will briefly review the literature with respect to (1) known ubiquitination signals, (2) factors that in addition to E2s and E3s are involved in protein ubiquitination, and (3) regulatory mechanisms.

4.2. Ubiquitination Signals

The first known ubiquitination signal was determined by studying the turnover of artificial substrates of the N-end rule pathway (reviewed in Varshavsky, 1992). As earlier mentioned, this pathway relates the half-life of a protein to the identity of its N-terminal residue. In addition to the identity of the N-terminal residue, however, one or more lysine residues that are properly located toward the N-terminal residue are necessary to allow ubiquitination of an N-end rule substrate. This has led to the hypothesis that ubiquitination signals are in general of bipartite structure (Bachmair *et al.*, 1989). One part of the signal can be regarded as the primary determinant that is directly recognized by the respective E2 or E3 or specificity factor ("recognition site"), whereas the other part is the site to which ubiquitin can be covalently attached ("attachment site"). These two parts of a ubiquitination signal may be partially overlapping or may be completely separated with respect to the primary sequence of a target protein. However, it seems likely that they have to be in close spatial proximity on the native protein for efficient ubiquitination.

Similar to the N-end rule, the identity of the penultimate N-terminal residue (i.e., proline) appears to play an important role in determining the metabolic fate of the *c-mos* proto-oncogene product, Mos, which plays a role in the meiotic cell cycle ("second codon rule") (Nishizawa *et al.*, 1992). Moreover, it has been reported that (auto)phosphorylation of the serine residue at position 3 (Ser-3) of Mos results in stabilization of Mos and that a lysine residue at position 34 constitutes the major ubiquitin attachment site (Nishizawa *et al.*, 1993). Whether proline at position 2 and Ser-3 alone constitute the recognition site for the ubiquitin-conjugation system, or whether they are part of a larger structure, is not clear at present. Similarly, the E2s and/or the E3s and/or specificity factors involved in regulation of Mos have not yet been determined.

Among the most prominent and best-studied substrates of the ubiquitin-conjugation system are the cyclins. Accordingly, regions of some of these proteins that are involved in half-life regulation have been determined (see also Section 3.3). The first described is the so-called "destruction box" of B-type cyclins (Glotzer *et al.*, 1991). The destruction box is a 9-amino-acid motif that appears to be conserved throughout evolution. It is not clear, however, whether the destruction box on its own represents a recognition site or whether it has to be presented within a larger structure for recognition (to our knowledge the minimal region tested that could confer a short half-life onto an otherwise stable protein was approximately 60 amino acids in length). Interestingly, uracil permease (see Section 3.43) appears to contain a motif similar to the destruction box and it is thought that it plays a role in degradation of the permease (Galan *et al.*, 1996, and references therein). This may suggest that at least some recognition sites of the ubiquitin-conjugation system are shared by otherwise unrelated target proteins. The G1 cyclins do not appear to contain a destruction box motif. Deletion analyses as well as experiments with fusion proteins have shown that a major recognition site(s) of both CLN2 and CLN3 is present within a C-terminal region of approximately 100–200 amino acids in length enriched in proline, glutamic acid, serine, and threonine residues (Salama *et al.*, 1994; Deshaies *et al.*, 1995; Yaglom *et al.*, 1995). This amino acid signature constitutes a so-called "PEST motif," which has ben suggested to represent a degradation signal (Rogers *et al.*, 1986). However, studies on CLN2 and CLN3 indicate that it may not be the PEST region *per se* but rather the phosphorylation of a specific serine or threonine residue, within this region, that is the crucial event required to trigger destruction (Salama *et al.*, 1994; Yaglom *et al.*, 1995; Lanker *et al.*, 1996).

Regions involved in determining the turnover rate have been identified in three transcription factors, namely MATα2, GCN4, and c-Jun (Hochstrasser and Varshavsky, 1990; Kornitzer *et al.*, 1994; Treier *et al.*, 1994). MATα2 and GCN4 contain at least two signals that can function independently of each other. This hypothesis is supported by the observation that different UBCs, and therefore presumably different pathways, are involved in the recognition of these signals as discussed above (Chen *et al.*, 1993; Kornitzer *et al.*, 1994). The proto-oncogene product c-Jun, and its cognate oncogene v-Jun, differ in their turnover in that c-Jun is normally a short-lived protein whereas v-Jun is a constitutively long-lived protein. The difference in their half-lives could be attributed to the so-called "δ domain" of c-Jun, which is 27 amino acids in length and is missing in v-Jun (Treier *et al.*, 1994).

Using β-galactosidase fusion proteins, to which short randomized peptide sequences had been fused, Finley and co-workers were able to identify short stretches of amino acids that can serve as signal for the ubiquitin-conjugation system in yeast (Sadis *et al.*, 1995). Three signals, termed class I, II, and III signals, could be identified. Class I signals are degraded via the N-end rule pathway involving the action of UBC2 and UBR1. It appears that class I signals are not recognition signals *per se* but, rather, promote proteolytic cleavage by an unknown protease near the N-terminus of the fusion protein to provide an amino acid that is recognized by the N-end rule pathway. Turnover of class II signals requires UBC4/5 as well as UBC6 and UBC7, whereas only UBC4/5 are involved in the turnover of class III signals. In contrast to the identified naturally occurring signals, class II and III signals are short (approximately 10 amino acids) and are less efficient signals. Taken together, this may indicate in general that short stretches of amino acids, such as class II and III signals, may serve as direct contact sites for E2s, E3s, or specificity factors, but for efficient recognition they have to be presented in a proper structural context. To support this hypothesis it will be necessary to determine more ubiquitination signals and, in particular, the factors that directly recognize these signals.

4.3. Ancillary Factors

Within the frame of this chapter, ancillary factors are defined as proteins that in addition to El, E2s, and at least in some cases E3s, are involved in the turnover of a particular protein. Based on their proposed role in ubiquitin conjugation, ancillary factors can be presently classified into three categories: specificity factors, targeting factors, and regulatory factors. This classification, of course, is not mutually exclusive.

The existence of specificity factors has been suggested by studies examining the ubiquitination of natural substrates *in vitro*. As already mentioned, characterization of the HPV E6 oncoprotein-facilitated ubiquitination of p53 revealed that the E6 oncoprotein forms a stable complex with the E3 enzyme E6-AP and that only the complex of E6 and E6-AP can detectably bind to p53 (Huibregtse *et al.*, 1991, 1993). Although the exact role of the E6 oncoprotein is not completely understood (for instance, it is not known whether both proteins contact p53 directly or whether binding of E6 induces a conformational change in E6-AP so that E6-AP can directly bind to p53), it is clear that the specificity of E6-AP for p53 is provided by interaction with the E6 oncoprotein. Similarly, it has been shown that the specificity in the ubiquitination of B-type cyclins is provided by a protein complex termed APC (anaphase promoting complex) or cyclosome (King *et al.*, 1995; Sudakin *et al.*, 1995; Irniger *et al.*, 1995). It is currently not clear whether APC functions as a specificity factor for certain E2(s), or whether similar to the E6/E6-AP complex APC consists of both specificity factors and a protein with E3 activity (for a discussion of APC, see Chapter 12).

Efficient turnover of particular proteins may require proper localization of certain E2s and/or E3s and this may in part be achieved by targeting factors that direct certain E2s and/or E3s to a distinct subcellular structure. This has been suggested by the observations that UBC2/RAD6 can complex with RAD18 (Bailly *et al.*, 1994) and human UBC9 with RAD51 (Kovalenko *et al.*, 1996) and mouse UBC9 with adenovirus E1A (Hateboer *et al.*, 1996). As discussed in

Section 3.3, it is thought that RAD18 targets UBC2 to sites of DNA damage. Similarly, RAD51 and E1A may target mammalian UBC9 to specific intracellular sites and/or substrates. What these substrates are, however, is not known at present.

Regulatory factors can be considered as proteins that influence either the accessibility of a ubiquitination signal, or the activity of proteins involved in substrate recognition (i.e., E2s, E3s, specificity and targeting factors). Good candidates for regulatory factors are obviously enzymes such as protein kinases that reversibly modify their target proteins. For instance, ubiquitination of the G1 cyclins CLN2 and CLN3, as well as of the CLB-CDC28 kinase inhibitor SIC1, involves the CDC28 kinase (Salama *et al.*, 1994; Deshaies *et al.*, 1995; Yaglom *et al.*, 1995; Lanker *et al.*, 1996; Schneider *et al.*, 1996). Mos appears to be both a substrate of the ubiquitin-conjugating system and a regulatory factor of its own ubiquitination. It has been shown that phosphorylation of Ser-3 (see above) inhibits the turnover of Mos and this phosphorylation is most likely catalyzed by Mos itself (Nishizawa *et al.*, 1992).

4.4. Regulation

Over the last few years it has become clear that phosphorylation plays an important regulatory role in the ubiquitination of a number of proteins, including CLN2, CLN3, cyclin B, Mos, IκB, SIC1, and GCN4. In the case of CLN2, CLN3, IκB, and GCN4, phosphorylation of these proteins appears to be necessary to induce ubiquitination of the respective protein (Kornitzer *et al.*, 1994; Yaglom *et al.*, 1995; Chen *et al.*, 1995; Lanker *et al.*, 1996), whereas in the case of Mos, phosphorylation of Mos presumably prevents ubiquitination (Nishizawa *et al.*, 1992). This suggests that the phosphorylation status of these substrates regulates the accessibility of either the recognition site or the attachment site of the ubiquitination signal. Alternatively, the amino acid residues that are modified by phosphorylation may serve as direct contact sites for recognition factors. Interestingly, the phosphorylation sites in CLN3 have been mapped to the so-called PEST regions (Yaglom *et al.*, 1995). As previously mentioned, this may indicate that PEST regions *per se* do not constitute primary degradation signals as previously assumed but, rather, provide phosphorylation sites to regulate the accessibility of such signals. Regulation does not necessarily have to be on the level of the substrate, although it is equally likely that the activity of factors that are involved in substrate recognition can be modulated by phosphorylation. In support of this, it has been suggested that the phosphorylation status of APC determines the cell cycle-specific ubiquitination of cyclin B (King *et al.*, 1995; Lahav-Baratz *et al.*, 1995). Various extracellular, as well as intracellular stimuli have been reported to influence the ubiquitin-dependent turnover of certain proteins. In higher plants, red light induces the turnover of phytochrome (Shanklin *et al.*, 1987). The half-life of several proteins is influenced by the nutritional status of a cell (e.g., glucose induces the degradation of fructose biphosphatase and amino acid starvation results in stabilization of GCN4) (Schork *et al.*, 1994; Kornitzer *et al.*, 1994). Of particular importance for the ubiquitination of cell surface receptors may be the finding that ubiquitination of the IgE receptor is induced on engagement of the receptor with its ligand (regulation by complex formation) (Paolini and Kinet, 1993). Because the ligand binds to the extracellular portion of the IgE receptor, whereas ubiquitination obviously occurs at the intracellular portion, the ligand has to be considered as a regulatory factor (as opposed to a specificity factor). Finally, the tumor suppressor p53 is stabilized on treatment of cells with DNA-damaging agents (Maltzman and Czyzyk, 1984; Kastan *et al.*, 1992). It should be noted that it is not known how the regulation occurs on the molecular level for any of these substrates. It seems likely, however, that in some (if not most) of the cases the respective stimuli result in modification of either the substrate proteins, or the recognition components of the ubiquitin-conjugating system. Regulation of the ubiquitin-conjugating system by extracellular and intracellular stimuli has been described not only on the protein level but also on the level of gene expression. In yeast, heat shock induces the expression of UBC4 and UBC5 and exposure to heavy metals leads to an induction of *UBC5* and *UBC7* mRNAs (Seufert and Jentsch, 1990; Jungmann *et al.*, 1993; see above). Similarly, expression of genes for certain E2s can vary during differentiation or development (Wefes *et al.*, 1995; Zhen *et al.*, 1996). Although not yet established, the different availability of certain E2s (and presumably E3s and ancillary factors) during differentiation (or development or in distinct tissues) should, therefore, be reflected in a concomitant regulated turnover of their respective target proteins.

5. PERSPECTIVE

The basic mechanism by which ubiquitin is transferred to the substrate is largely understood. However, the identification of a family of E3s that accept ubiquitin as a thioester necessitates studies to determine which E2 and E3 enzymes function together. Processive multiubiquitination of substrates is assumed to require complex formation between E1, E2, and E3 enzymes. It is not known, however, how a specific association between these enzymes is achieved *in vivo* and whether special tethering factors (analogous to the yeast STE5 protein MAP kinase cascades) are involved. The identification of a number of physiological substrates of the ubiquitin system will permit a detailed analysis of the degradation signals and the enzymes and ancillary factors that recognize the different signals.

The existence of distinct multiubiquitin chains that differ in their ubiquitin–ubiquitin linkage poses several interesting questions. Do they serve different cellular functions and are they recognized by specific receptors for each chain type? How does the cell decide whether a ubiquitinated substrate is targeted to the proteasome or, as in the case of the STE2 protein (Hicke and Riezman, 1996), the lysosome? In this context it is important to stress that the precise *biochemical* function of ubiquitin remains poorly understood. Ubiquitin is unlikely to function as a mere tag for proteolysis: A single ubiquitin moiety attached to a substrate should suffice for such a role. The ubiquitin moieties of multiubiquitin chains may rather possess additional activities such as an "unfoldase" activity for the conjugated substrate, or the chain may function as a tether by binding the conjugate to the proteasome. X-ray crystallography studies of ubiquitin–protein conjugates are expected to yield the answers.

Acknowledgments. We thank Drs. Noel J. Whitaker and Manfred Koegl for comments on the manuscript. M.S. was supported by grants from Deutsche Forschungsgemeinschaft, S.J. by grants from Deutsche Forschungsgemeinschaft, Human Frontier Science Program, German–Israeli Foundation for Research and Development, and Fonds der Chemischen Industrie. Parts of this chapter have been adapted from Smith *et al.* (1996) with permission from Walter DeGruyter Verlag, Berlin.

6. REFERENCES

Alagramam, K., Naider, F., and Becker, J. M., 1995, A recognition component of the ubiquitin system is required for peptide transport in *Saccharomyces cerevisiae*, *Mol. Microbiol.* **15:**225–234.

Amon, A., Tyers, M., Futcher, B., and Nasmyth, K., 1993, Mechanisms that help the yeast cell cycle clock tick: G2 cyclins transcriptionally activate G2 cyclins and repress G1 cyclins, *Cell* **74:**993–1007.

Amon, A., Irniger, S., and Nasmyth, K., 1994, Closing the cell cycle circle in yeast: G2 cyclin proteolysis initiated at mitosis persists until the activation of G1 cyclins in the next cycle, *Cell* **77:**1037–1050.

Aristarkhov, A., Eytan, E., Moghe, A., Admon, A., Hershko, A., and Ruderman, J. V., 1996, E2-C, a cyclin-selective ubiquitin carrier protein required for the destruction of mitotic cyclins, *Proc. Natl. Acad. Sci. USA* **93:**4294–4299.

Arnason, T., and Ellison, M. J., 1994, Stress resistance in *Saccharomyces cerevisiae* is strongly correlated with assembly of a novel type of multiubiquitin chain, *Mol. Cell. Biol.* **14:**7876–7883.

Ayusawa, D., Kaneda, S., Itoh, Y., Yasuda, H., Muramaki, Y., Sugasawa, K., Hanaoka, F., and Seno, T., 1992, Complementation by cloned human ubiquitin-activating enzyme El of the S-phase-arrested mouse *FM3A* cell mutant with thermolabile El, *Cell Struct. Funct.* **17:**113–122.

Bachmair, A., and Varshavsky, A., 1989, The degradation signal in a short-lived protein, *Cell* **56:**1019–1032.

Bachmair, A., Finley, D., and Varshavsky, A., 1986, *In vivo* half-life of a protein is a function of its amino-terminal residue, *Science* **234:**179–186.

Bailly, V., Lamb, J., Sung, P., Prakash, S., and Prakash, L., 1994, Specific complex formation between yeast RAD6 and RAD18 proteins: A potential mechanism for targeting RAD6 ubiquitin-conjugating activity to DNA damage sites,*Genes Dev.* **8:**811–820.

Ball, E., Karlik, C. C., Beall, C. J., Saville, D. L., Sparrow, J. C., Bullard, B., and Fryberg, E. A., 1987, Arthrin, a myofibrillar protein of insect flight muscle, is an actin–ubiquitin conjugate, *Cell* **51:**221–228.

Bartel, B., Wünning, I., and Varshavsky, A., 1990, The recognition component of the N-end rule pathway, *EMBO J.* **9:**3179–3189.

Berleth, E. S., and Pickart, C. M., 1996, Mechanism of ubiquitin conjugating enzyme E2-230K: Catalysis involving a thiol relay? *Biochemistry* **35:**1664–1671.

Biederer, T., Volkwein, C., and Sommer, T., 1996, Degradation of subunits of the Sec61p complex, an integral component of the ER membrane, by the ubiquitin-proteasome pathway, *EMBO J.* **15:**2069–2076.

Chang, F., and Herskowitz, I., 1990, Identification of a gene necessary for cell cycle arrest by a negative growth factor of yeast: FAR1 is an inhibitor of a G1 cyclin, CLN2, *Cell* **63:**999–1011.

Chau, V., Tobias, J. W., Bachmair, A., Marriott, D., Ecker, D. J., Gonda, D. K., and Varshavsky, A., 1989, A multiubiquitin chain is confined to specific lysine in a targeted short-lived protein, *Science* **243:**1576–1583.

Chen, P., Johnson, P., Sommer, T., Jentsch, S., and Hochstrasser, M., 1993, Multiple ubiquitin-conjugating enzymes participate in the in vivo degradation of the yeast MATα2 repressor, *Cell* **74:**357–369.

Chen, Z., Hagler, J., Palombella, V. J., Melandri, F., Scherer, D., Ballard, D., and Maniatis, T., 1995, Signal-induced site-specific phosphorylation targets IκB α to the ubiquitin-proteasome pathway, *Genes Dev.* **9:**1586–1597.

Ciechanover, A., 1994, The ubiquitin-proteasome proteolytic pathway, *Cell* **79:**13–21.

Ciechanover, A., Elias, S., Heller, H., and Hershko, A., 1982, "Covalent affinity" purification of ubiquitin-activating enzyme, *J. Biol. Chem.* **257:**2537–2542.

Cook, W. J., Jeffrey, L. C., Sullivan, M. L., and Vierstra, R. D., 1992, Three dimensional structure of a ubiquitin-conjugating enzyme (E2), *J. Biol. Chem.* **267:**15116–15121.

Cook, W. J., Jeffrey, L. C., Xu, Y., and Chau, V., 1993, Tertiary structures of class I ubiquitin-conjugating enzymes are highly conserved: Crystal structure of yeast Ubc4, *Biochemistry* **32:** 13809–13817.

Deshaies, R. J., Chau, V., and Kirschner, M., 1995, Ubiquitination of the G1 cyclin Cln2p by a Cdc34p-dependent pathway, *EMBO J.* **14:**303–312.

Dohmen, R. J., Madura, K., Bartel, B., and Varshavsky, A., 1991, The N-end rule is mediated by the UBC2(RAD6) ubiquitin-conjugating enzyme, *Proc. Natl. Acad. Sci. USA* **88:**7351–7355.

Dohmen, R. J., Stappen, R., McGrath, J. P., Forrova, H., Kolarov, J., Goffeau, A., and Varshavsky, A., 1995, An essential yeast gene encoding a homolog of ubiquitin-activating enzyme, *J. Biol. Chem.* **270:**18099–18109.

Finley, D., and Chau, V., 1991, Ubiquitination, *Annu. Rev. Cell Biol.* **7:**25–69.

Finley, D., Özkaynak, E., and Varshavsky, A., 1987, The yeast polyubiquitin gene is essential for resistance to high temperatures, starvation, and other stresses, *Cell* **48:**1035–1046.

Finley, D., Bartel, B., and Varshavsky, A., 1989, The tails of ubiquitin precursors are ribosomal proteins whose fusion to ubiquitin facilitates ribosome biogenesis, *Nature* **338:**394–401.

Finley, D., Sadis, S., Monia, B. P., Boucher, P., Ecker, D. J., Crook, S. T., and Chau, V., 1994, Inhibition of proteolysis and cell cycle progression in a multiubiquitination-deficient yeast mutant, *Mol. Cell. Biol.* **14:**5501–5509.

Fitch, I., Dahmann, C., Surana, U., Amon, A., Nasmyth, K., Goetsch, L., Byers, B., and Futcher, B., 1992, Characterization of four B-type cyclin genes of the budding yeast *Saccharomyces cerevisiae*, *Mol. Biol. Cell.* **3:**805–818.

Galan, J. M., Moreau, V., Andre, B., Volland, C., and Haguenauer-Tsapis, R., 1996, Ubiquitination mediated by the Npi1p/Rsp5p ubiquitin-protein ligase is required for endocytosis of the yeast uracil permease, *J. Biol. Chem.* **271:**10946–10952.

Glotzer, M., Murray, A. W., and Kirschner, M. W., 1991, Cyclin is degraded by the ubiquitin pathway, *Nature* **349:**132–138.

Goebl, M. G., Yochem, J., Jentsch, S., McGrath, J. P., Varshavsky, A., and Byers, B., 1988, The yeast cell cycle gene *CDC34* encodes a ubiquitin-conjugating enzyme, *Science* **241:**1331–1335.

Goldknopf, I. L., and Busch, H., 1977, Isopeptide linkage between nonhistone and histone 2A polypeptides of chromosomal conjugate-protein A24, *Proc. Natl. Acad. Sci. USA* **74:**864–868.

Haas, A. L., and Rose, I. A., 1982, The mechanism of ubiquitin activating enzyme. A kinetic and equilibrium analysis, *J. Biol. Chem.* **257:**10329–10337.

Haas, A. L., Warms, J. V., Hershko, A., and Rose, I. A., 1982, Ubiquitin-activating enzyme. Mechanism and role in protein–ubiquitin conjugation, *J. Biol. Chem.* **257:**2543–2548.

Hadwiger, J. A., Wittenberg, C., Richardson, H. E., de Barros-Lopes, M., and Reed, S. I., 1989, A family of cyclin homologs that control the G1 phase in yeast, *Proc. Natl. Acad. Sci. USA* **86:**6255–6259.

Handley, P. M., Mueckler, M., Siegel, N. R., Ciechanover, A., and Schwarz, A. L., 1991, Molecular cloning, sequence, and tissue distribution of the human ubiquitin-activating enzyme E1, *Proc. Natl. Acad. Sci. USA* **88:**258–262. (Erratum appeared in *Proc. Natl. Acad. Sci. USA* **88:**7456)

Hateboer, G., Hijmans, E. M., Nooij, J. B. D., Schlenker, S., Jentsch, S., and Bernads, R., 1996, mUBC9, a novel adenovirus E1A-interacting protein that complements a yeast cell cycle defect, *J. Biol. Chem.* **271:**25906–25911.

Hatfield, P. M., and Vierstra, R. D., 1992, Multiple forms of ubiquitin-activating enzyme *E*1 from wheat. Identification of an essential cysteine by *in vitro* mutagenesis, *J. Biol. Chem.* **267:**14799–14803.

Hatfield, P. M., Callis, J., and Vierstra, R. D., 1990, Cloning of ubiquitin activating enzyme from wheat and expression of a functional protein in *Escherichia coli*, *J. Biol. Chem.* **265:**15813–15817.

Heichman, K. A., and Roberts, J. M., 1996, The yeast *CDC16* and *CDC27* genes restrict DNA replication to once per cell cycle, *Cell* **85:**39–48.

Hein, C., Springael, J.-Y., Volland, C., Haguenauer-Tsapis, R., and Andre, B., 1995, *NPI1*, an essential yeast gene involved in induced degradation of Gap1 and Fur4 permeases, encodes the Rsp5 ubiquitin-protein ligase, *Mol. Microbiol.* **18:**77–87.

Hershko, A., and Ciechanover, A., 1992, The ubiquitin system for protein degradation, *Annu. Rev. Biochem.* **61:**761–807.

Hershko, A., Heller, H., Elias, S., and Ciechanover, A., 1983, Components of ubiquitin-protein ligase system. Resolution, affinity purification, and role in protein breakdown, *J. Biol. Chem.* **258:**8206–8214.

Hicke, L., and Riezman, H., 1996, Ubiquitination of a yeast plasma membrane receptor signals its ligand stimulated endocytosis, *Cell* **84:**277–287.

Hingamp, P. M., Arnold, J. E., Mayer, R. J., and Dixon, L. K., 1992, A ubiquitin conjugating enzyme encoded by African swine fever virus, *EMBO J.* **11:**361–366.

Hinnebusch, A. G., 1988, Mechanisms of gene regulation in the general control of amino acid biosynthesis in *Saccharomyces cerevisiae*, *Microb. Rev.* **52:**248–273.

Hinnebusch, A. G., 1994, Translational control of GCN4: An *in vivo* barometer of initiation-factor activity, *Trends Biochem. Sci.* **19:**409–414.

Hochstrasser, M., 1995, Ubiquitin, proteasomes, and the regulation of intracellular protein degradation, *Curr. Opin. Cell Biol.* **7:**215–223.

Hochstrasser, M., and Varshavsky, A., 1990, In vivo degradation of a transcriptional regulator: The yeast $\alpha 2$ repressor, *Cell* **61:**697–708.

Huibregtse, J. M., Scheffner, M., and Howley, P. M., 1991, A cellular protein mediates association of p53 with the E6 oncoprotein of human papillomavirus types 16 or 18, *EMBO J.* **10:**4129–4135.

Huibregtse, J. M., Scheffner, M., and Howley, P. M., 1993, Localization of the E6-AP regions that

direct human papillomavirus E6 binding, association with p53, and ubiquitination of associated proteins, *Mol. Cell. Biol.* **13:**4918–4927.
Huibregtse, J. M., Scheffner, M., Beaudenon, S., and Howley, P. M., 1995, A family of proteins structurally and functionally related to the E6-AP ubiquitin-protein ligase, *Proc. Natl. Acad. Sci. USA* **92:**2563–2567.
Hunt, L. T., and Dayhoff, M. O., 1977, Amino-terminal sequence identity of ubiquitin and the nonhistone component of nuclear protein A24, *Biochem. Biophys. Res. Commun.* **74:**650–655.
Imhof, M. O., and McDonnell, D., 1996, Yeast RSP5 and its human homolog hRPF1 potentiate hormone-dependent activation of transcription by human progesterone and glucocorticoid receptors, *Mol. Cell. Biol.* **16:**2594–2605.
Irniger, S., Piatti, S., Michaelis, C., and Nasmyth, K., 1995, Genes involved in sister chromatid separation are needed for B-type cyclin proteolysis in budding yeast, *Cell* **81:**269–278.
Jensen, J. P., Bates, P. W., Yang, M., Vierstra, R. D., and Weissman, A. M., 1995a, Identification of a family of closely related human ubiquitin conjugating enzymes, *J. Biol. Chem.* **51:**30408–30414.
Jensen, T.J., Loo, M. A., Pind, S., Williams, D. B., Goldberg, A. L., and Riordan, J. R., 1995b, Multiple proteolytic systems, including the proteasome, contribute to CFTR processing, *Cell* **83:**129–135.
Jentsch, S., 1992a, The ubiquitin-conjugation system, *Annu. Rev. Genet.* **26:**179–207.
Jentsch, S., 1992b, Ubiquitin-dependent protein degradation: A cellular perspective, *Trends Cell Biol.* **2:**98–103.
Jentsch, S., and Schlenker, S., 1996, Selective protein degradation: A journey's end within the proteasome, *Cell* **82:**881–884.
Jentsch, S., McGrath, J. P., and Varshavsky, A., 1987, The yeast DNA repair gene *RAD6* encodes a ubiquitin-conjugating enzyme, *Nature* **329:**131–134.
Jentsch, S., Seufert, W., Sommer, T., and Reins, H.-A., 1990. Ubiquitin-conjugating enzymes: Novel regulators of eukaryotic cells, *Trends Biochem. Sci.* **15:**195–198.
Jentsch, S., Seufert, W., and Hauser, H.-P., 1991, Genetic analysis of the ubiquitin system, *Biochim. Biophys. Acta* **1089:**127–139.
Johnson, E. S., Gonda, D. K., and Varshavsky, A., 1990, *Cis–trans* recognition and subunit-specific degradation of short-lived proteins, *Nature* **346:**287–291.
Johnson, E. S., Ma, P. C., Ota, I. M., and Varshavsky, A., 1995, A proteolytic pathway that recognizes ubiquitin as a degradation signal, *J. Biol. Chem.* **270:**17442–17456.
Jones, J. S., Weber, S., and Prakash, L., 1988, The *Saccharomyces cerevisiae RAD18* gene encodes a protein that contains potential zinc finger domains for nucleic acid binding and a putative nucleotide binding sequence, *Nucleic Acids Res.* **16:**7119–7131.
Jungmann, J., Reins, H.-A., Schobert, C., and Jentsch, S., 1993, Resistance to cadmium mediated by ubiquitin-dependent proteolysis, *Nature* **361:**369–371.
Kang, X. L., Yadao, F., Gietz, R. D., and Kunz, B. A., 1992, Elimination of the yeast RAD6 ubiquitin conjugase enhances base-pair transitions and G·C–T·A transversions as well as transposition of the Ty element: Implications for the control of spontaneous mutation, *Genetics* **130:**285–294.
Kastan, M. B., Zhan, Q., El-Deiry, W. S., Carrier, F., Jacks, T., Walsh, W. V., Plunkett, B. S., Vogelstein, B., and Fornace, A. J., Jr., 1992, A mammalian cell cycle checkpoint pathway utilizing p53 and GADD45 is defective in ataxia-telangiectasia, *Cell* **71:**587–597.
Kay, G. F., Ashworth, A., Penny, G. D., Dunlop, M., Swift, S., Brockdorff, N., and Rastan, S., 1991, A candidate spermatogenesis gene on the mouse Y chromosome is homologous to ubiquitin-activating enzyme E1, *Nature* **354:**486–489.
King, R. W., Peters, J. M., Tugendreich, S., Rolfe, M., Hieter, P., and Kirschner, M. W., 1995, A 20S complex containing CDC27 and CDC16 catalyzes the mitosis-specific conjugation of ubiquitin to cyclin B, *Cell* **81:**279–288.
Kok, K., Hofstra, R., Pilz, A., van den Berg, A., Terpstra, P., Buys, C. H. C. M., and Carrit, B., 1993, A gene in the chromosomal region 3p21 with greatly reduced expression in lung cancer is similar to the gene for ubiquitin-activating enzyme, *Proc. Natl. Acad. Sci. USA* **90:**6071–6075.

Koken, M. H. M., Reynolds, P., Jaspers-Dekker, I., Prakash, L., Prakash, S., Bootsma, D., and Hoeijmakers, J. H. J., 1991, Structural and functional conservation of two human homologs of the yeast DNA repair gene *RAD6*, *Proc. Natl. Acad. Sci. USA* **88:**8865–8869.

Kolman, C. J., Toth, J., and Gonda, D. K., 1992, Identification of a portable determinant of cell cycle function within the carboxyl-terminal domain of the yeast CDC34 (UBC3) ubiquitin-conjugating (E2) enzyme, *EMBO J.* **11:**3081–3090.

Kornitzer, D., Raboy, B., Kulka, R. G., and Fink, G. R., 1994, Regulated degradation of the transcription factor Gcn4, *EMBO J.* **13:**6021–6030.

Kovalenko, O. V., Plug, A. W., Haaf, T., Gonda, D. K., Ashley, T., Ward, D. C., Radding, C. M., and Golub, E. I., 1996, Mammalian ubiquitin-conjugating enzyme Ubc9 interacts with Rad51 recombination protein and localizes in synaptonemal complexes, *Proc. Natl. Acad. Sci. USA* **93:**2958–2963.

Lahav-Baratz, S., Sudakin, V., Ruderman, J. V., and Hershko, A., 1995, Reversible phosphorylation controls the activity of cyclosome-associated cyclin-ubiquitin ligase, *Proc. Natl. Acad. Sci. USA* **92:**9303–9307.

Lanker, S., Valdivieso, M. H., and Wittenburg, C., 1996, Rapid degradation of the G1 cyclin Cln2 induced by CDK-dependent phosphorylation, *Science* **271:**1597–1601.

Lawrence, C., 1994, The RAD6 DNA repair pathway in *Saccharomyces cerevisiae*: What does it do, and how does it do it? *Bioessays* **16:**253–258.

Leyser, H. M. O., Lincoln, C. A., Timpte, C., Lammer, D., Turber, J., and Estelle, M., 1993, Arabidopsis auxin-resistance gene *AXR1* encodes a protein related to ubiquitin-activating enzyme E1, *Nature* **364:**161–164.

Madura, K., and Varshavsky, A., 1994, Degradation of Gα by the N-end rule pathway, *Science* **265:** 1454–1458.

Madura, K., Dohmen, R. J., and Varshavsky, A., 1993, N-recognin/Ubc2 interactions in the N-end rule pathway, *J. Biol. Chem.* **268:**12046.

Maltzman, W., and Czyzyk, L., 1984, UV irradiation stimulates levels of p53 cellular tumor antigen in nontransformed mouse cells, *Mol. Cell. Biol.* **4:**1689–1694.

Marsh, L., Neiman, A. M., and Herskowitz, I., 1991, Signal transduction during pheromone response in yeast, *Annu. Rev. Cell Biol.* **7:**699–728.

Matuschewski, K., Hauser, H.-P., Treier, M., and Jentsch, S., 1996, Identification of a novel family of ubiquitin-conjugating enzymes with distinct amino-terminal extensions, *J. Biol. Chem.* **271:** 2789–2794.

McGrath, J. P., Jentsch, S., and Varshavsky, A., 1991, *UBA1*: An essential yeast gene encoding ubiquitin-activating enzyme, *EMBO J.* **10:**227–236.

McKinney, J. D., Chang, F., Heintz, N., and Cross, F. R., 1993, Negative regulation of FAR1 at the Start of the yeast cell cycle, *Genes Dev.* **7:**833–843.

Mitchell, M. J., Woods, D. R., Tucker, P. K., Opp, J. S., and Bishop, C. E., 1991, Homology of a candidate spermatogenic gene from the mouse Y chromosome to the ubiquitin-activating enzyme E1, *Nature* **354:**483–486.

Mitchell, M. J., Woods, D. R., Wilcox, S. A., Graves, J. A. M., and Bishop, C. E., 1992, Marsupial Y chromosome encodes a homologue of the mouse Y-linked candidate spermatogenesis gene *Ubely*, *Nature* **359:**528–531.

Morrison, A., Miller, E. J., and Prakash, L., 1988, Domain structure and functional analysis of the carboxyl-terminal polyacidic sequence of the RAD6 protein of *Saccharomyces cerevisiae*, *Mol. Cell. Biol.* **8:**1179–1185.

Muralidhar, M. G., and Thomas, J. B., 1993, The *Drosophila bendless* gene encodes a neural protein related to ubiquitin-conjugating enzymes, *Neuron* **11:**253–266.

Nasmyth, K., 1993, Control of the yeast cell cycle by the Cdc28 protein kinase, *Curr. Opin. Cell Biol.* **5:**166–179.

Nefsky, B., and Beach, D., 1996, Pub1 acts as an E5-AP-like protein ubiquitin ligase in the degradation of cdc25, *EMBO J.* **15:**1301–1312.

Nishizawa, M., Okazaki, K., Furuno, N., Watanabe, N., and Sagata, N., 1992, The 'second codon rule' and autophosphorylation govern the stability and activity of Mos during the meiotic cell cycle in *Xenopus oocytes*, *EMBO J.* **11:**2433–2446.

Nishizawa, M., Furuno, N., Okazaki, K., Tanaka, H., Ogawa, Y., and Sagata, N., 1993, Degradation of mos by the N-terminal proline (pro2)-dependent ubiquitin pathway on fertilization of Xenopus eggs: Possible significance of natural selection for pro2 in mos, *EMBO J.* **12:**4021–4027.

Nuber, U., Schwarz, S., Kaiser, P., Schneider, R., and Scheffner, M., 1996, Cloning of human ubiquitin-conjugating enzymes UbcH6 and UbcH7 (E2-F1) and characterization of their interaction with E6-AP and RSP5, *J. Biol. Chem.* **271:**2795–2800.

Nugroho, T. T., and Mendenhall, M. D., 1994, An inhibitor of yeast cyclin-dependent protein kinase plays an important role in ensuring the genomic integrity of daughter cells, *Mol. Cell. Biol.* **14:** 3320–3328.

Oh, C. E., McMahon, R., Benzer, S., and Tanouye, M., 1994, *bendless*, a *Drosophila* gene affecting neuronal connectivity, encodes a ubiquitin-conjugating enzyme homolog, *J. Neurosci.* **14:**3166–3179.

Özkaynak, E., Finley, D., Solomon, M. J., and Varshavsky, A., 1987, The yeast ubiquitin genes: A family of natural gene fusions, *EMBO J.* **6:**1429–1439.

Paolini, R., and Kinet, J.-P., 1993, Cell surface control of the multiubiquitination and deubiquitination of high affinity immunoglobulin E receptors, *EMBO J.* **12:**779–786.

Peter, M., and Herskowitz, I., 1994, Direct inhibition of the yeast cyclin-dependent kinase Cdc28-Cln by Far1, *Science* **265:**1228–1231.

Peter, M., Gartner, A., Horecka, J., Ammerer, G., and Herskowitz, I., 1993, FAR1 links the signal transduction pathway to the cell cycle machinery in yeast, *Cell* **73:**747–760.

Pickart, C. M., and Rose, I. A., 1985, Functional heterogeneity of ubiquitin carrier proteins, *J. Biol. Chem.* **260:**1573–1581.

Picologlou, S., Brown, N., and Liebman, S. W., 1990, Mutations in *RAD6*, a yeast gene encoding a ubiquitin-conjugating enzyme, stimulate retrotransposition, *Mol. Cell. Biol.* **10:**1017–1022.

Prendergast, J. A., Ptak, C., Kornitzer, D., Steussy, C. N., Hodgins, R., Goebl, M., and Ellison, M. J., 1996, Identification of a positive regulator of the cell cycle ubiquitin-conjugating enzyme Cdc34 (Ubc3), *Mol. Cell. Biol.* **16:**677–684.

Reynolds, P., Weber, S., and Prakash, L., 1985, *RAD6* gene of *Saccharomyces cerevisiae* encodes a protein with 13 consecutive aspartates, *Proc. Natl. Acad. Sci. USA* **82:**168–172.

Richardson, H. E., Wittenberg, C., Cross, F., and Reed, S. I., 1989, An essential G1 function for cyclin-like proteins in yeast, *Cell* **59:**1127–1133.

Rodriguez, J. M., Salas, M. L., and Viñuela, E., 1992, Genes homologous to ubiquitin-conjugating proteins and eukaryotic transcription factor SII in African swine fever virus, *Virology* **186:**40–52.

Rogers, S., Wells, R., and Rechsteiner, M., 1986, Amino acid sequences common to rapidly degraded proteins: The PEST hypothesis, *Science* **234:**364–368.

Sadis, S., Atienza, C., Jr., and Finley, D., 1995, Synthetic signals for ubiquitin-dependent proteolysis, *Mol. Cell. Biol.* **15:**4086–4094.

Salama, S. R., Hendricks, K. B., and Thorner, J., 1994, G1 cyclin degradation: The PEST motif of yeast Cln2 is necessary, but not sufficient, for rapid protein turnover, *Mol. Cell. Biol.* **14:**7953–7966.

Scheffner, M., Huibregtse, J. M., Vierstra, R. D., and Howley, P. M., 1993, The HPV-16 E6 and E6-AP complex functions as a ubiquitin-protein ligase in the ubiquitination of p53, *Cell* **75:**495–505.

Scheffner, M., Huibregtse, J. M., and Howley, P. M., 1994, Identification of a human ubiquitin-conjugating enzyme that mediates the E6-AP-dependent ubiquitination of p53, *Proc. Natl. Acad. Sci. USA* **91:**8797–8801.

Scheffner, M., Nuber, U., and Huibregtse, J. M., 1995, Protein ubiquitination involving an E1-E2-E3 enzyme ubiquitin thioester cascade, *Nature* **373:**81–83.

Schneider, B. L., Yang, Q-H., and Futcher, A. B., 1996, Linkage of replication to Start by the Cdk inhibitor Sic1, *Science* **272:**560–562.

Schneider, R., Eckerskorn, C., Lottspeich, F., and Schweiger, M., 1990, The human ubiquitin carrier protein E2(M_r = 17 000) is homologous to the yeast DNA repair gene *RAD6*, *EMBO J.* **9:**1431–1435.
Schork, S. M., Bee, G., Thumm, M., and Wolf, D. H., 1994, Catabolite inactivation of fructose-1,6-biphosphatase in yeast is mediated by the proteasome, *FEBS Lett.* **349:**270–274.
Schwob, E., and Nasmyth, K., 1993, CLB5 and CLB6, a new pair of B cyclins involved in DNA replication in *Saccharomyces cerevisiae*, *Genes Dev.* **7:**1160–1175.
Schwob, E., Bohm, T., Mendenhall, M. D., and Nasmyth, K., 1994, The B-type cyclin kinase inhibitor p40^{SIC1} controls the Gl to S transition in *S. cerevisiae*, *Cell* **79:**233–244. (Erratum appeared in *Cell* **84:** Jan. 12, 1996)
Seufert, W., and Jentsch, S., 1990, Ubiquitin-conjugating enzymes UBC4 and UBC5 mediate selective degradation of short-lived and abnormal proteins, *EMBO J.* **9:**543–550.
Seufert, W., McGrath, J. P., and Jentsch, S., 1990, *UBC1* encodes a novel member of an essential subfamily of yeast ubiquitin-conjugating enzymes involved in protein degradation, *EMBO J.* **9:**4535–4541.
Seufert, W., Futcher, B., and Jentsch, S., 1995, Role of a ubiquitin-conjugating enzyme in degradation of S- and M-phase cyclins, *Nature* **373:**78–81.
Shanklin, J., Jabben, M., and Vierstra, R. D., 1987, Red light-induced formation of ubiquitin–phytochrome conjugates: Identification of possible intermediates of phytochrome degradation, *Proc. Natl. Acad. Sci. USA* **84:**359–363.
Silver, E. T., Gwozd, T. J., Ptak, C., Goebl, M., and Ellison, M. J., 1992, A chimeric ubiquitin-conjugating enzyme that combines the cell cycle properties of CDC34 (UBC3) and the DNA repair properties of RAD6 (UBC2): Implications for the structure, function and evolution of the E2s, *EMBO J.* **11:**3091–3098.
Smith, S. E., Koegl, M., and Jentsch, S., 1996, Role of the ubiquitin/proteasome system in regulated protein degradation in *Saccharomyces cerevisiae*, *Biol. Chem.* **377:**437–446.
Sommer, T., and Jentsch, S., 1993, A protein translocation defect linked to ubiquitin conjugation at the endoplasmic reticulum, *Nature* **365:**176–179.
Spence, J., Sadis, S., Haas, A. L., and Finley, D., 1995, A ubiquitin mutant with specific defects in DNA repair and multiubiquitination, *Mol. Cell. Biol.* **15:**1265–1273.
Staub, O., Dho, S., Henry, P. C., Correa, J., Ishikawa, T., McGlade, J., and Rotin, D., 1996, WW domains of Nedd4 bind to the proline-rich PY motifs in the epithelial Na^+ channel deleted in Liddle's syndrome, *EMBO J.* **15:**2371–2380.
Sudakin, V., Ganoth, D., Dahan, A., Heller, H., Hershko, J., Luca, F., Ruderman, J. V., and Hershko, A., 1995, The cyclosome, a large complex containing cyclin-selective ubiquitin ligase activity, targets cyclins for destruction at the end of mitosis, *Mol. Biol. Cell* **6:**185–198.
Sudol, M., Bork, P., Einbond, A., Kastury, K., Druck, T., Negrini, M., Huebner, K., and Lehman, D., 1995, Characterization of the mammalian *YAP* (Yes-associated protein) gene and its role in defining a novel protein module, the WW domain, *J. Biol. Chem.* **270:**14733–14741.
Sung, P., Prakash, S., and Prakash, L., 1988, The RAD6 protein of *Saccharomyces cerevisiae* polyubiquitinates histones, and its acidic domain mediates this activity, *Genes Dev.* **2:**1476–1485.
Sung, P., Prakash, S., and Prakash, L., 1991a, Stable ester formation between the Saccharomyces cerevisiae RAD6 protein and ubiquitin has no biological activity, *J. Mol. Biol.* **221:**745–749.
Sung, P., Berleth, E., Pickart, C., Prakash, S., and Prakash, L., 1991b, Yeast *RAD6* encoded ubiquitin conjugating enzyme mediates protein degradation dependent on the N-end-recognizing E3 enzyme, *EMBO J.* **10:**2187–2193.
Surana, U., Robitsch, H., Price, C., Schuster, T., Fitch, I., Futcher, A. B., and Nasmyth, K., 1991, The role of CDC28 and cyclins during mitosis in the budding yeast *S. cerevisiae*, *Cell* **65:**145–161.
Treier, M., Seufert, W., and Jentsch, S., 1992, *Drosophila UbcD1* encodes a highly conserved ubiquitin-conjugating enzyme involved in selective protein degradation, *EMBO J.* **11:**367–372.
Treier, M., Staszewski, L. M., and Bohmann, D., 1994, Ubiquitin-dependent c-Jun degradation in vivo is mediated by the δ domain, *Cell* **78:**787–798.

van der Hoeven, F., Schimmang, T., Volkmann, A., Mattei, M.-G., Kyewski, B., and Rüther, U., 1994, Programmed cell death is affected in the mouse mutant *Fused toes* (*Ft*), *Development* **120:**2601–2607.

Varshavsky, A., 1992, The N-end rule, *Cell* **69:**725–735.

Varshavsky, A., 1996, The N-end rule, *Cold Spring Harbor Symp. Quant. Biol.* **60:**461–478.

Ward, C. L., Omura, S., and Kopito, R. R., 1995, Degradation of CFTR by the ubiquitin-proteasome pathway, *Cell* **83:**121–127.

Wefes, I., Mastrandrea, L. D., Haldeman, M., Koury, S. T., Tamburlin, J., Pickart, C. M., and Finley, D., 1995, Induction of ubiquitin-conjugating enzymes during terminal erythroid differentation, *Proc. Natl. Acad. Sci. USA* **92:**4982–4986.

Wiebel, F. F., and Kunau, W.-H., 1992, The Pas2 protein essential for peroxisome biogenesis is related to ubiquitin-conjugating enzymes, *Nature* **359:**73–76.

Yaglom, J., Linskens, M. H., Sadis, S., Rubin, D. M., Futcher, B., and Finley, D., 1995, $p34^{Cdc28}$-mediated control of Cln3 cyclin degradation, *Mol. Cell. Biol.* **15:**731–741.

Yu, H., King, R. W., Peters, J.-M., and Kirschner, M. W., 1996, Identification of a novel ubiquitin-conjugating enzyme involved in mitotic cyclin degradation, *Curr. Biol.* **6:**455–466.

Zacksenhaus, E., and Sheinin, R., 1990, Molecular cloning, primary structure and expression of the human X linked A1S9 gene cDNA which complements the ts A1S9 mouse L cell defect in DNA replication, *EMBO J.* **9:**2923–2929.

Zhen, M., Heinlein, R., Jones, D., Jentsch, S., and Candido, E. P. M., 1993, The *ubc2* gene of *Caenorhabditis elegans* encodes a ubiquitin-conjugating enzyme involved in selective protein degradation, *Mol. Cell. Biol.* **13:**1371–1377.

Zhen, M., Schein, J. E., Baillie, D. L., and Candido, E. P. M., 1996, An essential ubiquitin-conjugating enzyme with tissue and developmental specificity in the nematode *Caenorhabditis elegans*, *EMBO J.* **15:**3229–3237.

CHAPTER 4

The Deubiquitinating Enzymes

Keith D. Wilkinson and Mark Hochstrasser

1. INTRODUCTION

As detailed elsewhere in this volume, modification of proteins by the 76-residue ubiquitin polypeptide is involved in many aspects of protein metabolism. Among the cellular processes affected by ubiquitin-dependent reactions are chromosome structure and segregation, cell-cycle progression, receptor-mediated signal transduction, gene expression, protein localization, organelle biogenesis, antigen presentation, viral pathogenesis, and the stress response (reviewed in Hochstrasser, 1995, 1996a; Rubin and Finley, 1995; Wilkinson, 1995; Ciechanover, 1994; Hershko and Ciechanover, 1992; Finley and Chau, 1991). One type of ubiquitination, attachment of a polyubiquitin chain(s) to a protein, targets the modified protein for proteolysis by the proteasome (see references above). The ubiquitin molecules in these polyubiquitin chains are most often linked to one another by isopeptide bonds between the C-terminus of one ubiquitin and the ε-amino group of lysine 48 of the next ubiquitin (Chau *et al.*, 1989; Gregori *et al.*, 1990; Hochstrasser *et al.*, 1991). There is also evidence that polyubiquitin chains can be formed with isopeptide linkages involving lysines 6, 11 (Baboshina and Haas, 1996), 29 (Arnason and Ellison, 1994), or 63 (Arnason and Ellison, 1994; Spence *et al.*, 1995) of ubiquitin. In addition, a variety of ubiquitinlike proteins have been described

Keith D. Wilkinson • Department of Biochemistry, Emory University, Atlanta, Georgia 30322. **Mark Hochstrasser** • Department of Biochemistry and Molecular Biology, University of Chicago, Chicago, Illinois 60637.

Ubiquitin and the Biology of the Cell, edited by Peters *et al.* Plenum Press, New York, 1998.

(Toniolo *et al.*, 1988; Banerji *et al.*, 1990; Meyers *et al.*, 1991; Kumar *et al.*, 1993; Olvera and Wool, 1993; Linnen *et al.*, 1993; Haas *et al.*, 1996; Nakamura *et al.*, 1996; Narasimhan *et al.*, 1996; Biggins *et al.*, 1996; Shen *et al.*, 1996). The variety of possible polyubiquitin structures and the existence of a family of ubiquitinlike proteins suggest that considerable diversity exists in the functions of ubiquitin and related molecules, presumably as covalently linked regulators of other proteins.

Ubiquitination of proteins is reversible. Deubiquitination is catalyzed by processing proteases called *deubiquitinating enzymes*. These enzymes specifically hydrolyze peptide (or isopeptide) bonds after Gly76 of ubiquitin (see below). As a group, deubiquitinating enzymes comprise the largest known family of enzymes in the ubiquitin system. For instance, there are 17 genes for such enzymes in the yeast *Saccharomyces cerevisiae*. However, comparatively little is known about them at present–in yeast or any other organism. It seems very likely that the deubiquitinating enzymes serve distinct functions at discrete times and places in the cell, although there may well be overlap in specificity among members of this group of enzymes.

In this chapter, we review what is currently known about the functions of the deubiquitinating enzymes and about their structures and mechanisms of action. It is already clear that deubiquitination reactions are central to the dynamic control of the ubiquitin system and, as such, provide attractive targets for cellular regulation. We will begin with a brief structural description of ubiquitin and its polymeric forms, which will be followed by a short summary of the currently known consequences of protein ubiquitination. This will provide the context for a detailed discussion of the deubiquitinating enzymes, their catalytic mechanisms, and their various functions in cellular metabolism.

2. UBIQUITIN AND POLYUBIQUITIN CHAINS

Ubiquitin exists in both monomeric and polymeric forms and is also found conjugated to various proteins, either as single ubiquitin molecules linked to different substrate lysines or as polyubiquitin chains.

2.1. Ubiquitin Precursors

All ubiquitin genes code for fusion proteins (e.g., Özkaynak *et al.*, 1987) which must be accurately processed by deubiquitinating enzymes. The primary translation products of one class of ubiquitin genes consist of head-to-tail fusions of from 3 to as many as 52 ubiquitin molecules (Özkaynak *et al.*, 1984; Swindle *et al.*, 1988; and database entries). No structural or physical studies have been done on such "linear" peptide-bond linked ubiquitin chains. The remaining ubiquitin (and many ubiquitinlike) genes also encode precursor proteins (Finley *et al.*,

1989), with the ubiquitin (or ubiquitinlike protein) fused N-terminal to one of several proteins or peptides.

2.2. Polyubiquitin Chains

Another polymeric form of ubiquitin is found on proteins destined for proteolysis by the 26 S proteasome. In this case, the ubiquitin molecules in each chain are linked by isopeptide bonds in which the C-terminus of one ubiquitin is linked to the ε-amino group of the next ubiquitin in the chain. The most commonly observed isopeptide linkage in such chains involves Lys48 of ubiquitin. To date, this is the only ubiquitin–ubiquitin linkage verified by direct physical mapping (Chau *et al.*, 1989; Gregori *et al.*, 1990) and is the only type of ubiquitin chain definitely associated with targeting to the proteasome (Chau *et al.*, 1989; Hochstrasser *et al.*, 1991; Finley *et al.*, 1994). Lys48-linked ubiquitin chains also are found unattached to protein substrates *in vivo* (Haldeman *et al.*, 1995; Van Nocker and Vierstra, 1993); the levels of such unanchored ubiquitin chains in the cell are often substantial (see below).

From various genetic data, it has been demonstrated that ubiquitin–ubiquitin linkages other than those involving Lys48 can also be formed. Experiments with yeast suggest that the stress response (Arnason and Ellison, 1994) and error-prone DNA repair pathways (Spence *et al.*, 1995) require the formation of Lys63-linked ubiquitin chains. *In vivo* formation of ubiquitin–ubiquitin linkages involving Lys29 have also been reported (Arnason and Ellison, 1994; Johnson *et al.*, 1995). Whether ubiquitin chains with mixed isopeptide linkages are ever synthesized *in vivo* has not been determined. *In vitro* experiments involving ubiquitin mutants with different Lys-to-Arg substitutions suggest that at least two additional kinds of ubiquitin chains can be synthesized. A purified keratinocyte E2, E2-EPF, assembles a Lys11-linked ubiquitin chain on itself in an autoubiquitination reaction, and recombinant Rad6 protein carries out a similar reaction making Lys6-linked ubiquitin chains (Baboshina and Haas, 1996). The physiological significance of the alternative linkages observed *in vitro* is not known. In the presence of substrate proteins and crude mixtures of E3 ligases, Rad6 has been implicated only in the assembly of Lys48-linked ubiquitin chains on substrate proteins (Baboshina and Haas, 1996).

These forms of ubiquitin are in a dynamic steady state in the cell, with approximately 50 to 80% of total ubiquitin generally being present as ubiquitin–protein conjugates at any time (Riley *et al.*, 1988; Haas and Bright, 1985, 1987). The levels of total ubiquitin conjugates change rapidly during rat muscle degeneration caused by starvation, denervation, sepsis, or apoptosis (Riley *et al.*, 1988; Haas *et al.*, 1995; Medina *et al.*, 1991; Wing *et al.*, 1995), in molt-induced lobster claw atrophy (Shean and Mykles, 1995), in response to serum factors in cultured cells (Haas and Bright, 1987), during synthesis of abnormal proteins (Hershko *et al.*, 1982), and during certain stages of development (Haas *et al.*, 1995; Wunsch and Haas, 1995). In general, immunochemical studies demonstrate that pools of free and conjugated ubiquitin are in rapid exchange, and thus point to the importance of deubiquitinating enzymes in cellular ubiquitin metabolism.

2.3. Nonproductive Derivatives of Ubiquitin

Another potential role for deubiquitinating enzymes is in the salvage of adventitiously trapped derivatives of ubiquitin. The formation of isopeptide-linked ubiquitin conjugates involves the intermediacy of a number of thioesters between enzymes of the ubiquitin conjugation system and the C-terminus of ubiquitin (Scheffner *et al.*, 1995, and reviews above). These intermediates are subject to at least two "nonproductive" fates. First, high concentrations of cellular thiols or amines can trap the catalytic intermediates. Ubiquitin sequestered by these reactions must be regenerated by the action of deubiquitinating enzymes capable of cleaving small leaving groups from the C-terminus of ubiquitin (Pickart and Rose, 1985; Rose and Warms, 1983). A second, apparently nonproductive fate of such thioester intermediates is autoubiquitination. Under some conditions, the catalytic intermediates formed by some ubiquitinating enzymes are able to transfer ubiquitin from the active site thiol to a nearby lysine (Gwozd *et al.*, 1995; Baboshina and Haas, 1996). In some cases, additional ubiquitin molecules may be added to build a polyubiquitin chain on this enzyme site (Banerjee *et al.*, 1993). Although a function for this reaction is unknown, it is possible that this is a strategy for downregulation wherein the absence of substrate leads to ubiquitination and degradation of the ubiquitinating enzyme.

2.4. Structural Features of Polymeric Ubiquitin

There are several structural features of ubiquitin and its polymers that can, in principle, be recognized by deubiquitinating enzymes. The X-ray crystal structures of human, yeast, and oat ubiquitins have been determined and all three structures are very similar (Vijay-Kumar *et al.*, 1987); the solution structure of ubiquitin has been solved by NMR techniques (Weber *et al.*, 1987; Di Stefano and Wand, 1987) and is fully congruent with the crystallographic data. The protein folds into a compact, stable structure with a central five-stranded β-sheet, one α-helix, and one short 3-10 helix. Ubiquitin is unusually stable to proteases, with the exception of a trypsin-sensitive cleavage site near the C-terminus; trypsin releases the terminal Gly–Gly dipeptide (Wilkinson and Audhya, 1981). The C-terminus of ubiquitin projects from the body of the protein, and is the site of its covalent attachment to other proteins. The globular domain of ubiquitin exhibits three distinct "faces": a basic face, an acidic face, and a hydrophobic face (Wilkinson, 1988). Well over 20 examples of ubiquitinlike sequences are now also known; many of the sequence differences cluster near the acidic face and the N-terminal end of the α-helix.

The structure of a Lys48-linked tetraubiquitin has been reported (Cook *et al.*, 1994). The chain is slightly elliptical in cross section, resembling a flattened tube. The structure has a pseudo-twofold symmetry with even-numbered subunits related by a 180° rotation along the screw axis to the odd-numbered subunits. A prominent feature of the chain is a hydrophobic stripe extending along both flattened faces. Part of this stripe is formed by the exposed side chains of Leu8, Ile44, and Val70. These residues have been shown to be important for binding of the ubiquitin chain to a regulatory subunit of the 26 S proteasome, called S5a, and for degradation by the proteasome of proteins modified by Lys48-linked ubiquitin chains (Beal *et al.*, 1996).

In summary, a wide variety of ubiquitin polymers, ubiquitin–small molecule adducts, and ubiquitin–protein conjugates are found in eukaryotic cells. It follows that both the maintenance of adequate free ubiquitin pools and the observed dynamic exchange of ubiquitin between free and conjugated ubiquitin pools require a range of deubiquitinating activities. This demand for cleavage of diverse ubiquitinated species may explain, at least in part, the diversity and number of deubiquitinating enzymes found in all eukaryotes (see below).

3. THE DEUBIQUITINATING ENZYME FAMILIES

It is apparent that ubiquitination of proteins is a versatile covalent modification with multiple roles in cellular regulation. Ubiquitination is known to be a highly dynamic protein modification (Ellison and Hochstrasser, 1991). All deubiquitination reactions require accurate proteolytic processing at the C-terminal glycine of ubiquitin. The enzymes responsible for these reactions have been called *isopeptidases* (Matsui *et al.*, 1982), *ubiquitin carboxyl-terminal hydrolases* (Pickart and Rose, 1985; Rose, 1988), *ubiquitin thiolesterases* (Rose and Warms, 1983), or *ubiquitin-specific processing proteases* (Tobias and Varshavsky, 1991). We propose as a general acronym *DUB enzymes* (DUBs) (for *deub*iquitinating enzymes). The term *deubiquitinating enzymes* appears to have been first used to describe this class of enzymes in 1993 (Papa and Hochstrasser, 1993), and the DUB acronym later used to identify several specific murine deubiquitinating enzymes (Zhu *et al.*, 1996a,b). Two distinct classes of DUB enzymes are known (Fig. 1): The Family 1 enzymes are a small group with significant sequence similarity to the neuron-specific human protein PGP 9.5 (UCH-L1) (Wilkinson *et al.*, 1989). Most of the known Family 1 of UCH enzymes are relatively small (<40 kDa). Family 2 enzymes form a much larger group of thiol proteases seemingly unrelated to the Family 1 enzymes. These are referred to as *ub*iquitin-specific

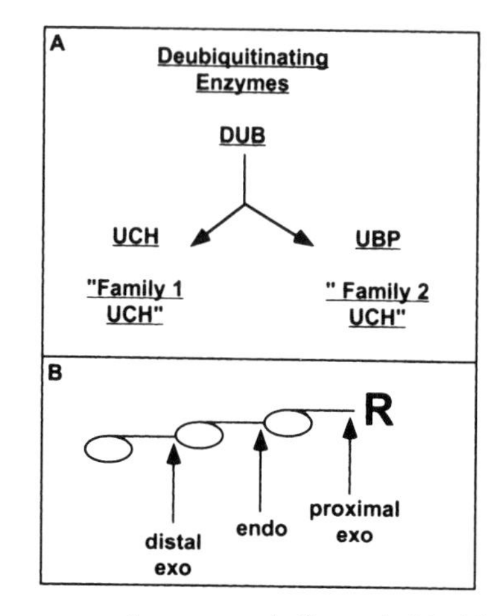

Figure 1. (A) These enzymes are known generically as deubiquitinating enzymes (DUB) (EC 3.1.2.15). Two gene families are currently recognized. The UCH Family 1 enzymes are members of the peptidase family C12 (Rawlings and Barrett, 1994), are evolutionarily related to YUH1 in yeast, and are commonly referred to as UCHs. The UCH Family 2 enzymes are members of the peptidase family C19 (Rawlings and Barrett, 1994), and are related to the yeast UBP gene products. (B) These enzymes can be further characterized as to the substrate and type of bond hydrolyzed, when known. Polyubiquitin chains can be described by the type of linkage (e.g., K48, K63). Fusion proteins, including the ubiquitin polyprotein precursor and the ubiquitin–ribosomal protein fusions, would be referred to by their trivial names. The distal isopeptide bond is the one most distant from the site of chain attachment to the substrate R, or from the single free C-terminus in the case of free chains. Trimming of one subunit at a time from the distal end would be classified as distal-exo-mono-deubiquitination. Trimming of two subunits at a time would be classified as distal-endo-di-deubiquitination, etc. A similar nomenclature could be applied for the proximal trimming.

processing proteases (UBP) (Tobias and Varshavsky, 1991). The UBPs vary widely in size, ranging from 50 to 250 kDa. Each family of enzymes will be discussed in more detail below.

The first example of a deubiquitinating enzyme was characterized over 15 years ago as an activity that deubiquitinated ubiquitin-histone H2A, a monoubiquitin conjugate also known as A24 (Matsui *et al.*, 1982; Andersen *et al.*, 1981). This thiol protease was reported to have an apparent molecular mass of 38 kDa and was termed A24 lyase or isopeptidase. It has never been purified, so it is not clear to which DUB enzyme family it belongs.

Using a screen for cleavage of a ubiquitin fusion peptide (Miller *et al.*, 1989) or hydrolysis of ubiquitin ethyl ester (Mayer and Wilkinson, 1989; Wilkinson *et al.*, 1989), the first deubiquitinating enzymes (which turned out to be members of the UCH family) were purified and their corresponding genes cloned and characterized. These proteins generally hydrolyze ubiquitin fusion proteins or adducts with small, unfolded leaving groups, although this generalization is not absolute.

There are at least three mammalian isozymes and several are tissue-specific, with particularly high levels of expression in nervous tissue for one of them (Wilkinson *et al.*, 1992; Wilkinson, 1995). There is only one such protein in *S. cerevisiae* and deletion of this gene is without obvious consequence for growth under standard laboratory conditions (Miller *et al.*, 1989). Clearly, other enzymes must function in the place of YUH1, and these additional enzymes are likely to be UBPs, first cloned by virtue of their ability to hydrolyze ubiquitin–protein fusions (Tobias and Varshavsky, 1991; Baker *et al.*, 1992).

4. GENERAL CHARACTERISTICS OF THE DEUBIQUITINATING ENZYMES

Both DUB gene families encode thiol proteases (Mayer and Wilkinson, 1989) based on their sensitivity to inhibition by iodoacetamide, *N*-ethyl maleimide, and other thiol reagents. The active site thiol has been identified in both gene families by site-directed mutagenesis (Papa and Hochstrasser, 1993; Larsen *et al.*, 1996), as have catalytically important histidines (Huang *et al.*, 1995; Larsen *et al.*, 1996). Most reports fail to note any effect of classical serine protease inhibitors, although at least one partially purified activity was inhibited by such inhibitors (Rose, 1988). It is not clear in this last case whether such inhibition reflects the existence of a novel type of DUB enzyme as the original observation was never pursued.

The second general feature of the DUB enzymes is the presence of a specific, tight binding site for ubiquitin. Ubiquitin binds to members of the UCH enzyme family with a micromolar biding constant (Larsen *et al.*, 1996) and to isopeptidase T (a UBP enzyme) with similar affinity (Hadari *et al.*, 1992; Wilkinson *et al.*, 1995; Stein *et al.*, 1995; Melandri *et al.*, 1996). The binding of ubiquitin to UCH-L3 is inhibited by salt, whereas increased salt strengthens the binding to isopeptidase T. These results suggest that in the former case, ionic interactions are predominant, while hydrophobic interactions may be important in the binding of ubiquitin to isopeptidase T. Whether these properties can be used generally to distinguish members of the UCH and UBP families is not yet known. Isopeptidase T, and possibly other UBP isozymes, bind to ubiquitin chains and have specific binding interactions with up to four subunits of the polyubiquitin substrates (Hadari *et al.*, 1992; Wilkinson *et al.*, 1995; Stein *et al.*, 1995; Melandri *et al.*, 1996). Finally, several of these enzymes bind to ubiquitin affinity resins, albeit with different affinities and specificity. For instance, UCH-L3, a hematopoietic cell-specific enzyme, and isopeptidase T both bind tightly to a ubiquitin affinity column in which ubiquitin is coupled via lysine side chains to the resin (Pickart and Rose, 1985; Hadari *et al.*, 1992). UCH-L1, a neuron-specific UCH isozyme, does not bind to this resin well, but does bind tightly to a resin where ubiquitin is linked via its C-terminus to the support (Duerksen-Hughes *et al.*, 1989). Similarly, ubiquitin linked through a cysteine residue placed at different positions in the primary sequence exhibits selective binding of DUB proteins (G. Kapp and K. D. Wilkinson, unpublished data).

The presence of a strong ubiquitin-binding site, and knowledge of the active site chemistry afford opportunities to develop inhibitors of these enzymes. A potent and specific inhibitor of many deubiquitinating enzymes is the C-terminal aldehyde of ubiquitin (Pickart and Rose, 1986). Large-scale syntheses have been reported (Mayer and Wilkinson, 1989; Dunten and Cohen, 1989; Melandri *et al.*, 1996), making it possible to use ubiquitin aldehyde in a variety of *in vitro* experiments. The effect of this inhibitor is usually to increase the steady-state levels of bulk ubiquitin conjugates, suggesting a fairly general effect on cellular deubiquitinating activities. The affinity of this inhibitor for UCH enzymes has been estimated to be in the nanomolar range, at least three orders of magnitude tighter than the binding of free ubiquitin (Mayer and Wilkinson, 1989). This is most likely related to the formation of a thiol hemiacetal between the aldehyde and the active site cysteine of these enzymes. Indeed, ubiquitin aldehyde was first synthesized by borohydride reduction of a small equilibrium amount of ubiquitin thioester formed by dehydration of the binary ubiquitin–UCH-L3 complex (Pickart and Rose, 1986). This is an unusual reaction and indicates that the active site cysteine is a very strong nucleophile.

The effectiveness of ubiquitin aldehyde as an inhibitor of deubiquitinating enzymes is variable (Hershko and Rose, 1987; Mayer and Wilkinson, 1989; Sokolik and Cohen, 1991; Hadari *et al.*, 1992; Eytan *et al.*, 1993; Schaeffer and Cohen, 1996). For isopeptidase T, normal cleavage of polyubiquitin occurs by hydrolysis of the peptide bond linking ubiquitin bound in the S1 (substrate binding) site and the proximal ubiquitin (leaving group) bound in the S1′ side. Inhibition of isopeptidase T requires micromolar concentrations of ubiquitin aldehyde (Melandri *et al.*, 1996), but inhibition is much more efficient in the presence of ubiquitin. Binding of free ubiquitin to the S1′ leaving group site enhances the binding of the aldehyde to the S1 substrate binding site (Hadari *et al.*, 1992; Wilkinson *et al.*, 1995; Melandri *et al.*, 1996). With the UCH enzymes, free ubiquitin has little effect on the inhibition by ubiquitin aldehyde, as the aldehyde binds 1000-fold more tightly than ubiquitin and there is only a single ubiquitin binding site (Mayer and Wilkinson, 1989).

5. POTENTIAL ROLES FOR UBPS AND UCHS

All ubiquitinated proteins are potential substrates for deubiquitinating enzymes, but there is generally a balance between deubiquitination and other fates for the modified protein. The exact path taken may depend on the localization, abundance, and activity of appropriate DUB enzymes; the ability of specific substrate-binding proteins to modulate accessibility to DUB enzymes; the rate at which the ubiquitinated proteins are targeted to the organelles or subcellular particles that metabolize ubiquitin conjugates; and the type of ubiquitin chain or ubiquitin–protein linkages that are formed on the substrate.

A major challenge in the study of the DUB enzymes is to assign specific physiological roles to members of this large and diverse group of enzymes. A variety of such roles have been suggested. It seems likely that one or more of the known deubiquitinating enzymes will be involved with each of the following metabolic and regulatory processes. In most cases, the identity of the enzymes involved has not been established.

5.1. Processing of Ubiquitin Precursors

Ubiquitin (Özkaynak *et al.*, 1987) and several ubiquitinlike proteins (Haas *et al.*, 1987; Olvera and Wool, 1993; Tautz *et al.*, 1993) are synthesized as fusion proteins that must be accurately processed at the C-terminus of the ubiquitin segment(s) to generate functional monomers. Multiple deubiquitinating enzymes are likely to participate in processing these precursors (cleaving an α-amide bond). This inference is based on the finding that several different UCH and UBP enzymes have been shown to cleave ubiquitin–ribosome protein precursors *in vitro* and on the observation that the yeast *ubp1 ubp2 ubp3 yuh1* quadruple mutant continues to process a ubiquitin-β-galactosidase fusion protein, albeit with a slightly reduced efficiency (Baker *et al.*, 1992). Ubiquitin-β-galactosidase fusion proteins can be processed by many different UBP family deubiquitinating enzymes (Hochstrasser, 1996b).

Normally, ubiquitin precursor processing is very rapid both in crude extracts and *in vivo*. However, individual DUB enzymes differ in the rate at which they can process these proproteins. Both the polyubiquitin primary translation product and Lys48-linked ubiquitin chains can be processed by isopeptidase T (Falquet *et al.*, 1995; Wilkinson *et al.*, 1995). The ubiquitin–ribosomal S27a fusion protein is efficiently processed by UCH-L3, but the ubiquitin–ribosomal L40 fusion protein is only slowly processed by this same enzyme. When substrate and enzyme are coexpressed in bacteria, UCH-L1 and yeast Ubp1 can cotranslationally process the ubiquitin polyprotein precursor (Larsen *et al.*, 1996). FAU, a fusion between a ubiquitinlike protein and the ribosomal S30 protein, is processed by unknown enzymes (Kas *et al.*, 1992). Similarly, mammalian ubiquitin cross-reactive protein (UCRP) is synthesized in precursor form, but the enzyme(s) responsible for removing the eight-residue propeptide at its C-terminus has not yet been identified molecularly (although it has recently been purified; A. Haas, personal communication).

5.2. Recycling of Adventitiously Trapped Catalytic Intermediates

The activation of ubiquitin for conjugation to other proteins involves formation of thiol esters with the ubiquitin activating enzyme, the ubiquitin conjugating enzymes, and, in some cases, the ubiquitin-protein ligase (Scheffner *et al.*, 1995). Each of these intermediates can, in principle, react with thiols or amines leading to relatively stable C-terminal ubiquitin thiol esters or amides. To the extent to which this modification occurs in the cell, it must be reversed by deubiquitinating enzymes. It has been estimated that such adducts could consume the pool of free ubiquitin in less than a minute in mammalian cells were they not rapidly broken down again (Pickart and Rose, 1985).

5.3. Proofreading of Protein Ubiquitination

Deubiquitinating enzymes can stimulate or inhibit ubiquitin-dependent processes, depending on their time and place of action. Inhibition could result from the deconjugation of ubiquitin from either mono- or polyubiquitinated protein substrates prior to their commitment to other fates, e.g., proteolysis by the proteasome. These reactions would reverse the conjugation of ubiquitin and define a biochemical "futile cycle." Such a futile cycle could function as a proofreading mechanism designed to ensure that only appropriately targeted proteins are actually degraded (Cox *et al.*, 1986; Wilkinson and Mayer, 1986; Ellison and Hochstrasser, 1991). The prevailing dogma suggests that specificity is obtained by specific E2/E3 combinations controlling the rate of ubiquitination and by relatively nonspecific DUB enzymes modulating this modification. Alternatively, conjugation of ubiquitin to proteins might be a relatively promiscuous process, and deubiquitination more selective. These hypotheses are of course not mutually exclusive. In either case, protein stability would be strongly dependent on the relative rate of ubiquitin removal by DUB enzymes. This in turn implies that DUB enzymes might be able to distinguish among different ubiquitinated proteins. Such a discriminatory capacity would be based on the direct recognition either of determinants in the nonubiquitin portion of the ubiquitin–protein conjugate or in ubiquitin itself (or both). For this latter mechanism, the conformation of conjugated ubiquitin must differ for different ubiquitinated proteins. For instance, ubiquitin may retain its native conformation when conjugated to a normal, long-lived protein, which would then be rapidly deubiquitinated, but if the conjugated protein were damaged or denatured, the exposed hydrophobic regions could cause a conformational change in ubiquitin, inhibiting deubiquitination and allowing further ubiquitin additions (Cox *et al.*, 1986; Wilkinson and Mayer, 1986).

In principle, polyubiquitinated proteins could be deubiquitinated by hydrolysis of the isopeptide bond joining the chain to the target protein, by trimming in from the distal end of the ubiquitin chain (an "exo-isopeptidase" activity), or by cleaving within the chain (an "endo-isopeptidase" activity) (see Fig. 1). Hydrolysis of the isopeptide bond joining the chain to the target protein would immediately release the protein from its polyubiquitin chain. There is no direct evidence for such activity, but it would be expected to be tightly regulated in order to avoid removal of the polyubiquitin signal before degradation of the target protein and to avoid the accumulation of unanchored ubiquitin chains that might act as inhibitors of the proteasome. There is some evidence for trimming of polyubiquitin chains from the distal end. Treatment of ubiquitin–lysozyme conjugates with reticulocyte lysate or partially purified UBP preparations leads to release of ubiquitin and a decrease in the molecular weight of the ubiquitinated protein (Hadari *et al.*, 1992). Recent unpublished studies (R. Cohen, personal communication) suggest that a "trimming" activity copurifies with PA700, the 19 S regulatory complex of the 26 S proteasome (DeMartino *et al.*, 1994). This may serve as an editing process whereby slowly degraded substrates are deubiquitinated before total proteolysis can occur (cf. NFκB), as a proofreading mechanism that would bias proteasomal degradation toward polyubiquitinated substrates, or simply as an intermediate step in the catabolism of polyubiquitin-linked substrates (see below).

5.4. Recycling of Ubiquitin from Polyubiquitinated Proteins following Commitment to Degradation

Ubiquitin is a stable protein and must therefore be recycled after targeting of the ubiquitinated protein to the proteasome (Haas and Bright, 1987), indicating the proteasome contains or can recruit deubiquitinating enzymes. It is not known when in the degradative cycle ubiquitin is normally released from the substrate. A mechanism that removes ubiquitin only after the substrate is broken down into peptides has the advantage of keeping incompletely digested substrate fragments from dissociating prematurely. An apparent disadvantage is the topological complication of having part(s) of the protein bound to the 19 S particle while the protein is being unfolded and fed into the proteasome core. This may not be a serious problem if the ubiquitin chain tether is flexible or if the chain undergoes repeated cycles of partial release and rebinding. Such a tether may even help to constrain the unfolding polypeptide chain near the outer 20 S proteasome pore such that the probability of entry of an end or loop of the polypeptide into the channel will be increased. The link between substrate and the ubiquitin chain would be severed only late in the degradation of substrate, perhaps because the DUB enzyme(s) involved only efficiently cleaves ubiquitin chains attached to small adducts. The point at which a substrate is removed from its ubiquitin tether(s) may also be substrate-dependent, e.g., early release may allow a partly degraded protein to be released, as appears to occur in NFκB p105 processing (Palombella *et al.*, 1994).

At least two distinct DUB enzymes are thought to participate in the processing of late degradation intermediates and will be discussed in more detail below. First, an *S. cerevisiae* protein called Doa4 (Ubp4) appears to be responsible for releasing the polyubiquitin chain from residual peptides produced by proteasome-dependent proteolysis of ubiquitinated proteins (Papa and Hochstrasser, 1993).

Yeast cells lacking the Doa4 enzyme accumulate what appear to be peptides attached to either ubiquitin or short ubiquitin oligomers (proposed to be proteolytic remnants generated by proteasomes) (Papa and Hochstrasser, 1993). These data suggest that most proteins are substantially degraded prior to complete deubiquitination (by Doa4). The small size of the putative peptides and the fairly large distance between the positions of the active sites and the outer pores of the 20S proteasome further suggest that if ubiquitin is indeed tethered to the proteasome during degradation, the tethering site(s) must be very close to the pores and, possibly, that ubiquitin can extend (unfold?) into the proteasome channel. An attractive idea is that following Doa4 action, isopeptidase T (Ubp14 in *S. cerevisiae*) degrades the remaining ubiquitin chain by hydrolysis of one ubiquitin unit at a time from the end of the chain previously attached to the substrate (Hadari *et al.*, 1992; Wilkinson *et al.*, 1995; Stein *et al.*, 1995; Amerik *et al.*, 1997).

5.5. Maintaining Free Ubiquitin Levels and Keeping Proteasomes Free of Ubiquitin Chains

There are several ways, in addition to those discussed above, by which deubiquitinating enzymes may function to enhance protein ubiquitination and/or degradation. Long ubiquitin chains are often assembled onto intracellular proteins, in some cases perhaps even on inappropriately targeted proteins. A failure to regenerate ubiquitin from these conjugates could deplete the intracellular pool of free ubiquitin. In addition, proteasomes and certain enzymes of the ubiquitin conjugation system, e.g., E3α, are known to bind ubiquitin chains avidly (Reiss *et al.*, 1989; Deveraux *et al.*, 1994). Such chains, either generated *de novo* or in the course of substrate proteolysis, may need to be rapidly disassembled to prevent extensive binding to and therefore inhibition of proteasomes or other ubiquitin system enzymes. Clearly, it will be important to understand which DUB enzymes catalyze these reactions and to determine their distribution and regulation.

6. ANALYSIS OF SPECIFIC DEUBIQUITINATING ENZYMES

6.1. The UCH Family

6.1.1. Structure and Specificity of the UCH Isozymes

Using antibodies to the major thymus isozyme, UCH-L3, a cDNA encoding this protein was cloned and sequenced (Wilkinson *et al.*, 1989). The UCH-L3 sequence showed considerable similarity to a yeast UCH, Yuh1, that is also small (26 kDA) and had been shown to cleave preferentially ubiquitin conjugates with small leaving groups (Liu *et al.*, 1989). UCH-L3 also bears strong sequence similarity to human PGP 9.5 (UCH-L1), a neuron-specific protein of remarkable abundance (~2% of soluble cell protein, Schofield *et al.*, 1995). A recently identified *Drosophila* protein prominent in nurse cells (Zhang *et al.*, 1993) is also related to the UCHs. Somewhat surprisingly, however, the *Drosophila* UCH is able to deubiquitinate polyubiquitinated proteins quite efficiently (Roff *et al.*, 1996), indicating that the preference of some UCH isozymes for small ubiquitin adducts is not absolute.

The UCHs are unrelated to any other known proteases. The sequences are conserved as strongly across species as they are among isozymes (~40% identity), suggesting significant selective pressure for the maintenance of their structures throughout eukaryotic evolution. More UCH-related sequences are now known from the various genome sequencing projects (Fig. 2). Interestingly, despite the presence of multiple UCH isozymes in different multicellular eukaryotes, only one UCH, Yuh1, appears to be encoded in the *S. cerevisiae* genome (Hochstrasser,

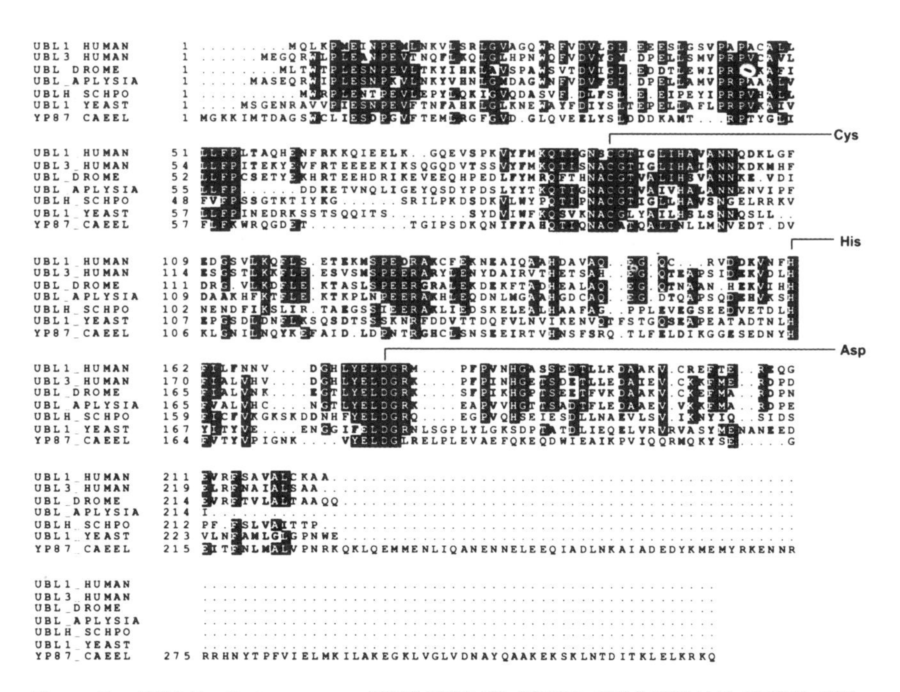

Figure 2. UCH Family 1 sequences. UBL1-HUMAN, P09936; UBL3-HUMAN, P15374, UBL-DROME, P35122; UBL-APLYSIA, A. Hegde and J. Schwartz, personal communication; UBLH-SCHPO, Q10171; UBL1-YEAST, P35127; YP87-CAEEL, Q09444. Identities are indicated by black boxes and similarities by gray boxes. The residues comprising the catalytic triad (C95, H169, D184 using the human L3 numbering) are indicated.

1996b). No defects have yet been associated with deletion of *YUH1* (Liu *et al.*, 1989).

All known thiol proteases use the side chain of a cysteine for peptide bond cleavage (by definition) and utilize histidine as the general base. In some, but not all, of these enzymes, an Asp residue is also critical for polarizing the His group (Rawlings and Barrett, 1994). The predicted protein sequences of the UCHs reveal a single conserved Cys residue as well as conserved His and Asp residues. The importance of these residues for catalysis has been verified by site-specific mutagenesis (Larsen *et al.*, 1996). Mutation of either Cys90 or His161 or UCH-L1 eliminated measurable catalytic activity. Substitution of Asp176 with Asn reduced V_m/K_m 250-fold.

The X-ray crystal structure of the seleno-methionine substituted human UCH-L3 has recently been solved (Johnston *et al.*, 1997). The assignments of catalytic residues based on site-directed mutagenesis are supported by the structure data. Parts of the overall fold and the arrangement of catalytic residues closely resemble those of papain (Fig. 3). Of particular interest is the location of the active site cysteine. It lies at the bottom of a narrow groove, which readily explains the inability of UCH-L3 to cleave ubiquitin from large peptides or proteins. Moreover, only small side chains in the terminal two residues of ubiquitin (both normally glycines) could be accommodated in this groove. Even though a portion of the peptide chain that forms the S1′ leaving group site of the enzyme is disordered in the crystal structure, it is apparent that only small leaving groups and/or extended conformations can be accommodated in the substrate structure.

The crystallographic data help to rationalize the observed substrate specificity. Recombinant ubiquitin–ribosomal L40 protein fusion (UbCEP52) is not processed by UCH-L3 in crude bacterial lysates. However, the purified fusion protein is readily processed by the same enzyme; addition of RNA or DNA to the purified fusion protein protects it from processing (Larsen *et al.*, 1998). Based on the crystal structure, these data could be explained as follows: When the fusion protein is purified away from nucleic acid, the first few amino acids of the ribosomal protein segment are disordered and can be bound productively in the active site of UCH-L3. When nucleic acid is bound (by the putative zinc-binding motif in L40), a conformational change prevents UCH-L3 binding by folding the ribosomal protein more tightly against the ubiquitin. This suggests that the ability of a fusion protein to be processed by these enzymes will depend more on the conformation or flexibility of the region following the ubiquitin moiety of the fusion than on the size of the leaving group.

Although these descriptive studies do not define physiological functions for the different UCHs, they do suggest that the roles of these isozymes will be distinguishable. Studies on the substrate specificities of UCH-L1 and UCH-L3 provide important constraints on what their functions could be (Wilkinson, 1994). When the leaving group at the C-terminus of ubiquitin is a large folded domain,

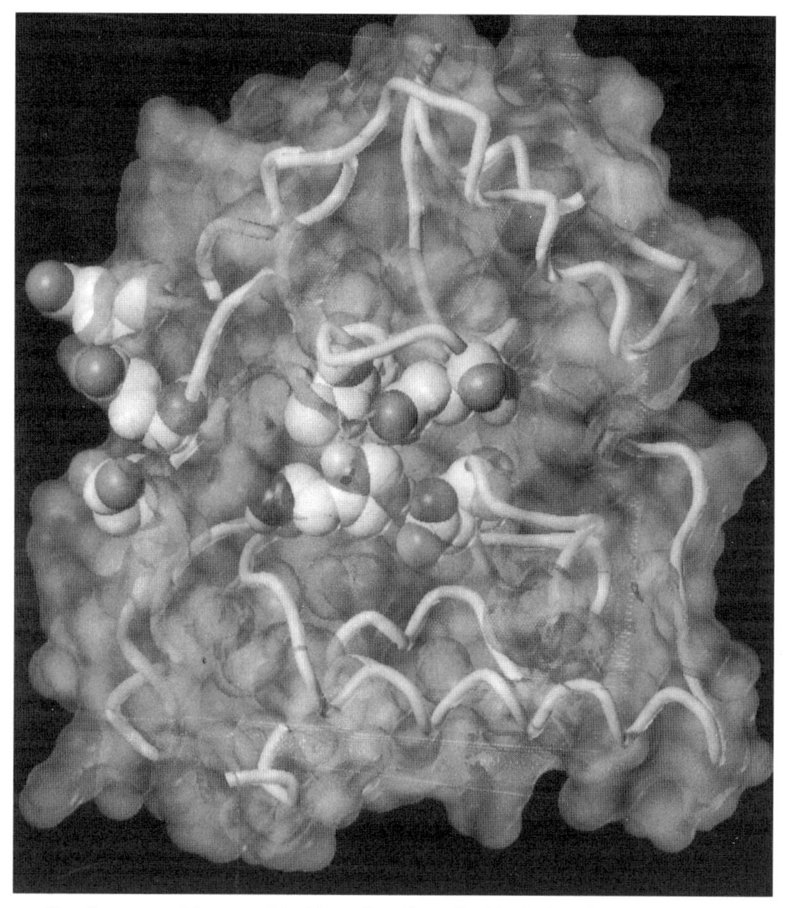

Figure 3. Structure of human ubiquitin carboxyl-terminal hydrolase isozyme L3. The α-carbon backbone is represented by a tube, and the solvent-accessible surface is shown as a translucent surface. The three highly conserved acidic residues (E10, E14, and D33) are shown in a space-filling representation at the left center of the molecule (9 o'clock). The residues forming the catalytic center are shown in a space-filling representation in the center of the molecule. The catalytic triad comprises residues of C95, H169, and D184, while Q89 forms an oxyanion hole analogous to other thiol proteases.

none of these enzymes will hydrolyze the amide bond, perhaps because of steric exclusion from the active site. The UCH enzymes may contribute to the cotranslational processing of ubiquitin precursor proteins if folding of the N-terminal ubiquitin precedes the stable folding of the downstream peptide sequence following their emergence from the ribosome. Alternatively, UCH enzymes may function in the regeneration of ubiquitin from small, adventitious adducts and/or in removing peptide remnants from ubiquitin (chains) following proteolysis by the proteasome.

6.1.2. Regulation of UCH Gene Expression

Using both immunological and biochemical approaches (Wilkinson *et al.*, 1992), considerable tissue specificity in the distribution of UCH isozymes L1 and L3 has been observed. Isozyme L1 is strongly expressed in neuronal, neuroendocrine, and perhaps some fetal cells. Isozyme L3 is present mainly in hematopoietic cells. Many tissues and cells contain significant amounts of isozyme L2, which may be a constitutive isozyme. The expression of UCH isozymes is also developmentally regulated, although in tissue culture, conditions that modulate the levels of expression significantly have not been found (Wilkinson *et al.*, 1992). The appearance of UCH-L1 (PGP 9.5) immunoreactivity in developing mouse brain correlates with the arrival of the neuronal precursor cells at the neural plate and the elaboration of neural processes (Schofield *et al.*, 1995). It is induced in the gonads of fish undergoing the sexual transition from female to male (Fujiwara *et al.*, 1994) and in experimentally induced axonal dystrophy (Bacci *et al.*, 1994). Interestingly, the enzyme is found in several neural inclusion bodies in humans (Lowe *et al.*, 1990). UCH-L1 levels are strongly downregulated on viral transformation of lung fibroblasts (Honore *et al.*, 1991).

In *Drosophila* the mRNA for uch-D is strongly expressed in nurse cells, the ovary, and the testis (Zhang *et al.*, 1993). The transcripts are also easily identified during the first few hours of embryonic development. By 4 to 6 hr of development, transcript levels drop markedly, consistent with the pattern usually seen for maternal transcripts. The uch-D transcripts are preferentially localized to the ventral surface of both the oocyte and the nurse cell. Inasmuch as dorsal/ventral gradients of certain proteins are required for proper embryogenesis in *Drosophila*, it is conceivable that the establishment or maintenance of the dorsal/ventral embryonic axis depends on a spatial gradient of proteolytic potential. The asymmetric distribution of uch-D mRNA in *Drosophila*, as well as the graded distribution of proteasomes in *Xenopus* embryos, are consistent with this possibility (Ryabova *et al.*, 1994).

6.2. The UBP Family

6.2.1. Structure of UBPs

The UBP family of DUB enzymes is extremely divergent, but all members contain several short consensus sequences that are likely to help form the catalytic domains (Fig. 4) (Baker *et al.*, 1992; Papa and Hochstrasser, 1993). Although the Cys and His boxes shown in Fig. 4 are the most conserved elements, additional

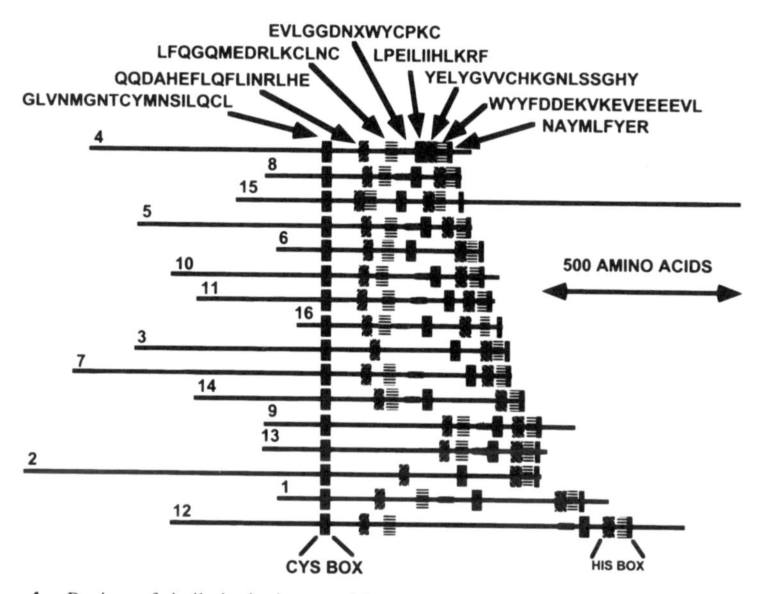

Figure 4. Regions of similarity in the yeast UBP family. Eight blocks of similarity were identified by submitting the yeast protein sequences to BLOCKS analysis (Blockmaker, http://www.blocks.fhc.org/blockmkr/make_blocks.html) and the consensus sequence is indicated at the top. The sequences are aligned on the active site cysteine and ordered by the size of the putative catalytic domain which is thought to be terminated near the His box. UBP1 (swissprot UBP1_YEAST, p25037); UBP2 (swissprot UBP2_YEAST, q01476); UBP3 (swissport UBP3_YEAST, q01477); UBP4 (swissprot UBP4_YEAST, p32571); UBP5 (swissprot UBC5_YEAST, p39944); UBP6 (swissprot UBPF_YEAST, P43593): UBP7 (swissprot UBP7_YEAST, p40453); UBP8 (swissport UBPN_YEAST, p50102); UBP9 (swissprot UBP9_YEAST, p39967); UBP10 (swissprot UBPM_YEAST, p50101); UBP11 (swissprot UBPB_YEAST, p36026); UBP12 (swissprot UBPC_YEAST, P39538); UBP13 (swissprot UBPD_YEAST, p38187); UBP14 (swissprot UBPE_YEAST, p38237); UBP15 (genbank, gp_all:u41849); UBP16 (genbank, gp_all:z71462);)

short sequences show some conservation as well (Papa and Hochstrasser, 1993; Wilkinson *et al.*, 1995). Interestingly, the conserved stretches that define the UBP enzymes are not obviously related to the regions surrounding the conserved Cys and His residues of the UCHs. Pairwise comparisons of certain isozymes of each class around the active site cysteines suggest some similarity, but the possibility that the UCH and UBP enzyme classes diverged from a common DUB forebear seems unlikely.

Analysis of sequences from the various protein sequence databases has revealed that the UBP enzyme family is remarkably large. In yeast there are 16 genes encoding potential UBP enzymes (Fig. 4), which exceeds the number of E2 ubiquitin-conjugating enzymes in this organism (Hochstrasser, 1996b). Many yeast UBP mutants do not display striking phenotypic abnormalities, which means either that many UBPs function in a very restricted set of metabolic processes (that were not tested by the standard phenotypic assays) or that there is considerable overlap in Ubp functions (Baker *et al.*, 1992; Papa and Hochstrasser, 1993; our unpublished data).

6.2.2. Role of UBPs in Development

Genetic analysis of a UBP-type deubiquitinating enzyme in *Drosophila*, the product of the *fat facets* (*faf*) gene, provides evidence in favor of the notion of substrate-specific modulation of degradation rates by UBPs (Huang *et al.*, 1995). The *faf* gene is required for normal eye development, and *faf* mutations also have a maternal effect on embryogenesis, with embryos from homozygous mutant mothers dying at an early stage in development. Mutations of the conserved Cys and His boxes in Faf behave as null mutations in transgenic flies, suggesting that the deubiquitinating activity of Faf is critical to its biological function. Most interestingly, Huang *et al.* (1995) demonstrated that several different mutant alleles of a 20 S proteasome subunit gene could dramatically suppress the defect in eye development seen in *faf* flies. These data suggest that Faf functions to reverse the ubiquitination of a regulatory protein(s) and thereby prevent or slow its degradation by the proteasome. Ubiquitin-dependent degradation of specific proteins may be regulated by changes in activity of UBPs such as Faf and/or by concomitant changes in the activity of ubiquitinating enzymes, which would shift the dynamic balance between ubiquitination and deubiquitination.

Another recent report suggests that substrate specificity of DUBs may be controlled in part by controlling access of the DUB enzyme to the substrate. Ubp3 has been shown to bind tightly to Sir4, a protein involved in transcriptional silencing of genes near telomeres and in the silent mating type loci of yeast (Moazed and Johnson, 1996). Silencing requires the maintenance of an altered state of chromatin, probably similar to heterochromatin in higher eukaryotes. A silencing complex assembled at a specific subset of chromosome sites contains the origin recognition complex (ORC), Sir2, Sir3, and Sir4. Deletion of the *UBP3* gene enhances silencing. Exactly how Ubp3 may interfere with silencing is unknown, as is the role of proteasome-mediated protein degradation in this process.

Interestingly, a mutation in another UBP, called D-Ubp-64E, has been identified as an enhancer of position effect variegation (PEV) in *Drosophila*. Extra copies of the *D-Ubp-64E* gene suppress PEV (Henchoz *et al.*. 1996). PEV is a form of transcriptional silencing involving chromosome rearrangements that place euchromatic genes next to a heterochromatin boundary. The presence of the same mutated proteasome subunit alleles used in the faf studies discussed above did not modify the level of PEV seen in the D-Ubp-64E mutant. However, this result does not rule out a potential function for the proteasome in D-Ubp-64E-dependent PEV.

6.2.3. UBPs Involved in Growth Control

Further evidence that deubiquitinating enzymes can have regulatory functions comes from recent work in mammalian cells. Several proteins implicated in tumorigenesis or growth control have recently been shown to be deubiquitinating enzymes. Interestingly, for the human *tre-2* oncogene, it appears that it is an inactive form of the deubiquitinating enzyme that is tumorigenic (Papa and Hochstrasser, 1993). The inactive protein may act in a dominant-negative fashion, interfering with *tre-2* enzyme-mediated degradation of one or more positive regulators of cell proliferation, e.g., the G1 cyclins. Alternatively, the *tre-2* enzyme may normally limit the degradation of a negative regulator(s) of growth, such as p53, by rapidly disassembling ubiquitinated intermediates.

Several additional mammalian growth regulators have been found to be deubiquitinating enzymes of the UBP class. One such protein is a mouse enzyme called DUB-1. DUB-1 is an erythroid cell-specific, immediate-early gene induced by the cytokines interleukin 3, interleukin 5, and GM-CSF. Induction requires the common β subunit of the corresponding cytokine receptors, βc (Zhu *et al.*, 1996a,b). *DUB-1* rapidly disappears after a short burst of expression. Interestingly, cells engineered to express DUB-1 continuously were found to arrest in the G1 phase of the cell cycle, suggesting a growth regulatory role for this deubiquitinating enzyme (Zhu *et al.*, 1996a). Under the same conditions, an active-site mutant that has lost DUB activity does not arrest growth.

Recently, additional UBP enzymes highly related to DUB-1 have been identified. Southern blotting experiments suggest the existence of a subfamily of enzymes with four to six members. DUB-2, the first of these other DUB-1-like enzymes to be analyzed in detail, is 88% identical to DUB-1. Interestingly, DUB-2 is also a cytokine-inducible, immediate-early gene, but unlike DUB-1, DUB-2 is induced by IL-2. Like DUB-1, it is ubiquitinated and degraded shortly after its synthesis (A. D. D'Andrea, personal communication). It is proposed that this subfamily of cytokine-inducible UBP enzymes regulates different growth regulatory factors, resulting in cytokine-specific growth responses. Because the active site thiol of DUB-1 is required for its effect on growth, the enzyme presumably functions by altering the degradation rate or ubiquitination state of one or more regulatory factors.

6.2.4. Metabolism of Degradation Intermediates

Evidence is now accumulating for several different deubiquitinating enzymes that indicates they function *in vivo* to facilitate ubiquitin-dependent degradation by preventing the excessive accumulation of inhibitory ubiquitin oligomers. Mammalian isopeptidase T is a Ubp-type enzyme that acts largely, if not exclusively, on unanchored ubiquitin chains, i.e., polyubiquitin with a free ubiquitin C-terminus (Wilkinson *et al.*, 1995). *In vitro* studies suggested that isopeptidase T could facilitate proteasomal degradation, possibly by preventing the accumulation of ubiquitin chains generated as intermediates in substrate degradation (Hadari *et al.*, 1992). A number of E2 enzymes can also synthesize unanchored ubiquitin chains from free ubiquitin (see Chapter 2) and it may be important that chains formed in this way also be processed.

Genetic experiments in yeast have recently provided results that extend the *in vitro* data on isopeptidase T (A. Amerik, *et al.*, 1997). Human isopeptidase T is the functional homologue of yeast Ubp14. The two proteins are about 31% identical and share similar enzymological properties. A yeast mutant lacking Ubp14 has defects in the ubiquitin-dependent degradation of a number of distinct proteins and shows a striking accumulation of unanchored ubiquitin oligomers. The defects of the yeast *ubp14* mutant can be reversed by expression of human isopeptidase T. That it is the excess unanchored ubiquitin chains that cause the proteolytic defects is suggested by experiments in which an overabundance of ubiquitin chains are formed on overexpression of a chain terminator, desGlyGly ubiquitin; these cells show defects in proteolysis as well.

Another yeast deubiquitinating enzyme, Doa4, has been shown to have a very broad role in ubiquitin-dependent degradation *in vivo* (Papa and Hochstrasser, 1993). Mutant *doa4* cells have a variety of phenotypic defects, ranging from an inability to sporulate (Papa and Hochstrasser, 1993) to a failure to properly coordinate replication of different regions of the genome (Singer *et al.*, 1996). Molecular and genetic analyses strongly suggest that Doa4 works in conjunction with the 26 S proteasome, and purified preparations of the yeast 26 S proteasome contain Doa4 (F. R. Papa and M. Hochstrasser, unpublished data). In *doa4* mutants, small ubiquitinated species accumulate; these species are all slightly larger than unanchored ubiquitin chains, leading to the suggestion that they may be the ubiquitinated proteolytic remnants of 26 S proteasome action (Papa and Hochstrasser, 1993). Such remnants may concentrate on the proteasome in *doa4* cells, preventing its recycling for further rounds of protein degradation.

Several points concerning isopeptidase T/Ubp14 and Doa4 function *in vivo* are worthy of elaboration. First, given the abundance of different deubiquitinating enzymes in the cell, it is somewhat surprising that any ubiquitinated species should accumulate in *ubp14* or *doa4* mutants. This may be related to very high substrate specificity among the Ubps, but *in vitro* cleavage experiments (as well as the minimal phenotypic abnormalities of many *ubp* mutants noted above) suggest there will be significant overlap in specificities among Ubps. Potentially, compensating enzymes do exist that can hydrolyze ubiquitin chains, but their activity against these substrates is lower than that of the missing enzymes. Loss of a large fraction of enzyme activity may have only minimal phenotypic effects unless the increase in ubiquitin chains caused proteolysis to become rate-limiting. For instance, if Ubp14 were responsible for cleavage of 80% of unanchored ubiquitin chains in a wild-type cell, then its loss would increase the steady-state concentration of chains by fivefold, which may not affect the degradation of most proteins too severely (Amerik *et al.*, 1997).

An alternative explanation is based on a compartmentalization model. Specifically, ubiquitinated species may accumulate in compartments where they are inaccessible to the bulk of the deubiquitinating enzymes in the cell. As already mentioned, the ubiquitinated species that build up in *doa4* cells were proposed to bind the 26 S proteasome where they might be sequestered from cellular isopeptidases other than Doa4. An analogous explanation may account for the accumulation of unanchored ubiquitin chains in *ubp14* cells. This does not require that Ubp14 be an integral component of the proteasome but simply that it, unlike most other UBPs (including Doa4), can access specific sites of unanchored ubiquitin chain binding on the proteasome.

A second interesting point about Doa4 and Ubp14/isopeptidase T is that both may have as their primary substrates late intermediates of proteolysis. These products may provide a natural brake on proteasome activity in the cell, which could have several important consequences. First, by restraining the action of the 26 S proteasome complex, it could confer on the protease a degree of substrate discrimination that it would not have if rates of proteolysis greatly exceeded rates of substrate ubiquitination. In fact, elimination of Doa4 or Ubp14 from yeast cells stabilizes various proteins to very different degrees, as might be predicted if ubiquitinated substrates were competing for limiting amounts of active protease. A second, related effect of such a feedback loop on the 26 S proteasome is that changes in *relative* protein degradation rates could be engendered by varying the level of Doa4 or Ubp14. For instance, Doa4 levels are partially rate-limiting for α2 degradation but not for N-end rule substrates (Papa and Hochstrasser, 1993). Hence, an increase in Doa4 levels, such as occurs when cells enter stationary phase, could change the relative degradation rates of different proteins.

Finally, as noted earlier, a DUB enzyme that is an integral component of the 26 S proteasome has recently been described; there is one DUB active site per 19 S regulatory subunit (R. Cohen, personal communication). Interestingly, this enzyme has precisely the opposite specificity to that of isopeptidase T/Ubp14 in that it cleaves ubiquitin chains from the distal end in, one ubiquitin at a time, toward the site of attachment to substrate (a distal-exo-monodeubiquitination mechanism). Cohen and colleagues have proposed that this PA700-associated DUB enzyme serves an editing function, biasing proteasome action toward substrates with longer ubiquitin chains. This could help explain why monoubiquitinated substrates, such as histone H2A or actin in insect flight muscles, can be metabolically stable.

7. CONCLUSIONS AND FUTURE DIRECTIONS

The deubiquitinating or DUB enzymes comprise the largest and most diverse family of proteins currently known within the ubiquitin system. Although this statement rests largely on data from systematic genome sequencing projects, many of the predicted proteins have been shown to have DUB activity. Thus, the predictive value of the short consensus sequences found in the UBP family, for instance, is very high. Nevertheless, only a very limited set of ubiquitinated substrates have been tested with these enzymes, and in many cases, activity measurements have been at best semiquantitative. For an understanding of how these many different DUB enzymes function, it will be necessary to get much more precise kinetic data using a variety of substrates. Extensive genetic analyses are also needed to complement such enzymological investigations so as to elucidate how the enzymatic specificity of each enzyme is matched to its physiological function.

The study of the DUB enzymes is in its infancy. Already it has become apparent that these enzymes will be key regulatory factors in the ubiquitin system. They could, for instance, determine the probability of a particular ubiquitinated protein's being targeted to the proteasome rather than some alternative metabolic fate, e.g., triggering of plasma membrane endocytosis and trafficking to the vacuole/lysosome. There are indications that ubiquitin conjugation to cellular proteins may be fairly promiscuous. The DUB enzymes could therefore provide a proofreading function that affords ubiquitin-dependent proteolysis sufficient specificity *in vivo*. Like the kinase/phosphatase couples that orchestrate the phosphorylation and dephosphorylation of thousands of cell proteins, ubiquitin ligase/ DUB enzyme couples could furnish a similarly rich set of molecular switches for a wide range of biological regulatory mechanisms.

8. REFERENCES

Amerik, A. Y., Swaminathan, S., Krantz, B. A., Wilkinson, K. D., and Hochstrasser, M., 1997, In vivo disassembly of free polyubiquitin chains by yeast Ubp14 modulates rates of protein degradation by the proteasome. *EMBO J.* **16:**4826–4838.

Anderson, M. W., Ballal, N. R., Goldknopf, I. L., and Busch, H., 1981, Protein A24 lyase activity in nucleoli of thioacetamide-treated rat liver releases histone 2A and ubiquitin from conjugated protein A24, *Biochemistry* **20:**1100–1104.

Arnason, T., and Ellison, M. J., 1994, Stress resistance in Saccharomyces cerevisiae is strongly correlated with assembly of a novel type of multiubiquitin chain, *Mol. Cell. Biol.* **14:**7876–7883.

Baboshina, O. V., and Haas, A. L., 1996, Novel multiubiquitin chain linkages catalyzed by the conjugating enzymes E2EPF and RAD6 are recognized by 26 S proteasome subunit 5, *J. Biol. Chem.* **271:**2823–2831.

Bacci, B., Cochran, E., Nunzi, M. G., Izeki, E., Mizutani, T., Patton, A., Hite, S., Sayre, L. M., Autilio

Gambetti, L., and Gambetti, P., 1994, Amyloid beta precursor protein and ubiquitin epitopes in human and experimental dystrophic axons. Ultrastructural localization, *Am. J. Pathol.* **144:** 702–710.

Baker, R. T., Tobias, J. W., and Varshavsky, A., 1992, Ubiquitin-specific proteases of Saccharomyces cerevisiae. Cloning of UBP2 and UBP3, and functional analysis of the UBP gene family, *J. Biol. Chem.* **267:**23364–23375.

Banerjee, A., Gregori, L., Xu, Y., and Chau, V., 1993, The bacterially expressed yeast CDC34 gene product can undergo autoubiquitination to form a multiubiquitin chain-linked protein, *J. Biol. Chem.* **268:**5668–5675.

Banjerji, J., Sands, J., Strominger, J. L., and Spies, T., 1990, A gene pair from the human major histocompatibility complex encodes large proline-rich proteins with multiple repeated motifs and a single ubiquitin-like domain, *Proc. Natl. Acad. Sci. USA* **87:**2374–2378.

Beal, R., Deveraux, Q., Xia, G., Rechsteiner, M., and Pickart, C., 1996, Surface hydrophobic residues of multiubiquitin chains essential for proteolytic targeting, *Proc. Natl. Acad. Sci. USA* **93:**861–866.

Biggins, S., Ivanovska, I., and Rose, M. D., 1996, Yeast ubiquitin-like genes are involved in duplication of the microtubule organizing center, *J. Cell Biol.* **133:**1331–1346.

Chau, V., Tobias, J. W., Bachmair, A., Marriott, D., Ecker, D. J., Gonda, D. K., and Varshavsky, A., 1989, A multiubiquitin chain is confined to specific lysine in a targeted short-lived protein, *Science* **243:**1576–1583.

Ciechanover, A., 1994, The ubiquitin-proteasome proteolytic pathway, *Cell* **79:**13–21.

Cook, W. J., Jeffrey, L. C., Kasperek, E., and Pickart, C. M., 1994, Structure of tetraubiquitin shows how multiubiquitin chains can be formed, *J. Mol. Biol.* **236:**601–609.

Cox, M. J., Haas, A. L., and Wilkinson, K. D., 1986, Role of ubiquitin conformations in the specificity of protein degradation: Iodinated derivatives with altered conformation and activities, *Arch. Biochem. Biophys.* **250:**400–409.

DeMartino, G. N., Moomaw, C. R., Zagnitko, O. P., Proske, R. J., Chu-Ping, M., Afendis, S. J., Swaffield, J. C., and Slaughter, C. A., 1994, PA700, an ATP-dependent activator of the 20 S proteasome, is an ATPase containing multiple members of a nucleotide-binding protein family, *J. Biol. Chem.* **269:**20878–20884.

Deveraux, Q., Ustrell, V., Pickart, C., and Rechsteiner, M., 1994, A 26 S protease subunit that binds ubiquitin conjugates, *J. Biol. Chem.* **269:**7059–7061.

Di Stefano, D. L., and Wand, A. J., 1987, Two-dimensional 1H NMR study of human ubiquitin: A main chain directed assignment and structure analysis, *Biochemistry* **26:**7272–7281.

Duerksen-Hughes, P. J., Williamson, M. M., and Wilkinson, K. D., 1989, Affinity chromatography using protein immobilized via arginine residues: Purification of ubiquitin carboxyl-terminal hydrolases, *Biochemistry* **28:**8530–8536.

Dunten, R. L., and Cohen, R. E., 1989, Recognition of modified forms of ribonuclease A by the ubiquitin system, *J. Biol. Chem.* **264:**16739–16747.

Ellison, M. J., and Hochstrasser, M., 1991, Epitope-tagged ubiquitin. A new probe for analyzing ubiquitin function, *J. Biol. Chem.* **266:**21150–21157.

Eytan, E., Armon, T., Heller, H., Beck, S., and Hershko, A., 1993, Ubiquitin C-terminal hydrolase activity associated with the 26 S protease complex, *J. Biol. Chem.* **268:**4668–4674.

Falquet, L., Paquet, N., Frutiger, S., Hughes, G. J., Hoang-Van, K., and Jaton, J. C., 1995, A human deubiquitinating enzyme with both isopeptidase and peptidase activities in vitro, *FEBS Lett.* **359:**73–77.

Finley, D., and Chau, V., 1991, Ubiquitination, *Annu. Rev. Cell Biol.* **7:**25–69.

Finley, D., Bartel, B., and Varshavsky, A., 1989, The tails of ubiquitin precursors are ribosomal proteins whose fusion to ubiquitin facilitates ribosome biogenesis, *Nature* **338:**394–401.

Finley, D., Sadis, S., Monia, B. P., Boucher, P., Ecker, D. J., Crooke, S. T., and Chau, V., 1994, Inhibition of proteolysis and cell cycle progression in a multiubiquitination-deficient yeast mutant, *Mol. Cell. Biol.* **14:**5501–5509.

Fujiwara, Y., Hatano, K., Hirabayashi, T., and Miyazaki, J. I., 1994, Ubiquitin C-terminal hydrolase as a putative factor involved in sex differentiation of fish (temperate wrasse, Halichoeres poecilopterus), *Differentiation* **56:**13–20.

Gregori, L., Poosch, M. S., Cousins, G., and Chau, V., 1990, A uniform isopeptide-linked multiubiquitin chain is sufficient to target substrate for degradation in ubiquitin-mediated proteolysis, *J. Biol. Chem.* **265:**8354–8357.

Gwozd, C. S., Arnason, T. G., Cook, W. J., Chau, V., and Ellison, M. J., 1995, The yeast UBC4 ubiquitin conjugating enzyme monoubiquitinates itself in vivo: Evidence for an E2-E2 homointeraction, *Biochemistry* **34:**6296–6302.

Haas, A. L., and Bright, P. M., 1985, The immunochemical detection and quantitation of intracellular ubiquitin–protein conjugates, *J. Biol. Chem.* **260:**12464–12473.

Haas, A. L., and Bright, P. M., 1987, The dynamics of ubiquitin pools within cultured human lung fibroblasts, *J. Biol. Chem.* **262:**345–351.

Haas, A. L., Ahrens, P., Bright, P. M., and Ankel, H., 1987, Interferon induces a 15-kilodalton protein exhibiting marked homology to ubiquitin, *J. Biol. Chem.* **262:**11315–11323.

Haas, A. L., Baboshina, O., Williams, B., and Schwartz, L. M., 1995, Coordinated induction of the ubiquitin conjugation pathway accompanies the developmentally programmed death of insect skeletal muscle, *J. Biol. Chem.* **270:**9407–9412.

Haas, A. L., Katzung, D. J., Reback, P. M., and Guarino, L. A., 1996, Functional characterization of the ubiquitin variant encoded by the baculovirus Autographa californica, *Biochemistry* **35:**5385–5394.

Hadari, T., Warms, J. V., Rose, I. A., and Hershko, A., 1992, A ubiquitin C-terminal isopeptidase that acts on polyubiquitin chains. Role in protein degradation, *J. Biol. Chem.* **267:**719–727.

Haldeman, M. T., Finley, D., and Pickart, C. M., 1995, Dynamics of ubiquitin conjugation during erythroid differentiation in vitro, *J. Biol. Chem.* **270:**9507–9516.

Henchoz, S., De Rubertis, F., Pauli, D., and Spierer, P., 1996, The dose of a putative ubiquitin-specific protease affects position-effect variegation in Drosophila melanogaster, *Mol. Cell. Biol.* **16:**5717–5725.

Hershko, A., and Ciechanover, A., 1992, The ubiquitin system for protein degradation, *Annu. Rev. Biochem.* **61:**761–807.

Hershko, A., and Rose, I. A., 1987, Ubiquitin-aldehyde: A general inhibitor of ubiquitin-recycling processes, *Proc. Natl. Acad. Sci. USA* **84:**1829–1833.

Hershko, A., Eytan, E., Ciechanover, A., and Haas, A. L., 1982, Immunochemical analysis of the turnover of ubiquitin–protein conjugates in intact cells. Relationship to the breakdown of abnormal proteins, *J. Biol. Chem.* **257:**13964–13970.

Hochstrasser, M., 1995, Ubiquitin, proteasomes, and the regulation of intracellular protein degradation, *Curr. Opin. Cell Biol.* **7:**215–223.

Hochstrasser, M., 1996a, Protein degradation or regulation: Ub the judge, *Cell* **84:**813–815.

Hochstrasser, M., 1996b, Ubiquitin-dependent protein degradation, *Annu. Rev. Genet.* **30:**405–439.

Hochstrasser, M., Ellison, M. J., Chau, V., and Varshavsky, A., 1991, The short-lived Matα2 transcriptional regulator is ubiquitinated in vivo, *Proc. Natl. Acad. Sci. USA* **88:**4606–4610.

Honore, B., Rasmussen, H. H., Vandekerckhove, J., and Celis, J. E., 1991, Neuronal protein gene product 9.5 (IEF SSP 6104) is expressed in cultured human MRC-5 fibroblasts of normal origin and is strongly down-regulated in their SV40 transformed counterparts, *FEBS Lett.* **280:**235–240.

Huang, Y., Baker, R. T., and Fischer-Vize, J. A., 1995, Control of cell fate by a deubiquitinating enzyme encoded by the fat facets gene, *Science* **270:**1828–1831.

Johnson, E. S., Ma, P. C. M., Ota, I. M., and Varshavsky, A., 1995, A proteolytic pathway that recognizes ubiquitin as a degradation signal, *J. Biol. Chem.* **270:**17442–17456.

Johnston, S. C., Larsen, C. N., Cook, W. J., Wilkinson, K. D., and Hill, C. P., 1997, Crystal structure of a deubiquitinating enzyme (human UCH-L3) at 1.8 A resolution. *EMBO J.* **16:**3787–3796.

Kas, K., Michiels, L., and Merregaert, J., 1992, Genomic structure and expression of the human fau gene: Encoding the ribosomal protein S30 fused to a ubiquitin-like protein, *Biochem. Biophys. Res. Commun.* **187:**927–933.

Kumar, S., Yoshida, Y., and Noda, M., 1993, Cloning of a cDNA which encodes a novel ubiquitin-like protein, *Biochem. Biophys. Res. Commun.* **195:**393–399.

Larsen, C. N., Price, J. S., and Wilkinson, K. D., 1996, Substrate binding and catalysis by ubiquitin C-terminal hydrolases: Identification of two active site residues, *Biochemistry* **35:**6735–6744.

Linnen, J. M., Bailey, C. P., and Weeks, D. L., 1993, Two related localized mRNAs from Xenopus laevis encode ubiquitin-like fusion proteins, *Gene* **128:**181–188.

Liu, C. C., Miller, H. I., Kohr, W. J., and Silber, J. I., 1989, Purification of a ubiquitin protein peptidase from yeast with efficient in vitro assays, *J. Biol. Chem.* **264:**20331–20338.

Lowe, J., McDermott, H., Landon, M., Mayer, R. J., and Wilkinson, K. D., 1990, Ubiquitin carboxyl-terminal hydrolase (PGP 9.5) is selectively present in ubiquitinated inclusion bodies characteristic of human neurodegenerative diseases, *J. Pathol.* **161:**153–160.

Matsui, S., Sandberg, A. A., Negoro, S., Seon, B. K., and Goldstein, G., 1982, Isopeptidase: A novel eukaryotic enzyme that cleaves isopeptide bonds, *Proc. Natl. Acad. Sci. USA* **79:**1535–1539.

Mayer, A. N., and Wilkinson, K. D., 1989, Detection, resolution, and nomenclature of multiple ubiquitin carboxyl-terminal esterases from bovine calf thymus, *Biochemistry* **28:**166–172.

Medina, R., Wing, S. S., Haas, A., and Goldberg, A. L., 1991, Activation of the ubiquitin-ATP-dependent proteolytic system in skeletal muscle during fasting and denervation atrophy, *Biomed. Biochim. Acta* **50:**347–356.

Melandri, F., Grenier, L., Plamondon, L., Huskey, W. P., and Stein, R. L., 1996, Kinetic studies on the inhibition of isopeptidase T by ubiquitin aldehyde, *Biochemistry* **35:**12893–12900.

Meyers, G., Tautz, N., Dubovi, E. J., and Thiel, H. J., 1991, Viral cytopathogenicity correlated with integration of ubiquitin-coding sequences, *Virology* **180:**602–616.

Miller, H. I., Henzel, W. J., Ridgeway, J. B., Kuang, W., Chisholm, V., and Liu, C., 1989, Cloning and expression of a yeast ubiquitin-protein cleaving activity in Escherichia coli, *Biotechnology* **7:**698–704.

Moazed, D., and Johnson, D., 1996, A deubiquitinating enzyme interacts with SIR4 and regulates silencing in *S. cerevisiae*, *Cell* **86:**667–677.

Nakamura, M., Xavier, R. M., and Tanigawa, Y., 1996, Ubiquitin-like moiety of the monoclonal nonspecific suppressor factor beta is responsible for its activity, *J. Immunol.* **156:**532–538.

Narasimhan, J., Potter, J. L., and Haas, A. L., 1996, Conjugation of the 15-kDa interferon-induced ubiquitin homolog is distinct from that of ubiquitin, *J. Biol. Chem.* **271:**324–330.

Olvera, J., and Wool, I. G., 1993, The carboxyl extension of a ubiquitin-like protein is rat ribosomal protein S30, *J. Biol. Chem.* **268:**17967–17974.

Özkaynak, E., Finley, D., and Varshavsky, A., 1984, The yeast ubiquitin gene: Head-to-tail repeats encoding a polyubiquitin precursor protein, *Nature* **312:**663–666.

Özkaynak, E., Finley, D., Solomon, M. J., and Varshavsky, A., 1987, The yeast ubiquitin genes: A family of natural gene fusions, *EMBO J.* **6:**1429–1439.

Palombella, V. J., Rando, O. J., Goldberg, A. L., and Maniatis, T., 1994, The ubiquitin-proteasome pathway is required for processing the NF-kappa B1 precursor protein and the activation of NF-kappa B, *Cell* **78:**773–785.

Papa, F. R., and Hochstrasser, M., 1993, The yeast DOA4 gene encodes a deubiquitinating enzyme related to a product of the human tre-2 oncogene, *Nature* **366:**313–319.

Pickart, C. M., and Rose, I. A., 1985, Ubiquitin carboxyl-terminal hydrolase acts on ubiquitin carboxyl-terminal amides, *J. Biol. Chem.* **260:**7903–7910.

Pickart, C. M., and Rose, I. A., 1986, Mechanism of ubiquitin carboxyl-terminal hydrolase. Borohydride and hydroxylamine inactive in the presence of ubiquitin, *J. Biol. Chem.* **261:**10210–10217.

Rawlings, N. D., and Barrett, A. J., 1994, Families of cysteine peptidases, *Methods Enzymol.* **244:** 461–485.

Reiss, Y., Heller, H., and Hershko, A., 1989, Binding sites of ubiquitin-protein ligase. Binding of ubiquitin-protein conjugates and of ubiquitin-carrier protein, *J. Biol. Chem.* **264:**10378–10383.

Riley, D. A., Bain, J. L., Ellis, S., and Haas, A. L., 1988, Quantitation and immunocytochemical localization of ubiquitin conjugates within rat red and white skeletal muscles, *J. Histochem. Cytochem.* **36:**621–632.

Roff, M., Thompson, J., Rodriguez, M. S., Jacque, J. M., Baleux, F., Arenzana-Seisdedos, F., and Hay, R. T., 1996, Role of IκBα ubiquitination in signal-induced activation of NFκB in vivo, *J. Biol. Chem.* **271:**7844–7850.

Rose, I. A., 1988, Ubiquitin carboxyl-terminal hydrolases, in *Ubiquitin* (M. Rechsteiner, ed.), pp. 135–155, Plenum Press, New York.

Rose, I. A., and Warms, J. V., 1983, An enzyme with ubiquitin carboxy-terminal esterase activity from reticulocytes, *Biochemistry* **22:**4234–4237.

Rubin, D. M., and Finley, D., 1995, Proteolysis. The proteasome: A protein-degrading organelle? *Curr. Biol.* **5:**854–858.

Ryabova, L. V., Virtanen, I., Olink-Coux, M., Scherrer, K., and Vassetzky, S. G., 1994, Distribution of prosome proteins and their relationship with the cytoskeleton in oogenesis of Xenopus laevis, *Mol. Reprod. Dev.* **37:**195–203.

Schaeffer, J. R., and Cohen, R. E., 1996, Differential effects of ubiquitin aldehyde on ubiquitin and ATP-dependent protein degradation, *Biochemistry* **35:**10886–10893.

Scheffner, M., Nuber, U., and Huibregtse, J. M., 1995, Protein ubiquitination involving an E1-E2-E3 enzyme ubiquitin thioester cascade, *Nature* **373:**81–83.

Schofield, J. N., Day, I. N., Thompson, R. J., and Edwards, Y. H., 1995, PGP9.5, a ubiquitin C-terminal hydrolase; pattern of mRNA and protein expression during neural development in the mouse, *Brain Res. Dev. Brain Res.* **85:**229–238.

Shean, B. S., and Mykles, D. L., 1995, Polyubiquitin in crustacean striated muscle: Increased expression and conjugation during molt-induced claw muscle atrophy, *Biochim. Biophys. Acta* **1264:** 312–322.

Shen, Z., Pardington-Purtymun, P. E., Cary, R. B., Peterson, S. R., Comeaux, J. C., and Chen, D. J., 1996, hRAP12, a human ubiquitin-like nuclear protein that associates with human RAD51/RAD52 proteins, *FASEB J.* **10:**A963.

Singer, J. D., Manning, B. M., and Formosa, T., 1996, Coordinating DNA replication to produce one copy of the genome requires genes that act in ubiquitin metabolism, *Mol. Cell. Biol.* **16:**1356–1366.

Sokolik, C. W., and Cohen, R. E., 1991, The structures of ubiquitin conjugates of yeast iso-2-cytochrome c, *J. Biol. Chem.* **266:**9100–9107.

Spence, J., Sadis, S., Haas, A. L., and Finley, D., 1995, A ubiquitin mutant with specific defects in DNA repair and multiubiquitination, *Mol. Cell. Biol.* **15:**1265–1273.

Stein, R. L., Chen, Z., and Melandri, F., 1995, Kinetic studies of isopeptidase T: Modulation of peptidase activity by ubiquitin, *Biochemistry* **34:**12616–12623.

Swindle, J., Ajioka, J., Eisen, H., Sanwal, B., Jacquemot, C., Browder, Z., and Buck, G., 1988, The genomic organization and transcription of the ubiquitin genes of Trypanosoma cruzi, *EMBO J.* **7:**1121–1127.

Tautz, N., Meyers, G., and Thiel, H. J., 1993, Processing of poly-ubiquitin in the polyprotein of an RNA virus, *Virology* **197:**74–85.

Tobias, J. W., and Varshavsky, A., 1991, Cloning and functional analysis of the ubiquitin-specific protease gene UBP1 of Saccharomyces cerevisiae, *J. Biol. Chem.* **266:**12021–12028.

Toniolo, D., Persico, M., and Alcalay, M., 1988, A "housekeeping" gene on the X chromosome encodes a protein similar to ubiquitin, *Proc. Natl. Acad. Sci. USA* **85:**851–855.

Van Nocker, S., and Vierstra, R. D., 1993, Multiubiquitin chains linked through lysine 48 are abundant in vivo and are competent intermediates in the ubiquitin proteolytic pathway, *J. Biol. Chem.* **268:**24766–24773.

Vijay-Kumar, S., Bugg, C. E., Wilkinson, K. D., Vierstra, R. D., Hatfield, P. M., and Cook, W. J., 1987,

Comparison of the three-dimensional structures of human, yeast, and oat ubiquitin, *J. Biol. Chem.* **262:**6396–6399.

Weber, P. L., Brown, S. C., and Mueller, L., 1987, Sequential 1H NMR assignments and secondary structure identification of human ubiquitin, *Biochemistry* **26:**7282–7290.

Wilkinson, K. D., 1988, Purification and structural properties of ubiquitin, in *Ubiquitin* (M. Rechsteiner, ed.), pp. 5–38, Plenum Press, New York.

Wilkinson, K. D., 1994, Cellular roles of ubiquitin, in *Heat Shock Proteins in the Nervous System* (R. J. Mayer and I. R. Brown, eds.), pp. 191–234, Academic Press, London.

Wilkinson, K. D., 1995, Roles of ubiquitinylation in proteolysis and cellular regulation, *Annu. Rev. Nutr.* **15:**161–189.

Wilkinson, K. D., and Audhya, T. K., 1981, Stimulation of ATP-dependent proteolysis requires ubiquitin with the COOH—terminal sequence Arg-Gly-Gly, *J. Biol. Chem.* **256:**9235–9241.

Wilkinson, K. D., and Mayer, A. N., 1986, Alcohol-induced conformational changes of ubiquitin, *Arch. Biochem. Biophys.* **250:**390–399.

Wilkinson, K. D., Lee, K. M., Deshpande, S., Duerksen-Hughes, P. J., Boss, J. M., and Pohl, J., 1989, The neuron-specific protein PGP 9.5 is a ubiquitin carboxyl-terminal hydrolase, *Science* **246:** 670–673.

Wilkinson, K. D., Deshpande, S., and Larsen, C. N., 1992, Comparisons of neuronal (PGP 9.5) and non-neuronal ubiquitin C-terminal hydrolases, *Biochem. Soc. Trans.* **20:**631–637.

Wilkinson, K. D., Tashayev, V. L., O'Connor, L. B., Larsen, C. N., Kasperek, E., and Pickart, C. M., 1995, Metabolism of the polyubiquitin degradation signal: Structure, mechanism, and role of isopeptidase T, *Biochemistry* **34:**14535–14546.

Wing, S. S., Haas, A. L., and Goldberg, A. L., 1995, Increase in ubiquitin–protein conjugates concomitant with the increase in proteolysis in rat skeletal muscle during starvation and atrophy denervation, *Biochem. J.* **307:**639–645.

Wunsch, A. M., and Haas, A. L., 1995, Ubiquitin–protein conjugates selectively distribute during early chicken embryogenesis, *Dev. Dyn.* **204:**118–132.

Zhang, N., Wilkinson, K. D., and Bownes, M., 1993, Cloning and analysis of expression of a ubiquitin carboxyl terminal hydrolase expressed during oogenesis in Drosophila melanogaster, *Dev. Biol.* **157:**214–223.

Zhu, Y., Carroll, M., Papa, F. R., Hochstrasser, M., and D'Andrea, A. D., 1996a, DUB-1, a novel deubiquitinating enzyme with growth-suppressing activity, *Proc. Natl. Acad. Sci. USA* **93:**3275–3279.

Zhu, Y., Pless, M., Inhorn, R., Mathey-Prevot, B., and D'Andrea, A. D., 1996b, The murine DUB-1 gene is specifically induced by the beta-c subunit of the interleukin-3 receptor, *Mol. Cell. Biol.* **16:**4808–4817.

CHAPTER 5

The 20 S Proteasome

Andrei Lupas and Wolfgang Baumeister

1. INTRODUCTION

The 26 S proteasome—discussed in Chapter 6—is the central protease of the ubiquitin pathway of protein degradation. The core of this 2-MDa enzyme is formed by the 20 S proteasome (Peters *et al.*, 1993), a barrel-shaped protease of about 700 kDa, which is the subject of this chapter (Fig. 1). Whereas the 26 S proteasome degrades folded proteins in an ATP-dependent manner, the 20 S proteasome is ATP-independent and only degrades entirely unfolded polypeptides.

Since its first sighting in erythrocyte extracts in 1968 (Harris, 1968), the 20 S proteasome has accumulated more than 20 different names—including *cylindrin* (Harris, 1988), *alkaline protease* (Hase *et al.*, 1980), *multicatalytic proteinase* MCP (Orlowski and Wilk, 1988), *ingensin* (Ishiura *et al.*, 1985), *prosome* (Schmid *et al.*, 1984), *low-molecular-weight protein* LMP (Monaco and McDevitt, 1984), *macropain* (McGuire and DeMartino, 1986)—and almost as many functions, illustrating clearly its important role in many cellular processes. Not all functions have been substantiated by subsequent research; for example, the association of proteasomes with RNA into "prosomes" (Schmid *et al.*, 1984) is now widely considered to be an artifact of purification, given the heterogeneity and substoichiometric amounts of the RNA that can be copurified with proteasomes (Pamnani *et*

Andrei Lupas and Wolfgang Baumeister • Molekulare Strukturbiologie, Max-Planck-Institute für Biochemie, D-82152 Martinsried, Germany.

Ubiquitin and the Biology of the Cell, edited by Peters *et al.* Plenum Press, New York, 1998.

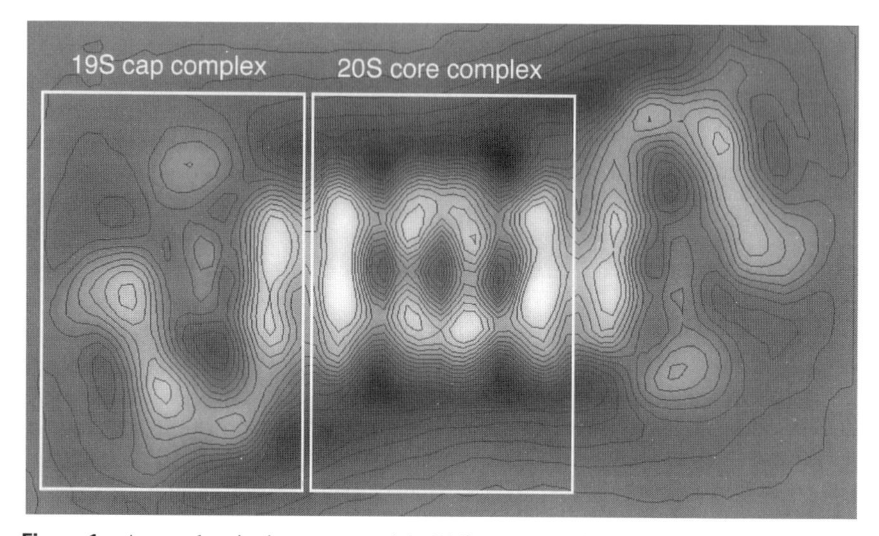

Figure 1. Averaged projection structure of the 26 S proteasome from *Xenopus laevis* oocytes at a resolution of 1.5 nm, showing the location of the 19 S caps and the 20 S core.

al., 1994). Except for a report on an RNAse activity for one of its subunits (Pouch *et al.*, 1995), which awaits confirmation, the only clearly established activity of the 20 S proteasome is proteolytic and accounts for all known functions of the complex. The name *proteasome* was coined in 1988 (Arrigo *et al.*, 1988) and quickly replaced the previous names because it united in one term the two most salient features of the particle: its proteolytic activity and its complex nature.

The 20 S proteasome is ubiquitous in eukaryotes. In archaebacteria, it was initially only found in *Thermoplasma acidophilum* (Dahlmann *et al.*, 1989; Pühler *et al.*, 1994), but recent genome sequencing projects have shown that it is widespread, at least in euryarchaeotes. In eubacteria, the 20 S proteasome has so far only been identified in the actinomycetes *Mycobacterium* (Lupas *et al.*, 1994), *Rhodococcus* (Tamura *et al.*, 1995), *Streptomyces* (I. Nagy, T. Tamura, R. De Mot, and W. Baumeister, unpublished), and possibly *Frankia* (Benoist *et al.*, 1992); the completed genome sequences of *Escherichia coli*, *Synechocystis* PCC6803, *Mycoplasma pneumoniae*, and *Haemophilus influenzae* show that it is absent from these organisms. Instead, *E. coli* and many other eubacteria contain a simpler protease, HslV (ClpQ), related to the 20 S proteasome (Lupas *et al.*, 1994; Kessel *et al.*, 1996; Rohrwild *et al.*, 1997). It seems conceivable that the antinomycetes have acquired the proteasome genes by horizontal gene transfer after their separation from other gram-positive bacteria. If so, the assembly pathway and the sensitivity to inhibitors indicate that the source was eukaryotic rather than an archaebacterium.

The main difference between prokaryotic and eukaryotic proteasomes is one of complexity (Fig. 2). Prokaryotic proteasomes generally contain two different but related subunits, α and β (Dahlmann *et al.*, 1989; Zwickl *et al.*, 1991, 1992), except for the proteasome of *Rhodococcus erythropolis*, which contains 2 α-type and 2 β-type subunits, probably as a result of a recent operon transfer (Tamura *et al.*, 1995). The proteasome-related protease of eubacteria, HslV, is simpler and contains only a β-type subunit (Rohrwild *et al.*, 1996). In contrast, eukaryotic proteasomes contain 7 α-type and 7 β-type subunits (reviewed by Hilt and Wolf, 1995; Tanaka, 1995). Vertebrates have achieved an even higher degree of com-

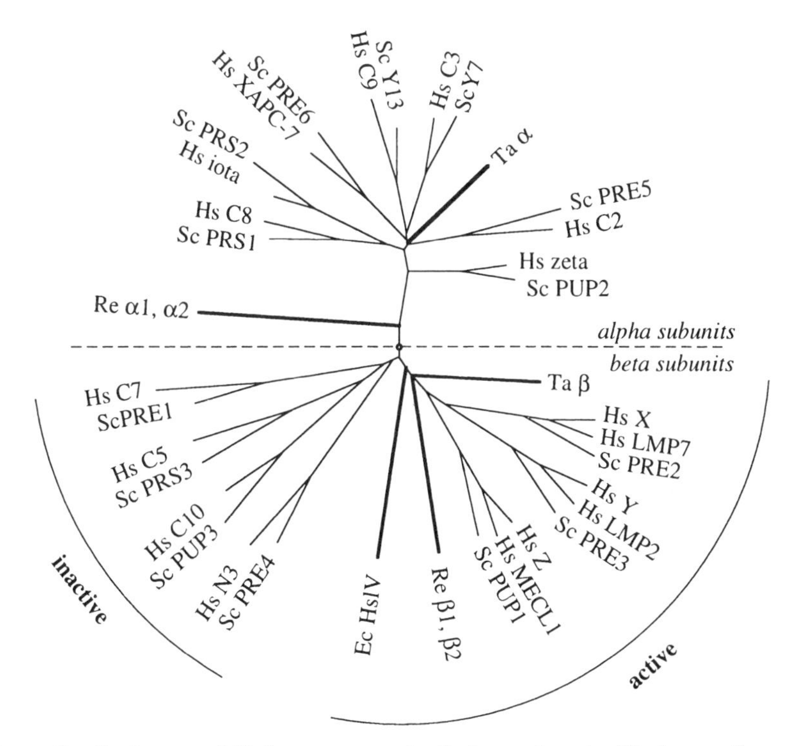

Figure 2. Dendrogram of 20 S proteasome subunits from eukaryotes (Hs, human; Sc, yeast), archaebacteria (Ta, *Thermoplasma acidophilum*), and eubacteria (Re, *Rhodococcus erythropolis*), and of the HslV protease (Ec, *Escherichia coli*). Prokaryotic branches are shown in bold. For the three branches of active human β-type subunits, the γ-interferon-inducible alternates (MECL1, LMP2, and LMP7) are also shown.

plexity: In addition to the 14 constitutive subunits, they contain 3 immune-encoded, γ-interferon-inducible β-type subunits, which can replace their constitutive counterparts to further modify the substrate specificity of the proteasome. These subunits and their role in the immune response are discussed in Chapter 7.

2. QUATERNARY STRUCTURE

Under the electron microscope, 20 S proteasomes from prokaryotes and eukaryotes are essentially indistinguishable. They appear as barrel-shaped particles 11 nm wide and 15 nm long, composed of four stacked, seven-membered rings of subunits (Tamura *et al.*, 1995; Pühler *et al.*, 1992; Baumeister *et al.*, 1988; Hase *et al.*, 1980). Immunoelectron microscopy of *Thermoplasma* and eukaryotic proteasomes showed that the outer rings are formed by α-type subunits and the inner rings by β-type subunits (Kopp *et al.*, 1993, 1995; Schauer *et al.*, 1993; Grziwa *et al.*, 1991). Overall, proteasomes have the appearance of complex dimers with *C*2 symmetry.

To date, the structures of two 20 S proteasome complexes have been determined by X-ray crystallography (Fig. 3); the *Thermoplasma* proteasome at 3.4-Å resolution (Löwe *et al.*, 1995) and the yeast proteasome at 2.4-Å resolution (Groll *et al.*, 1997). In the *Thermoplasma* proteasome, the two outer rings are rotated counterclockwise by 21° relative to the inner rings, and the two inner rings are rotated counterclockwise by 9° relative to each other. Collectively, the four rings form an elongated cylinder 14.8 nm long and 11.3 nm wide, which is traversed from end to end by a channel and contains three large inner cavities separated by narrow constrictions. The two cavities between the α- and β-subunit rings (the "antechambers") have a height of 4.0 nm and a maximum diameter of 5.0 nm. The third cavity at the center of the complex has a height of 3.8 nm and a maximum diameter of 5.5 nm. Access to the cavities is controlled by four constrictions: two in the α-subunit rings with a diameter of 1.3 nm, and two in the β-subunit rings with a diameter of 2.2 nm. The N-terminal 12 residues of the α-subunits, lying directly above the α-ring constrictions, are not visible in the electron density and thus presumably have a disordered conformation (Löwe *et al.*, 1995).

Experiments with Nanogold™-labeled insulin show that substrates penetrate into the proteasome via the central channel (Wenzel and Baumeister, 1995). As expected from the dimensions of the α-ring constrictions, only entirely unfolded polypeptide chains can pass. Disulfide bonds or residual secondary structure elements (as are found for example in the molten globule state) are sufficient to prevent substrate uptake. By a still unknown mechanism (possibly involving a molecular ratchet), substrates wind their way through the outer cavities to the proteolytic active-site clefts in the central cavity. Here, they are degraded to

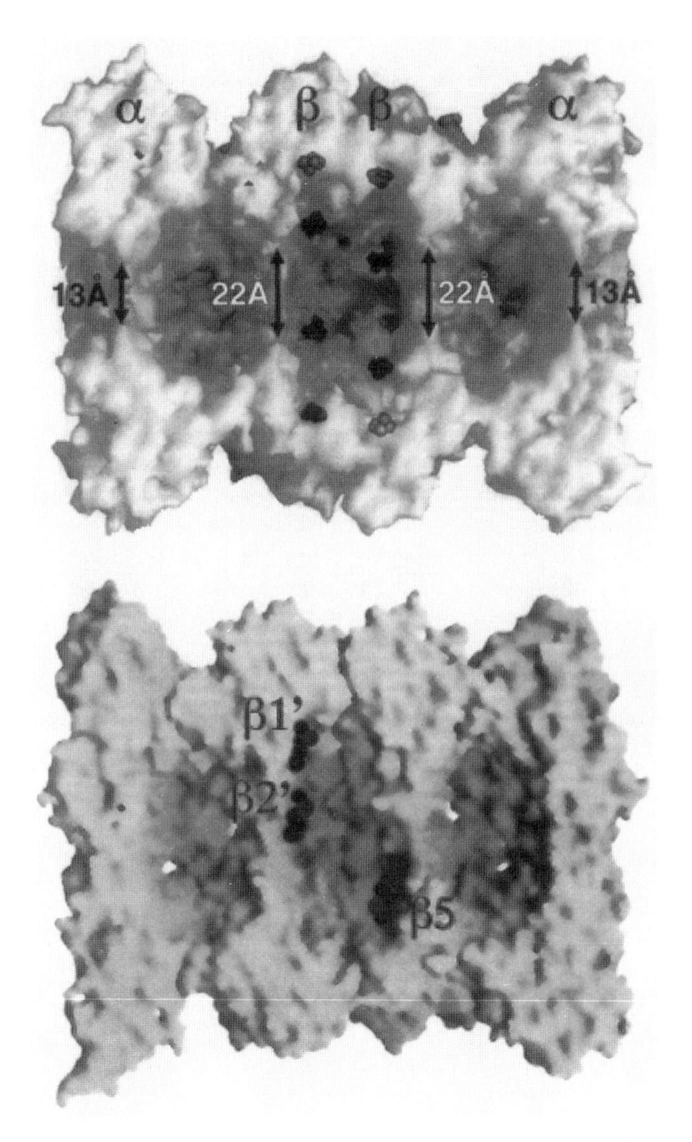

Figure 3. Space-filling representation of the proteasomes from *Thermoplasma* (upper panel) and yeast (lower panel) with bound inhibitor, clipped along the cylinder axis. The upper panel is modified from Fig. 5 of Stock *et al.* (1996) and the lower panel is from Fig. 8 of Groll *et al.* (1997).

peptides of a narrow size range (mostly between 4 and 10 residues). This property, which is shared between prokaryotic and eukaryotic proteasomes (Ehring *et al.*, 1996; Wenzel *et al.*, 1994), has led to the proposal of the "molecular ruler" hypothesis (Wenzel *et al.*, 1994). Although neighboring active sites were found to be 2.8 nm apart in the *Thermoplasma* proteasome structure—a distance that can be bridged by a hepta- or octapeptide in extended conformation—it is not yet clear whether the "molecular ruler" is determined by the distance between neighboring active sites. It is also not clear of what general use (if any) this "ruler" would be to the cell. In vertebrates, the focused product length of the proteasome has been exploited by the immune system for the production of peptides displayed by the class I major histocompatibility complex; this is discussed in Chapter 7.

The yeast proteasome quaternary structure is very similar to that of *Thermoplasma*, with one significant difference: The N-terminal residues of the α-subunits, which are disordered in the structure of the *Thermoplasma* proteasome, are visible and seen to tightly close the α-ring constrictions with several layers of interdigitating subunits (Fig. 3). This provides a possible structural explanation for the observation that eukaryotic 20 S proteasomes, but not the *Thermoplasma* proteasome, are purified in a latent state and can be activated by mild chaotropic agents such as SDS or heat. Presumably, these selectively denature the N-terminal sequences, resulting in an opening of the channel. Several small openings are visible in the side walls of the "antechambers" in yeast but not in *Thermoplasma*. Although these may serve as exit sites for product peptides, it seems unlikely that they represent a significant uptake route given that the sequences surrounding these openings are poorly conserved even among eukaryotes, suggesting that these openings are heterogeneous in size and hydrophilicity even between yeast and humans. The sequences lining the central pore, on the other hand, contain the most highly conserved parts of the α-subunits, as would be expected for the main substrate uptake route. A final difference in quaternary structure results from the higher subunit complexity of the yeast proteasome. Thus, while the *Thermoplasma* proteasome has multiple axes of symmetry, the yeast proteasome has a single *C*2 axis with all 14 subunits found exactly twice within the complex, and in defined locations. This had been anticipated, down to a surprising level of detail, by immunoelectron microscopy (Kopp *et al.*, 1993, 1995, 1997). The location of subunits relative to the axis of *C*2 symmetry has prompted a new, systematic nomenclature for eukaryotic proteasome subunits (Table I; Groll *et al.*, 1997).

In eubacteria, the proteasome has not yet been studied to high structural detail. An initial uncertainty on the number of subunits in the rings forming the complex has now been resolved: As in all other proteasomes, the rings are seven-membered (Zühl *et al.*, 1997a). *In vitro* reassembly studies indicate that in the *Rhodococcus* proteasome, the two α-type and two β-type subunits occur randomly within their respective rings (F. Zühl *et al.*, 1997b). Such a lack of

Table I
Systematic Names of Eukaryotic Proteasome Subunits and Their Former Names in Yeast and Mammals[a]

Subunit	Yeast names	Mammalian names
α1	C7α, PRS2, Y8, PRC2, SCL1	ι, PROS-27
α2	PRS4, Y7, PRE*	C3
α3	PRS5, Y13, PRE9	C9
α4	PRE6	XAPC-7, C6
α5	PUP2, DOA5	ζ
α6	PRE5	C2, PROS-30
α7	C1, PRS1, PRC1, PRE10	C8
β1*PGPH	PRE3	Y, δ, N5
β1i*		LMP2, RING12, N7
β2*tryp	PUP1	Z, α, N1
β2i*		MECL-1, N2
β3	PUP3, RAD51-ORF1	C10-II, θ
β4	C11, PRE1	C7-I
β5*chym	PRE2, PRG1, DOA3	X, MB1, ε
B5i*		LMP7E1, C1, C13 LMP7E2, RING10
β6	PRS3, PTS1, PRE7	C5, γ, N4
β7	PRE4	N3, PROS-26, β

[a]The new systematic names result from the location of subunits relative to the axis of *C*2 symmetry (Groll *et al.*, 1997). The catalytically active subunits are marked with an asterisk. The subunits carrying the peptidylglutamyl-peptide hydrolyzing activity (PGPH), chymotrypsinlike activity (chym), and trypsinlike activity (tryp) are labeled. Subunits marked with "i" are the mammalian, γ-interferon-inducible homologues of subunits β1, β2, and β5. For β5i, two alternatively spliced products have been described.

differentiation agrees with a recent divergence of these subunits, as deduced from phylogenetic studies (Tamura *et al.*, 1995).

The proteasome-related HslV protease of eubacteria, for which a crystal structure has now also been determined (Bochtler *et al.*, 1997), has a simpler quaternary structure than the 20 S proteasome. It is formed by only two stacked rings of β-type subunits, with an overall height of 8 nm and a diameter of 11 nm. It essentially corresponds to the two central rings of the proteasome; however, HslV is built of six-membered rings (Kessel *et al.*, 1996; Rohrwild *et al.*, 1997). Also, the constrictions in the central channel are much narrower in HslV than in the β-subunit rings of the yeast and *Thermoplasma* proteasomes and only slightly larger than in the α-subunit rings, indicating that the gatekeeping function of the α-subunits has been integrated into the β-type subunits of HslV (Bochtler *et al.*, 1997).

3. SECONDARY AND TERTIARY STRUCTURE

As anticipated from their sequence similarity, proteasome α- and β-type subunits have the same fold (Fig. 4): a four-layer α+β structure with two central antiparallel β-sheets, flanked on either side by α-helices (Löwe *et al.*, 1995; Groll *et al.*, 1997; Bochtler *et al.*, 1997). Helices 1 and 2 mediate the association of α- and β-rings by intercalating in a wedgelike fashion between two subunits of the facing ring; helices 3 and 4 mediate the association of β-rings. In the yeast proteasome, additional contacts are formed by highly variable extensions at the C-termini of subunits as well as by insertions. The central β-sheets have a very low twist and an unusual dihedral angle of 30° (the usual value is around −30°). The β-sheet sandwich is open at one end to form the active-site cleft, and is closed at the other end by four hairpin loops. Proteasome subunits share this general fold with a newly described family of proteins, the Ntn hydrolases, which include aspartylglucosaminidase, glutamine PRPP amidotransferase, and penicillin acylase (Branningan *et al.*, 1995). The main difference between α- and β-type subunits lies in a highly conserved N-terminal extension of α-subunits (fig. 5). This extension forms an α-helix across the top of the central β-sandwich, thus filling the active-site cleft, and extends into the central channel, which it closes in the yeast proteasome. Proteolytic activity is limited to β-type subunits. In place of the N-terminal α-helix, these contain a prosequence (Lee *et al.*, 1990; Lilley *et al.*,

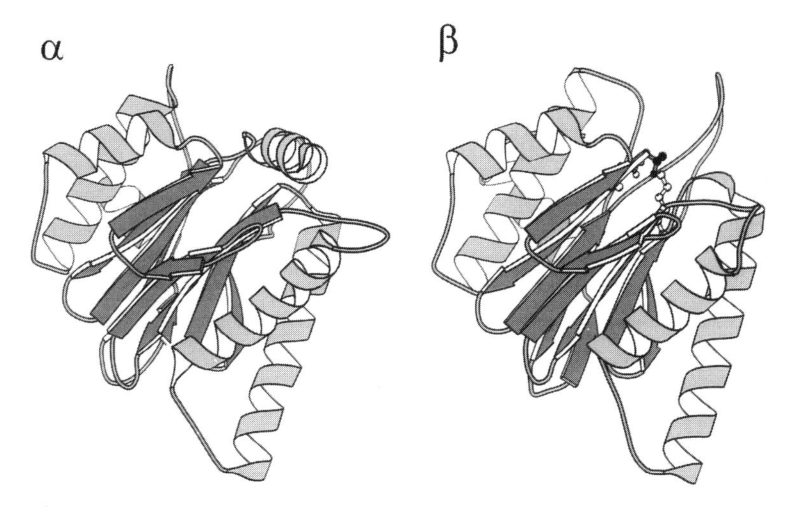

Figure 4. Structure of the α and β subunits of the *Thermoplasma* 20 S proteasome. The active-site residues of the β subunit are shown in ball-and-stick representation (Thr1 in black, and Glu17 and Lys33 in white).

1990; Zwickl *et al.*, 1992; Schauer *et al.*, 1993; Tamura *et al.*, 1995; Chen and Hochstrasser, 1995), which is cleaved off during proteasome assembly to free the active site. Although all residues required for the catalytic activity of one active site are located in one subunit, part of each active-site pocket in the assembled β-subunit rings is formed by a loop from a neighboring subunit, illustrating the close interaction of subunits in the central cavity and providing one explanation for the observation that processed but unassembled β subunits are essentially inactive (Zwickl *et al.*, 1994).

In addition to the N-terminal helix—whose function is still unclear but whose location at the top of the α-rings indicates that it may be important for interactions between the proteasome and its regulatory complexes (see Chapter 6)—α-type subunits contain two further conserved sequence motifs (Fig. 5): an RPxG motif, located at the base of the loop that constricts the proteasome channel at the level of the α-rings, and a GxxxD motif, which is almost completely conserved also in β-type subunits. The aspartate of the GxxxD motif makes multiple interactions with residues in the RPxG motif-containing loop, both of its own and of the neighboring subunit, and may play a role in determining the size and rigidity of the constriction. Mutation of this aspartate in the *Thermoplasma* α-subunit does not significantly impair proteasome activity, but the corresponding mutation in the β-subunit leads to both a strongly increased activity against peptides (Seemüller *et al.*, 1995a) and a simultaneous drop in activity against proteins. In the yeast β-subunit DOA3, mutation of the same residue leads to a destabilization of the interaction between the two central β-subunit rings (Chen and Hochstrasser, 1995). Besides the GxxxD motif, sequence similarity in β-type subunits is largely concentrated in the N-terminal region, which contains the active-site residues, and in a central GSG motif, which is located in a loop at one side of the active-site cleft. This serine is conserved in all active β-type subunits and is hydrogen-bonded to the catalytic threonine, Thr1.

Several α-type subunits, including the subunit from *Thermoplasma*, contain a nuclear location signal (NLS) in either one of two loops found at the upper, outer edge of α-rings (Nederlof *et al.*, 1995). This NLS sequence is not conserved between different α-type subunits and its role in *Thermoplasma*, which is a prokaryote and thus lacks a nucleus, is quite unclear. The function of the NLS sequence in eukaryotic subunits is discussed in Chapter 6.

4. MECHANISM

N-terminal nucleophile (Ntn) hydrolases have a unique, “single-residue” active site (Brannigan *et al.*, 1995; Duggleby *et al.*, 1995): Both the catalytic nucleophile and the primary proton acceptor are found in the same residue at the

```
Ta α      1        MQQGQMAYDRAITVFSPDGRLFQVEYAREAVKKGST-ALGMKFANGVLLISDKKVRS-RLIE
Sc α1     1    MSGAAAASAAGYDRHITIFSPEGRLYQVEYAFKATNQTNINSLAVRGKDCTVVISQKKVPD-KLLD
Sc α2     1           MTDRYSFSLTTFSPSGKLGQIDYALTAVKQGVT-SLGIKATNGVVIATEKKSSS-PLAM
Sc α3     1          MGSRRYDSRTTIFSPEGRLYQVEYALESISHAGT-AIGIMASDGIVLAAERKVTS-TLLE
Sc α4     1            MSGYDRALSIFSPDGHIFQVEYALEAVKRGTC-AVGVKGKNCVVLGCERRSTL-KLQD
Sc α5     1        MFLTRSEYDRGVSTFSPEGRLFQVEYSLEAIKLGST-AIGIATKEGVVLGVEKRATS-PLLE
Sc α6     1          MFRNNYDGDTVTFSPTGRLFQVEYALEAIKQGSV-TVGLRSNTHAVLVALKRNAD-ELSS
Sc α7     1        MTSIGTGYDLSNSVFSPDGRNFQVEYAVKAVENGTT-SIGIKCNDGVVFAVEKLITS-KLLV

Ec HslV   1                                           TTIVSVRRNGHVVIAGDGQATLGNTVM
                                                      *               *
Ta β      1                                           TTTVGITLKDAVIMATERRVTMENFIM
Sc β1     1                                           TSIMAVTFKDGVILGADSRTTTGAYIA
Sc β2     1                                           TTIVGVKFNNGVVIAADTRSTQGPIVA
Sc β3     1                                  MSDPSSINGGIVVAMTGKDCVAIACDLRLGSQSLGV
Sc β4     1                                          MDIILGIRVQDSVILASSKAVTRGISVL
Sc β5     1                                           TTTLAFRFQGGIIVAVDSRATAGNWVA
Sc β6     1                                  QFNPYGDNGGTILGIAGEDFAVLAGDTRNITDYSIN
Sc β7     1                                   TQQPIVTGTSVISMKYDNGVIIAADNLGSYGSLLR

Ta α     61   QN-SIEKIQLID-DYVAAVTSGLVADARVLVDFARISAQ-QEKVTYGSLVNIENLVKRVADQMQQY
Sc α1    66   P--TTVSYIFCISRTIGMVVNGPIPDARNAALRAKAEAA-EFRYKYGYDMPCDVLAKRMANLSQIY
Sc α2    58   SE-TLSKVSLLT-PDIGAVYSGMGPDYRVLVDKSRKVAHTSYKRIYGEYPPTKLLVSEVAKIMQEA
Sc α3    59   QDTSTEKLYKLN-DKIAVAVAGLTADAEILINTARIHAQ-NYLKTYNEDIPVEILVRRLSDIKQGY
Sc α4    57   TRITPSKVSKID-SHVVLSFSGLNADSRILIEKARVEAQ-SHRLTLEDPVTVEYLTRYVAGVQQRY
Sc α5    61   SD-SIEKIVEID-RHIGCAMSGLTADARSMIEHARTAAV-THNLYYDEDINVESLTQSVCDLALRF
Sc α6    59   YQ---KKIIKCD-EHMGLSLAGLAPDARVLSNYLRQQCN-YSSLVFNRKLAVERAGHLLCDKAQKN
Sc α7    61   PQ-KNVKIQVVD-RHIGCVYSGLIPDGRHLVNRGREEAA-SFKKLYKTPIPIPAFADRLGQYVQAH

Ec HslV  28   KG-NVKKVRRLYNDKVIAGFAGGTADAFTLFELFERKLE-MHQ------GHLVKAAVELAKDWRTD
                    *
Ta β     28   HK-NGKKLFQID-TYTGMTIAGLVGDAQVLVRYMKAELE-LYRLQRRVNMPIEAVATLLSNMLNQV
Sc β1    28   NR-VTDKLTRVH-DKIWCCRSGSAADTQAIADIVQYHLE-LYTSQYGTPSTETAASVFKELCYENK
Sc β2    28   DK-NCAKLHRIS-PKIWCAGAGTAADTEAVTQLIGSNIE-LHSLYTSREPRVVSALQMLKQHLFKY
Sc β3    37   SN-KFEKIFHYG--HVFLGITGLATDVTTLNEMFRYKTN-LYKLKEERAIEPETFTQLVSSSLYER
Sc β4    29   KD-SDDKTRQLS-PHTLMSFAGEAGDTVQFAEYIQANIQ-LYSIREDYELSPQAVSSFVRQELAKS
Sc β5    28   SQ-TVKKVIEIN-PFLLGTMAGGAADCQFWETWLGSQCR-LHELREKERISVAAASKILSNLVYQY
Sc β6    37   SR-YEPKVFDCG-DNIVMSANGFAADGDALVKRFKNSVK-WYHFDHNDKKLSINSAARNIQHLLYG
Sc β7    36   FN-GVERLIPVG-DNTVVGISGDISDMQHIERLLKDLVT-ENAYDNPLADAEEALEPSYIFEYLAT

Ta α    124   TQYGGV------RPYGVSLIFAGID-QIG-PRLFDCDPAGTINEYK--ATAIGSGKDAVVSFLERE
Sc α1   129   TQRAYM------RPLGVILTFVSVDEELG-PSIYKTDPAGYYVGYK--ATATGPKQQEITTNLENH
Sc α2   122   TQSGGV------RPFGVSLLIAGHDEFNG-FSLYQVDPSGSYFPWK--ATAIGKGSVAAKTFLEKR
Sc α3   123   TQHGGL------RPFGVSFIYAGYDDRYG-YQLYTSNPSGNYTGWK--AISVGANTSAAQTLLQMD
Sc α4   121   TQSGGV------RPFGVSTLIAGFDPRDDEPKLYQTEPSGIYSSWS--AQTIGRNSKTVREFLEKN
Sc α5   124   GEGASGEERLMSRPFGVALLIAGHDADDG-YQLFHAEPSGTFYRYN--AKAIGSGSEGAQAELLNE
Sc α6   120   TQSYGG------RPYGVGLLIIGYDKSGA--HLLEFQPSGNVTELY--GTAIGARSQGAKTYLERT
Sc α7   124   TLYNSV------RPFGVSTIFGGVDKNGA--HLYMLEPSGSYWGYK--GAATGKGRQSAKAELEKL

Ec HslV  86   RML---------RKLEALLAVADET------ASLIITGNGDVVQPENDLIAIGSGGPYAQAAARAL

Ta β     91   KY----------MPYMVQLLVGGID-TAP--HVFSIDAAGGSVEDI--YASTGSGSPFVYGVLESQ
Sc β1    91   ------------DNLTAGIIVAGYDDKNK--GEVYTIPLGGSVHKL-PYAIAGSGSTFIYGYCDKN
Sc β2    91   Q-----------GHIGAYLIVAGVDPTGS--HLFSIHAHGSTDVGY--YLSLGSGSLAAMAVLESH
Sc β3    99   RF----------GPYFVGPVVAGINSKSGKPFIAGFDLIGCIDEAK-DFIVSGTASDQLFGMCESL
Sc β4    92   IRSR--------RPYQVNVLIGGYDKKKNKPELYQIDYLGTKVELP--YGAHGYSGFYTFSLLDHH
Sc β5    91   KG----------AGLSMGTMICGYTRKEG-PTIYYVDSDGTRLKGD--IFCVGSGQTFAYGVLDSN
Sc β6   100   KRF---------FPYYVHTIIAGLD-EDGKGAVYSFDPVGSYEREQ--CRAGGAAASLIMPFLDNQ
Sc β7    99   VMYQRRSKM---NPLWNAIIVAGVQ-SNGDQFLRYVNLLGVTYSSP--TLATGFGAHMANPLLRKV
```

Figure 5. Alignment of proteasome α- and β-type subunits. The location of secondary structure elements is marked by open boxes (helices) and arrows (strands). Conserved residues are labeled in reverse type. Thr1, Glu17, and Lys33 of the β-subunit, which form the active site, are marked by asterisks.

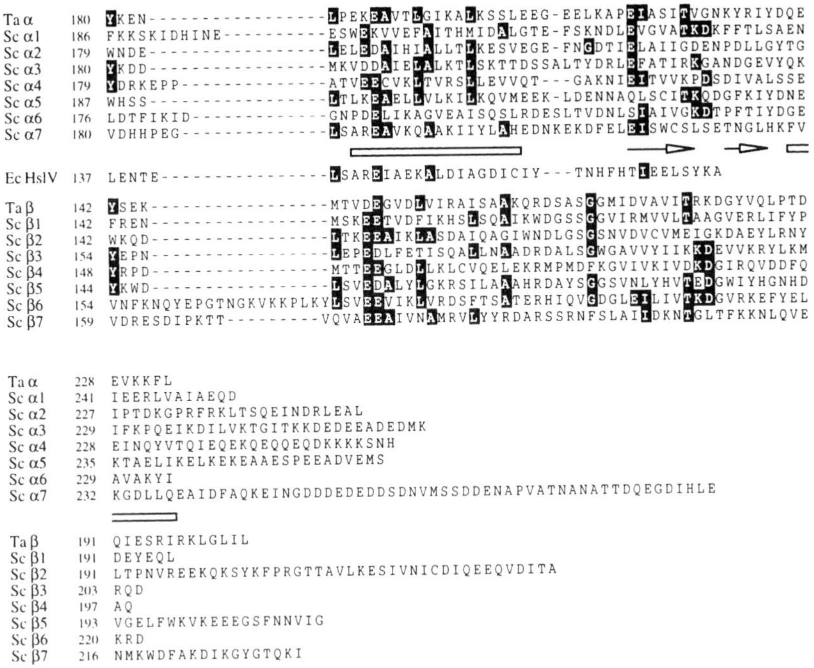

```
Ta α     180 YKEN-----------------LPEKEAVTLGIKALKSSLEEG-EELKAPEIASITVGNKYRIYDQE
Sc α1    186 FKKSKIDHINE----------ESWEKVVEFAITHMIDALGTE-FSKNDLEVGVATKDKFFTLSAEN
Sc α2    179 WNDE-----------------LELEDAIHIALLTLKESVEGE-FNGDTIELAIIGDENPDLLGYTG
Sc α3    180 YKDD-----------------MKVDDAIELALKTLSKTTDSSALTYDRLEFATIRKGANDGEVYQK
Sc α4    179 YDRKEPP--------------ATVEECVKLTVRSLLEVVQT---GAKNIEITVVKPDSDIVALSSE
Sc α5    187 WHSS-----------------LTLKEAELLVLKILKQVMEEK-LDENNAQLSCITKQDGFKIYDNE
Sc α6    176 LDTFIKID-------------GNPDELIKAGVEAISQSLRDESLTVDNLSIAIVGKDTPFTIYDGE
Sc α7    180 VDHHPEG--------------LSAREAVKQAAKIIYLAHEDNKEKDFELEISWCSLSETNGLHKFV

Ec HslV  137 LENTE----------------LSAREIAEKALDIAGDICIY---TNHFHTIEELSYKA

Ta β     142 YSEK-----------------MTVDEGVDLVIRAISAAKQRDSASGGMIDVAVITRKDGYVQLPTD
Sc β1    142 FREN-----------------MSKEETVDFIKHSLSQAIKWDGSSGGVIRMVVLTAAGVERLIFYP
Sc β2    142 WKQD-----------------LTKEEAIKLASDAIQAGIWNDLGSGSNVDVCVMEIGKDAEYLRNY
Sc β3    154 YEPN-----------------LEPEDLFETISQALLNAADRDALSGWGAVVYIIKKDEVVKRYLKM
Sc β4    148 YRPD-----------------MTTEEGLDLLKLCVQELEKRMPMDFKGVIVKIVDKDGIRQVDDFQ
Sc β5    144 YKWD-----------------LSVEDALYLGKRSILAAAHRDAYSGGSVNLYHVTEDGWIYHGNHD
Sc β6    154 VNFKNQYEPGTNGKVKKPLKYLSVEEVIKLVRDSFTSATERHIQVGDGLEILIVTKDGVRKEFYEL
Sc β7    159 VDRESDIPKTT---------VQVAEEAIVNAMRVLYYRDARSSRNFSLAIIDKNTGLTFKKNLQVE


Ta α     228 EVKKFL
Sc α1    241 IEERLVAIAEQD
Sc α2    227 IPTDKGPRFRKLTSQEINDRLEAL
Sc α3    229 IFKPQEIKDILVKTGITKKDEDEEADEDMK
Sc α4    228 EINQYVTQIEQEKQEQQEQDKKKKSNH
Sc α5    235 KTAELIKELKEKEAAESPEEADVEMS
Sc α6    229 AVAKYI
Sc α7    232 KGDLLQEAIDFAQKEINGDDDEDEDDSDNVMSSDDENAPVATNANATTDQEGDIHLE

Ta β     191 QIESRIRKLGLIL
Sc β1    191 DEYEQL
Sc β2    191 LTPNVREEKQKSYKFPRGTTAVLKESIVNICDIQEEQVDITA
Sc β3    203 RQD
Sc β4    197 AQ
Sc β5    193 VGELFWKVKEEEGSFNNVIG
Sc β6    220 KRD
Sc β7    216 NMKWDFAKDIKGYGTQKI
```

Figure 5. *(Continued)*

N-termini of the proteins. The nucleophilic attack is initiated when the free N-terminus (the primary proton acceptor) strips the proton off the catalytic side chain (the nucleophile), a transfer that—at least in the proteasome and in penicillin acylase—involves an intermediate water molecule (Fig. 6A). Three different N-terminal residues can act as the nucleophile in Ntn hydrolases: serine (penicillin acylase), cysteine (glutamine PRPP amidotransferase), or threonine (the proteasome). In this respect, Ntn hydrolases resemble inteins, the self-exision elements of protein splicing, but the mechanisms utilized by the two families of proteins appear to be different. In all Ntn hydrolases, the amide backbone group of a residue at the C-terminal end of strand β4 (Gly47 in *Thermoplasma* β) is involved in forming the oxyanion hole.

In addition to these generally shared features, the proteasome has some unique aspects:

1. Besides the N-terminal threonine, it requires two further residues for activity, whose exact role remains to be clarified (Seemüller *et al.*, 1995b, 1996): a lysine residue (Lys33 in *Thermoplasma* β) and a negatively charged residue (Glu17 in *Thermoplasma* β). These form a salt bridge across the bottom of the

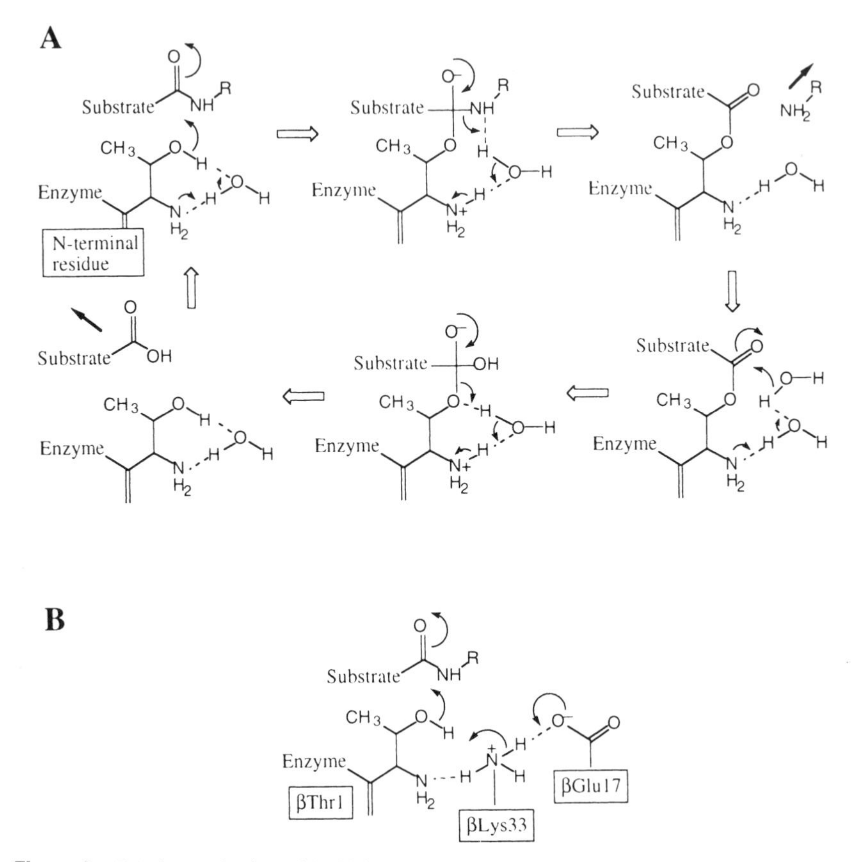

Figure 6. Putative mechanism of the 20 S proteasome. Panel A shows the general mechanism of Ntn hydrolases and panel B the potential involvement of Lys33 and Glu17 as a charge relay system in the delocalization of the Thr1 side-chain proton.

active site and either may lower the pK_a of the N-terminus by electrostatic effects, thus facilitating its function as a reversible proton acceptor, or may actually participate in the delocalization of the threonine side-chain proton by forming a charge relay system (Fig. 6B; Löwe *et al.*, 1995). A third residue (Asp166 in *Thermoplasma* β) appears to also be involved in formation of the active site, probably via electrostatic effects, as its mutation in *Thermoplasma* inactivates the proteasome (Seemüller *et al.*, 1996). This aspartate is conserved in all active β-type subunits except HslV.

2. Although all proteasome subunits have the Ntn hydrolase fold, only β-type subunits are active; in α-type subunits, the active-site cleft is blocked by the

N-terminal helical extension. Of the different β-type subunits, the prokaryotic ones are all active. The active-site residues of the eukaryotic ones, however, have a pattern of evolutionary conservation that indicates that of the seven β-type subunits, only three have proteolytic activity (Fig. 5) (Seemüller *et al.*, 1995b). This is confirmed by the observation that in the yeast crystal structure, these three subunits contain the proteasome inhibitor N-acetyl-Leu-Leu-norleucinal bound covalently to their N-terminal threonine (Groll *et al.*, 1997). The function of the inactive β-subunits is still unclear. Mutagenesis studies have shown that each of the three major peptidase activities can be abolished by mutation of either an active or an inactive β-subunit (Heinemeyer *et al.*, 1991, 1993; Dick *et al.*, 1992; Hilt *et al.*, 1993; Enenkel *et al.*, 1994; Fenteany *et al.*, 1995), and in the yeast crystal structure, these pairs of subunits are neighbors within their rings. This suggest that subunits are very sensitive to conformational changes in their neighbors, pointing to a regulatory role of inactive subunits via allosteric changes.

3. The proteasome is the only Ntn hydrolase that forms large oligomers and segregates its active sites to an inner compartment. It is a slow enzyme, which functions in a processive manner. After taking up a substrate protein, it degrades this entirely to peptides before attacking the next molecule. At V_{max}, digestion of a protein of average size, such as casein, requires approximately 1 to 2 min.

The determination of the yeast crystal structure has led to the proposal of a second kind of proteolytic active site in the proteasome, located in the loops forming the β-subunit ring constriction (Groll *et al.*, 1997). The proposal was made from the observation that the propeptides of the inactive β-type subunits β7 and β6, which are processed only to an intermediate form, bind fairly tightly to the constriction loop of β7, pointing to this as the site of endoproteolysis. This proposal must be viewed with great caution given that the constriction loops are highly variable in length and sequence, are largely solvent-exposed, and would have to process substrates in at least two orientations (because the two propeptides approach the β7 loop from different sides). Also, with the propeptides bound tightly to it, at least the β7 loop would be unable to further participate in proteolysis. A different explanation for the tight binding of the propeptides to the β7 constriction loop would be that they are processed by active subunits in their vicinity and must afterwards be immobilized at a different site, lest they continuously drift back into the active sites and hinder their activity.

5. SUBSTRATE SPECIFICITY

Eukaryotic proteasomes have three major peptidase activities that have been defined using fluorogenic peptides: a "chymotryptic" activity, which cleaves after large hydrophobic residues, a "tryptic" activity, which cleaves after basic residues, and a "peptidylglutamyl-hydrolyzing" activity, which cleaves after acidic residues (reviewed by Cardozo, 1993). These activities are the result of different active sites and are affected differently by inhibitors and mutations. Mutational studies in yeast have assigned the "peptidylglutamyl-hydrolyzing" activity to β1, the "tryptic" activity to β2, and the "chymotryptic" activity to β5 (Heinemeyer *et al.*, 1991, 1993; Hilt *et al.*, 1993; Enenkel *et al.*, 1994). Two additional specificities have been determined for mammalian proteasomes, involving cleavage after branched-chain amino acids (BrAAP activity) and between small neutral amino acids (SNAAP activity) (Orlowski *et al.*, 1993); these may be due to the γ-interferon-inducible subunits, which are present constitutively at various levels in various tissues. The proteasomes of *Thermoplasma* and of *Rhodococcus* have only one major, "chymotryptic" activity (Dahlmann *et al.*, 1992; Tamura *et al.*, 1995), in accordance with the fact that they have only one type of active site.

Although fluorogenic peptides have been useful for the characterization of the different proteolytic activities, it should be noted that they have little (if any) relevance for the cleavage specificities in nonsynthetic peptides and proteins. An analysis of the degradation products of oxidized insulin B chain by *Thermoplasma* proteasomes has shown that most peptide bonds can be cleaved by the proteasome, and that many of the preferred sites cannot be described in terms of a "chymotryptic" activity (Wenzel *et al.*, 1994). Thus, of the 29 different peptide bonds in insulin B chain, 7 were cleaved rapidly and 16 after prolonged incubation. The three bonds cleaved most frequently at brief incubation times were Gly–Ser, Glu–Ala, and Val–Cys (E. Seemüller and W. Baumeister, unpublished). Insulin B chain is also a much better substrate for the proteasome and is degraded with k_{cat}/K_m value of 25 $mM^{-1}s^{-1}$ at 60°C, pH 7.5, whereas the best fluorogenic peptide substrate, Suc-Leu-Leu-Val-Tyr-AMC, is degraded with a k_{cat}/K_m value of 0.35 $mM^{-1}s^{-1}$ (I. Dolenc and W. Baumeister, unpublished).

Similar results are obtained with eukaryotic proteasomes: Incubation of oxidized insulin B chain with purified human 20 S proteasomes led to cleavage at 20 different bonds after 15 min and at essentially every position after 4 hr (Ehring *et al.*, 1996). Again, many of the major cleavage sites were not interpretable in terms of the specificities described by experiments with fluorogenic peptides. The same is true for the cleavage patterns observed in human histone H3 (Ehring *et al.*, 1996), ovalbumin (Dick *et al.*, 1994), gonadotropin-releasing hormone (Leibovitz *et al.*, 1995), bradykinin (Zolfaghari *et al.*, 1987), and neurotensin (Cardozo *et al.*, 1992). The classification of cleavage specificities using the residue in the P1 position clearly falls short of reality. Rather, residues in the P3 and P4 positions are likely to contribute to the selection of a cleavage site by the proteasome (Cardozo *et al.*, 1994; Coux *et al.*, 1996; Tsubuki *et al.*, 1996; Groll *et al.*, 1997). The P2 position is solvent-exposed, unconstrained by the active-site cleft, and therefore unlikely to contribute to the cleavage specificity (Groll *et al.*, 1997). So far, nothing is known on the influence of the P′ positions in determining preferred cleavage sites.

6. PROCESSING AND ASSEMBLY

Like most proteases, proteasome β-type subunits are synthesized in an inactive form containing a propeptide. To prevent proteolytic damage to cellular proteins, processing is coupled to proteasome assembly. This ensures that, once processed, the activity of β-type subunits is confined to the innermost cavity of a molecular microcompartment, well segregated from the cellular environment. As in other (and possibly all) Ntn hydrolases, the processing step is autocatalytic.

The events that lead from individual subunits to mature proteasome complexes are best understood for the archaebacterial proteasome from *Thermoplasma* (Seemüller *et al.*, 1996; Zwickl *et al.*, 1994). Here, the α-subunits can associate into seven-membered rings in the absence of β-subunits, and can further aggregate into double rings, which mimic the interaction between α- and β-rings in the native proteasome (Zwickl *et al.*, 1994). The association of α-subunits is dependent on the N-terminal helical extension and deletion of this helix leads to monomeric subunits. Although α-rings may form the template on which the β-subunits are assembled, their role as an assembly intermediate *in vivo* remains unproven. In the absence of α-subunits, β-subunits fail to either assemble or cleave their propeptide. Indeed, there are indications that the β-subunits may not even fold properly without the assistance of α-subunits. In the presence of α-subunits, β-subunits can be processed and assembled both *in vivo* and *in vitro* (Seemüller *et al.*, 1996). The assembly requires neither the presence of the prosequence nor its cleavage, and subunits genetically deleted for the prosequence as well as unprocessed subunits can be efficiently incorporated into nascent proteasome complexes. Indeed, the assembly proceeds normally even if the prosequence is replaced genetically by an entirely unrelated sequence—the N-terminal extension of α-subunits. Nevertheless, processed subunits are preferentially incorporated into proteasomes, probably because the processing complex represents an assembly intermediate. The processing reaction requires the active-site residues of the mature proteasome (but tolerates the replacement of the N-terminal threonine by cysteine or serine), the glycine in the cleavage-site sequence Gly-Thr, and a partly unfolded site of the subunit to be processed. The reaction is relatively insensitive to partial deletion of the prosequence and even to its complete replacement with the N-terminal extension of α-subunits. Although it is not yet clear whether the processing reaction is intra- (*cis*) or intermolecular (*trans*) *in vivo*, the coexpression of an active with an inactive β-subunit leads to efficient processing of the inactive subunit, suggesting that the normal processing reaction in the wild-type case is also intermolecular.

In eukaryotes, processing and assembly of proteasomes present some differences, mainly related to the increase in complexity. Here, prosequences are generally much longer than the eight residues found in *Thermoplasma* and appear to fulfill other functions in addition to muzzling the proteasome β-subunits. This becomes clear from the fact that the proteasome from *Dictyostelium* cannot be reconstituted *in vitro* after dissociation (Schauer *et al.*, 1993), as the archaebacterial one can (Grziwa *et al.*, 1994), showing that, once processed, its subunits lose the ability to find their correct place in the complex. In yeast, expression of the DOA3 subunit without its prosequence results in a failure to incorporate this subunit into proteasomes, although, surprisingly, the prosequence can correct this defect if provided as a separate polypeptide (Chen and Hochstrasser, 1996). DOA3 also fails to be incorporated if its prosequence is replaced by the prosequence of another yeast β-type subunit, PRE3. It has been proposed that this is related to a subunit-specific chaperone activity of the prosequences (Chen and Hochstrasser, 1996; Schauer *et al.*, 1993), but the evidence for this remains scant. In contrast to *Thermoplasma* proteasomes, for which no assembly intermediates have been observed, eukaryotic proteasomes assemble via 15 S precursor particles, which may correspond to "half-proteasomes" (Frentzel *et al.*, 1994). These contain unprocessed β-subunits, indicating that processing is coupled to the association of these particles into complete proteasomes.

In eubacteria finally, the study of the assembly pathway has made great progress recently and has revealed a surprising similarity between the pathways in *Rhodococcus* and in eukaryotes (F. Zühl, *et al.*, 1997b). *In vitro*, the *Rhodococcus* proteasome assembles via half-proteasome intermediates consisting of an α- and a β-subunit ring. The β-subunits are still unprocessed at this stage and are activated by autocatalytic cleavage of their prosequence during the association of half-proteasomes to full complexes. The prosequences appear to play an active role in every step of the assembly pathway prior to their cleavage and are fully functional if provided as separate polypeptides. The assembly of proteasomes from any combination of *Rhodococcus* α- and β-subunits *in vitro* is essentially quantitative.

7. CONCLUDING REMARKS

The segregation of proteolytic activity into "molecular organelles," for which the proteasome has become a paradigm, is obviously an ancient evolutionary principle and has led to a convergence in molecular architecture of many proteolytic systems, such as ClpP (Kessel *et al.*, 1995), HslV (Rohrwild *et al.*,

1997; Bochtler *et al.*, 1997), bleomycin hydrolase (Joshua-Tor *et al.*, 1995), or tricorn protease (Tamura *et al.*, 1996). It provides a cell with the possibility to protect itself from random proteolysis while at the same time adding a further level of regulation to the degradation of cellular proteins. In addition, this architecture provides a functional coupling to the ATP-dependent unfolding systems, which, being under similar evolutionary pressures, have also converged toward a barrel-like shape.

8. REFERENCES

Arrigo, A. P., Tanaka, K., Goldberg, A. L., and Welch, W. J., 1988, Identity of the 19S 'prosome' particle with the large multifunctional protease complex of mammalian cells (the proteasome), *Nature* **331:**192–194.

Baumeister, W., Dahlmann, B., Hegerl, R., Kopp, F., Kuehn, L., and Pfeifer, G., 1988, Electron microscopy and image analysis of the multicatalytic protease, *FEBS Lett.* **241:**239–245.

Benoist, P., Muller, A., Diem, H. G., and Schwencke, J., 1992, High-molecular-mass multicatalytic proteinase complexes produced by the nitrogen-fixing actinomycete *Frankia* strain BR, *J. Bacteriol.* **174:**1495–1504.

Bochtler, M., Ditzel, L., Groll, M., and Huber, R., 1997, Crystal structure of heat shock locus V (HslV) from Escherichia coli, *Proc. Natl. Acad. Sci. USA* **94:**6070–6074.

Brannigan, J. A., Dodson, G., Duggleby, H. J., Moody, P. C. E., Smith, J. L., Tomchick, D. R., and Murzin, A. G., 1995, A protein catalytic framework with an N-terminal nucleophile is capable of self-activation, *Nature* **378:**416–419.

Cardozo, C., 1993, Catalytic components of the bovine pituitary multicatalytic proteinase complex (proteasome), *Enzyme Protein* **47:**296–305.

Cardozo, C., Vinitsky, A., Hidalgo, M. C., Michaud, C., and Orlowski, M., 1992, A 3,4-dichloroisocoumarin-resistant component of the multicatalytic proteinase complex, *Biochemistry* **31:**7373–7380.

Cardozo, C., Vinitsky, A., Michaud, C., and Orlowski, M., 1994, Evidence that the nature of amino acid residues in the P_3 position directs substrates to distinct catalytic sites of the pituitary multicatalytic proteinase complex (proteasome), *Biochemistry* **33:**6483–6489.

Chen, P., and Hochstrasser, M., 1995, Biogenesis, structure and function of the yeast 20S proteasome, *EMBO J.* **14:**2620–2630.

Chen, P., and Hochstrasser, M., 1996, Autocatalytic subunit processing couples active-site formation in the 20S proteasome to completion of assembly, *Cell* **86:**961–972.

Coux, O., Tanaka, K., and Goldberg, A. L., 1996, Structure and functions of the 20S and 26S proteasomes, *Annu. Rev. Biochem.* **65:**801–847.

Dahlmann, B., Kopp, F., Kuehn, L., Niedel, B., Pfeifer, G., Hegerl, R., and Baumeister, W., 1989, The multicatalytic proteinase (prosome) is ubiquitous from eukaryotes to archaebacteria, *FEBS Lett.* **251:**125–131.

Dahlmann, B., Kuehn, L., Grziwa, A., Zwickl, P., and Baumeister, W., 1992, Biochemical properties of the proteasome from *Thermoplasma acidophilum*, *Eur. J. Biochem.* **208:**789–797.

Dick, L. R., Moomaw, C. R., Pramanik, B. C., DeMartino, G. N., and Slaughter, C. A., 1992, Identification and localization of a cysteinyl residue critical for the trypsin-like catalytic activity of the proteasome, *Biochemistry* **31:**7347–7355.

Dick, L. R., Aldrich, C., Jameson, S. C., Moomaw, C. R., Pramanik, B. C., Doyle, C. K., DeMartino,

G. N., Bevan, M. J., Forman, J. M., and Slaughter, C. A., 1994, Proteolytic processing of ovalbumin and β-galactosidase by the proteasome to yield antigenic peptides, *J. Immunol.* **152:**3884–3894.

Duggleby, H. J., Tolley, S. P., Hill, C. P., Dodson, E. J., Dodson, G., and Moody, P. C. E., 1995, Penicillin acylase has a single-amino-acid catalytic centre, *Nature* **373:**264–265.

Ehring, B., Meyer, T. H., Eckerskorn, C., Lottspeich, F., and Tampe, R., 1996, Effects of major-histocompatibility-complex-encoded subunits on the peptidase and proteolytic activities of human 20S proteasomes—Cleavage of proteins and antigenic peptides, *Eur. J. Biochem.* **235:**404–415.

Enenkel, C., Lehmann, H., Kipper, J., Guckel, R., Hilt, W., and Wolf, D. H., 1994, PRE3, highly homologous to the human major histocompatibility complex-linked LMP2 (RING12) gene, codes for a yeast proteasome subunit necessary for the peptidylglutamyl-peptide hydrolyzing activity, *FEBS Lett.* **341:**193–196.

Fenteany, G., Standaert, R. F., Lane, W. S., Choi, S., Corey, E. J., and Schreiber, S. L., 1995, Inhibition of proteasome activities and subunit-specific amino-terminal threonine modification by lactacystin, *Science* **268:**726–731.

Frentzel, S., Pesold, H. B., Seelig, A., and Kloetzel, P. M., 1994, 20S proteasomes are assembled *via* distinct precursor complexes. Processing of LMP2 and LMP7 proproteins takes place in 13–16S preproteasome complexes, *J. Mol. Biol.* **236:**975–981.

Groll, M., Ditzel, L., Löwe, J., Stock, D., Bochtler, M., Bartunik, H. D., and Huber, R., 1997, Structure of the 20S proteasome from yeast at 2.4Å resolution, *Nature* **386:**463–471.

Grziwa, A., Baumeister, W., Dahlmann, B., and Kopp, F., 1991, Localization of subunits in proteasomes from *Thermoplasma acidophilum* by immunoelectron microscopy, *FEBS Lett.* **290:**186–190.

Grziwa, A., Maack, S., Puhler, G., Wiegand, G., Baumeister, W., and Jaenicke, R., 1994, Dissociation and reconstitution of the *Thermoplasma* proteasome, *Eur. J. Biochem.* **223:**1061–1067.

Harris, J. R., 1968, Release of a macromolecular protein component from human erythrocyte ghosts, *Biochim. Biophys. Acta* **150:**534–537.

Harris, J. R., 1988, Erythrocyte cylindrin: Possible identity with the ubiquitous 20S high molecular weight protease complex and the prosome particle, *Indian J. Biochem. Biophys.* **25:**459–466.

Hase, J., Kobashi, K., Nakai, N., Iwata, K., and Takadera, T., 1980, The quaternary structure of carp muscle alkaline protease, *Biochim. Biophys. Acta* **611:**205–213.

Heinemeyer, W., Kleinschmidt, J. A., Saidowsky, J., Escher, C., and Wolf, D. H., 1991, Proteinase yscE, the yeast proteasome/multicatalytic-multifunctional proteinase: Mutants unravel its function in stress induced proteolysis and uncover its necessity for cell survival, *EMBO J.* **10:**555–562.

Heinemeyer, W., Gruhler, A., Mohrle, V., Mahe, Y., and Wolf, D. H., 1993, PRE2, highly homologous to the human major histocompatibility complex-linked RING10 gene, codes for a yeast proteasome subunit necessary for chymotryptic activity and degradation of ubiquitinated proteins, *J. Biol. Chem.* **268:**5115–5120.

Hilt, W., and Wolf, D. H., 1995, Proteasomes of the yeast *Saccharomyces cerevisiae*—Genes, structure and functions, *Mol. Biol. Rep.* **21:**3–10.

Hilt, W., Enenkel, C., Gruhler, A., Singer, T., and Wolf, D. H., 1993, The PRE4 gene codes for a subunit of the yeast proteasome necessary for peptidylglutamyl-peptide hydrolyzing activity. Mutations link the proteasome to stress- and ubiquitin-dependent proteolysis, *J. Biol. Chem.* **268:**3479–3486.

Ishiura, S., Sano, M., Kamakura, K., and Sugita, H., 1985, Isolation of two forms of the high-molecular-mass serine protease, ingensin, from porcine skeletal muscle, *FEBS Lett.* **189:**119–123.

Joshua-Tor, L., Xu, H. E., Johnston, S. A., and Rees, D. C., 1995, Crystal structure of a conserved protease that binds DNA: The bleomycin hydrolase, Gal6, *Science* **269:**945–950.

Kessel, M., Maurizi, M. R., Kim, B., Kocsis, E., Trus, B. L., Singh, S. K., and Steven, A. C., 1995, Homology in structural organization between *Escherichia coli* ClpAP protease and the eukaryotic 26S proteasome, *J. Mol. Biol.* **250:**587–594.

Kessel, M., Wu, W., Gottesman, S., Kocsis, E., Steven, A. C., and Maurizi, M. R., 1996, Six-fold rotational symmetry of ClpQ, the *E. coli* homolog of the 20S proteasome, and its ATP-dependent activator, ClpY, *FEBS Lett.* **398:**274–278.

Kopp, F., Dahlmann, B., and Hendil, K. B., 1993, Evidence indicating that the human proteasome is a complex dimer, *J. Mol. Biol.* **229:**14–19.

Kopp, F., Kristensen, P., Hendil, K. B., Johnsen, A., Sobek, A., and Dahlmann, B., 1995, The human proteasome subunit hsn3 is located in the inner rings of the complex dimer, *J. Mol. Biol.* **248:** 264–272.

Kopp, F., Hendil, K. B., Dahlmann, B., Kristensen, B., Sobek, A., and Uerkvitz, W., 1997, Subunit arrangement in the human 20S proteasome, *Proc. Natl. Acad. Sci. USA* **94:**2939–2944.

Lee, L. W., Moomaw, C. R., Orth, K., McGuire, M. J., DeMartino, G. N., and Slaughter, C. A., 1990, Relationships among the subunits of the high molecular weight proteinase, macropain (proteasome), *Biochim. Biophys. Acta* **1037:**178–185.

Leibovitz, D., Koch, Y., Fridkin, M., Pitzer, F., Zwickl, P., Dantes, A., Daumeister, W., and Amsterdam, A., 1995, Archaebacterial and eukaryotic proteasomes prefer different sites in cleaving gonadotropin-releasing-hormone, *J. Biol. Chem.* **270:**11029–11032.

Lilley, K. S., Davison, M. D., and Rivett, A. J., 1990, N-terminal sequence similarities between components of the multicatalytic proteinase complex, *FEBS Lett.* **262:**327–329.

Löwe, J., Stock, D., Jap, B., Zwickl, P., Baumeister, W., and Huber, R., 1995, Crystal structure of the 20S proteasome from the archaeon *Thermoplasma acidophilum* at 3.4 Å resolution, *Science* **268:** 533–539.

Lupas, A., Zwickl, P., and Baumeister, W., 1994, Proteasome sequences in eubacteria, *Trends Biochem. Sci.* **19:**533–534.

McGuire, M. J., and DeMartino, G. N., 1986, Purification and characterization of a high molecular weight proteinase (macropain) from human erythrocytes, *Biochim. Biophys. Acta* **873:**279–289.

Monaco, J. J., and McDevitt, H. O., 1984, H-2-linked low-molecular weight polypeptide antigens assemble into an unusual macromolecular complex, *Nature* **309:**797–799.

Nederlof, P. M., Wang, H. R., and Baumeister, W., 1995, Nuclear-localization signals of human and *Thermoplasma* proteasomal α-subunits are functional *in vitro*, *Proc. Natl. Acad. Sci. USA* **92:** 12060–12064.

Orlowski, M., and Wilk, W., 1988, Multicatalytic proteinase complex or multicatalytic proteinase: A high M_r endopeptidase, *Biochem. J.* **255:**751.

Orlowski, M., Cardozo, C., and Michaud, C., 1993, Evidence for the presence of five distinct proteolytic components in the pituitary multicatalytic proteinase complex. Properties of two components cleaving bonds on the carboxyl side of branched chain and small neutral amino acids, *Biochemistry* **32:**1563–1572.

Pamnani, V., Haas, B., Puhler, G., Sanger, H. L., and Baumeister, W., 1994, Proteasome-associated RNAs are non-specific, *Eur. J. Biochem.* **225:**511–519.

Peters, J. M., Cejka, Z., Harris, J. R., Kleinschmidt, J. A., and Baumeister, W., 1993, Structural features of the 26S proteasome complex, *J. Mol. Biol.* **234:**932–937.

Pouch, M. N., Petit, F., Buri, J., Briand, Y., and Schmid, H. P., 1995, Identification and initial characterization of a specific proteasome (prosome) associated RNAse activity, *J. Biol. Chem.* **270:**22023–22028.

Pühler, G., Weinkauf, S., Bachmann, L., Müller, S., Engel, A., Hegerl, R., and Baumeister, W., 1992, Subunit stoichiometry and three-dimensional arrangement in proteasomes from *Thermoplasma acidophilum*, *EMBO J.* **11:**1607–1616.

Pühler, G., Pitzer, F., Zwickl, P., and Baumeister, W., 1994, Proteasomes—Multisubunit proteinases common to *Thermoplasma* and eukaryotes, *Syst. Appl. Microbiol.* **16:**734–741.

Rohrwild, M., Coux, O., Huang, H. C., Moerschell, R. P., Yoo, S. J., Seol, J. H., Chung, C. H., and Goldberg, A. L., 1996, HslV-HslU—A novel ATP-dependent protease complex in *Escherichia coli* related to the eukaryotic proteasome, *Proc. Natl. Acad. Sci. USA* **93:**5808–5813.

Rohrwild, M., Pfeifer, G., Santarius, U., Müller, S. A., Huang, H.-C., Engel, A., Baumeister, W., and

Goldberg, A. L., 1997, The ATP-dependent HslVU protease from *Escherichia coli* is a four-ring structure resembling the proteasome, *Nature Struct. Biol.* **4**:133–139.

Schauer, T. M., Nesper, M., Kehl, M., Lottspeich, F., Müller, T. A., Gerisch, G., and Baumeister, W., 1993, Proteasomes from *Dictyostelium discoideum*: Characterization of structure and function, *J. Struct. Biol.* **111**:135–147.

Schmid, H. P., Akhayat, O., Martins De Sa, C., Puvion, F., Koehler, K., and Scherrer, K., 1984, The prosome: A ubiquitous morphologically distinct RNP particle associated with repressed mRNPs and containing ScRNA and a characteristic set of proteins, *EMBO J.* **3**:29–34.

Seemüller, E., Lupas, A., Zuhl, F., Zwickl, P., and Baumeister, W., 1995a, The proteasome from *Thermoplasma acidophilum* is neither a cysteine nor a serine protease, *FEBS Lett.* **359**:173–178.

Seemüller, E., Lupas, A., Stock, D., Lowe, J., Huber, R., and Baumeister, W., 1995b, Proteasome from *Thermoplasma acidophilum*—A threonine protease, *Science* **268**:579–582.

Seemüller, E., Lupas, A., and Baumeister, W., 1996, Autocatalytic processing of the 20S proteasome, *Nature* **382**:468–470.

Stock, D., Nederlof, P., Seemüller, E., Baumeister, W., Huber, R., and Löwe, J., 1996, Proteasome: From structure to function, *Curr. Opin. Biotech.* **7**:376–385.

Tamura, T., Nagy, I., Lupas, A., Lottspeich, F., Cejka, Z., Schoofs, G., Tanaka, K., Demot, R., and Baumeister, W., 1995, The first characterization of a eubacterial proteasome—The 20s complex of rhodococcus, *Curr. Biol.* **5**:766–774.

Tamura, T., Tamura, N., Cejka, Z., Hegerl, R., Lottspeich, F., and Baumeister, W., 1996, Tricorn protease—the core of a modular proteolytic system, *Science* **274**:1385–1389.

Tanaka, K., 1995, Molecular biology of proteasomes, *Mol. Biol. Rep.* **21**:21–26.

Tsubuki, S., Saito, Y., Tomioka, M., Ito, H., and Kawashima, S., 1996, Differential inhibition of calpain and proteasome activities by peptidyl aldehydes of di-leucine and tri-leucine, *J. Biol. Chem.* **119**:572–576.

Wenzel, T., and Baumeister, W., 1995, Conformational constraints in protein degradation by the 20S proteasome, *Nature Struct. Biol.* **2**:199–204.

Wenzel, T., Eckerskorn, C., Lottspeich, F., and Baumeister, W., 1994, Existence of a molecular ruler in proteasomes suggested by analysis of degradation products, *FEBS Lett.* **349**:205–209.

Zolfaghari, R., Baker, C. R. F., Jr., Canizaro, P. C., Amirgholami, A., and Behal, F. J., 1987, A high-molecular-mass neutral endopeptidase-24.5 from human lung, *Biochem. J.* **241**:129–135.

Zühl, F., Tamura, T., Dolenc, I., Cejka, Z., Nagy, I., De Mot, R., and Baumeister, W., 1997a, Subunit topology of the *Rhodococcus* proteasome, *FEBS Lett.* **400**:83–90.

Zühl, F., Seemüller, E., Golbik, R., and Baumeister, W., 1997b, Dissecting the assembly pathway of the 20S proteasome. *FEBS Lett.* **418**:189–194.

Zwickl, P., Lottspeich, F., Dahlmann, B., and Baumeister, W., 1991, Cloning and sequencing of the gene encoding the large (α-) subunit of the proteasome from *Thermoplasma acidophilum*, *FEBS Lett.* **278**:217–221.

Zwickl, P., Grziwa, A., Pühler, G., Dahlmann, B., Lottspeich, F., and Baumeister, W., 1992, Primary structure of the *Thermoplasma* proteasome and its implication for the structure, function, and evolution of the multicatalytic proteinase, *Biochemistry* **31**:964–972.

Zwickl, P., Kleinz, J., and Baumeister, W., 1994, Critical elements in proteasome assembly, *Nature Struct. Biol.* **1**:765–770.

CHAPTER 6

The 26 S Proteasome

Martin Rechsteiner

1. INTRODUCTION

The 26 S proteasome is a large multisubunit complex that degrades intracellular proteins in a nucleotide-dependent reaction. The enzyme is composed of about 30 different proteins. Fourteen subunits are provided by the proteasome and an additional 16 or more subunits are present in a particle that I will call the *regulatory complex*. The 26 S enzyme is the only ATP-dependent protease so far identified in the nuclear and cytosolic compartments of eukaryotic cells—a surprising observation in view of the fact that at least four ATP-dependent proteases are present in *E. coli* cells (i.e., Lon, Clp, Hfl, and HslV). The 26 S proteasome has been shown to be responsible for the degradation of more than 20 specific proteins. Among the known substrates are a variety of important regulatory proteins including transcription factors, cyclins, and cdk inhibitors. The enzyme may well be primarily responsible for the *selective* degradation of eukaryotic proteins.

The 26 S proteasome was discovered by its ability to hydrolyze ubiquitin–lysozyme conjugates. The existence of such an enzyme was implicit in the ubiquitin-marking hypothesis (Hershko *et al.*, 1980). In 1978, Ciechanover *et al.*, demonstrated that ATP-dependent proteolysis in rabbit reticulocyte lysate re-

Martin Rechsteiner • Department of Biochemistry, University of Utah School of Medicine, Salt Lake City, Utah 84132.

Ubiquitin and the Biology of the Cell, edited by Peters *et al.* Plenum Press, New York, 1998.

quired APF1, a small protein factor. When added to reticulocyte lysate, radioiodinated APF1 (now known to be ubiquitin) became covalently attached to a number of proteins, including exogenous substrates (Ciechanover *et al.*, 1980). This finding led Hershko *et al.* (1980) to propose that ubiquitin marks proteins for destruction—a very important concept that provides the theme unifying many chapters in this book. Within 3 years of this proposal, Hershko *et al.* (1983) had identified the E1, E2, and E3 enzymes required to activate and couple ubiquitin to proteolytic substrates. An additional 3 years passed before the enzyme that degrades ubiquitylated proteins was identified (Hough *et al.*, 1986). Its discovery was facilitated by preparation of a specific substrate, ubiquitin–lysozyme conjugates. In 1981, Haas and Rose reported that hemin inhibited the ATP-dependent degradation of [^{125}I]RNase despite the fact that the substrate was ubiquitylated. Using their procedure, Hough and Rechsteiner (1986) prepared ubiquitin–lysozyme conjugates in hemin-inhibited reticulocyte lysate. The partially purified conjugates proved to be ideal substrates for a novel proteolytic activity present in rabbit reticulocyte extracts lacking hemin. The protease degraded the ubiquitin–lysozyme conjugates in the presence of ATP, but it did not cleave unmodified lysozyme (Hough *et al.*, 1986). Thus, its substrate specificity matched that of the protease predicted by the ubiquitin-marking hypothesis.

The ATP-dependent protease, purified a year later by Hough *et al.* (1987), was found to be an exceptionally large enzyme. Gel filtration and sedimentation on glycerol gradients indicated that the particle had an apparent S value of 26 and a mass greater than 10^6 Da. Throughout purification the 26 S proteasome was accompanied by an energy-independent protease sedimenting at 20 S. The smaller enzyme was identified as the multicatalytic protease described earlier by Wilk and Orlowski (1980)—an enzyme now most frequently called the *proteasome*. SDS–PAGE analyses of each protease revealed that the 20 S enzyme was composed of subunits with molecular masses between 20 and 30 kDa, whereas the 26 S proteasome contained at least 10 subunits with apparent molecular masses between 40 and 110 kDa in addition to the smaller subunits characteristic of the proteasome. Based on the gel patterns, Hough *et al.* (1987) suggested that the 26 S proteasome and the proteasome share subunits.

During the past 10 years a number of research groups have contributed to our current understanding of the 26 S proteasome. It is now widely accepted that the proteasome is a component of the larger 26 S enzyme. The 26 S complex has been examined by electron microscopy and cDNAs encoding almost all of the 26 S proteasome subunits have been isolated and sequenced. The recently published X-ray structure of the archaebacterial proteasome (Löwe *et al.*, 1995) is a major achievement that has generated a number of recent reviews (Hilt and Wolf, 1996; Coux *et al.*, 1996; Jentsch and Schlenker, 1995; Rubin and Finley, 1995; Weissman *et al.*, 1995; see Chapter 5). For this reason, I will not provide a comprehensive review here. Rather, I will focus on various structural and functional aspects of the

26 S proteasome. I will also speculate on possible biochemical mechanisms used in the degradation of intracellular proteins.

2. ELECTRON MICROSCOPIC IMAGES OF THE 26 S PROTEASOME

The 26 S proteasome was first visualized more than 25 years ago. In their studies on aminoacyltransferases, Shelton *et al.* (1970) used electron microscopy to examine particles obtained by sucrose density gradient centrifugation of HeLa cell extracts. They described images of a mushroom-shaped particle that sedimented at 26 S. Based on their published micrographs and the subunit stoichiometry of the rabbit reticulocyte 26 S proteasome, Hough *et al.* (1988) proposed a model for the 26 S proteasome in which a large "spherical" particle was attached to two rings of subunits provided by the multicatalytic protease or proteasome. It now seems that this initial model is incorrect as a number of subsequent electron microscopic analyses indicate that all four rings of the proteasome are present in the 26 S proteasome. For example, Ikai *et al.* (1991) described the rat liver 26 S proteasome as a barbell-shaped structure 40 nm in length. The central "bar" portion of the enzyme contains four rings that in their model are provided by the proteasome. Attached to each end of the proteasome is a larger structure 12 nm long and 20 nm wide. Peters *et al.* (1991) reported a similar model for the 26 S proteasome from *Xenopus* oocytes. That is, the enzyme is 40 nm long consisting of a central core formed by the 20 S proteasome to which large particles are attached at each end.

Follow-up studies by J. Peters in collaboration with W. Baumeister have provided a higher-resolution image of the *Xenopus* 26 S proteasome (Peters *et al.*, 1993). Using digital image analysis these authors describe a barbell-shaped structure in which the two particles attached at each end of the protease are highly asymmetric having the appearance of a Chinese dragon head motif. Interestingly, in those particles with two attached "dragon heads," the dragons face in opposite direction, implying that the regulatory complexes bind proteasome ends in a highly specific manner (see Fig. 1A). Equivalent images were also obtained with preparations of the rat liver 26 S proteasome (Yoshimura *et al.*, 1993). Thus, recent electron microscopic studies on the 26 S proteasome reveal a particle in which the ends of an intact proteasome associate with one or two regulatory complexes. Whether the functioning 26 S proteasome is capped at both ends by a regulatory complex has not been resolved. Particles in which an intact proteasome is attached to a single regulatory complex are seen in the electron microscope (Fig. 1C). Furthermore, the X-ray structure of the *Thermoplasma* proteasome does not indicate the presence of channels leading from the central proteolytic chamber formed by the β rings to the exterior of the particle (Löwe *et al.*, 1995). Thus, it is

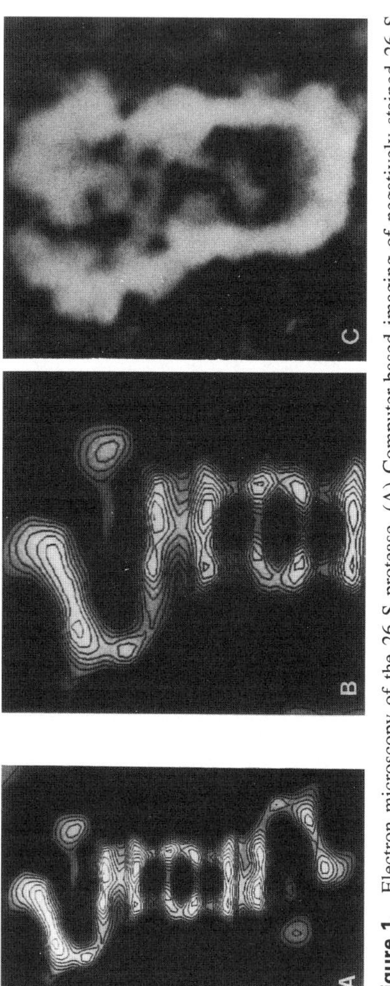

Figure 1. Electron microscopy of the 26 S protease. (A) Computer-based imaging of negatively stained 26 S proteases from *Xenopus laevis.* The particle, which is about 45 nm from top to bottom, consists of four central rings contributed by the proteasome with two regulatory complexes attached at each end in opposite orientation. (B) A cropped portion of the image in A used for comparison to the adjacent metal-shadowed 26 S protease. (C) A metal-shadowed 26 S protease from rat liver. Only one regulatory complex is attached to the proteasome. The micrograph in panels A and B was provided by Jan Peters; the micrograph in C was a gift from John Heuser.

difficult to see how peptide fragments could be released by a 26 S proteasome with regulatory complexes attached to both ends. For this reason, I favor the idea that the functional enzyme contains a single regulatory complex attached to the proteasome. It should be noted that the original model (Hough *et al.*, 1988) in which a regulatory complex binds only two rings of the proteasome cannot be rigorously excluded. In fact, that early model can readily explain product release.

3. THE PROTEASOME

The proteasome is an ~ 700-kDa complex possessing at least five endopeptidase activities that cleave bonds at the carboxyl sides of basic, hydrophobic, acidic, small neutral, and branched-chain amino acid residues (Orlowski *et al.*, 1993). The crystal structure of the proteasome from the archaebacterium *Thermoplasma acidophilum* reveals an assembly of four stacked rings each of which contains seven subunits (Löwe *et al.*, 1995). The *Thermoplasma* enzyme is composed of two related polypeptides called the α and β subunit. The outer rings are composed of α subunits and the inner rings of β subunits. Mutagenic data indicate that the catalytic activity resides in the β subunits (Seemüller *et al.*, 1995), and difference Fourier analysis of inhibitor complexes localizes the protease active site to the central chamber within the cylindrical structure (Löwe *et al.*, 1995). The archaebacterial and eukaryotic proteasomes appear to share the same α7-β7-β7-α7 ring structure, and they are very similar by electron microscopic analysis (Dahlmann *et al.*, 1989), although not all previous studies concluded that eukaryotic proteasomes possess sevenfold symmetry (Kleinschmidt *et al.*, 1983; Arrigo *et al.*, 1988; Baumeister *et al.*, 1988). Substrate access to the proteasome active sites is assumed to be through the central channel that passes along the sevenfold axis of the complex (Wenzel and Baumeister, 1995). This restricted access to the active sites can explain how indiscriminate degradation of cellular proteins is avoided and why the proteasome is inactive against folded proteins. Because an entire chapter is devoted to the proteasome, I will only discuss selected features of the enzyme from higher eukaryotes.

3.1. C-Terminal Extensions

Unlike the archaebacteria, which express only one type of α and one type of β subunit (Zwickl *et al.*, 1992), eukaryotes encode a number of different subunits that can be grouped as α-like or β-like based on homology to the archaebacterial subunits (Heinemeyer *et al.*, 1994). There are at least 7 unique α subunits and 10 β subunits in proteasomes from higher eukaryotes. Antibody decoration experiments analyzed by electron microscopy have provided strong evidence that the human proteasome is a complex dimer (Kopp *et al.*, 1993, 1995). Individual subunits are present in equal amounts, indicating that the assembled 20 S particle contains two copies of each subunit (Hendil *et al.*, 1993). It is reasonable to assume that individual subunits occupy defined positions in the α rings, and preliminary cross-linking studies support this assumption (K. Hendil, personal communication). Thus, assembled eukaryotic proteasomes may have a single quaternary structure with respect to the arrangement of individual α subunits. Whether the β subunits have specific nearest neighbors within the inner rings is an open question.

Amino acid sequence comparisons to the archaebacterial α and β subunits reveal the existence of C-terminal extensions in four α subunits and one β subunit from eukaryotic proteasomes (see Fig. 2). All four α-subunit extensions are highly

α-subunits C-terminal extensions

T.α... EVKKFL
C3....... YL. AAIA
C6....... YV. AEIEKEKEENEKKKQKKAS
C8....... YA. KESLKEEDESDDDNM
C9....... LI. KKHEEEEAKAEREKKEKEQKEKDK
C2....... FL. EGLEERPQRKAQPAQPADEPAEKADEPMEH

β-subunit C-terminal extension

T.β ... RKLGLIL
MECL-1... KRSGRYH. FVPGTTAVLTQTVKPLTELVEETVQAMEVE

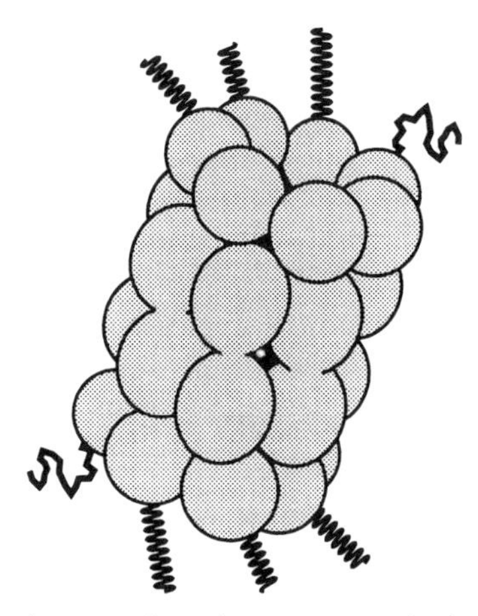

Figure 2. C-terminal extensions on eukaryotic proteasome subunits. Portions of the amino acid sequences of the *Thermoplasma* α and β proteasome subunits are compared to the longer C-termini from several human α subunits and one human β subunit, MECL-1. A schematic representation of possible conformations for the α-subunit extensions is shown below the sequences.

charged. The extensions on α subunits C6, C8, and C9 are predicted to be α-helical whereas the proline-rich extension on C2 is unlikely to form an α helix. Possible conformations and positions of these α-subunit extensions are depicted in the cartoon at the bottom of Fig. 2. The extensions on subunits C6 and C9 consist of "alternating" lysine (K) and glutamate (E) residues. These "KEKE motifs" are particularly interesting because, as discussed below, similar tracts of "alternating" glutamates and lysines are present in proteins known to associate with the proteasome, namely, the 11 S regulator (PA28) as well as five subunits in the regulatory complex (subunits 1, 2, S5a, 10, and 12). KEKE motifs are also found in various chaperones, including hsp90, which also associates with the proteasome (see Table I). It has been hypothesized that KEKE sequences mediate protein–protein interactions (Realini *et al.*, 1994a), and this idea is supported by the observation that KEKE fusions cause heterologous proteins to cosediment with recombinant 11 S regulators (Realini and Rechsteiner, unpublished).

The sequence of the single β-subunit extension is also presented in Fig. 2. This extension is found on MECL-1, and a very similar sequence occupies the C-terminus of subunit Z, which is replaced by MECL-1 on interferon-γ induction

Table I
KEKE Motifs[a]

Proteolytic components	
26 S protease S1	...KEKEKEKVSTAVLSITAKAKKKEKEKEKKEEEKMEVDEAEK KEEKEKKKE...
26 S protease S2	...KERRDAGDKDKEQELSEE...
	...KEKEEDKDKKEKKDKDKKE...
26 S protease S5a	...KDGKKDKKEEDKK
	...EDNEKDLVKLAKRLKKEK...
26 S protease S10	...KKANEDELKRLDEELE...
26 S protease S12	...EKKEGQEKEESKKDRKEDKEKDKDKEKSDVKKEEKK
Isopeptidase T	...KEELLEYEEKKRQAEEEK...
11 S regulator	...KEKEKEERKKQQEKEDKDEKKKGEDEDK...
Proteasome C6	...EIEKEKEENEKKKQK...
Proteasome C9	...KKHEEEEAKAEREKKEKEQKEKDK
HSPs/chaperonins	
HSP 90	...EEKEDKEEEKEKEEKESEDK...
Fli L	...EKKEEKKKEKKKEEKGDKKDAEK...
HSP 70	...EKLAAQRKAEAEKKEEKKDTE...
Calcium-binding proteins	
Calnexin	...EEEEKEEEKDKGDEE...
Calreticulin	...KDKQDEEQRLKEEEE...
Ca^{2+}-ATPase	...EEKKDEKKKEKK...

[a]The KEKE motifs shown are presented in the one-letter amino acid code. A KEKE motif is defined as a sequence of 13 amino acids or longer starting and ending with a Lys or Glu; greater than 60% of the residues are Lys or Glu/Asp; no more than 4 consecutive negative or positive residues are present; and the sequence is devoid of Trp, Tyr, Phe, Cys, or Pro. KEKE sequences so defined are present in 6% of the entries in an edited library of human proteins.

(Nandi *et al.*, 1996; Hisamatsu *et al.*, 1996). An interesting feature of the β-subunit extension is the presence of Gly-Thr-Thr-Ala. Several β subunits contain N-terminal extensions that are processed after a glycine followed by two or three threonines (Thomson and Rivett, 1996; Lilley *et al.*, 1990). Thus, a full-length C-terminal extension may not be present on MECL-1 or subunit Z following their assembly in the proteasome. Interestingly enough, the yeast β subunit most closely related to MECL-1 contains a C-terminal extension in which Gly-Thr-Thr-Ala is the region of greatest homology.

3.2. On the Nature of Proteasome Active Sites

The *Thermoplasma* proteasome contains 14 copies of a single α subunit and 14 copies of a unique β subunit. Although it can hydrolyze a wide range of peptide bonds (Wenzel *et al.*, 1994), the archaebacterial proteasome cleaves best small fluorogenic peptides with hydrophobic residues in the P1 position, and hence it is said to exhibit chymotrypsinlike activity (Dahlmann *et al.*, 1992). Eukaryotic proteasomes are composed of a number of different subunits, and the eukaryotic enzyme hydrolyzes a wider variety of fluorogenic peptides. There is substantial evidence that specific β subunits exhibit different substrate specificities. Soon after their initial description of the multicatalytic protease (proteasome), Wilk and Orlowski (1983) used inhibitors to identify trypsinlike, chymotrypsinlike, and peptidylglutamyl-preferring sites in the pituitary enzyme. Subsequent studies have expanded the number of sites to include branch chain-preferring (BRAAP) and small neutral-preferring (SNAAP) sites as well (Orlowski *et al.*, 1993). An extensive survey of the substrate specificities of human proteasomes from red blood cells, wild-type human lymphoblasts, and lymphoblasts lacking two β subunits, LMP2 and LMP7, has recently been published. Ustrell *et al.* (1995) found that the presence of LMP2 and LMP7 correlates with markedly greater cleavage of several fluorogenic peptides. Studies on yeast proteasomes with mutations in individual β subunits confirm that some subunits make major contributions to the hydrolysis of specific fluorogenic peptides (see Heinemeyer *et al.*, 1994, for review). Finally, the X-ray structure of the *Thermoplasma* enzyme bound to the inhibitor Ac-Leu-Leu-norleucinal places the proteolytic active site within a β subunit although it is clear that the active-site geometry is stabilized by a neighboring subunit. All of these observations provide evidence that the individual β subunits of eukaryotic proteasomes differ in their substrate specificities.

In a previous review (Rechsteiner *et al.*, 1993), we suggested that the proteasome would be found to be an atypical serine protease using a highly conserved lysine rather than histidine as the proton donor. We also proposed that Ser129 in the *Thermoplasma* β subunit might act as the nucleophile during peptide bond cleavage. The importance of the conserved lysine appears to have been correct.

Surprisingly enough, however, the proteasome uses a newly generated N-terminal threonine in catalysis. This has been shown in several ways. Baumeister and his colleagues have used site-directed mutagenesis to assess the importance of potential catalytic residues in the proteasome β subunit from *Thermoplasma* (Seemüller *et al.*, 1995). They find that mutation of the N-terminal threonine leads to folded, but inactive, β subunits. Fenteany *et al.* (1995) showed that lactacystin, a bacterial metabolite that inhibits proteasomes, forms covalent adducts to the N-terminal threonine of a human β subunit. And the X-ray structure of the *Thermoplasma* proteasome bound to the protease inhibitor, Ac-Leu-Leu-norleucinal, places the inhibitor very near the newly generated N-terminal threonine (Löwe *et al.*, 1995).

Although I do not question the importance of the N-terminal threonine in catalysis, I believe it is premature to exclude a catalytic role for Ser129 or amino acids at the equivalent position in eukaryotic proteasomes. In the first place, β subunits may well use two "catalytic centers" as they are likely to generate their mature N-termini by autoproteolysis. Consistent with this possibility, the X-ray structure of the *Thermoplasma* proteasome places Ser129 very near the N-terminal threonine (Löwe *et al.*, 1995). Also, an alignment of human β sequences reveals that the residue at "Ser129" invariably contains a hydroxyl group (i.e., it is Ser, Thr, or Tyr). Finally, it is conceivable that either or both the N-terminal threonine and "Ser129" may act as nucleophiles during proteolysis. Seemüller *et al.* (1996) have recently published a paper on autocatalytic processing of *Thermoplasma* β subunits. Based on site-directed mutagenesis and coexpression of active and inactive variants, the authors suggest that processing is intermolecular. Although their findings would seem to invalidate the arguments just presented, they did not address the role of Ser129 in self-processing. Moreover, their assumption that Lys33 is directly involved in self-processing may or may not be correct as mutation of Lys33 may cause structural changes that prevent processing.

The observation that only 6 of 10 eukaryotic β subunits have a consensus Gly-Thr-Thr site in their N-terminal extensions led Seemüller *et al.* (1995) to propose that the eukaryotic proteasome contains inactive subunits. For the reasons stated above, I suspect that this may not be the case. Two human β subunits lacking the GTT processing site, subunits C5 and N3, are nevertheless processed 8 or 9 residues N-terminal to a threonine virtually equivalent to the N-terminal threonine (Thomson and Rivett, 1996). It is possible that the catalytic threonine need not be at the N-terminus, and to my knowledge there is no evidence that C5 and N3 are inactive. Whether all β subunits are catalytically active is, of course, an important mechanistic issue that requires further study.

In summary, the overall structure of the proteasome from higher eukaryotes will almost certainly prove to be very similar to the *Thermoplasma* enzyme. Nonetheless, the presence of multiple α and β subunits and the presence of C-terminal extensions on some of these subunits should have two major consequences: First, the eukaryotic enzyme is likely to exhibit wider substrate specificity. Second, eukaryotic proteasomes probably associate with more proteins or protein complexes than can the *Thermoplasma* enzyme.

4. REGULATORY COMPLEX

4.1. Conjugate-Degrading Factors

Soon after the discovery of the 26 S proteasome, Ganoth *et al.* (1988) reported that degradation of ubiquitin–lysozyme conjugates required three components, the conjugate-degrading factors (CFs). CF1 and CF2 exhibited apparent molecular masses of 600 and 250 kDa, respectively, and ATP stabilized both components to heat inactivation. CF3 chromatographed on gel filtration columns with an apparent size of 650 kDa, and this component was more stable to high temperature. Mixtures of CF1, CF2, CF3, and ATP produced a conjugate degrading enzyme with a mass greater than 1000 kDa, indicating that the CFs assembled to form the 26 S proteasome. Subsequent studies (Eytan *et al.*, 1989; Driscoll and Goldberg, 1990) identified CF3 as the proteasome, thereby providing support for the shared subunit model of Hough *et al.* (1988).

More recently, several groups have identified a single particle that assembles with the proteasome to form the 26 S proteasome. This protein complex has been termed the *ball* (Hoffman *et al.*, 1992), the *μ particle* (Udvardy, 1993), *PA700* (Ma *et al.*, 1994), and the *19 S cap complex* (Peters *et al.*, 1994), SDS–PAGE analyses show that the particle contains the higher-molecular-weight subunits present in the 26 S proteasome. Subunits of the proteasome are absent, and the particle does not exhibit proteolytic activity when assayed with fluorogenic peptides. Because this subcomponent of the 26 S proteasome contains ATPases and substrate recognition components, I call it the *regulatory complex*. Almost everyone in the field would agree that the proteasome provides peptide cleavage functions to the 26 S proteasome, and there is a widespread belief that the regulatory complex confers substrate recognition, substrate unfolding, and the ability to transfer polypeptide chains to the protease active sites within the central axis of the proteasome. If this view is correct, then the regulatory complex is a sophisticated biochemical entity. The presence of more than 15 distinct subunits in the regulatory complex certainly supports such an idea.

4.2. Naming Subunits of the Regulatory Complex

Although there has been a claim that the 26 S proteasome contains subunits with apparent molecular masses larger than 110 kDa (Goldberg, 1992), everyone now seems to agree that the largest subunit migrates on an SDS–PAGE gel at about 110 kDa. And there is widespread agreement that the regulatory complex contains 15 subunits, give or take a few. In 1992, we introduced a simple numbering system that identified the largest subunit as S1 and the smallest as S15 (Dubiel *et al.*, 1992b). Unfortunately, the names for 26 S proteasome subunits have subsequently multiplied. In characterizing bovine PA700, which has proved to be comparable to the regulatory complex, DeMartino *et al.* (1994) introduced molecular weight designations for its subunits (e.g., p112, p97, p63). Peters *et al.* (1992) had already used p97 to name a hexameric ATPase related to the ATPases discussed below, so this molecular weight nomenclature is ambiguous. Also, the regulatory complex contains numerous subunits in the size range of 40 to 60 kDa, and designations such as p55.4 are not useful identifications given the limited resolution of PAGE gels. I believe that it is more sensible to designate the subunits S1, S2, and so on, from largest to smallest, and hope those in the field will adopt the original, simpler nomenclature.

4.3. The ATPases

4.3.1. Identifying Characteristics

Currently, we know the sequences of 14 subunits in the regulatory complex, and we have some idea of the functions of 7 of those. Six of the known subunits (S4, S6, S6′, S7, S8, and S10b) are members of a novel family of putative ATPases, called the AAA family (*A*TPases *a*ssociated with a variety of cellular *a*ctivities) (Confalonieri and Duguet, 1995). The larger family contains a diverse set of proteins that includes NSF, Pas1p, and p97. Some members of the family possess two nucleotide binding sites, others only one. AAA family members found in the 26 S proteasome are characterized by a length of ~ 400 residues, the presence of a single nucleotide binding site, and several characteristic sequence motifs. The glycine-rich P-loop is almost always GPPGXGKT; about 50 residues toward the C-terminus one finds a conserved DEID; a cysteine close to the C-terminus is present in all of the ATPases from the human 26 S proteasome. For the sake of discussion, we call the 26 S proteasome ATPases the *S4 subfamily* based on identification of its first member as a 26 S proteasome subunit (Dubiel *et al.*, 1992b).

Sequence alignment of S4 subfamily members reveals some interesting features. These proteins possess a central nucleotide binding domain that is about 60% identical between family members. The C-terminal 150 amino acids are clearly conserved but to a lesser extent (~ 40% identity). Interestingly, there is almost no homology in the first 150 amino acids (< 20% identity). Thus, sequences of the individual subfamily members can be considered tripartite with a variable region followed by a conserved nucleotide binding domain and a semiconserved C-terminus (see Fig. 3). Despite sequence differences among members of the S4 subfamily, the sequence of each ATPase has been highly conserved

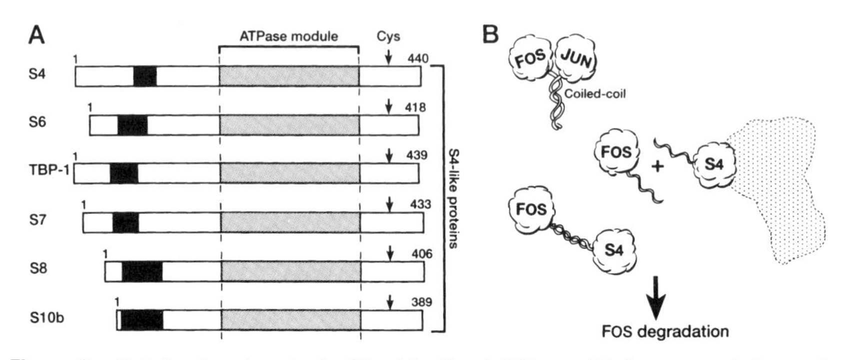

Figure 3. Coiled-coil regions in the S4 subfamily of ATPases. (A) Sequences of six putative ATPases of the regulatory complex are depicted as open lines proportional to their length. The ATPase module is represented by close-spaced diagonal lines and the coiled-coil regions are shown in black. The highly conserved cysteine residue is identified by the arrow. (B) Coiled coils may be used to target proteins for proteolytic degradation as in the hypothetical scheme shown. A mechanism for proteolytic substrate recognition by the 26 S protease is proposed by which an unpaired α helix in Fos, which normally dimerizes with a similar region on Jun, could dimerize with α helices present in ATPase subunits of the 26 S protease. It is proposed that interaction with the S4-like ATPase rather than Jun would result in degradation of Fos.

during evolution. The non-ATPase subunits, S1, S2, S5a, and so on, are only 30–45% identical between humans and yeast. By contrast, individual members of the S4 subfamily are almost 75% identical across this evolutionary gap. Moreover, the degree of conservation between human S4 and yeast S4, for example, is uniform along the sequence including the N-terminal regions that are so divergent between family members. This high degree of conservation clearly implies that the N-terminal variable regions serve important functions.

4.3.2. Roles for the N-Terminal Variable Regions

We have suggested that the N-terminal regions of the S4-like ATPases are variable because they are used for substrate selection (Hoffman and Rechsteiner, 1996; Rechsteiner *et al.*, 1993). One can imagine that some proteins are selected for destruction by binding S4 whereas others first bind S7 or S8. The presence of coiled-coil motifs in the N-terminal portion of each ATPase led us to suggest that these regions of the proteins might bind unassembled substrates through the latter's unpaired coiled-coil motifs (Hoffman and Rechsteiner, 1996; Rechsteiner *et al.*, 1993). For example, fos and jun normally form heterodimers via their leucine zipper motifs. If one of the two proteins were produced in excess, then the coiled-coil regions in an S4-like ATPase might bind the unassembled monomer and thereby promote its degradation. This idea, depicted in Fig. 3, recently received support in studies by Wang *et al.* (1996) who showed that the human homologue of SUG1 (S8) specifically binds c-fos through leucine zippers in each protein.

Whether the coiled-coil regions of the S4-like ATPases actually target proteins for destruction will only be known after further experimentation. There is additional evidence, albeit circumstantial, that supports this hypothesis. Several of the ATPases were identified by their ability to bind proteins that could well be proteolytic substrates. For example, TBP1 (S6) was discovered using biotinylated TAT protein encoded by the HIV virus (Nelbock *et al.*, 1990). Yeast two-hybrid screens identified S8 as a binding partner of the thyroid receptor (Lee *et al.*, 1995). In yeast, mutations in Sug1 affect the metabolic stability of Cdc68 protein (Xu *et al.*, 1995). And there is the interesting report that yeast cim3 and cim5 (S8 and S7) are required for the degradation of mouse ornithine decarboxylase whereas only cim3 is required for proteolysis of yeast ornithine decarboxylase (Mamroud-Kidron and Kahana, 1994). Finally, over the past 4 years I have been contacted by about 20 investigators who identified S4 as a binding partner of some intracellular protein or another not present in the 26 S proteasome.

The variable N-terminal regions might serve another purpose. They could promote interactions between members of the subfamily to ensure that each ATPase assembles with a specified neighbor. In fact, recent studies from my lab indicate that the coiled-coil regions mediate specific interactions between members of the S4 subfamily (Richmond *et al.*, 1997). Each of the six ATPases was synthesized in reticulocyte lysate containing [^{35}S] methionine and then used in binding assays to subunits of the regulatory complex separated on two-dimensional gels and transferred to nitrocellulose filters. S4 bound to S7; S6 bound to S8; S6′ (TBP1) bound S10b (SUG2). Progressive-terminal deletions of S4 up to Thr167 had no effect on its binding to S7. However, truncation of 85 N-terminal residues from S4 abolished its association with S7. The observation that specific interaction between the ATPases persists after transfer of the regulatory complex subunits to nitrocellulose and the fact that the inactivating N-terminal truncation disrupts the coiled-coil region of S4 led us to propose that the coiled-coil regions are responsible for the specific association between the ATPases. This apparent importance of the variable N-terminal regions in assembly does not preclude their playing a role in substrate selection as hypothesized above. Once the ATPases are incorporated into the regulatory complex, the N-terminal regions could well be free to bind potential substrates.

It has been suggested that some members of the S4 subfamily of ATPases are transcription factors. Several early studies connected TBP1, TBP7, MSS1, and SUG1 (subunits S6, S6′, S7, and S8, respectively) with transcriptional control by the TAT protein of HIV (Nelbock *et al.*, 1990; Ohana *et al.*, 1993; Shibuya *et al.*, 1992) or yeast GAL 4 (Swaffield *et al.*, 1992). More recently, Swaffield *et al.* (1995) presented evidence that overexpressed, His-tagged Sug1p was associated with the TATA-box binding protein and was specifically not present in the 26 S proteasome. This report appears to have been in error as Akiyama *et al.* (1995) subsequently demonstrated that the human homologue of Sug1p is, indeed, present in the 26 S proteasome, and Rubin *et al.* (1996) found Sug1p to be a component of the yeast 26 S proteasome and not present in transcription complexes. Based on these recent findings it is likely that members of the S4-like ATPases are present only in the 26 S proteasome. This is certainly the most economical hypothesis to entertain at the present time.

4.3.3. Arrangement of ATPases within the Regulatory Complex

Sequence analysis indicates that six S4-like ATPases are present in humans and in yeast. The Coomassie blue stain patterns of regulatory complexes from several species are consistent with the presence of one copy each of S1, S2, S3, and S4 in the 26 S proteasome. There is too much overlap among subunits S5 to S12 for accurate measurements of relative abundance. Nonetheless, two-dimensional gels indicate that other putative ATPases, S6, S7, S8, and S10b, are present at levels comparable to S4. From this it is reasonable to postulate that there is one copy of each S4-like ATPase in the regulatory complex. This assumption is also consistent with an apparent mass of 700 kDa for the regulatory complex, as the ATPases would contribute about 300 kDa, S1 and S2 would contribute 200 kDa, and the other subunits would add another 200 to 300 kDa.

Where are the ATPases located, and do they form a hexamer? If the S4-like subunits actually function by pumping the substrate polypeptide into the central proteolytic chamber formed by proteasome β subunits, it seems most reasonable to assume that they are attached directly to the terminal α rings of the proteasome. Recent proteolytic protection studies by Haracska and Udvardy (1996) support this arrangement. These investigators found that p50, p48, and p42 of the *Drosophila* 26 S proteasome were sensitive to trypsin cleavage in the unassembled regulatory complex, but protected when the 26 S proteasome was exposed to the intestinal protease. p48 and p42 are members of the S4 subfamily of ATPases; p50 is probably also an ATPase (S6). Interestingly, three other subunits, S2, S5a, and what is probably *Drosophila* S12, were trypsin sensitive both in the free regulatory complex and in the assembled 26 S proteasome (Haracska and Udvardy, 1996).

Direct attachment of the ATPases to the proteasome would also be consistent with the relationship of ClpA and ClpP. In the simpler *E. coli* protease, ClpA subunits form a hexameric ring that attaches to a heptameric ring of ClpP subunits (Flanagan *et al.*, 1995; Kessel *et al.*, 1995). Likewise, a member of the AAA family of nucleotidases, p97, is found to be a hexamer (Peters *et al.*, 1990). Thus, one might expect the six S4-like ATPases to form a hexameric ring. There are, however, some problems with this idea. First, if the ATPases sit atop the proteasome, then the image in Fig. 1A is difficult to reconcile with a *hexameric* ring of ATPases. Assuming the ATPases are spheres with diameters proportional to their mass, a hexameric ring of these 50-kDa proteins would have a diameter larger than the underlying proteasome and would extend beyond the edge of the central cylinder. By contrast, a tetramer of ATPases would have a diameter less than the proteasome, more consistent with the image in Fig. 1A. Second, DeMartino and his colleagues have identified a smaller particle, the modulator, that contains two members of the S4-like family (DeMartino *et al.*, 1996). Although it is not clear that the modulator exists as a separate component *in vivo*, it does seem that two of the S4-like ATPases may dissociate from the regulatory complex more readily than the other four. This observation might also be taken as evidence against a hexameric ring of S4-like ATPases. The available evidence is not compelling one way or the other. Still, I believe one should not exclude the possibility that four ATPases contact the proteasome with an additional two ATPases (S6′ and S10b) located peripherally. Electron microscopic immunolocalization experiments should eventually allow us to position the ATPases within the 26 S proteasome.

4.4. Non-ATPase Subunits

Although cDNAs for almost all 26 S proteasome subunits have been isolated and sequenced (see Table II), obvious functional motifs have yet to be found among the non-ATPases. That is, none of the sequences contain regions homologous to isopeptidases, protein disulfide isomerases, prolyl isomerases, or proteases. Certain motifs are, however, quite prevalent. Portions of almost all of the non-ATPase subunits score very high using the coiled-coil search algorithm of Lupas *et al.* (1991), and five of the proteins contain KEKE motifs (see Table II). These "structural" motifs may play an important role in assembly of the regulatory complex or they may be used in substrate selection. We can assign a function to only one of the non-ATPase subunits, S5a. This protease subunit binds polyubiquitin chains, and it presumably functions in substrate selection. Despite the absence of functional information, it may prove useful to summarize the properties of subunits other than S5a, which is discussed at some length in Section 6.

4.4.1. Subunits of the Regulatory Complex

4.4.1a. Subunit 1. The largest subunit in the regulatory complex has been identified in both yeast and humans. Yeast S1 was obtained in a screen for cells that express high levels of Sen1p, a rapidly degraded protein involved in tRNA splicing. Selection for increased levels of Sen1-fusion protein by DeMarini *et al.* (1995) led to the isolation of mutants with alterations in a 945-residue protein encoded by *SEN3* (i.e., subunit 1). Mutants in Sen3p are defective in degrading

Table II
Subunits of the Human 26 S Protease

Subunit	Amino acids	SDS–PAGE apparent MW (× 10^3)	Calcd MW	Calcd pI	Coiled-coils[a]	KEKE motifs[b]	Functions
1	953	110	105,878	5.2	−	+	Bind polyUb?
2	908	100	100,208	5.0	+	+	Bind polyUb?
3	534	65	60,983	8.4	+		—
4	440	58	49,275	6.2	+		ATPase
5a	377	53	40,811	4.6	−	+	Bind polyUb
5b	504	53	56,201	5.3	+		—
6	418	50	47,340	5.0	+		ATPase
6′	439	50	49,060	5.0	+		ATPase
7	433	49	48,638	5.7	+		ATPase
8	406	48	45,647	7.8	+		ATPase
9	422	46	47,453	6.3	+		—
10a	389	45	45,536	5.4	+	+	—
10b	389	45	44,178	7.4	+		ATPase
11	—	43	—	—			—
12	321	39	36,643	6.4	−	+	—
13	257	34	30,007	7.2	−		—
14	—	31	—	—			—
15	—	26	—	—			—

[a]Coiled-coils were calculated using version 2 of the program of Lupas *et al.* (1991). A window of 28 residues was employed and a cutoff of 85% probability.
[b]KEKE motifs are defined in Table I.

Sen1-fusion proteins, and they accumulate ubiquitin conjugates. A cDNA encoding human S1 was obtained by Yokota *et al.* (1996) using PCR protocols based on partial amino acid sequences. The 953-residue human protein is 42% identical to yeast S1. Both the yeast and human proteins contain a long KEKE motif (Table I). Yokota *et al.* (1996) also showed that mutations in yeast S1 accumulate ubiquitin conjugates and that high-copy expression of yeast S1 suppresses a mutation in subunit S13/S14 (*NIN1*) of the regulatory complex. Yeast S1 also shows interaction with *SON1* whose product is a component of the pathway required for the degradation of noncleavable ubiquitin fusion proteins (Johnson *et al.*, 1995). Whereas DeMarini *et al.* (1995) reported that disruption of *SEN3* is lethal, Yokota *et al.* (1996) found that its disruption had a modest effect on cell growth at permissive temperatures. Interestingly, disruption of *SEN3* resulted in defective nuclear translocation of a reporter protein, leading Yokota *et al.* (1996) to suggest that the regulatory complex might be involved in nuclear import. In summary, the difference in the outcomes of the deletion experiments may reflect differences in genetic backgrounds or the structures of the disrupted alleles.

4.4.1b. Subunit 2. A cDNA encoding most of human subunit 2 was isolated from a two-hybrid screen for proteins that bind the intracellular domain of the p55 tumor necrosis factor (TNF) receptor (Boldin *et al.*, 1995). Peptide sequences obtained by W. Dubiel in my laboratory allowed us to deduce the complete sequence of human S2 (Dubiel and Rechsteiner, in press), and Tsurumi *et al.* (1996) have obtained a full-length cDNA for subunit 2. This subunit is a 908-residue protein that contains two KEKE motifs and a region with high coiled-coil potential (see Tables I and II). SDS–PAGE gels indicate that one copy of S1 and S2 are present in the 26 S proteasome from rabbit reticulocytes, and a comparison of their sequences indicates that the two proteins may be evolutionarily related. Moreover, both S1 and S2 possess several regions that are identical or similar to the hydrophobic core of the polyubiquitin binding sites in subunit S5a (see below). Thus, S1 and S2 may also be able to bind polyubiquitin chains.

The physiological significance of the interaction of S2 with the TNF-receptor is unclear. It may be a two-hybrid screen artifact. On the other hand, it is conceivable that the intracellular domain of the receptor is a substrate for the 26 S proteasome. Boldin *et al.* (1995) noted that TNF receptors appear to have a very short half-life. In addition, the portion of the TNF receptor that binds S2 is part of its intracellular domain, and this region scores positive using the algorithm for PEST proteolytic signals (Rechsteiner and Rogers, 1996). Thus, S2 may select the TNF receptor or its intracellular domain for degradation. Alternatively, the binding of S2 to the TNF receptor could serve to localize some 26 S proteasome complexes to the plasma membrane.

4.4.1c. Subunit 3. S3 was identified as a 26 S proteasome component by Ma *et al.* (1994) based on partial peptide sequences from components of the bovine red blood cell regulatory complex. Their peptides matched the sequence of a mouse tumor transplantation antigen called P91A (Lurquin *et al.*, 1989). This protein, in turn, is homologous to a *Drosophila* gene, thought to encode a diphenol oxidase component (Pentz and Wright, 1991). However, the encoded protein is probably not a diphenol oxidase as the *Drosophila* protein is not found in purified preparations of diphenol oxidase (T. Wright, personal communication). Although S3 might be evolutionarily related to diphenol oxidase, I suspect that *Drosophila* S3 is involved in selecting diphenol oxidase for destruction, and this provides its connection to oxidative enzymes in fruit flies.

S3 is a 534-residue protein that migrates with an apparent mass of 65 kDa on SDS–PAGE gels. The sequence of S3 contains a number of the hydrophobic regions that are too short to span a lipid bilayer, but longer than hydrophobic clusters present in most globular proteins. The significance of these hydrophobic regions is unclear. S3 does not contain a KEKE region but an analogous sequence, EERREREQQD, is present near the C-terminus of S3 from mouse, carrots, and flies. S3 has been identified in yeast as a suppressor of *nin1-1* (S13/S14); yeast S3 has been called Sun2p (Kawamura *et al.*, 1996). Interestingly, Sun1p, the yeast homologue of the human multiubiquitin binding subunit 5a, is also a suppressor of *nin1-1*. Thus, S3, S5a, and S13/S14 may well share structural and/or functional properties.

4.4.1d. Subunit 5b. A cDNA encoding a 50-kDa subunit of the human 26 S proteasome was identified by Deveraux *et al.* (1995a) in their attempt to isolate a cDNA for a 50-kDa multiubiquitin binding subunit (S5a) discovered a year earlier (Deveraux *et al.*, 1994). S5b is somewhat unusual among 26 S proteasome components in that an obvious homologue is not present in *S. cerevisiae*. The protein is leucine-rich, and nine dileucine repeats are present in its sequence. Dileucine repeats have been implicated in protein sorting to Golgi cisternae, lysosomes and in the internalization of several transmembrane proteins.

4.4.1e. Subunit 9. A cDNA encoding subunit 9 was recently obtained in my laboratory by L. Hoffman. S9 has 45% similarity to a gene on *S. cerevisiae* chromosome IV. The sequence of S9 contains several dileucine repeats like those present in S5b. The function of S9 is unknown.

4.4.1f. Subunit 10. The sequence of S10 has been published (Dubiel *et al.*, 1995a). This component contains a short KEKE motif. Originally designated S10, this component should now be called S10a because a member of the ATPase subfamily (the human SUG2 homologue or S10b) comigrates with S10a on one-dimensional gels. The function of subunit 10a is unknown.

4.4.1g. Subunit 11. A cDNA encoding subunit 11 has been isolated in the laboratory of W. Dubiel (W. Dubiel, personal communication).

4.4.1h. Subunit 12. Dubiel *et al.* (1995b) isolated, sequenced, and expressed a cDNA for subunit 12 of the human erythrocyte 26 S proteasome. The cDNA contains an open reading frame that encodes a 36.6-kDa protein with a KEKE motif at its C-terminus. Antibodies produced against two S12 fragments react with S12 transferred to nitrocellulose from SDS–PAGE. In contrast, after transfer from native gels, the epitope(s) recognized on one of the fragments is exposed in the regulatory complex but appears to be masked in the 26 S proteasome. S12 may play a role in binding the regulatory complex to the proteasome.

4.4.1i. Subunit 13/14. Amino acid sequences for this subunit have been obtained from humans, *S. cerevisiae*, and *S. pombe*. The gene *NIN1* was identified as a temperature-sensitive yeast mutant arrested at G2 phase; nuclear disintegration eventually occurred (Kominami and Toh-e, 1994). Nin1p was later shown to be a component of the 26 S proteasome (Kominami *et al.*, 1995). A functional

Nin1p is needed for both G1/S and G2/M transitions in the cell cycle, perhaps by activating Cdc28p kinase via degradation of the inhibitor p40^{Sic1} (Kominami *et al.*, 1995). *NIN1* is the budding yeast homologue of the fission yeast mts3^{+} gene, and a conditional lethal mutant, mts3-1, has been isolated in *S. pombe*. mts3-1 cells are defective in the metaphase-to-anaphase transition at the restrictive temperature (Gordon *et al.*, 1996). The 26 S proteasome isolated from mts3-1 cells grown at the restrictive temperature does not contain mts3 protein and cannot degrade ubiquitin conjugates (W. Dubiel, personal communication).

4.4.2. The Non-ATPase Subunits—a Summary

Knowing the amino acid sequences of almost all 26 S proteasome subunits has not, with the exception of the ATPases, provided a great deal of insight into their functions. Even though clear homologies to known enzymes are absent, one cannot be certain that the subunits are not proteases, isopeptidases, or rotamases. The 26 S proteasome subunits might employ novel catalytic mechanisms, and in such cases, sequences characteristic of known enzyme active sites could well be absent. Nonetheless, I suspect that most, if not all, of the non-ATPase subunits will prove to be enzymatically inactive. Rather, they will be found to function in substrate recognition and perhaps in localizing the 26 S proteasome within cells.

4.5. Activities of the Regulatory Complex

4.5.1. Nucleotide Hydrolysis

Although it has been reported that the 20 S proteasome from skeletal muscle degrades proteins in an ATP-dependent fashion (Driscoll and Goldberg, 1989), most in the field would agree that nucleotide-dependent proteolysis by the proteasome requires the presence of the regulatory complex. The first demonstration of a relevant ATPase activity in the 26 S proteasome was provided by Armon *et al.* (1990), who observed that formation of the 26 S proteasome from CF1, CF2, and CF3 generated nucleotidase activity. ATPase activity in purified human 26 S proteasomes (Kanayama *et al.*, 1992) and in bovine PA700 (DeMartino *et al.*, 1994) has also been reported. Recently, L. Hoffman in my laboratory has characterized the NTPase activities of the rabbit reticulocyte 26 S proteasome and its regulatory complex. Both particles hydrolyze ATP, CTP, GTP, and UTP to the corresponding nucleoside diphosphate and inorganic phosphate. The K_m values for hydrolysis of nucleotides by the 26 S proteasome are 15 μM for ATP and CTP, 50 μM for GTP, and 100 μM for UTP; the values for hydrolysis by the regulatory complex are two- to fourfold higher for each nucleotide. The K_m for ATP hydrolysis by the 26 S proteasome is virtually identical to the observed K_m for the ATP concentration required for Ub-conjugate degradation. Although nucleotide hydrolysis is needed for protein degradation by the 26 S proteasome, nucleotide hydrolysis and peptide bond cleavage are not strictly coupled. Substrate specificity constants (k_{cat}/K_m) are similar for hydrolysis of each nucleotide, yet GTP and UTP support Ub-conjugate degradation less effectively (Hoffman and Rechsteiner, 1996).

In prokaryotes, 70–80% of the energy-dependent protein degradation is provided by two large multisubunit proteases, Clp and Lon (Maurizi, 1992). The Clp protease consists of two components: Small (~20 kDa) ClpP subunits contribute the proteolytic activity and larger (~ 80kDa) ClpA subunits contribute the ATPase activity. Only ATP or dATP support casein degradation by Clp, and the K_m for ATP is 210 μM. The ClpAP protease is composed of two heptameric rings of ClpP subunits (Flanagan *et al.*, 1995) presumably topped by a larger hexameric ring of ClpA subunits (Kessel *et al.*, 1995). The assembly of the two subcomplexes, the ATPase activity (ClpA) and the proteolytic activity (ClpP), into a larger ATP-dependent protease is highly reminiscent of regulatory complex and proteasome associating to form the 26 S proteasome.

Biochemically, however, the nucleotidase activities of the 26 S proteasome more closely resemble the *E. coli* Lon protease, which is composed of identical subunits that provide both the ATPase and the protease activities (Maurizi, 1992). Both Lon protease and the 26 S proteasome have broad nucleotide specificity, utilizing ATP, CTP, GTP, and UTP, but not ADP or AMP. Although the values for nucleotide concentrations needed for casein degradation by Lon vary among several reports (Waxman and Goldberg, 1982; Larimore *et al.*, 1982; Chung and Goldberg, 1981; Charette *et al.*, 1981), the pattern is consistently ATP > CTP > UTP > GTP, similar to the nucleotide dependence for Ub-conjugate degradation by the 26 S proteasome (Armon *et al.*, 1990; Hough *et al.*, 1986). The similarity in nucleotide preference between the eukaryotic 26 S proteasome and prokaryotic Lon protease suggests that these enzymes may use common reaction mechanisms. In this regard, the sequence of Lon contains the motifs, GPPGXGKT and DEID, that are hallmarks of the S4 subfamily of ATPases.

4.5.2. Isopeptidases

An entire chapter is devoted to these enzymes, so I will limit my discussion to the isopeptidase activity associated with the 26 S proteasome, to isopeptidase T, and to the eukaryotic elongation factor, EFIα. For our purposes, the most relevant of these three enzymes is the one inherent to the 26 S proteasome. The prevailing idea that polypeptide substrates are threaded down the proteasome's narrow central axis requires that ubiquitin or multiubiquitin chains be removed from the protein substrate. Thus, one expects to find a 26 S proteasome component that can cleave the isopeptide bond between substrate lysines and the C-terminus of ubiquitin.

4.5.2a. Isopeptidases Associated with the 26 S Complex. In 1993, Eytan *et al.* identified just such an activity in the 26 S proteasome complex from rabbit reticulocytes. They showed that purified preparations of 26 S complex release free Ub from lysozyme conjugated to reductively methylated Ub. The activity that releases free Ub comigrated with the 26 S proteasome complex on glycerol gradients, and the complex-associated hydrolase also acted on a bacterially expressed construct in which the C-terminus of Ub was fused to the α-NH_2 group of a fragment of Ub that contains 60% of its N-terminal region (a protein called 1.6 Ub). ATP produced a marked stimulation of Ub release from 1.6 Ub. ATP-dependent Ub release from the Ub-carboxyl extension protein, UbRP52, was slower, and Ub was not cleaved from Ub-metallothionein. Thus, a 26 S proteasome-associated, nucleotide-stimulated hydrolase acts on mono-Ub-protein adducts in either α- or ε-NH_2 linkage, and its action depends on the nature of the protein to which Ub is attached. The fact that Ub release was stimulated by ATP, CTP, and GTP in that order suggests that the S4 ATPases are involved in the isopeptidase step as the nucleotide preference matches that of the regulatory complex itself. The involvement might be indirect with substrate polypeptide engagement being required prior to the action of an energy-independent isopeptidase. Alternatively, members of the S4 subfamily could themselves possess isopeptidase activity. In this regard, we have already noted that the unique C-terminal regions of these proteins contain a conserved cysteine that could serve this purpose (Rechsteiner *et al.*, 1993).

4.5.2b. Isopeptidase T. Attachment of polyUb chains to a protein substrate promotes its degradation much more than ligation of a Ub monomer. Presumably this results from recognition and binding of polyUb chains by subunit S5a and perhaps other components of the regulatory complex. It is not clear how binding of the polyUb chains to the 26 S proteasome is reversed once their linkage to the substrate has been broken. Two possibilities come to mind. Nucleotide hydrolysis by the S4 ATPases could produce a conformational change in S5a and/or additional subunits that reduce their affinity for polyUb chains. Alternatively, the polyUb chains might be converted to Ub monomers while still bound to the 26 S proteasome. Isopeptidase T, an enzyme capable of catalyzing polyUb chain degradation, was first described by Hadari *et al.* (1992) as an abundant monomeric Ub-binding protein. They showed that preincubation of conjugates with the 26 S proteasome increased release of free Ub on subsequent incubation with isopeptidase T. These investigators proposed that isopeptidase T serves to remove polyUb chains following the degradation of conjugates by the 26 S proteasome. Consistent with this idea, Wilkinson *et al.* (1995) have shown that isopeptidase T exhibits high discrimination against chains that are blocked or modified at the proximal end, indicating that the enzyme acts after release of the chains from conjugated proteins or degradation intermediates. Presumably this ensures that the proteolytic degradation signal is not disassembled by isopeptidase T before the ubiquitylated protein is degraded. In principle, Ub chain shortening could occur on the 26 S proteasome complex or after the chains have dissociated. Concerning the former possibility, it is interesting that isopeptidase T contains an 18-residue KEKE-like region that might promote its association with the 26 S proteasome (Wilkinson *et al.*, 1995; Falquet *et al.*, 1995).

4.5.2c. EF1α. Ciechanover and his colleagues have published three papers advancing the idea that the protein synthesis factor EF1α is required for the degradation of some ubiquitylated substrates. In 1991 Gonen *et al.* isolated a factor required for the degradation of N-α-acetylated proteins by the Ub pathway. The protein (factor H) was an apparent dimer of identical 46-kDa subunits. Factor H stimulated proteolysis by interacting with Ub conjugates prior to their degradation by the 26 S proteasome. Three years later, Gonen *et al.* (1994) reported the factor H was, surprisingly enough, identical to EF1α, a protein that binds GTP and tRNA in its role as a protein synthesis elongation factor. Just recently, this same group has proposed that EF1α and its bacterial counterpart EF-Tu are isopeptidases (Gonen *et al.*, 1996). This proposition may turn out to be correct, but the idea that a small, 46-kDa protein can catalyze GTP hydrolysis, bind tRNA, and break isopeptide bonds is difficult to accept. In my opinion, additional work is needed to exclude the presence of a bacterial endoprotease in the EF-Tu and EF1α preparations. Such an enzyme might nick the ubiquitylated substrate to provide "free" ends. Subsequent transfer of these ends into the proteasome could account for the stimulation of proteolysis by "EF1α."

4.5.3. Proteasome Activation

Two groups have shown that the regulatory complex can activate hydrolysis of fluorogenic peptides by the proteasome. Using rabbit reticulocyte components, Hoffman and Rechsteiner (1994) found that addition of the regulatory complex to the proteasome stimulated hydrolysis of several fluorogenic peptides by 2- to 3-fold. Different results were obtained by Ma *et al.* (1994) using components obtained from bovine erythrocytes. They observed markedly different degrees of stimulation depending on the peptide. For example, cleavage of sLLVY-MCA was stimulated 30-fold, whereas hydrolysis of z-GGL-MCA was stimulated only 3-fold. The reason for these divergent results is not clear. Methods for isolating bovine and rabbit regulatory complexes differ between the two groups. Conceivably, Ma *et al.* (1994) have isolated either a novel proteasome activator or PA28 in association with the bovine regulatory complex. Alternatively, Hoffman and Rechsteiner (1994) may have purified the regulatory complex missing some components. In any event, whether binding of the regulatory complex to the proteasome results in uniform or selective activation of peptide hydrolysis is an important issue. It is well established that PA28 or the 11 S regulators of the proteasome (Ma *et al.*, 1992; Dubiel *et al.*, 1992a; Realini *et al.*, 1994b; Gray *et al.*, 1994) stimulate hydrolysis of fluorogenic peptides differentially, e.g., some peptides are hydrolyzed 3-fold faster, others 50-fold. Presumably, PA28 or 11 S regulators transfer substrates, remove products, or impart conformational changes to the β subunits in a selective fashion. The results obtained using the regulatory complex from rabbit reticulocytes indicate the absence of selectivity in peptidase activation. It is as though substrate access and/or product release were simply increased in a uniform way. By contrast, the results obtained with bovine components suggest that binding of the regulatory complex to the proteasome differentially activates proteasome β subunits or leads to differential transfer of substrates/products. Because we want to know how the regulatory complex stimulates proteolysis, this issue should be reexamined, perhaps using proteasomes and regulatory complexes from yeast or flies.

4.6. The Regulatory Complex as a Polypeptide Pump

Although there is the prevailing idea that a major function of the regulatory complex is to transfer or "pump" substrate polypeptides into the central cavity of the proteasome, other mechanisms may be used by the 26 S proteasome to degrade substrates. For example, nucleotide hydrolysis could cause the symmetric structure shown in Fig. 1A to open at the interface of the β subunits to permit contact between protease active sites and substrate. In this context, Ishii *et al.* (1995) showed that ATP causes the *Thermus thermophilus* holochaperonin to dissociate into two halves. Still, the hypothesis that the regulatory complex pumps polypeptide strands down the cylinder axis is a most attractive one. If correct, it raises immediate questions. Is the pump directional? Does a substrate enter the cylinder carboxyl end first? Vice versa? Can it be threaded in either orientation? Or is it even possible that β-hairpin loops are sent into the proteasome?

As far as I know, these questions have not been answered. And there is not much to favor one orientation over another. I might hazard the guess, however, that proteins are degraded by the 26 S proteasome from their C- to N-terminal ends. Several observations and a reasonable assumption provide hints for choosing C to N over the other possibilities. First, I assume that the interaction of attached polyUb chains with S5a (and possibly S1 or S2 or other subunits) must persist until the substrate polypeptide is engaged by the S4-like ATPases, which are the only components likely to serve as the pump(s). Interestingly, the known attachment sites for polyUb chains reside near substrate N-termini. For example, IκBα is ubiquitylated at Lys21 or 22 (Baldi *et al.*, 1996; Scherer *et al.*, 1995), various Ub-β-galactosidase N-end constructs at Lys15 or 17 in the N-terminal extension (Bachmair and Varshavsky, 1989), or on the very N-terminal Ub in the case of noncleavable Ub-β-galactosidase fusions (Johnson *et al.*, 1995). Lys13 is the most probable site for the attachment of Ub chains to lysozyme (Hill *et al.*, 1993). With the known locations of polyUb chains at or near substrate N-termini, it seems reasonable to speculate that C-termini are available for capture by the ATPases. The interesting report that the NFκB precursor p105 is processed by the 26 S proteasome is also consistent with proteolysis from the carboxyl end. Maniatis and his colleagues have presented evidence that p105 is ubiquitylated and partially degraded to release the N-terminal NFκB subunit p50 (Palombella *et al.*, 1994). If p105 degradation is mediated by the 26 S proteasome, then proteolysis proceeds C to N or by a loop mechanism. It should be noted, however, that recent results by Lin and Ghosh (1996) are difficult to reconcile with a mechanism that is strictly C to N as they found that both N- and C-terminal domains were liberated from fusion proteins containing a central portion of p105.

The available evidence is certainly meager, and my speculations on directional transfer of the substrate should be considered tentative at best. On the other hand, this is certainly an interesting problem that awaits solution. Over 35 years ago, Dintzis (1961) showed that the synthesis of a protein proceeds from its N to C ends. There is a pleasing symmetry in the thought that degradation of many proteins may proceed in the opposite direction.

5. ASSEMBLY OF THE 26 S PROTEASOME

5.1. *In Vitro* Studies

There is ample evidence that either the regulatory complex of CF1 plus CF2 can associate with the proteasome to produce the 26 S proteasome. All four ribonucleotides support Ub-conjugate degradation by the 26 S proteasome, but only hydrolyzable ATP or CTP promotes its assembly from the proteasome and regulatory complex (DeMartino *et al.*, 1994; Armon *et al.*, 1990). Beyond this simple fact, we know very little about the biochemistry of assembly.

Hoffman and Rechsteiner (1994) found that the regulatory complex can outcompete PA28 or the 11 S regulator described above (Section 4.5.3) for limiting amounts of proteasome. They also noted that assembly of the 26 S proteasome *in vitro* was unusual in that titration of proteasomes with increasing amounts of regulatory complex did not result in the graded appearance of active 26 S proteasome. Rather, 26 S proteasome activity required equal amounts of regulatory complex and proteasome. The molecular basis for the almost all-or-none appearance of an active 26 S proteasome at a 1:1 ratio of regulatory complex to proteasome is not understood. It is possible that two regulatory complexes are required for each proteasome, as formation of active 26 S proteasomes from 1:1 mixtures consumed all of the regulatory complexes and only half of the proteasomes. In this regard, assembly of the 26 S proteasome differs markedly from similar titrations of proteasomes with increasing levels of the 11 S regulator. Levels of activated proteasomes were proportional to the amount of 11 S regulator added to the assembly reaction.

5.2. *In Vivo* Studies

There is a single published study on the *in vivo* assembly of the 26 S proteasome. Yang *et al.* (1995) examined assembly of the proteasome and the 26 S proteasome in murine lymphoma cells using a metabolic labeling approach. They found that all newly synthesized regulatory complex components were associated with proteasomes. That is, there did not appear to be a substantial pool of free regulatory complexes. By contrast, there was a significant pool of "free" proteasome, perhaps three times as much as present in the 26 S proteasome. The studies also revealed uniform labeling of the individual subunits within the regulatory complex implying that the particle assembled from newly synthesized components and then associated with the proteasome to produce the 26 S enzyme. The uniform labeling pattern argues against differential metabolic stabilities among individual subunits in the regulatory complex. It seems that once the regulatory complex is assembled, its subunits generally remain together or, at least, have similar half-lives. These authors also implicated phosphorylation in 26 S proteasome assembly by showing that the protein kinase inhibitor, staurosporine, resulted in a complete dissociation of the 26 S proteasome to free regulatory complexes and proteasomes within living lymphoma cells.

5.3. Dynamics of the 26 S Proteasome: Ribosome versus Solid-State Models

The combined *in vitro* and *in vivo* studies leave little doubt that regulatory complex and proteasome reversibly associate to form the 26 S proteasome. This raises the interesting possibility that association/dissociation reactions are required for the degradation of ubiquitylated proteins. Perhaps, much like the 40 S subunit of ribosomes binds mRNAs, the regulatory complex binds substrates and then associates with the proteasome to initiate proteolysis *per se* (see Fig. 4). On the other hand, it has been demonstrated that ClpA and ClpP remain together through multiple rounds of substrate degradation (Maurizi, 1992). Thus, each 26 S proteasome complex may be able to capture and funnel substrates to its central proteolytic chamber without ever dissociating.

5.4. Subcomponents of the Regulatory Complex

5.4.1. CF1 and CF2 Revisited

As noted earlier, Hershko and his colleagues identified three components, CF1, CF2, and CF3, required for the degradation of Ub conjugates (Ganoth *et al.*,

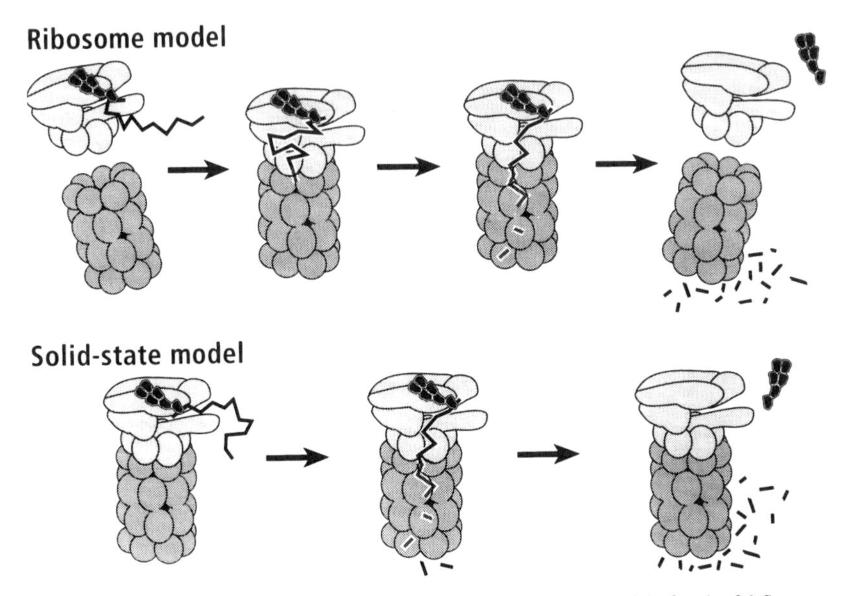

Figure 4. Schematic representation of ribosome versus solid-state models for the 26 S protease enzymatic cycle. In both models, the proteasome is depicted as a cylinder, the ATPase complex is shown as a ball, the kinked heavy black line represents the polypeptide substrate, and the black pear-shaped structures are ubiquitin. The models differ principally with regard to whether association–dissociation reactions are required for the degradation of ubiquitin conjugates. In the ribosome model, it is assumed that regulatory complexes and proteasomes must cycle between assembled and unassembled state for degradation of substrates. In the solid-state model, a preformed 26 S protease can degrade ubiquitin conjugates without dissociation.

1988). With CF3 clearly shown to be the proteasome, it is reasonable to hypothesize that CF1 and CF2 combine to form the regulatory complex. However, in his studies on the *Drosophila* μ particle, Udvardy (1993) reported that assembly of the *Drosophila* 26 S proteasome required a third factor, but this factor was not incorporated into the conjugate degrading enzyme. Thus, it is possible that CF2 is an enzyme or assembly factor not permanently associated with the 26 S proteasome.

Two groups have identified smaller protein assemblages that may be CF2. Driscoll *et al.* (1992) described an ATP- stabilized proteasome inhibitor composed of 40-kDa subunits. They reported that this 250-kDa native complex is CF2 and that it is equivalent to an inhibitor identified earlier by Etlinger and his colleagues (Li *et al.*, 1991). According to Driscoll *et al.* (1992), the 250-kDa inhibitor is incorporated into the 26 S proteasome. This conclusion is surprising as the 26 S enzyme does not contain many 40-kDa subunits. Moreover, the 250-kDa inhibitor of Etlinger and his colleagues was subsequently identified as δ-aminolevulinic acid dehydratase (Guo *et al.*, 1994); to my knowledge, no one has identified this enzyme as a component of the 26 S proteasome.

DeMartino *et al.* (1996) identified and purified a heteromeric complex called the *modulator*. By itself, the modulator does not stimulate peptidase hydrolysis by the proteasome, but in the presence of PA700 (the bovine regulatory complex) it enhances peptide hydrolysis almost 10-fold. This coactivator of the proteasome has a native molecular mass of 300 kDa by gel filtration and is composed of three subunits with apparent molecular masses of 50, 42, and 27 kDa. The two larger subunits are members of the S4 subfamily of ATPases (see above), and they are also present in the 26 S proteasome. The 27-kDa subunit, however, is not a component of the 26 S proteasome or its regulatory complex. Although DeMartino *et al.* (1996) did not favor the idea that the modulator is equivalent to CF2, this remains a distinct possibility.

5.4.2. Free Subunits and Other Complexes

In their studies on the regulatory complex from *Drosophila*, Haracska and Udvardy (1995) noted that besides being present in the 26 S proteasome, *Drosophila* S5a was also found as a monomer or perhaps in a smaller protein complex. Likewise, Peters *et al.* (1994) observed that p50, a subunit of the *Xenopus* 26 S proteasome, was present in the 26 S enzyme, in the 19 S regulatory complex, and in a 10 S complex. Finally, the labeling studies of Yang *et al.* (1995) show three subunits presumed to be components of the regulatory complex that sediment slower than the 20 S proteasome. Although the relationship of these "subparticles" to the modulator and/or CF2 is not known, it seems clear that some components of the 26 S proteasome either readily dissociate during fractionation or exist as separate entities *in vivo*.

5.5. One 26 S Proteasome or Many?

The regulatory complex is said to have an apparent molecular mass of 700 kDa. If one assumes that each of its subunits is present once, and this seems reasonable from staining patterns, then the data in Table II predict a molecular mass of 945 kDa for the particle. There are a number of potential explanations for the rather substantial difference between predicted and apparent molecular masses. I will consider just one. Perhaps these are several varieties of the 26 S proteasome with slightly different subunit compositions. In this regard, Wang *et al.* (1996) reported that the SUG1 homologue is more abundant in nuclear 26 S complexes than in 26 S proteasomes from the cytoplasm of HeLa cells. Likewise, Palmer *et al.* (1996) found an enrichment of proteasome subunit Z in nuclear proteasomes. I suspect that all 26 S complexes will contain at least four of the six S4-like ATPases. It is not unreasonable to think that some 26 S proteasome complexes may contain S5b, for example, with others containing say S9 in its place. So concerning the question, "How many 26 S proteasomes?" the answer may well be several. Further experiments will be required to determine whether ATPase and non-ATPase subunits substitute for one another in subpopulations of the 26 S proteasome.

6. SUBSTRATE RECOGNITION BY THE 26 S PROTEASOME

6.1. Widespread Occurrence of Substrates

An increasing number of intracellular proteins are being identified as 26 S proteasome substrates. In 1991, there were less than a handful of natural substrates of the Ub-dependent pathway (Rechsteiner, 1991). A recent review lists almost 20 (Hilt and Wolf, 1996). I assume that most Ub-dependent substrates are necessarily 26 S proteasome substrates as it is the only cytosolic enzyme known to be capable of degrading ubiquitylated proteins. I recognize, however, that this assumption is difficult to prove. Substrates of the 26 S proteasome are found in the nucleus, the cytosol, and surprisingly in cellular membranes (see Chapters 10–15). Most known substrates are short-lived proteins with important regulatory functions, e.g., the cyclins, cdk inhibitors, IκBα, and transcription factors such as jun, MATα2 or GCN4, and so on. It is not clear that more stable cellular proteins, like lactate dehydrogenase, are degraded by the 26 S proteasome. (See Jennissen, 1995, for a comprehensive review.) Nonetheless, eukaryotic cells express so many rapidly degraded proteins that it should soon be impossible to present an updated list of 26 S proteasome substrates in a review of reasonable length. Rather than do this, I focus on issues relating to substrate recognition and the mechanism of polypeptide degradation by the 26 S enzyme.

6.2. Ubiquitin Dependence

Some proteins are degraded by the 26 S proteasome without being marked by Ub. In 1989, Rosenberg-Hasson *et al.* used antibodies to remove the Ub-activating enzyme, E1, from reticulocyte lysate. They observed that p53 was metabolically stable in the depleted extract, but the short lived enzyme, ornithine decarboxylase (ODC), was still rapidly degraded. This was the first piece of evidence that ODC is degraded in a Ub-independent reaction. Murakami *et al.* (1992) extended this observation significantly by demonstrating that the purified 26S proteasome degrades ODC in a reaction requiring ATP and the polyamine-induced protein antizyme, but not Ub or Ub-activating enzymes. In further support of the idea that

ODC degradation does not require ubiquitylation, Mahaffey *et al.* (manuscript in preparation) have found that proteolysis of full-length ODC in reticulocyte lysate is not inhibited by addition of S5a. This 26 S proteasome component has been shown to bind Ub–lysozyme conjugates, and it is a powerful inhibitor of cyclin degradation in *Xenopus* egg extracts as well as the degradation of preformed Ub conjugates by the human 26 S proteasome (Deveraux *et al.*, 1995b). Thus, three independent studies indicate that ODC is degraded in a Ub-independent fashion, and one report clearly implicates the 26 S proteasome. Coffino provides further discussion concerning recognition of ODC–antizyme complexes and antizyme fusion proteins by the 26 S proteasome (see Chapter 14).

The transcription factor c-jun provides a second example of a Ub-independent substrate for the 26 S proteasome. Although Ub can be conjugated to c-jun (Hermida-Matsumoto *et al.*, 1996; Treier *et al.*, 1994), Piechaczyk and his colleagues presented evidence that this modification is not required for degradation of c-jun by the 26 S proteasome *in vitro* (Jariel-Encontre *et al.*, 1995). They found that recombinant c-jun was degraded in an ATP-dependent reaction that required the 26 S proteasome; additional cofactors, such as Ub or antizyme, were not needed.

6.3. The Role of Ubiquitin in Substrate Recognition

The 26 S proteasome was identified by its ability to degrade Ub–lysozyme conjugates. This substrate consists of lysozyme molecules attached to polyUb chains, and Chau *et al.* (1989) showed that long Ub polymers markedly increase the rate of β-galactosidase degradation. Some cellular proteins, such as arthrin, a Ub conjugate of insect actin (Ball *et al.*, 1987), are found conjugated to one or a few Ub moieties. In many cases, these proteins are not rapidly degraded. Their metabolic stability could reflect the fact that they are sequestered from the 26 S proteasome; e.g., Ub–histone H2A by binding to DNA or arthrin by its assembly in flight muscle sarcomeres. Alternatively monoubiquitylated proteins may simply not be recognized by the 26 S proteasome. A recent paper by Shaeffer suggests, however, that this second possibility is not correct. Shaeffer and Kania (1995) isolated [^{125}I]α-globin molecules conjugated to one, two, or a mixture of three and four Ubs. Each of these conjugates was then incubated with the 26 S proteasome or with the proteasome. Whereas α-globin itself was not degraded by the 26 S proteasome, the monoUb adduct was proteolyzed at 24% per hr, and α-globin molecules attached to two or more Ubs were degraded threefold faster. Degradation of the globin conjugates by the 26 S proteasome was ATP-dependent. A surprisingly high rate of proteolysis of the globin conjugates by the proteasome alone raises some concerns, as the 20 S enzyme does not normally degrade intact proteins. Nonetheless, the presented evidence supports the view that Ub_2-α-globin is a 26 S proteasome substrate; Ub_1-α-globin may be one as well.

Several studies have shown that polyUb chains *per se* are not required for substrate recognition by the 26 S proteasome. Hershko and Heller (1985) blocked the amino groups on Ub by reductive methylation and showed that the modified Ub produced lysozyme conjugates of much lower molecular weight. They concluded that high-molecular-weight conjugates are produced by the formation of polyUb chains [This proposal was confirmed several years later in an elegant series of experiments by Chau *et al.* (1989), who showed that polyUb chains formed on β-galactosidase bearing an N-terminal extension.] Despite the fact that polyUb chains did not form, the MeUb–lysozyme conjugates were degraded in the reconstituted extract at half the rate of lysozyme conjugates formed from unmodified Ub. Studies by Haas *et al.* (1990) using histone H3 as substrate confirmed that conjugates formed from MeUb are degraded by the 26 S proteasome. A subsequent paper from Hershko's group in which five proteins were used as substrates showed that the requirement for polyUb chains differ significantly among proteins (Hershko *et al.*, 1991). For example, MeUb conjugates of lysozyme, S-protein, and β-lactoglobulin were degraded at almost half the rate seen with conjugates formed using native Ub. By contrast, the degradation of α-lactalbumin, oxidized ribonuclease, and clam cyclin was significantly inhibited when MeUb was added to the reaction mixtures. The requirement for polyUb chains thus appears to be substrate specific.

6.4. Polyubiquitin Chains Are Not Sufficient Degradation Signals

Two recent papers demonstrate that a protein may be ligated to a polyUb chain and yet not be rapidly degraded. Johnston *et al.* (1995) observed that methotrexate inhibited the degradation of dihydrofolate reductase (DHFR) despite the fact that the enzyme was ubiquitylated. A second example involving polyUb chains has recently been published by Klotzbücher *et al.* (1996). They transferred the cyclin destruction box from *Xenopus* cyclin A to cyclin B and found that the chimeric protein was ubiquitylated but not rapidly degraded. These two papers show that attachment of a polyUb chain to a protein may be necessary for its destruction, but is not always sufficient.

The findings of Johnston *et al.* (1995) are reminiscent of an earlier, somewhat vexing observation obtained in my laboratory. Löetscher *et al.* (1991) found that PEST-DHFR fusion proteins purified on methotrexate columns were metabolically stable unless the DHFR core was radioiodinated. This indicated that PEST sequences were not sufficient proteolytic signals. Presumably, an additional perturbation in the protein was needed for degradation. A possible explanation for these observations is as follows. PEST sequences or polyUb chains promote association of the proteolytic substrate with the 26 S proteasome. Once the substrate is bound to the complex, a second event must occur before the substrate dissociates. This second event might be engagement of the substrate by the ATPases. In this scenario, rapid proteolysis will require two signals: a binding event and the irreversible capture of the substrate by the proteolytic complex. If the marked protein is tightly folded, e.g., DHFR bound to methotrexate, the second step may not occur prior to dissociation.

6.5. Recognition of Polyubiquitin Chains by the 26 S Proteasome

It has been known for more than a decade that high-molecular-weight conjugates are preferred substrates for the 26 S proteasome. The molecular basis for this preference is beginning to emerge from the recent discovery of a 26 S proteasome subunit (S5a) that can bind multiUb chains. Using far Western blots, Deveraux *et al.* (1994) demonstrated that radiolabeled Ub–lysozyme conjugates bind the regulatory complex and the 26 S proteasome but not the proteasome. Surprisingly, the ^{125}I-labeled conjugates bound a 50-kDa component of the regulatory complex even after the latter had been subjected to SDS–PAGE. This subunit, now known as S5a, also binds free multiUb chains. Moreover, when incubated with Ub polymers of various lengths, S5a selected for species containing four or more Ub monomers and exhibited increased affinity for longer chains (Deveraux *et al.*, 1994). These two properties of S5a, length-dependent binding and insensitivity to denaturation, coupled with the X-ray structure of the Ub tetramer (Cook *et al.*, 1994), suggested a model in which tetramers and longer chains of Ub bound "loops" on S5a (Deveraux *et al.*, 1995a). Increased affinity for longer chains could be explained by the presence of multiple loops on the protease subunit and resistance to denaturation could be explained by assuming the "loops" were short enough to readily renature after SDS–PAGE and transfer of S5a to membranes. The model envisions that repeating "grooves" in the multiUb chain would interact with repeated "loops" on S5a.

Several recent publications and unpublished observations from my laboratory support the "loop–groove" model. The sequences of subunit S5a obtained from three higher eukaryotes reveal repeated motifs in the C-terminal half of S5a (Ferrell *et al.*, 1996; Van Nocker *et al.*, 1996b; Haracska and Udvardy, 1995). These motifs are characterized by five hydrophobic residues followed by a conserved serine and flanked by several aspartates and/or glutamates. The hydrophobic cluster consists of three large (Leu, Met, Ile, Tyr) and two small (Ala) residues that alternate, e.g., LALAL. The presence of repeated hydrophobic stretches in S5a is relevant to substrate recognition in view of the observations of Beal *et al.* (1996), showing that the surface hydrophobic residues L8, I44, and V70 in Ub are critical for targeting. Mutation of pairs of these residues to alanines had little effect on attachment of Ub to substrates but severely inhibited degradation of the resulting conjugates. The same mutations blocked the binding of the Ub chains to S5a. The side chains implicated in this binding—L8, I44, and V70—form repeating patches on the chain surface. Thus, hydrophobic interactions between these patches and the hydrophobic clusters in S5a may make major contributions to enhanced proteolytic targeting by multiUb chains. Recent deletion analyses of recombinant S5a have identified two independent regions of the protein that bind multiUb chains (Young *et al.*, manuscript in preparation). These regions correspond to the hydrophobic clusters in S5a, supporting the idea that hydrophobic regions in S5a bind to hydrophobic patches on multiUb chains.

To some extent, we are beginning to understand molecular details of the recognition of ubiquitylated substrates by the 26 S proteasome. However, S5a cannot be the whole story as this subunit does not exhibit substantial affinity for Ub monomers, dimers, or trimers and as mentioned Ub_1-globin may be a good substrate. More compelling is the finding that deletion of the gene for S5a in yeast is not lethal (Van Nocker *et al.*, 1996a; Toh-e, personal communication). This indicates that other 26 S proteasome components can recognize ubiquitylated proteins. In this regard, subunits 1 and 2 each contain several regions of alternating large and small hydrophobic amino acids, such as LGLGL or LALAL. Moreover, like S5a, both S1 and S2 possess robust KEKE motifs. The two largest 26 S proteasome subunits are, therefore, excellent candidates for being additional polyUb chain binding components.

7. INTRACELLULAR LOCALIZATION OF PROTEASOMES AND THE 26 S PROTEASOME

7.1. Light Microscopic Analyses

The first studies on the intracellular location of proteasomes were published well over a decade ago. Hügle *et al.* (1983) prepared antibodies to 22 S cylinder particles (proteasomes) from *Xenopus laevis* and used them to localize the proteasome in various frog tissues. They found proteasomes to be present in both nucleoplasm and cytosol but at higher concentration in nuclei of transcriptionally active cells. The nucleolus did not stain nor did metaphase chromosomes. Kleotzel *et al.* (1987) reported that proteasomes were largely cytoplasmic in *Drosophila* salivary gland cells. They described a speckled distribution of immunostaining within the cytoplasm. Tanaka *et al.* (1989) found proteasomes to be present in both rat liver nuclei and cytoplasm in accord with the earlier studies by Hügle *et al.* (1983). They observed intense nuclear staining in some cells, but not others. They also commented on the speckled appearance of immunostaining in the nuclei of some cells.

Three groups have focused on the distribution of proteasomes in dividing cells. Kawahara and Yokosawa (1992) reported rather dramatic changes in the location of proteasomes during mitosis of ascidian eggs. During interphase, proteasomes were present in the nucleoplasm but disappeared in early prophase only to reappear on metaphase chromosomes and the mitotic spindle. At anaphase, proteasomes were once again absent from chromosomes. Amsterdam *et al.* (1993) obtained different distributions in dividing ovarian granulosa cells. They found clustered immunofluorescence in both nucleus and cytoplasm of interphase cells. In accord with the studies of Kawahara and Yokosawa (1992), they observed intense staining of the spindle apparatus. They did not, however, see proteasomes associated with metaphase chromosomes. Amsterdam *et al.* (1993) suggest that proteasome accumulation in spindles reflects the abundance of proteolytic substrates, such as cyclins and centrosomal components, that need to be rapidly degraded during mitosis.

Palmer *et al.* (1996) have examined the intracellular location of proteasomes during the cell cycle in human lung cells and in the kangaroo rat line PtK_2. Proteasomes were present in both nucleus and cytoplasm throughout interphase. However, the intensity and pattern of staining changed with phase of the cell cycle. In the nucleus, the intensity of staining in early S phase was low and showed a punctate distribution, which changed to a more diffuse and intense labeling during S to G1. Double-label immunofluorescence studies using antiproteasome and anticytokeratin (TROMA-1) antibodies showed that proteasomes colocalize with intermediate filaments of the cytokeratin type, mainly during G2. In mitosis, proteasomes were found by immunogold electron microscopy to be localized around the chromosomes in both PtK_2 and L-132 cells.

Studies by Machiels *et al.* (1995) raise a note of caution concerning the apparent changes in proteasome location. Using two monoclonal antibodies, they observed that fixation conditions profoundly affected the apparent distribution of proteasomes between nucleus and cytoplasm as cells depleted their growth medium. In cells growing under favorable conditions (1 or 2 days in fresh medium), proteasomes were detected mainly in the nuclei, whereas when the medium became depleted of nutrients (4- or 5-day-old medium), the staining pattern changed to one with a much less pronounced nuclear staining. However, in immunofluorescence studies on cells grown under similar conditions, but fixed in ethanol (12°C) for 15 min, the changes in proteasome localization pattern were not detected during medium depletion. Using this fixation protocol the proteasomes were detected mainly in the nuclei at all stages of the medium exhaustion experiment.

The intracellular distribution of one 26 S proteasome subunit has also been examined by immunofluorescence. Kominami and Toh-e (1994) observed that Nin1p (S13/S14) was predominantly cytoplasmic in *S. cerevisiae*. Small bright dots were present reminiscent of the speckled patterns observed for proteasomes. These investigators examined the distribution of Nin1p by Western blotting subcellular fractions and observed a uniform distribution of the protein between low-speed and high-speed supernatants and pellets.

7.2. Ultrastructural Studies

Erwin Knecht and his colleagues have examined the distribution of proteasomes using electron microscopic techniques. In 1992, Rivett *et al.* reported an apparent partitioning of proteasomes between nucleus and cytoplasm in rat liver cells of 20 and 80%, respectively. About 17% of the cytoplasmic proteasomes were associated with the cytosolic facet of the endoplasmic reticulum membrane, and the remainder of the cytoplasmic proteasomes were present in the cytosol; mitochondria, lysosomes, peroxisomes, and Golgi were devoid of specific immunoreactivity. The authors described the cytosolic proteasomes as diffusely distributed; "speckles" were apparently not obvious at the ultrastructural level.

In a more recent publication, Palmer *et al.* (1996) have combined cellular fractionation and microscopic techniques to reexamine the distribution of proteasomes in rat liver cells. These studies generally confirm their earlier findings and provide several new pieces of information. As mentioned earlier, they found significant (13-fold) enrichment of the β subunit Z in nuclear fractions and a 6-fold enrichment of the β subunit LMP2 in microsomal fractions. They also localized the microsomal proteasomes to the smooth endoplasmic reticulum and *cis*-Golgi; proteasomes were not found in the rough endoplasmic reticulum.

In summary, there seems to be little doubt that proteasomes and 26 S proteasome complexes are found in both nucleus and cytoplasm. The microscopic approaches are supported by cell fractionation studies (Wang *et al.*, 1996; Palmer *et al.*, 1996; Peters *et al.*, 1994) that place the 26 S proteasome in both compartments. Whether these two enzymes are present in highly concentrated regions ("speckles") or are uniformly distributed is a matter for debate. Also, it is likely, but by no means certain, that proteasomes and for 26 S proteasomes change their location at the various stages of mitosis.

7.3. Signals for Nuclear Import

Tanaka *et al.* (1990) noted the presence of potential nuclear targeting signals in several proteasome subunits. They suggested that these regions may regulate the nuclear/cytoplasmic distribution of proteasomes during various growth states. Recently, Nederlof *et al.* (1995) showed that the regions identified by Tanaka *et al.* (1990) can serve as independent nuclear targeting signals. Nederlof *et al.* (1995) generated synthetic peptides containing these putative NLS sequences and conjugated them to fluorescent reporter proteins. All three putative NLS sequences from human proteasomal subunits were able to direct the reporter proteins to the nucleus in both cell types, although differences in efficiency were observed. The putative NLS sequence found in *T. acidophilum* was also functional as a nuclear targeting sequence.

8. SUMMARY AND PERSPECTIVES

In the 10 years since the 26 S proteasome was discovered, it has become clear that the enzyme plays an important role in regulation. It degrades key proteins involved in signaling pathways, in cell cycle control, and in general metabolism. In my coverage of the enzyme, I omitted discussion of several pertinent issues, such as regulation of 26 S proteasome activity, its impact on developmental processes, and its postulated role in antigen presentation. These omissions were necessary because this essay, even limited largely to structure–function aspects of the 26 S proteasome, stretched over 65 pages in manuscript form. Hopefully, the length reflects our increased knowledge of the 26 S proteasome rather than a diffuse writing style. Despite recent advances, a number of important structural questions remain unanswered. For example: Are there various 26 S proteasome isoforms? Where are the individual subunits within the 26 S proteasome? How dynamic is the enzyme complex? How are substrates captured by the 26 S proteasome? Are there fundamental differences in the hydrolysis of ubiquitylated and nonubiquitylated substrates? Do the ATPases actually unfold substrates? Do they pump? In a polar fashion? Given the progress over the past decade, I predict that many, if not all, of these questions will be answered in the third book on ubiquitin likely to appear early in the next century.

9. REFERENCES

Akiyama, K., Yokota, K., Kagawa, S., Shimbara, N., DeMartino, G. N., Slaughter, C. A., Noda, C., and Tanaka, K., 1995, cDNA cloning of a new putative ATPase subunit p45 of the human 26S proteasome, a homolog of yeast transcriptional factor Sug1p, *FEBS Lett.* **363:**151–156.

Amsterdam, A., Pitzer, F., and Baumeister, W., 1993, Changes in intracellular localization of proteasomes in immortalized ovarian granulosa cells during mitosis associated with a role in cell cycle control, *Proc. Natl. Acad. Sci. USA* **90:**99–103.

Armon, T., Ganoth, D., and Hershko, A., 1990, Assembly of the 26S complex that degrades proteins ligated to ubiquitin is accompanied by the formation of ATPase activity, *J. Biol. Chem.* **265:** 20723–20726.

Arnason, T., and Ellison, M. J., 1994, Stress resistance in *saccharomyces cerevisiae* is strongly correlated with assembly of a novel type of multiubiquitin chain, *Mol. Cell. Biol.* **14:**7876–7883.

Arrigo, A.-P., Tanaka, K., Goldberg, A. L., and Welch, W. J., 1988, Identity of the 19S 'prosome' particle with the large multifunctional protease complex of mammalian cells (the proteasome), *Nature* **331:**192–194.

Baboshina, O. V., and Haas, A. L., 1996, Novel multiubiquitin chain linkages catalyzed by the conjugating enzymes $E2_{EPF}$ and RAD6 are recognized by 26S proteasome subunit 5, *J. Biol. Chem.* **271:**2823–2831.

Bachmair, A., and Varshavsky, A., 1989, The degradation signal in a short-lived protein, *Cell* **56:**1019–1032.

Baldi, L., Brown, K., Franzoso, G., and Siebenlist, U., 1996, Critical role for lysines 21 and 22 in signal-induced, ubiquitin-mediated proteolysis of IκB-α, *J. Biol. Chem.* **271:**376–379.

Ball, E., Karlik, C. C., Beall, C. J., Saville, D. L., Sparrow, J. C., Bullard, B., and Fyrberg, E. A., 1987, Arthrin, a myofibrillar protein of insect flight muscle, is an actin–ubiquitin conjugate, *Cell* **51:** 221–218.

Baumeister, W., Dahlmann, B., Hegerl, R., Kopp, F., Kuehn, L., and Pfeifer, G., 1988, Electron microscopy and image analysis of the multicatalytic proteinase, *FEBS Lett.* **241:**239–245.

Beal, R., Deveraux, Q., Gang, X., Rechsteiner, M., and Pickart, C., 1996, Surface hydrophobic residues essential for proteolytic targeting by ubiquitin, *Proc. Natl. Acad. Sci. USA* **93:**861–866.

Boldin, M. P., Mett, I. L., and Wallach, D., 1995, A protein related to a proteasomal subunit binds to the intracellular domain of the p55 TNF receptor upstream to its "death domain," *FEBS Lett.* **367:**39–44.

Charette, M. F., Henderson, G. W., and Markovitz, A., 1981, ATP hydrolysis-dependent protease activity of the lon (capR) protein of *Escherichia coli* K-12, *Proc. Natl. Acad. Sci. USA* **78:**4728–4732.

Chau, V., Tobias, J. W., Bachmair, A., Marriott, D., Ecker, D. J., Gonda, D. K., and Varshavsky, A., 1989, A multiubiquitin chain is confined to specific lysine in a targeted short-lived protein, *Science* **243:**1576–1583.

Chung, C. H., and Goldberg, A. L., 1981, The product of the lon (capR) gene in *Escherichia coli* is the ATP-dependent protease, protease La, *Proc. Natl. Acad. Sci. USA* **78:**4931–4935.

Ciechanover, A., Hod, Y., and Hershko, A., 1978, A heat-stable polypeptide component of an ATP-dependent proteolytic system from reticulocytes, *Biochem. Biophys. Res. Commun.* **81:**1100–1105.

Ciechanover, A., Heller, H., Elias, S., Haas, A. L., and Hershko, A., 1980, ATP-dependent conjugation of reticulocyte proteins with the polypeptide required for protein degradation, *Proc. Natl. Acad. Sci. USA* **77:**1365–1368.

Confalonieri, F., and Duguet, M., 1995, A 200-amimo acid ATPase module in search of a basic function, *BioEssays* **17:**639–650.

Cook, W. J., Jeffrey, L. C., Kasperek, E., and Pickart, C. M., 1994, Structure of tetraubiquitin shows how multiubiquitin chains can be formed, *J. Mol. Biol.* **236:**601–609.

Coux, O., Tanaka, K., and Goldberg, A. L., 1996, Structure and functions of the 20S and 26S proteasomes, *Annu. Rev. Biochem.* **65:**801–847.

Dahlmann, B., Kopp, F., Kuehn, L., Niedel, B., Pfeifer, G., Hegerl, R., and Baumeister, W., 1989, The multicatalytic proteinase (prosome) is ubiquitous from eukaryotes to archaebacteria, *FEBS Lett.* **251:**125–131.

Dahlmann, B., Kuehn, L., Grziwa, A., Zwickl, P., and Baumeister, W., 1992, Biochemical properties of the proteasome from *Thermoplasma acidophilum*, *Eur. J. Biochem.* **208:**789–797.

DeMarini, D. J., Papa, F. R., Swaminathan, S., Ursic, D., Rasmussen, T. P., Culbertson, M. R., and Hochstrasser, M., 1995, The yeast SEN3 gene encodes a regulatory subunit of the 26S proteasome complex required for ubiquitin-dependent protein degradation *in vivo*, *Mol. Cell. Biol.* **15:**6311–6321.

DeMartino, G.N., Moomaw, C. R., Zagnitko, O. P., Proske, R. J., Ma, C.-P., Afendis, S. J., Swaffield, J. C., and Slaughter, C. A., 1994, PA700, an ATP-dependent activator of the 20S proteasome, is an ATPase containing multiple members of a nucleotide-binding protein family, *J. Biol. Chem.* **269:**20878–20884.

DeMartino, G. N., Proske, R. J., Moomaw, C. R., Strong, A. A., Song, X., Hisamatsu, H., Tanaka, K., and Slaughter, C. A., 1996, Identification, purification, and characterization of a PA700-dependent activator of the proteasome, *J. Biol. Chem.* **271:**3112–3118.

Deveraux, Q., Ustrell, V., Pickart, C., and Rechsteiner, M., 1994, A 26S protease subunit that binds ubiquitin conjugates, *J. Biol. Chem.* **269:**7059–7061.

Deveraux, Q., Jensen, C., and Rechsteiner, M., 1995a, Molecular cloning and expression of a 26S protease subunit enriched in dileucine repeats, *J. Biol. Chem.* **270:**23726–23729.

Deveraux, Q., van Nocker, S., Mahaffey, D., Vierstra, R., and Rechsteiner, M., 1995b, Inhibition of ubiquitin-mediated proteolysis by the *Arabidopsis* 26S protease subunit S5a, *J. Biol. Chem.* **270:** 29660–29663.

Dintzis, H. M., 1961, Assembly of the peptide chains of hemoglobin, *Proc. Natl. Acad. Sci. USA* **47:** 247–261.

Driscoll, J., and Goldberg, A. L., 1989, Skeletal muscle proteasome can degrade proteins in an ATP-dependent process that does not require ubiquitin, *Proc. Natl. Acad. Sci. USA* **86:**787–791.

Driscoll, J., and Goldberg, A. L., 1990, The proteasome (multicatalytic protease) is a component of the 1500 kDa proteolytic complex which degrades ubiquitin-conjugated proteins, *J. Biol. Chem.* **265:**4789–4792.

Driscoll, J., Frydman, J., and Goldberg, A. L., 1992, An ATP-stabilized inhibitor of the proteasome is a component of the 1500-kDa ubiquitin conjugate-degrading complex, *Proc. Natl. Acad. Sci. USA* **89:**4986–4990.

Dubiel, W., and Rechsteiner, M., The 19S regulatory complex of the 26S proteasome, in: *Advances in Molecular Cell Biology* (A. J. Rivett, ed.), JAI Press, Greenwich, CT, in press.

Dubiel, W., Pratt, G., Ferrell, K., and Rechsteiner, M., 1992a, Purification of an 11S regulator of the multicatalytic protease, *J. Biol. Chem.* **267:**22367–22377.

Dubiel, W., Ferrell, K., Pratt, G., and Rechsteiner, M., 1992b, Subunit 4 of the 26S protease is a member of a novel eukaryotic ATPase family, *J. Biol. Chem.* **267:**22699–22702.

Dubiel, W., Ferrell, K., and Rechsteiner, M., 1993, Peptide sequencing identifies MSS1, a modulator of HIV Tat-mediated transactivation, as subunit 7 of the 26S protease, *FEBS Lett.* **323:**276–278.

Dubiel, W., Ferrell, K., and Rechsteiner, M., 1994, Tat-binding protein 7 is a subunit of the 26S protease, *Biol. Chem. Hoppe-Seyler* **375:**237–240.

Dubiel, W., Ferrell, K., and Rechsteiner, M., 1995a, Subunits of the regulatory complex of the 26S protease, *Mol. Biol. Rep.* **21:**27–34.

Dubiel, W., Ferrell, K., Dumdey, R., Standera, S., Prehn, S., and Rechsteiner, M., 1995b, Molecular cloning and expression of subunit 12: A non-MCP and non-ATPase subunit of the 26S protease, *FEBS Lett.* **363:**97–100.

Eytan, E., Ganoth, D., Armon, T., and Hershko, A., 1989, ATP-dependent incorporation of 20S protease into the 26S complex that degrades proteins conjugated to ubiquitin, *Proc. Natl. Acad. Sci. USA* **86:**7751–7755.

Eytan, E., Armon, T., Heller, H., Beck, S., and Hershko, A., 1993, Ubiquitin C-terminal hydrolase activity associated with the 26S protease complex, *J. Biol. Chem.* **268:**4668–4674.

Falquet, L., Paquet, N., Frutiger, S., Hughes, G. J., Khan, H.-V., and Jaton, J.-C., 1995, cDNA cloning of a human 100 kDa de-ubiquitinating enzyme: The 100 kDa human de-ubiquitinase belongs to the ubiquitin C-terminal hydrolase family 2 UCH2, *FEBS Lett.* **376:**233–237.

Fenteany, G., Standaert, R. F., Lane, W. S., Choi, S., Corey, E. J., and Schreiber, S. L., 1995, Inhibition of proteasome activities and subunit-specific amino-terminal threonine modification by lactacystin, *Science* **268:**726–731.

Ferrell, K., Deveraux, Q., Van Nocker, S., and Rechsteiner, M., 1996, Molecular cloning and expression of human S5a, a multiubiquitin chain binding subunit of the 26S protease, *FEBS Lett.* **381:**143–148.

Flanagan, J. M., Wall, J. S., Capel, M. S., Schneider, D. K., and Shanklin, J., 1995, Scanning transmission electron microscopy and small-angle scattering provide evidence that native *Escherichia coli* ClpP is a tetradecamer with an axial pore, *Biochemistry* **34:**10910–10917.

Ganoth, D., Leshinsky, E., Eytan, E., and Hershko, A., 1988, A multicomponent system that degrades proteins conjugated to ubiquitin, *J. Biol. Chem.* **263:**12412–12419.

Goldberg, A. L., 1992, The mechanism and functions of ATP-dependent proteases in bacterial and animal cells, *Eur. J. Biochem.* **203:**9–23.

Gonen, H., Schwartz, A. L., and Ciechanover, A., 1991, Purification and characterization of a novel protein that is required for degradation of N-α-acetylated proteins by the ubiquitin system, *J. Biol. Chem.* **266:**19221–19231.

Gonen, H., Smith, C. E., Siegel, N. R., Kahana, C., Merrick, W. C., Chakraburtty, K., Schwartz, A. L., and Ciechanover, A., 1994, Protein synthesis elongation factor EF-1α is essential for ubiquitin-dependent degradation of certain N^{α}-acetylated proteins and may be substituted for by the bacterial elongation factor EF-Tu, *Proc. Natl. Acad. Sci. USA* **91:**7648–7652.

Gonen, H., Dickman, D., Schwartz, A. L., and Ciechanover, A., 1996, Protein synthesis elongation factor EF-1α is an isopeptidase essential for ubiquitin-dependent degradation of certain proteolytic substrates, in *Intracellular Protein Catabolism* (K. Suzuki and J. Bond, eds.), pp. 209–219, Plenum Press, New York.

Gordon, C., McGurk, G., Wallace, M., and Hastie, N. D., 1996, A conditional lethal mutant in the fission yeast 26S protease subunit *mts3+* is defective in metaphase to anaphase transition, *J. Biol. Chem.* **271:**5704–5711.

Gray, C. W., Slaughter, C. A., and DeMartino, G. N., 1994, PA28 activator protein forms regulatory caps on proteasome stacked rings, *J. Mol. Biol.* **236:**7–15.

Guo, G. G., Gu, M., and Etlinger, J. D., 1994, 240 kDa proteasome inhibitor (CF-2) is identical to δ-aminolevulinic acid dehydratase, *J. Biol. Chem.* **269:**12399–12402.

Haas, A., and Rose, I., 1981, Hemin inhibits ATP-dependent ubiquitin-dependent proteolysis, *Proc. Natl. Acad. Sci. USA* **78:**6845–6848.

Haas, A., Reback, P. M., Pratt, G., and Rechsteiner, M., 1990, Ubiquitin-mediated degradation of histone H3 does not require the substrate-binding ubiquitin protein ligase, E3, or attachment of polyubiquitin chains, *J. Biol. Chem.* **265:**21664–21669.

Hadari, T., Warms, J. V. B., Rose, I. A., and Hershko, A., 1992, A ubiquitin C-terminal isopeptidase that acts on polyubiquitin chains, *J. Biol. Chem.* **267:**719–727.

Haracska, L., and Udvardy, A., 1995, Cloning and sequencing a non-ATPase subunit of the regulatory complex of the *Drosophila* 26S protease, *Eur. J. Biochem.* **231:**720–725.

Haracska, L., and Udvardy, A., 1996, Dissection of the regulator complex of the *Drosophila* 26S protease by limited proteolysis, *Biochem. Biophys. Res. Commun.* **220:**166–170.

Heinemeyer, W., Trondle, N., Albrecht, G., and Wolf, D. H., 1994, PRE5 and PRE6, the last missing genes encoding 20S proteasome subunits from yeast? Indication for a set of 14 different subunits in the eukaryotic proteasome core, *Biochemistry* **33:**12229–12237.

Hendil, K. B., Welinder, K. G., Pedersen, D., Uerkvitz, W., and Kristensen, P., 1993, Subunit stoichiometry of human proteasomes, *Enzyme Protein* **47:**232–240.

Hermida-Matsumoto, M.-L., Chock, P. B., Curran, T., and Yang, D. C. H., 1996, Ubiquitinylation of transcription factors c-Jun and c-Fos using reconstituted ubiquitinylating enzymes, *J. Biol. Chem.* **271:**4930–4936.

Hershko, A., and Heller, H., 1985, Occurrence of a polyubiquitin structure in ubiquitin–protein conjugates, *Biochem. Biophys. Res. Commun.* **128:**1079–1086.

Hershko, A., Ciechanover, A., Heller, H., Haas, A. L., and Rose, I. A., 1980, Proposed role of ATP in protein breakdown: Conjugation of proteins with multiple chains of the polypeptide of ATP-dependent proteolysis, *Proc. Natl. Acad. Sci. USA* **77:**1783–1786.

Hershko, A., Ganoth, D., Pehrson, J., Palazzo, R. E., and Cohen, L. H., 1991, Methylated ubiquitin inhibits cyclin degradation in clam embryo extracts, *J. Biol. Chem.* **266:**16376–16379.

Hill, C. P., Johnston, N. L., and Cohen, R. E., 1993, Crystal structure of a ubiquitin-dependent degradation substrate: A three-disulfide form of lysozyme, *Proc. Natl. Acad. Sci. USA* **90:**4136–4140.

Hilt, W., and Wolf, D. H., 1996, Proteasomes: Destruction as a programme, *Trends Biochem. Sci.* **21:**96–102.

Hisamatsu, H., Shimbara, H., Saito, Y., Kristensen, P., Hendil, K. B., Fujiwara, T., Takahashi, E.,

Tanahashi, N., Tamura, T., Ichihara, A., and Tanaka, K., 1996, Newly identified pair of proteasomal subunits regulated reciprocally by interferon γ, *J. Exp. Med.* **183:**1807–1816.

Hoffman, L., and Rechsteiner, M., 1994, Activation of the multicatalytic protease, *J. Biol. Chem.* **269:**16890–16895.

Hoffman, L., and Rechsteiner, M., 1996, Regulatory feature of multicatalytic and 26 S proteases, *Curr. Top. Cell. Regul.* **34:**1–32.

Hoffman, L., and Rechsteiner, M., 1997, Nucleotidase activities of the 26 S protease and its regulatory complex, *J. Biol. Chem.* **271:**32538–32545.

Hoffman, L., Pratt, G., and Rechsteiner, M., 1992, Multiple forms of the 20 S multicatalytic and the 26 S ubiquitin/ATP-dependent proteases from rabbit reticulocyte lysate, *J. Biol. Chem.* **267:** 22362–22368.

Hough, R., and Rechsteiner, M., 1986, Ubiquitin–lysozyme conjugates: Purification and susceptibility to proteolysis, *J. Biol. Chem.* **261:**2391–2399.

Hough, R., Pratt, G., and Rechsteiner, M., 1986, Ubiquitin–lysozyme conjugates: Identification of an ATP-dependent protease from reticulocyte lysate, *J. Biol. Chem.* **261:**2400–2408.

Hough, R., Pratt, G., and Rechsteiner, M., 1987, Purification of two high molecular weight proteases from rabbit reticulocyte lysate, *J. Biol. Chem.* **262:**8303–8311.

Hough, R., Pratt, G., and Rechsteiner, M., 1988, ATP/ubiquitin-dependent proteases, in *Ubiquitin* (M. Rechsteiner, ed.), pp. 101–134, Plenum Press, New York.

Hügle, B., Kleinschmidt, J. A., and Franke, W.W., 1983, The 22S cylinder particles of *Xenopus laevis*. Immunological characterization and localization of their proteins in tissues and cultured cells, *Eur. J. Cell Biol.* **32:**157–163.

Ikai, A., Nishigai, M., Tanaka, K., and Ichihara, A., 1991, Electron microscopy of 26S complex containing 20S proteasome, *FEBS Lett.* **292:**21–24.

Ishii, N., Taguchi, H., Sasbe, H., and Yoshida, M., 1995, Equatorial split of holo-chaperonin from *Thermus thermophilus* by ATP and K^+, *FEBS Lett.* **362:**121–125.

Jariel-Encontre, I., Pariat, M., Martin, F., Carillo, S., Salvat, C., and Piechaczyk, M., 1995, Ubiquitinylation is not an absolute requirement for degradation of c-Jun protein by the 26S proteasome, *J. Biol. Chem.* **270:**11623–11627.

Jennissen, H. P., 1995, Ubiquitin and the enigma of intracellular protein degradation, *Eur. J. Biochem.* **231:**1–30.

Jentsch, S., and Schlenker, S., 1995, Selective protein degradation: A journey's end within the proteasome, *Cell* **82:**881–884.

Johnson, E. S., Ma, P. C. M., Ota, I. M., and Varshavsky, A., 1995, A proteolytic pathway that recognizes ubiquitin as a degradation signal, *J. Biol. Chem.* **270:**17442–17456.

Johnston, J. A., Johnson, E. S., Waller, P. R. H., and Varshavsky, A., 1995, Methotrexate inhibits proteolysis of dihydrofolate reductase by the N-end rule pathway, *J. Biol. Chem.* **270:**8172–8178.

Kanayama, H.-O., Tamura, T., Ugai, S., Kagawa, S., Tanahashi, N., Yoshimura, T., Tanaka, K., and Ichihara, A., 1992, Demonstration that the human 26S proteolytic complex consists of a proteasome and multiple associated protein components and hydrolyzes ATP and ubiquitin-ligated proteins by closely linked mechanisms, *Eur. J. Biochem.* **206:**567–578.

Kawahara, H., and Yokosawa, H., 1992, Cell cycle-dependent change of proteasome distribution during embryonic development of the ascidian *Halocynthia roretzi*, *Dev. Biol.* **151:**27–33.

Kawamura, M., Kominami, K.-I., Takeuchi, J., and Toh-e, A., 1996, A multicopy suppressor of *nin*1-1 of the yeast *Saccharomyces cerevisiae* is a counterpart of the *Drosophila melanogaster* diphenol oxidase A2 gene, *Dox-A2*, *Mol. Gen. Genet.* **251:**146–152.

Kessel, M., Maurizi, M. R., Kim, B., Kocsis, E., Trus, B. L., Singh, S. K., and Steven, A. C., 1995, Homology in structural organization between *E. coli* ClpAP protease and the eukaryotic 26S proteasome, *J. Mol. Biol.* **250:**587–594.

Kleinschmidt, J. A., Hugle, B., Grund, C., and Franke, W. W., 1983, The 22S cylinder particles of

Xenopus laevis. I. Biochemical and electron microscopic characterization, *Eur. J. Cell Biol.* **32:** 143–156.

Kloetzel, P.-M., Falkenburg, P. E., Hössl, P., and Glätzer, K. H., 1987, The 19S ring-type particles of *Drosophila*, *Exp. Cell Res.* **170:**204–213.

Klotzbücher, A., Stewart, E., Harrison, D., and Hunt, T., 1996, The "destruction box" of cyclin A allows B-type cyclins to be ubiquitinated, but not efficiently destroyed, *EMBO J.* **15:**3053–3064.

Kominami, K.-i., and Toh-e, A., 1994, Characterization of the function of the NIN1 gene product of *Saccharomyces cerevisiae*, *Exp. Cell Res.* **211:**203–211.

Kominami, K.-i., DeMartino, G. N., Moomaw, C. R., Slaughter, C. A., Shimbara, N., Tujimoto, M., Yokosawa, H., Hisamatsu, H., Tanahasi, N., Shimizu, Y., Tanaka, K., Toh-e, A., 1995, Nin1p, a regulatory subunit of the 26S proteasome, is necessary for activation of Cdc28p kinase of *Saccharomyces cerevisiae*, *EMBO J.* **14:**3105–3115.

Kopp, F., Dahlmann, B., and Hendil, K. B., 1993, Evidence indicating that the human proteasome is a complex dimer, *J. Mol. Biol.* **229:**14–19.

Kopp, F., Kristensen, P., Hendil, K. B., Johnsen, A., Sobeck, A., and Dahlmann, B., 1995, The human proteasome subunit HsN3 is located in the inner rings of the complex dimer, *J. Mol. Biol.* **248:** 264–272.

Larimore, F. S., Waxman, L., and Goldberg, A. L., 1982, Studies of the ATP-dependent proteolytic enzyme, protease La, from *Escherichia coli*, *J. Biol. Chem.* **257:**4187–4195.

Lee, J. W., Ryan, F., Swaffield, J. C., Johnston, S. A., and Moore, D. D., 1995, Interaction of thyroid-hormone receptor with a conserved transcriptional mediator, *Nature* **374:**91–94.

Li, X., Gu, M., and Etlinger, J. D., 1991, Isolation and characterization of a novel endogenous inhibitor of the proteasome, *Biochemistry* **30:**9709–9715.

Lin, L., and Ghosh, S., 1996, A glycine-rich region in NK-κB p105 functions as a processing signal for the generation of the p50 subunit, *Mol. Cell. Biol.* **16:**2248–2254.

Löetscher, P., Pratt, G., and Rechsteiner, M., 1991, The C-terminus of mouse ornithine decarboxylase confers rapid degradation on dihydrofolate reductase: Support for the PEST hypothesis, *J. Biol. Chem.* **266:**11213–11220.

Löwe, J., Stock, D., Jap, B., Zwickl, P., Baumeister, W., and Huber, R., 1995, Crystal structure of the 20S proteasome from the archaeon *T. acidophilum* at 3.4Å resolution, *Science* **268:**533–539.

Lupas, A., Van Dyke, M., and Stock, J., 1991, Predicting coiled coils from protein sequences, *Science* **252:**1162– 1164.

Lurquin, C., Van Pel, A., Mariamé, B., De Plaen, E., Szikora, J.-P., Janssens, C., Reddehase, M. J., Lejeune, J., and Boon, T., 1989, Structure of the gene of Tum$^-$ transplantation antigen P91A: The mutated exon encodes a peptide recognized with L^d by cytolytic T cells, *Cell* **58:**293–303.

Ma, C.-P., Slaughter, C. A., and DeMartino, G. N., 1992, identification, purification and characterization of a protein activator (PA28) of the 20S proteasome macropain, *J. Biol. Chem.* **267:**10515–10523.

Ma, C.-P., Vu, J. H., Proske, R. J., Slaughter, C. A., and DeMartino, G. N., 1994, Identification, purification, and characterization of a high molecular weight, ATP-dependent activator (PA700) of the 20S proteasome, *J. Biol. Chem.* **269:**3539–3547.

Machiels, B. M., Henfling, M. F. R., Broers, J. L. V., Hendil, K. B., and Ramaekers, F. C. S., 1995, Changes in immunocytochemical detectability of proteasome epitopes depending on cell growth and fixation conditions of lung cancer cell lines, *Eur. J. Cell Biol.* **66:**202–292.

Mamroud-Kidron, E., and Kahana, C., 1994, The 26S proteasome degrades mouse and yeast ornithine decarboxylase in yeast cells, *FEBS Lett.* **356:**162–164.

Maurizi, M., 1992, Proteases and protein degradation in *Escherichia coli*, *Experientia* **48:**178–201.

Murakami, Y., Matsufuji, S., Kameji, T., Hayashi, S.-I., Igarashi, K., Tamura, T., Tanaka, K., and Ichihara, A., 1992, Ornithine decarboxylase is degraded by the 26S proteasome without ubiquitination, *Nature* **360:**597–599.

Nandi, D., Jiang, H., and Monaco, J. J., 1996, Identification of MECL-1 (LMP-10) as the third IFN-γ-inducible proteasome subunit, *J. Immunol.* **156:**2361–2364.

Nederlof, P. M., Wang, H.-R., and Baumeister, W., 1995, Nuclear localization signals of human and *Thermoplasma* proteasomal α subunits are functional *in vitro*, *Proc. Natl. Acad. Sci. USA* **92:** 12060–12064.

Nelbock, P., Dillon, P. J., Perkins, A., and Rosen, C. A., 1990, A cDNA for a protein that interacts with the human immunodeficiency virus Tat transactivator, *Science* **248:**1650–1653.

Ohana, B., Moore, P. A., Ruben, S. M., Southgate, C. D., Green, M. R., and Rosen, C. A., 1993, The type 1 human immunodeficiency virus Tat binding protein is a transcriptional activator belonging to an additional family of evolutionarily conserved genes, *Proc. Natl. Acad. Sci. USA* **90:**138–142.

Orlowski, M., Cardozo, C., and Michaud, C., 1993, Evidence for the presence of five distinct proteolytic components in the pituitary multicatalytic proteinase complex. Properties of two components cleaving bonds on the carboxyl side of branched chain and small neutral amino acids, *Biochemistry* **32:**1563–1572.

Palmer, A., Mason, G. G. F., Paramio, J. M., Knecht, E., and Rivett, A. J., 1994, Changes in proteasome localization during the cell cycle, *Eur. J. Cell Biol.* **64:**163–175.

Palmer, A., Rivett, A. J., Thomson, S., Hendil, K. B., Butcher, G. W., Fuertes, G., and Knecht, E., 1996, Subpopulations of proteasomes in rat liver nuclei microsomes and cytosol, *Biochem. J.* **316:** 401–407.

Palombella, V. J., Rando, O. J., Goldberg, A. L., and Maniatis, T., 1994, The ubiquitin-proteasome pathway is required for processing the NF-κB1 precursor protein and the activation of NF-κB, *Cell* **78:**773–785.

Pentz, E. S., and Wright, T. R. F., 1991, *Drosophila melanogaster* diphenol oxidase A2: Gene structure and homology with the mouse mast-cell tum$^-$ transplantation antigen, P91A, *Gene* **103:**239–242.

Peters, J.-M., Walsh, M. J., and Franke, W. W., 1990, An abundant and ubiquitous homo-oligomeric ring-shaped ATPase particle related to the putative vesicle fusion proteins Sec18p and NSF, *EMBO J.* **9:**1757–1767.

Peters, J.-M., Harris, J. R., and Kleinschmidt, J. A., 1991, Ultrastructure of the ~26S complex containing the ~20S cylinder particle multicatalytic proteinase/proteasome, *Eur. J. Cell Biol.* **56:** 422–432.

Peters, J.-M., Harris, J. R., Lustig, A., Müller, S., Engel, A., Volker, S., and Franke, W. W., 1992, Ubiquitous soluble Mg^{2+}-ATPase complex. A structural study, *J. Mol. Biol.* **223:**557–571.

Peters, J.-M., Cejka, Z., Harris, J. R., Kleinschmidt, J. A., and Baumeister, W., 1993, Structural features of the 26S proteasome complex, *J. Mol. Biol.* **234:**932–937.

Peters, J.-M., Franke, W. W., and Kleinschmidt, J. A., 1994, Distinct 19S and 20S subcomplexes of the 26S proteasome and their distribution in the nucleus and the cytoplasm, *J. Biol. Chem.* **269:**7709–7718.

Realini, C., Rogers, S. W., and Rechsteiner, M., 1994a, KEKE motifs: Proposed roles in protein–protein association and presentation of peptides by MHC class I receptors, *FEBS Lett.* **348:**109–113.

Realini, C., Dubiel, W., Pratt, G., Ferrell, K., and Rechsteiner, M., 1994b, Molecular cloning and expression of a gamma-interferon inducible activator of the multicatalytic protease, *J. Biol. Chem.* **269:**20727–20732.

Rechsteiner, M., 1987, Ubiquitin-mediated pathways for intracellular proteolysis, *Annu. Rev. Cell Biol.* **3:**1–30.

Rechsteiner, M., 1991, Natural substrates for the ubiquitin proteolytic pathway, *Cell* **66:**615–618.

Rechsteiner, M., and Rogers, S., 1996, PEST sequences and regulation by proteolysis, *Trends Biochem. Sci.* **21:**267–271.

Rechsteiner, M., Hoffman, L., and Dubiel, W., 1993, The multicatalytic and 26S proteases, *J. Biol. Chem.* **268:**6065–6068.

Rivett, A. J., Palmer, A., and Knecht, E., 1992, Electron microscopic localization of the multicatalytic proteinase complex in rat liver and in cultured cells, *J. Histochem. Cytochem.* **40:**1165–1172.

Rosenberg-Hasson, Y., Bercovich, Z., Ciechanover, A., and Kahana, C., 1989, Degradation of ornithine decarboxylase in mammalian cells is ATP dependent but ubiquitin independent, *Eur. J. Biochem.* **185:**469–474.

Rubin, D. M., and Finley, D., 1995, The proteasome: A protein-degrading organelle? *Curr. Biol.* **5:** 854–858.

Rubin, D. M., Coux, O., Wefes, I., Hengartner, C., Young, R. A., Goldberg, A. L., and Finley, D., 1996, Identification of the gal4 suppressor Sug1 as a subunit of the yeast 26S proteasome, *Nature* **379:** 655–657.

Scherer, D. C., Brockman, J. A., Chen, Z., Maniatis, T., and Ballard, D. W., 1995, Signal-induced degradation of IκBa requires site-specific ubiquitination, *Proc. Natl. Acad. Sci. USA* **92:**11259–11263.

Seemüller, E., Lupas, A., Stock, D., Löwe, J., Huber, R., and Baumeister, W., 1995, Proteasome from *Thermoplasma acidophilum*: A threonine protease, *Science* **268:**533–539.

Seemüller, E., Lupas, A., and Baumeister, W., 1996, Autocatalytic processing of the 20S proteasome, *Nature* **382:**468–470.

Shaeffer, J. R., and Kania, M. A., 1995, Degradation of monoubiquitinated α-globin by 26S proteasomes, *Biochemistry* **34:**4015–4021.

Shelton, E., Kuff, E. L., Maxwell, E. S., and Harrington, J. T., 1970, Cytoplasmic particles and aminoacyl transferase I activity, *J. Cell Biol.* **45:**1–8.

Shibuya, H., Irie, K., Ninomiya-Tsuji, J., Goebl, M., Taniguchi, T., and Matsumoto, K., 1992, New human gene encoding a positive modulator of HIV Tat-mediated transactivation, *Nature* **357:** 700–702.

Stancovski, I., Gonen, H., Orian, A., Schwartz, A. L., and Ciechanover, A., 1995, Degradation of the proto-oncogene product c-fos by the ubiquitin proteolytic system *in vivo* and *in vitro*: identification and characterization of the conjugating enzymes, *Mol. Cell. Biol.* **15:**7106–7116.

Swaffield, J. C., Bromberg, J. F., and Johnston, S. A., 1992, Alterations in a yeast protein resembling HIV Tat-binding protein relieve requirement for an acidic activation domain in GAL4, *Nature* **357:**698–704.

Swaffield, J. C., Melcher, K., and Johnston, S. A., 1995, A highly conserved ATPase protein as a mediator between acidic activation domains and the TATA-binding protein, *Nature* **374:**88–91.

Tanaka, K., Kumatori, A., Ii, K., and Ichihara, A., 1989, Direct evidence for nuclear and cytoplasmic colocalization of proteasomes (multiprotease complexes) in liver, *J. Cell. Physiol.* **139:**34–41.

Tanaka, K., Yoshimura, T., Tamura, T., Fujiwara, T., Kumatori, A., and Ichihara, A., 1990, Possible mechanism of nuclear translocation of proteasomes, *FEBS Lett.* **271:**41–46.

Thomson, S., and Rivett, A. J., 1996, Processing of N3, a mammalian proteasome beta-type subunit, *Biochem. J.* **315:**733–738.

Treier, M., Staszewski, L. M., and Bohmann, D., 1994, Ubiquitin-dependent c-Jun degradation *in vivo* is mediated by the δ domain, *Cell* **78:**787–798.

Tsurumi, C., Shimizu, Y., Saeki, M., Kato, S., DeMartino, G. N., Slaughter, C. A., Fujimuro, M., Yokosawa, H., Yamasaki, M., Hendil, K. B., Toh-E, A., Tanahashi, N., and Tanaka, K., 1996, cDNA cloning and functional analysis of the p97 subunit of the 26 S proteasome, a polypeptide identical to the type-1 tumor-necrosis-factor-receptor-associated protein-2/55.11, *Eur. J. Biochem.* **239:**912–921.

Udvardy, A., 1993, Purification and characterization of a multiprotein component of the *Drosophila* 26S (1500 kDa) proteolytic complex, *J. Biol. Chem.* **268:**9055–9062.

Ustrell, V., Realini, C., Pratt, G., and Rechsteiner, M., 1995, Human lymphoblast and erythrocyte multicatalytic proteases: Differential peptidase activities and responses to the 11S regulator, *FEBS Lett.* **376:**155–158.

Van Nocker, S., Sadis, S., Rubin, D. M., Glickman, M., Fu, H., Coux, O., Wefes, I., Finley, D., and Vierstra, R. D., 1996a, The multiubiquitin chain-binding protein MCB1 is a component of the 26S proteasome in *Saccharomyces cerevisiae* and plays a nonessential, substrate-specific role in protein turnover, *Mol. Cell. Biol.* **16:**6020–6028.

Van Nocker, S., Deveraux, Q., Rechsteiner, M., and Vierstra, R., 1996b, *Arabidopsis MBP1* gene encodes a conserved ubiquitin recognition component of the 26S proteasome, *Proc. Natl. Acad. Sci. USA* **93:**856–860.

Wang, W., Chevray, P. M., and Nathans, D., 1996, Mammalian Sug1 and c-Fos in the nuclear 26S proteasome, *Proc. Natl. Acad. Sci. USA* **93:**8236–8240.

Waxman, L., and Goldberg, A. L., 1982, Protease La from *Escherichia coli* hydrolyzes ATP and proteins in a linked fashion, *Proc. Natl. Acad. Sci. USA* **79:**4883–4887.

Weissman, J. S., Sigler, P. B., and Horwich, A. L., 1995, From the cradle to the grave: Ring complexes in the life of a protein, *Science* **268:**523–524.

Wenzel, T., and Baumeister, W., 1995, Conformational constraints in protein degradation by the 20S proteasome, *Nature Struct. Biol.* **2:**199–204.

Wenzel, T., Eckerskorn, C., Lottspeich, F., and Baumesiter, W., 1994, Existence of a molecular ruler in proteasomes suggested by analysis of degradation products, *FEBS Lett.* **349:**205–209.

Wilk, S., and Orlowski, M., 1980, Cation-sensitive neutral endopeptidase: Isolation and specificity of the bovine pituitary enzyme, *J. Neurochem.* **35:**1172–1882.

Wilk, S., and Orlowski, M., 1983, Evidence that pituitary cation-sensitive neutral endopeptidase is a multicatalytic protease complex, *J. Neurochem.* **40:**842–849.

Wilkinson, K. D., Tashayev, V. L., O'Connor, L. B., Larsen, C. N., Kasperek, E., and Pickart, C. M., 1995, Metabolism of the polyubiquitin degradation signal: Structure, mechanism, and role of isopeptidase T, *Biochemistry* **34:**14535–14546.

Xu, Q., Singer, R. A., and Johnston, G. C., 1995, Sug1 modulates yeast transcription activation by cdc68, *Mol. Cell. Biol.* **15:**6025–6035.

Yang, Y., Früh, K., Ahn, K., and Peterson, P. A., 1995, *In vivo* assembly of the proteasomal complexes, implications for antigen processing, *J. Biol. Chem.* **270:**27687–27694.

Yokota, K.-Y., Kagawa, S., Shimizu, Y., Akioka, H., Tsurumi, C., Noda, C., Fujimoro, M., Yokosawa, H., Fujiwara, T., Takahashi, E., Ohba, M., Yamasaki, M., DeMartino, G. N., Slaughter, C. A., Toh-e, A., Tanaka, K., 1996, cDNA cloning of p112, the largest regulatory subunit of the human 26S proteasome, and functional analysis of its yeast homologue, Sen3p, *Mol. Biol. Cell* **7:** 853–870.

Yoshimura, T., Kameyama, K., Takagi, T., Ikai, A., Tokunaga, F., Koide, T., Tanahashi, N., Tamura, T., Cejka, Z., Baumeister, W., Tanaka, K., and Ichihara, A., 1993, Molecular characterization of the "26S" proteasome complex from rat liver, *J. Struc. Biol.* **111:**200–211.

Yukawa, M., *et al.*, 1991, Proteasome and its novel endogenous activator in human platelets, *Biochem. Biophys. Res. Commun.* **178:**256–262.

Zwickl, P., Grziwa, A., Pühler, G., Dahlmann, B., Lottspeich, F., and Baumeister, W., 1992, Primary structure of the *Thermoplasma* proteasome and its implications for the structure, function, and evolution of the multicatalytic proteinase, *Biochemistry* **31:**964–972.

Zwickl, P., Kleinz, J., and Baumeister, W., 1994, Critical elements in proteasome assembly, *Nature Struct. Biol.* **1:**765–770.

CHAPTER 7

Function of the Proteasome in Antigen Presentation

Jochen Beninga and Alfred L. Goldberg

1. INTRODUCTION

Recent studies using selective inhibitors of the proteasome have established that these particles catalyze the degradation of most proteins in mammalian cells, including both the rapid breakdown of abnormal and regulatory proteins and the slower degradation of the bulk of cytosolic proteins.

2. THE TWO PATHWAYS FOR ANTIGEN PRESENTATION

The cellular immune system constantly screens both the extracellular milieu and the intracellular compartment for small polypeptides that differ from those present normally. The monitoring of intracellular and endocytosed extracellular antigens involves two distinct arms of the immune system: the MHC class I and the MHC class II pathways. These two pathways use different proteolytic systems, different vesicular transport machinery, and distinct MHC molecules for antigen presentation (Goldberg and Rock, 1992; Germain and Margulies, 1993; York and

Jochen Beninga and Alfred L. Goldberg • Department of Cell Biology, Harvard Medical School, Boston, Massachusetts 02115.

Ubiquitin and the Biology of the Cell, edited by Peters *et al.* Plenum Press, New York, 1998.

Rock, 1996). MHC molecules are highly polymorphic membrane glycoproteins that function in presentation of small antigenic peptides to T lymphocytes.

The MHC class II pathway functions in presenting antigens derived from the extracellular environment on class II molecules, which are found on specialized "antigen-presenting cells." Extracellular proteins (e.g., bacterial antigens) taken up by endocytosis are degraded in the endosomal and lysosomal compartments by acid-optimal proteases (cathepsins B, H, L, S, or O) (Germain and Margulies, 1993). These enzymes are also involved in the normal turnover of many membrane receptors, and in the accelerated degradation in "autophagic vacuoles" of many cytosolic and organellar proteins in poor nutritional conditions (Dice, 1987). There is strong evidence that these lysosomal/endosomal proteases also degrade endocytosed proteins for antigen presentation on MHC class II molecules (Ziegler and Unanue, 1982; Germain and Margulies, 1993). For example, the presentation of exogenous antigens to T cells can be blocked by raising the pH in lysosomes with weak bases or by using inhibitors of lysosomal proteases (Ziegler and Unanue, 1982; Takahashi *et al.*, 1989). In antigen-presenting cells (e.g., macrophages), some of the peptide fragments generated during lysosomal/endosomal digestion of the endocytosed proteins bind to MHC class II molecules. The association between peptides and the MHC class II molecules occurs in a specialized endosomal compartment, MIIC (also CIIV or CPL), and requires the invariant chain (Ii) and the HLA-DM gene product, whose precise function remains unclear (Busch and Mellius, 1996). The antigenic peptides, which are heterogenous and generally range between 12 and 20 residues, are then transported together with the class II molecules to the plasma membrane. Their recognition by $CD4^+$ helper T lymphocytes leads to the initiation of inflammatory and antibody responses (Germain and Margulies, 1993).

The presentation of intracellular cytosolic antigens on MHC class I molecules occurs on most cells of the body. The proteolytic pathway that generates peptides for MHC class I presentation is a soluble, nonlysosomal ATP-dependent system present in the nucleus and cytosol (Goldberg, 1992). The initial step in the hydrolysis of most, but not all, such proteins is their covalent conjugation to the polypeptide ubiquitin. In this process, the C-terminus of ubiquitin becomes attached by an isopeptide bond to ε-amino groups on lysine residues on the protein substrate and then on other ubiquitins (Ciechanover, 1994; Finley and Chau, 1991). The attachment of long chains of ubiquitin to the protein substrate marks it for rapid hydrolysis by the very large ATP-dependent proteolytic complex, the 26 S proteasome (Goldberg, 1992; Rechsteiner *et al.*, 1993; Rivett, 1993). Although most of the peptides generated in the cytosol by this pathway during the breakdown of intracellular proteins (e.g., viral proteins) are hydrolyzed further by exoproteases to free amino acids, some are fed into the MHC class I pathway. These oligopeptides (Goldberg and Rock, 1992; Germain and Margulies, 1993; York and Rock, 1996) are then transported into the lumen of the ER by the MHC-

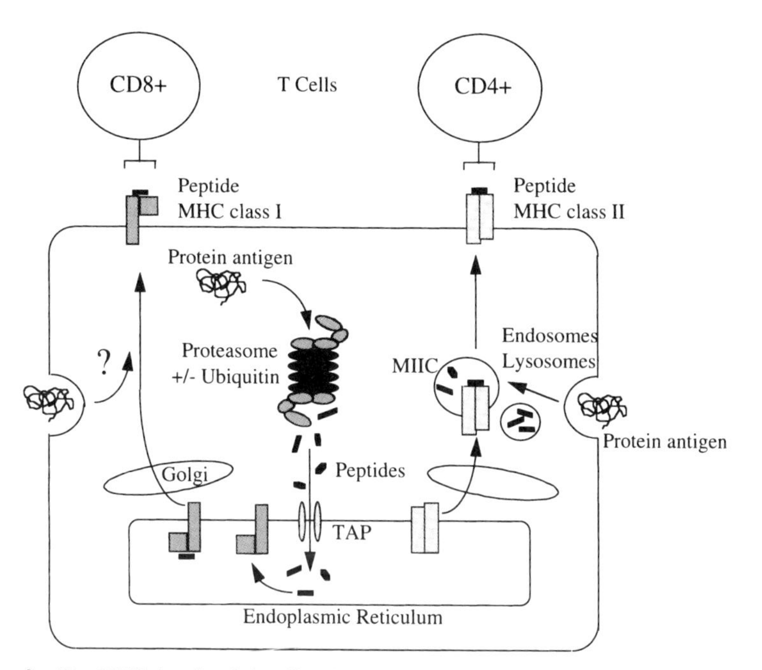

Figure 1. The MHC class I and class II pathways for presentation of intracellular and extracellular antigens.

encoded TAP complex (Fig. 1). The TAP complex, which is essential for MHC class I presentation, belongs to a group of homologous members of the ATP-dependent (ABC) family of transporter molecules. The TAP transporter, a heterodimer containing the TAP1 and TAP2 gene products, transports peptides by an ATP-dependent mechanism from the cytosol into the ER (York and Rock, 1996). In the ER, the transported peptides possibly may be trimmed further (Snyder *et al.*, 1994; Elliott *et al.*, 1995) and finally only peptides of eight or nine residues and appropriate sequence become bound to MHC class I molecules. These peptide–class I complexes are then transported through the Golgi network to the cell surface. If a peptide is displayed on class I molecules to which the immune system is not tolerant (e.g., a viral or a somatically mutated sequence), the cell is destroyed by cytolytic $CD8^+$ T lymphocytes (Germain and Margulies, 1993; York and Rock, 1996).

Over the last few years, it has become obvious that the MHC class I and class II pathways under certain conditions are not truly distinct, but are functionally connected to each other and are coordinately regulated, especially by γ-interferon (γ-IFN). For example, macrophages, which are specialized antigen-presenting cells, are able to take up and degrade particular exogenous antigens (e.g., proteins associated with cell debris or bound to beads) by the endosomal/lysosomal compartments and to feed the generated peptides into the class I pathway for presentation to $CD8^+$ cytotoxic T cells. What components of each pathway are involved in this "cross talk" are not yet clear and will require further investigation. However, these interactions between the class I and class II pathways seem to be exceptional, rather than a general immune mechanism (Rock, 1996).

3. PROTEIN UBIQUITINATION AND MHC CLASS I PRESENTATION

The first step in ubiquitin conjugation is an ATP-dependent reaction catalyzed by the enzyme El, which activates the C-terminus of ubiquitin to form a thioester (Ciechanover, 1994; Finley and Chau, 1991). To test whether the ubiquitin-dependent pathway is required for the generation of antigenic peptides, Michalek *et al.* (1993) used a Chinese hamster cell line (ts20) (Kulka *et al.*, 1988) that is temperature-sensitive for ubiquitin conjugation as a result of a defect in El. At the nonpermissive temperature (41°C), the mutant El enzyme is irreversibly inactivated. When a model antigenic protein, ovalbumin, was introduced into the cytosol at the permissive temperature (Moore *et al.*, 1988), an octapeptide comprising residues 257–264 (SIINFEKL) of ovalbumin appeared bound to surface MHC class I molecules (Michalek *et al.*, 1993). However, at 41°C, this process was blocked almost completely in the TS mutant, in contrast to the wild-type parent. Several control experiments indicated that this defect in antigen presentation was the result of a failure of the cells to generate this peptide in the absence of ubiquitin conjugation. For example, when the antigenic peptide, SIINFEKL, was expressed in the cells on a minigene, this peptide was presented on MHC molecules in the wild type and the mutant cells even at 41°C. Therefore, ubiquitination, and by extension the 26 S proteasome complex, must be required for the proteolytic generation of the antigenic peptide and not for some subsequent step in this pathway (e.g., peptide transport into the ER, binding to the MHC molecules, or vesicular transport to the surface).

Because other studies have failed to find a requirement for ubiquitination in antigen presentation (Cox *et al.*, 1995), these results were reinvestigated with an additional mouse cell line, ts85, which also has a temperature-sensitive El enzyme (Finley *et al.*, 1984). The ts85 cell line showed a similar ubiquitin dependence for SIINFEKL presentation from microinjected ovalbumin, as found with the ts20 hamster line (Michalek *et al.*, 1996), Surprisingly, in this set of experiments, antigen presentation from vaccinia virus-expressed ovalbumin and from microinjected, denatured (reduced and acylated) ovalbumin were not inhibited by heat inactivation of El to prevent ubiquitination. Nevertheless, their presentation was still sensitive to inhibitors of the proteasome (Michalek *et al.*, 1996). Possibly, these species of ovalbumin differ in tertiary structure, and therefore are degraded through a different pathway than the native protein. These findings raise the possibility that the structure of a protein can influence whether ubiquitin conjugation is required for its breakdown and its efficient presentation by MHC class I. In fact, when ovalbumin was denatured and methylated to block ubiquitination sites (i.e., the lysines, ε-amino groups) and were microinjected, its presentation was largely unaffected by heat inactivation of El in the ts85 cells.

This finding and those described below suggest that although ubiquitination is essential for generation of many antigenic peptides, some can be generated by proteasomes in a ubiquitin-independent process (Michalek *et al.*, 1996). However, it remains generally unclear to what extent 20 or 26 S proteasomes may function in protein breakdown *in vivo* without ubiquitin conjugation to substrates. Although rapid, regulated degradation of several (perhaps many) polypeptides clearly requires their ubiquitination, degradation of many polypeptides by isolated 20 and 26 S proteasomes does not require ubiquitination, and eubacteria and archaea clearly contain 20 S proteasomes, but lack ubiquitin (Coux *et al.*, 1996).

4. INFLUENCE OF THE RATE OF UBIQUITIN-DEPENDENT DEGRADATION ON ANTIGEN PRESENTATION

To further analyze the involvement of the ubiquitin–proteasome pathway in antigen presentation of certain proteins, experiments were undertaken to investigate systematically how rates of ubiquitin conjugation and proteasomal degradation influence the extent of antigen presentation. To analyze the influence of protein half-lives on class I presentation, Grant *et al.* (1995) built on the elegant work of Varshavsky and co-workers (Bachmair *et al.*, 1986; Gonda *et al.*, 1989) that demonstrated that proteins with abnormal N-terminal residues (ones terminating in large hydrophobic, basic, or acidic residues) undergo rapid ubiquitination and proteolysis (the "N-end rule") in yeast and reticulocytes (Bachmair and Varshavsky, 1989). Using β-galactosidase variants with or without destabilizing N-termini, Grant *et al.* (1995) showed that altering rates of ubiquitination markedly affected the presentation of antigenic peptides. The β-galatosidase variants were degraded with very different half-lives in cell extracts in accord with the reported N-end rule (Fig. 2). These proteins were then introduced into lymphoblasts, and antigen presentation was measured (Grant *et al.*, 1995). Townsend *et al.* (1988) had earlier shown that an N-terminal modification of a viral protein, which should stimulate ubiquitin-dependent degradation via the ubiquitin pathway (Bachmair and Varshavsky, 19889), enhanced its presentation on MHC class I. However, this phenomenon and the role of ubiquitination were not studied further.

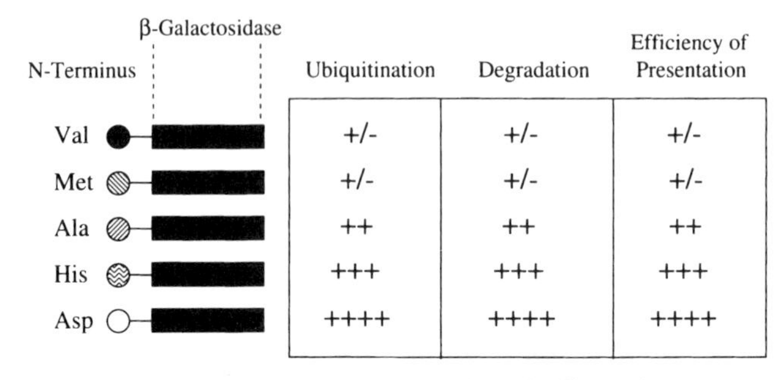

Figure 2. Rates of ubiquitin conjugation and degradation of different β-galactosidase constructs determine rates of presentation on MHC class I molecules. Summary of findings by Grant *et al.* (1995) comparing the rates of ubiquitination and degradation of β-galactosidase substrates in cell lysates with their rates of MHC class I-restricted presentation by LB27.4 cells.

In accord with the observations of Gonda *et al.* (1989), a striking correlation was found between the rate of degradation of these different variants by the N-end rule ubiquitin pathway in the extracts and the extent of class I-restricted presentation *in vivo* (Grant *et al.*, 1995). For example, 30 min after being microinjected into cells, the short-lived species (Asp-β-galactosidase or His-β-galactosidase) were presented on MHC class I molecules, unlike the stable species, Met-β-galactosidase or Val-β-galactosidase (Fig. 2). Moreover, the degradation of these proteins *in vivo* also seemed to require ubiquitin and the proteasome, as the presentation was blocked by inhibitors of proteasome function (see below) or by methylation of amino groups on the substrate, which prevents ubiquitin conjugation (Michalek *et al.*, 1996). These findings argue strongly that the rate of ubiquitin conjugation is a critical factor determining the rate of antigen presentation.

However, these results, obtained at short times after introducing the antigenic protein, do not mean that long-lived proteins are never presented on the surface. On the contrary, at longer times after introduction of the β-galactosidase variants, peptides from the short-lived variant progressively were lost from the cell surface, whereas there was a progressive increase in the appearance of the epitope from the much more stable wild-type β-galactosidase. For example, by 16 hr, antigen presentation was demonstrable only from the stable species, presumably because all of the short-lived protein (Asp-β-galactosidase) in the cell had been degraded, and the peptide–MHC molecules cleared from the surface. These findings strongly suggest that the appearance of antigenic peptides on the surface reflects the absolute rate of digestion of the specific protein by the proteasome, and that for maximal presentation, a protein should be degraded rapidly and also be abundant in the cell. In contrast to these observations, Shastri and colleagues were unable to demonstrate an increase in the presentation of antigenic peptides when the half-life of their endogenously synthesized precursor proteins was diminished (Groth *et al.*, 1996). The reasons for this difference are presently unclear.

Recently, Villanueva *et al.* (1996) directly investigated whether the abundance of an antigenic protein influences its capacity for class I antigen presentation. These investigators used *Listeria monocytogenes*, an intracellular bacterial pathogen, as a source for antigenic proteins. This organism secretes into the cytosol of infected cells an antigenic protein, the murine hydrolase, p60, from which a known 9mer antigenic peptide is generated. Three different strains of *L. monocytogenes*, with different capacities to secret p60 into the cytosol, were compared. These workers demonstrated a strong positive correlation between intracellular levels of the p60 proteins and the amount of MHC class I-presented p60 peptides. Moreover, they consistently observed a ratio of approximately 30 p60 protein molecules to each molecule of antigenic peptide independent of the cytosolic level of the p60 protein. These results indicate that the cytosolic abundance of an antigenic protein along with its rate of degradation are critical factors determining the extent of class I presentation via the MHC class I pathway. In other words, the presentation of peptides on the surface constitutes an "instant replay" of how much of a protein was being degraded by the cell's proteasomal machinery in the preceding several hours, and the duration of this record depends on the half-life of class I–peptide complexes (generally about several hours).

5. SELECTIVE INHIBITORS OF PROTEASOME ACTIVITY

The lack of specific inhibitors of the ubiquitin–proteasome pathway has long been a major factor limiting our understanding of its function *in vivo*. Recently, we have identified several peptide aldehydes that can block the activities of purified 20 and 26 S proteasomes, and thereby can reduce proteolysis in intact mammalian cells (Rock *et al.*, 1994). We showed that several compounds of this class could not only inhibit reversibly the chymotryptic and peptidylglutamyl activities of purified 20 and 26 S particles, but could also block the degradation of casein and ubiquitin-conjugated lysozyme by the 26 S complex purified from mammalian muscle. Moreover, these agents can readily enter cultured cells and inhibit overall proteolysis, without being acutely toxic at least for many hours (as shown by the normal morphological appearance of the cells, normal rates of protein synthesis and protein secretion, and so on).

Peptide aldehyde inhibitors of the proteasome are proving very useful in studies clarifying the role of the proteasome in different intracellular degradative processes. As expected from prior work, inhibition of proteasome function caused almost a complete block in the rapid breakdown of abnormal polypeptides (e.g., ones that have incorporated the amino acid analogue canavanine) and short-lived

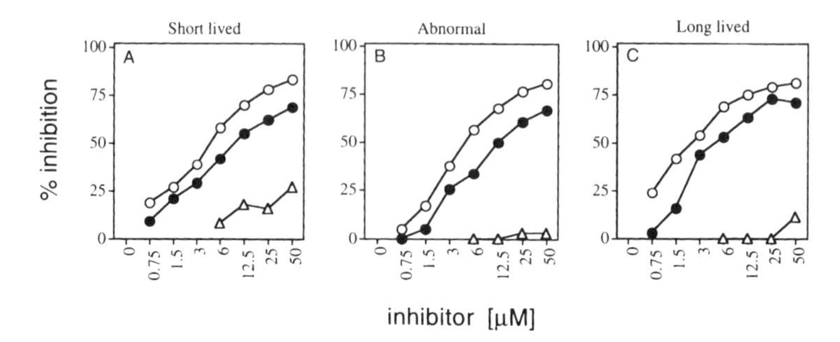

Figure 3. Effects of peptide aldehydes on protein catabolism in lymphoblasts. Representative experiments showing the effect of peptide aldehydes on the degradation of (A) short-lived proteins, (B) amino acid analogue (canavanine)-containing proteins, and (C) long-lived proteins without pre-incubation with inhibitors. The potencies of these inhibitors against these different classes of proteins are very similar and correlate with their relative potencies against the 20 and 26 S proteasome. Closed circles, LLnl; open circles, MG115; open triangles, LLM. Modified from Rock *et al.* (1994).

normal cell proteins (Rock *et al.*, 1994) (Fig. 3). It had long been recognized that such short-lived polypeptides are degraded by the cytosolic ATP-dependent pathway (Etlinger and Goldberg, 1977; Goldberg, 1992). By contrast, the bulk of cell proteins are relatively stable (i.e., they show an average half-life of 20–30 hr, which corresponds to a rate of about 2–3%/hr). It has been generally assumed that such polypeptides are not degraded by the ubiquitin–proteasome pathway, even though several studies have indicated that the responsible proteolytic system is a nonlysosomal, ATP-dependent process (Gronostajski *et al.*, 1985). Of particular interest was the finding that the peptide aldehyde inhibitors reduced almost completely the breakdown of these long-lived components in lymphoblasts and fibroblasts (Rock *et al.*, 1994) (Fig. 3). Moreover, with several different peptide aldehyde inhibitors, the potency in blocking the degradation of the most long-lived cell proteins was indistinguishable from that for the breakdown of abnormal or short-lived normal proteins (Fig. 3).

In addition to inhibiting the proteasome, these peptide aldehydes are also potent inhibitors of the cysteine proteases in the lysosome and the calpains (Rock *et al.*, 1994); however, inactivation of these proteolytic systems (e.g., with weak bases or inhibitors of cysteine proteases) does not reduce the bulk of intracellular proteolysis (Gronostajski *et al.*, 1984, 1985) or inhibit MHC class I antigen presentation (Goldberg and Rock, 1992). Moreover, the relative ability of different peptide aldehydes to inhibit intracellular proteolysis correlated precisely with their potencies in inhibiting the function of purified proteasomes. These findings clearly indicate that proteasomes are the primary sites for degradation of the bulk of proteins in mammalian cells, although the involvement of ubiquitin in this process remains to be established. These results (Rock *et al.*, 1994) were obtained in lymphoblasts under specific growth conditions. Under other nutritional conditions or with other cells, the lysosome may also be an important site for the breakdown of long-lived components; for example, in fibroblast cultures at high cell densities, overall protein degradation increases up to twofold by activation of a lysosomal mechanism (autophagic vacuole formation) (Gronostajski *et al.*, 1984).

One of the more potent of these peptide aldehyde inhibitors, MG132 (CbzLLLal), has proven particularly useful as an inhibitor of the ubiquitin–proteasome pathway in eukaryotes and mammalian cells. For example, at low concentrations (5 μM), the compound can also prevent the rapid degradation of the inhibitor of gene activation, IκB, and the proteolytic processing of the transcription factor, NF-κB, from its precursor, pl05, both of which require ubiquitination and the 26 S proteasome (Palombella *et al.*, 1994; Read *et al.*, 1995). In fact, selective inhibitors of the proteasome appear to have therapeutic potential as anti-inflammatory agents (Read *et al.*, 1995), because of their ability to prevent in intact animals the NF-κB-dependent production of inflammatory mediators (e.g., TNF, IL-1, IL-6) and adhesion molecules (e.g., ICAM-1 and VCAM-1 (Read *et al.*, 1995).

Another very useful and highly specific inhibitor of the proteasome is lactacystin. Fenteany *et al.* (1995) found that this agent reacts specifically and irreversibly with the active-site threonine of the β-subunits on proteasomes and does not affect other types of proteases. This agent was initially reported to inhibit multiple peptidase activities of the mammalian proteasome and covalently modify its X subunit. Recent experiments, however, indicate that it reacts with all of the catalytic active β-subunits as well as with their homologues induced by γ-IFN (Craiu *et al.*, 1997a). Also, L. R. Dick *et al.* (1996) showed that the spontaneous hydrolysis product of lactacystin, β-lactone, is actually the active species that reacts with the proteasome *in vivo*, and not lactacystin itself. Systematic studies have shown that both lactacystin and β-lactone are potent inhibitors of the degradation of short-lived and long-lived cell proteins (Craiu *et al.*, 1997).

6. BLOCKING PROTEASOME FUNCTION PREVENTS ANTIGEN PRESENTATION

These inhibitors have been used to determine whether the proteasome is the dominant site for generating peptides for MHC class I presentation. When ovalbumin was introduced by electroporation into the cytosol of lymphoblasts, the MHC class I presentation of the ovalbumin-derived peptide, SIINFEKL, could be blocked completely with the peptide aldehydes (MG115, Cbz-LLnVaL, MG101,

NAcLLnLal). However, these inhibitors did not reduce MHC class I expression when the cells were injected with the SIINFEKL peptide directly (Rock *et al.*, 1994). Thus, the inhibitors prevent the generation of the antigenic peptides without affecting the cell's capacity to transport peptides into the ER or to deliver the MHC–peptide complexes to the surface. These findings are also consistent with the ubiquitin requirement for class I presentation from ovalbumin discussed above (Michalek *et al.*, 1993, 1996). Furthermore, the relative efficacies of several inhibitors in blocking antigen presentation correlated with their relative potencies in reducing intracellular proteolysis and the activity of purified proteasomes.

In similar studies, Harding *et al.* (1995) used novel dipeptide aldehyde inhibitors to confirm that the proteasome is the proteolytic enzyme responsible for the production of class I-presented antigenic peptides from ovalbumin. Because these compounds are predominantly inhibitors of the chymotrypsin-like activity of the proteasome, the authors also argued that this activity is of primary importance in antigen processing. However, because multiple active sites must function together in the processive degradation of protein substrates by the 20 S particle, it is likely that inhibition of one or more active sites influences the function of other proteolytic sites.

The peptide aldehyde inhibitors can also affect the lysosomal proteases and calpains. However, specific inhibition of these cysteine proteases does not affect antigen presentation. To further investigate the role of the proteasome, Craiu *et al.* (1997b) tested whether the more specific inhibitors, lactacystin and β-lactone, also inhibit antigen presentation from ovalbumin. Irreversible inhibition of the proteasome by lactacystin or β-lactone results in a complete block of MHC class I presentation for ovalbumin, when this protein is expressed in cells using the vaccinia virus T7-expression system. By contrast, presentation of the SIINFEKL peptide, when expressed directly from a minigene construct, was not affected by lactacystin or β-lactone. As these agents block MHC class I antigen presentation in a similar way as MG101 (NAc-LLnLe) or (Cbz-LLLal), this process must involve the proteasome.

The peptide aldehyde inhibitors were also used to determine whether the proteasome is the source of most of the peptides presented on MHC class I molecules (Rock *et al.*, 1994). The stable assembly of MHC class I heterodimers in the ER and their movement to the cell surface require peptide binding (Germain and Margulies, 1993; York and Rock, 1996). Therefore, blocking the production of antigenic peptides should reduce the assembly of these complexes on the cell surface. In cells treated with these inhibitors, there was a marked reduction in the generation of MHC class I heterodimers without any decrease in the synthesis of unassembled MHC chains. This blockage of MHC heterodimer formation was completely reversed either when the peptide aldehyde inhibitors were removed and proteolysis was reinitiated, or when antigenic peptides were added to the cells (Rock *et al.*, 1994). In addition, lactacystin and β-lactone can block the generation of antigenic peptides from cellular proteins. By preventing the generation of antigenic peptides, these agents also block the assembly and transport of MHC class I molecules to the cell surface in a similar way as the peptide aldehydes (Craiu *et al.*, 1997a). These results constitute strong evidence that the great majority of peptides presented on MHC class I molecules are produced by proteasomes, although it is unclear to what extent they are functioning in a ubiquitin-dependent (26 S proteasome) or -independent (20 or 26 S proteasome) mode. In either case, it is very likely that proteasomes generate the antigenic peptides during the continuous degradation of cell proteins and that these fragments, unlike most proteasomal products, somehow escape further hydrolysis to amino acids. The capacity of the peptide aldehydes to inhibit the degradation of cell proteins correlated precisely with their capacity to block antigen presentation.

7. REGULATION OF PROTEASOME COMPOSITION BY γ-INTERFERON

Over 10 years ago, Monaco and McDevitt first described protein complexes termed *low-molecular-weight proteins* (LMPs), which they speculated played a role in antigen processing (Monaco and McDevitt, 1986). This suggestion was made because two LMP subunits, LMP2 and LMP7, were encoded in the MHC region and because their expression was enhanced by γ-IFN, a potent stimulator of immune responses (Monaco and McDevitt, 1986; Glynne *et al.*, 1991; Kelly *et al.*, 1991). Although many properties of the LMPs resembled those of the 20 S proteasome (e.g., both are complexes of about 600 kDa and contain multiple subunits ranging between 20 and 30 kDa (Brown *et al.*, 1991), there was until recently little or no communication between immunologists and biochemists studying intracellular proteolysis, and these important clues went unrecognized.

The first observations implicating the proteasome in class I antigen presentation came from studies of the genes in the MHC class II locus that encode LMP2 and LMP7 proteins, and adjacent to the TAP1 and 2 gene subunits of the peptide transporter (Glynne *et al.*, 1991; Kelly *et al.*, 1991; Brown *et al.*, 1991; Germain and Margulies, 1993; York and Rock, 1996). The sequences of the LMP2 and LMP7 genes were found to resemble closely those known β-subunits of the 20 S proteasome. Monaco's (Brown *et al.*, 1991) and Trowsdale's (Glynne *et al.*, 1991; Kelly *et al.*, 1991) laboratories then showed that the LMP complex and 20 S proteasome complex share most, if not all, of their subunits, and that LMP2 and LMP7 were actually components of the 20 S proteasome (Glynne *et al.*, 1991; Kelly *et al.*, 1991; Brown *et al.*, 1991; Aki *et al.*, 1992). These polypeptides correspond to the β-type subunits, which comprise the two central seven-member rings of this particle (Kopp *et al.*, 1993; Heinemeier *et al.*, 1994), and appear to contain the proteolytic active sites (for reviews, see Chapter 5; Goldberg, 1995; Coux *et al.*, 1996).

Subsequent work has established that γ-IFN, which is secreted by activated T lymphocytes and natural killer cells, induces multiple changes in the composition of the 20 S proteasome (Yang *et al.*, 1995). Three new subunits are expressed (LMP2, LMP7, and MECL-1) and incorporated into newly synthesized particles in place of normal subunits (X, Y, and Z, respectively) (Akiyama *et al.*, 1994b; Früh *et al.*, 1994; Hisamatsu *et al.*, 1996; Nandi *et al.*, 1996) (Fig. 4). These IFN-induced substitutions appear to favor specific antigen presentation as described below (Gaczynska *et al.*, 1993a; Goldberg *et al.*, 1995). Tanaka and co-workers have proposed that the proteasomes containing these γ-IFN-induced subunits be named *immunoproteasomes*. Unlike γ-IFN, the other major cytokines—α-interferon, β-interferon (β-IFN), and tumor necrosis factor α (TNF-α)—had no effect on the proteasomal subunit composition. However, addition of β-IFN and TNF-α together with γ-IFN further enhanced the induction of LMP2 and LMP7 subunits (Aki *et al.*, 1994).

8. BIOCHEMICAL EFFECTS OF THE γ-INTERFERON-INDUCED SUBUNITS

Several early studies showed that LMP2 and LMP7 are not essential for class I antigen presentation (Arnold *et al.*, 1992; Momburg *et al.*, 1992; Yewdell *et al.*, 1994; Zhou *et al.*, 1994), and these observations were widely interpreted to indicate that these subunits, and by association the proteasome, were not impor-

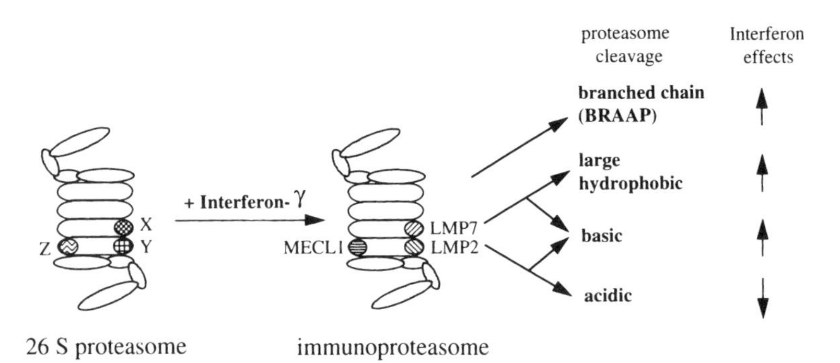

Figure 4. Model for the action of γ-interferon in altering proteasome function. A summary of the recent findings on the effects of the three γ-IFN-induced subunits, LMP2, LMP7, and MECL-1, and the resulting changes in the peptidase activities of proteasomes.

tant for class I antigen processing (Robertson, 1991). Subsequent studies, however, established clear biochemical effects of these γ-IFN-induced subunits on proteasome function that are important in antigen presentation (Goldberg *et al.*, 1995). Mammalian 20 S proteasome particles exhibit at least five distinct peptidase activities, and when associated with the 700-kDa (19 S) regulatory particle in the 26 S complex, these particles can degrade ubiquitin-conjugated proteins in an ATP-dependent process (Goldberg, 1992; Rechsteiner *et al.*, 1993). Three of these peptidase activities have been characterized most extensively: a "chymotrypsinlike" activity, which cleaves after large hydrophobic residues, a "trypsinlike" activity, which cleaves after basic groups, and a peptidylglutamyl activity, which cleaves after acidic residues (Orlowski, 1990). Mammalian proteasomes have also been shown to cleave peptides preferentially after branched-chain amino acids ["branched-chain amino acid preferring activity (BRAAP)] and after small neutral amino acids ["small neutral amino acid preferring" activity (SNAAP)], but these activities are less well characterized (Orlowski *et al.*, 1992).

To understand the functional significance of the γ-IFN-induced subunits, Gaczynska *et al.* (1993) and Driscoll *et al.* (1993) isolated proteasomes from control and γ-IFN-treated macrophages or lymphoblasts and compared their capacities to degrade different fluorogenic peptides or protein substrates. Interferon treatment did not change the total amount of 20 or 26 S complexes nor the rates of degradation of ubiquitinated lysozyme or casein into acid-soluble fragments *in vitro* (Gaczynska *et al.*, 1993), nor did it alter the rates of hydrolysis of proteins in cultured cells (Nie and Goldberg, unpublished observations).

γ-IFN did, however, alter in very interesting ways the capacity of proteasomes to hydrolyze certain fluorogenic peptide substrates. Proteasome fractions as well as purified 20 and 26 S particles from γ-IFN-treated cells cleaved hydrophobic and basic peptides faster than similar preparations from control cells. For example, treatment of macrophages with γ-IFN for several days caused a two- to sixfold increase in the maximal capacity (V_{max}) of the proteasomes to hydrolyze substrates after hydrophobic or basic residues in the P1 position. In contrast, their capacity to hydrolyze acidic substrates actually decreased after γ-IFN treatment (Gaczynska *et al.*, 1993a) (Fig. 5). In addition, in some recent studies, proteasomes from γ-IFN-treated macrophages showed a twofold increase in the cleavage of the peptide substrates preferentially cleaved by the BRAAP-active site (Beninga *et al.*, unpublished observations) (Fig. 5). Similar increases in the chymotrypsin- and trypsinlike activities were also reported by Driscoll *et al.* (1993), Aki *et al.* (1994), Ehring *et al.* (1996), and finally by Ustrell *et al.* (1995b), who had earlier questioned these findings (Ustrell *et al.*, 1995a). An opposite pattern of changes was reported by Kloetzel and co-workers (Boes *et al.*, 1994) for reasons that are unclear. Although these effects have now been observed in multiple cell lines by several laboratories, some variations in the observed changes have been seen, perhaps related to the different cell lines and experimental conditions used. It also

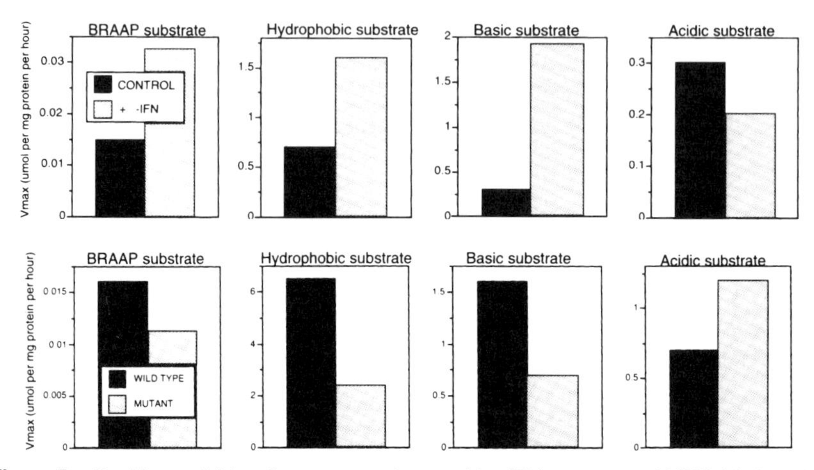

Figure 5. Peptidase activities of proteasomes change with γ-IFN treatment and MHC deletion. V_{max} values for degradation of BRAAP, hydrophobic, basic, and acidic substrates by proteasomes were determined. (Top) Proteasomes were isolated from control and γ-IFN-treated U937 macrophage cells; (bottom) proteasomes from wild-type and MHC-deficient mutant lymphoblasts (data from Gaczynska *et al.*, 1993a; Beninga *et al.*, unpublished observation).

has been shown that differences in the activation state of the proteasome can contribute to variations in the magnitude of the γ-IFN-induced effects (Ehring *et al.*, 1996).

While altering these peptidase activities, γ-IFN treatment increased several-fold the amounts of LMP2 and LMP7 in the proteasome (Gaczynska *et al.*, 1993a; Driscoll *et al.*, 1993; Aki *et al.*, 1994). To test whether the expression of these genes was responsible for the functional changes, Gaczynska *et al.* (1993a) compared the activities of proteasomes from a wild-type human lymphoblastoid line (721) with those from 721.174 mutant cells, which carry a homozygous deletion in the MHC region encoding LMP2 and LMP7 (DeMars *et al.*, 1985). The 20 and 26 S proteasomes from the mutant cells degraded basic and hydrophobic peptides more slowly, and the acidic substrate faster than particles from wild-type cells (Gaczynska *et al.*, 1993a) (Fig. 5). In addition, particles from the 721.174 mutant cells also have a lower BRAAP activity than those from wild-type cells (Beninga *et al.*, unpublished observations) (Fig. 5). Thus, as expected, the deletion of these genes has effects on proteasome function opposite those induced by γ-IFN. In wild-type lymphoblasts, LMP2 and LMP7 are constitutively present in proteasomes, and their levels increase further after γ-IFN treatment, concomitantly with the enhanced activity against hydrophobic and basic peptides. In contrast, this cytokine has little or no stimulatory effect on these activities in the LMP-deleted cells (Gaczynska *et al.*, 1993a; see below).

Several catalytic functions of proteasomes are thus regulated in distinct fashions by γ-IFN, and these adaptations require the function of MHC-encoded genes. It remains to be determined whether MHC-encoded genes regulate the reported ability of proteasomes to cleave after small neutral amino acids (Condors *et al.*, 1994) and whether the other γ-IFN-inducible subunit, MECL-1, which is not encoded in the MHC (Hisamatsu *et al.*, 1996; Nandi *et al.*, 1996), plays any role in altering the catalytic properties of this particle. MECL-1 replaces subunit Z when LMP2 and LMP7 are incorporated. Both subunits MECL-1 and Z bear a threonine at their N-termini and are modified by lactacystin (see below), which implies that they are catalytically active subunits of the "immunoproteasome" (Hisamatsu *et al.*, 1996; Nandi *et al.*, 1996).

9. SPECIFIC FUNCTIONS OF LMP2, LMP7, AND THE HOMOLOGOUS SUBUNITS

To define more precisely the biochemical effects of the LMP2 and LMP7 subunits, the cDNAs for these genes were transfected into the LMP-deficient mutant lymphoblasts or HeLa cells (Gaczynska *et al.*, 1994), and the peptidase activities and subunit composition of proteasomes isolated from the parental and transfectant cells were then analyzed. Transfection of LMP7 into mutant cells specifically increased the content of this subunit in proteasomes and enhanced the cleavage after hydrophobic and basic residues. LMP2 transfection caused a marked suppression of the hydrolysis of the acidic substrate (Gaczynska *et al.*, 1994) (Fig. 6). Moreover, in both cell types, the enhancement of the hydrophobic peptidase activity was proportional to the increase in LMP subunits actually incorporated into the proteasome fraction (Gaczynska *et al.*, 1994). Thus, the incorporation of LMPs reversed, at least partially, the functional defects reported in proteasomes from the MHC-deletion strain (Gaczynska *et al.*, 1993a, 1994), and incorporation of LMPs appears to account for most of the changes in proteasome activity induced by γ-IFN (Fig. 4). Using a similar approach, Kloetzel and co-workers (Kuckelkorn *et al.*, 1995) were also able to confirm the incorporation of the LMPs as responsible for the γ-IFN-induced changes in the hydrolytic activities of the proteasome. However, in their experiments, they were unable to correlate the changes in the peptide hydrolytic activities to specific LMP subunits.

The incorporation of LMP2 and LMP7 into proteasomes on γ-IFN treatment does not represent a simple addition of two new subunits to yield a larger particle, but instead the LMPs replace homologous subunits that are normally present in proteasomes (Früh *et al.*, 1994; Akiyama *et al.*, 1994b; Belich *et al.*, 1994). The constitutive subunit named either δ (DeMartino *et al.*, 1991), 2 (Früh *et al.*, 1994), or Y (Aki *et al.*, 1994; Akiyama *et al.*, 1994a), has about 61% identity with LMP2,

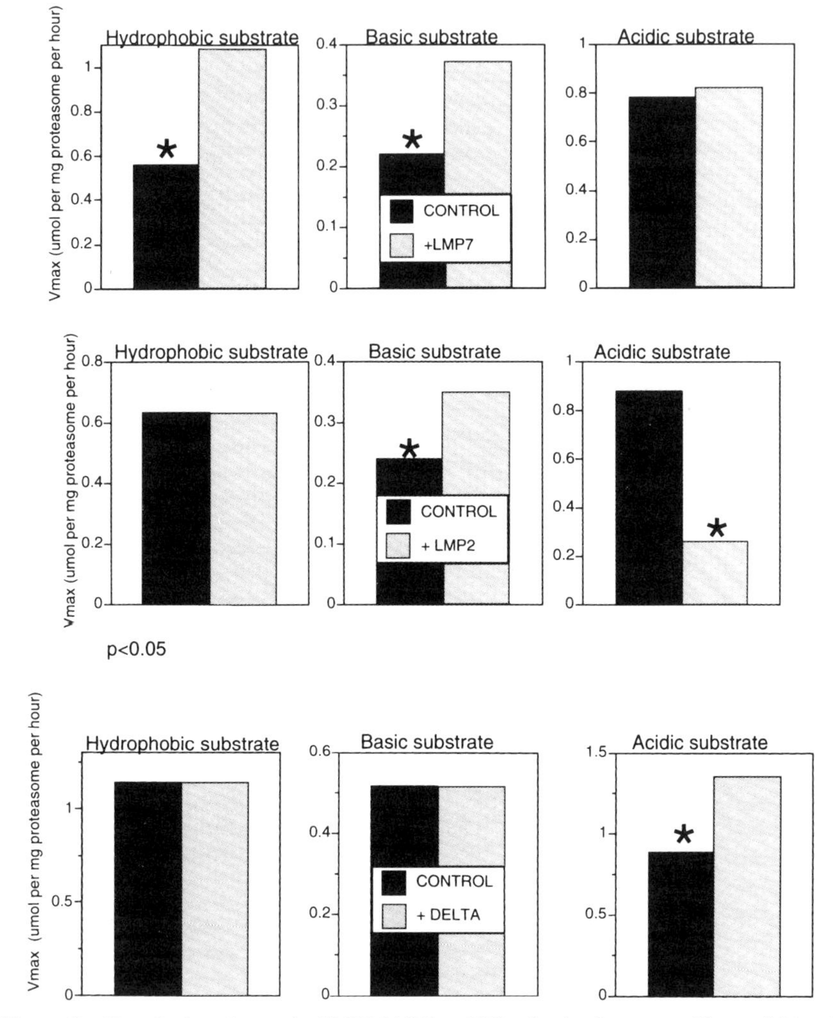

Figure 6. Transfection of genes for LMP7, LMP2, or Y(d) subunits changes peptidase activities of proteasomes. V_{max} values were determined for degradation of hydrophobic, basic, and acidic substrates by proteasomes isolated from parent cells and cells transfected with various cDNAs. (Top) MHC-deficient 0.174 mutant lymphoblasts and ones transfected with LMP7; (center) HeLa cells and ones transfected with LMP2; (bottom) HeLa cells and ones transfected with Y (d) subunit (data from Gaczynska *et al.*, 1994, 1996). Asterisk indicates significant difference in V_{max}, $p < 0.05$ (T test).

while the ε-subunit (also termed 10 or X) has about 69% identity to LMP7 (Aki *et al.*, 1994; Früh *et al.*, 1994; Akiyama *et al.*, 1994a; Belich *et al.*, 1994). Because LMP2 replaces the δ-subunit and suppresses cleavages after acidic residues, Gaczynska *et al.* (1996) tested whether the δ-subunit might stimulate cleavages after acidic residues. Transfection of cells with cDNA for the δ-subunit increased both the incorporation of this subunit into proteasomes and their peptidylglutamyl peptidase activity (Fig. 6). Furthermore, in the MHC-deletion mutant, unlike in wild-type cells, interferon treatment caused an increase in the peptidylglutamyl activity and also increased the proteasomal content of delta (presumably because LMP incorporation cannot occur in this strain).

Thus, the capacity to cleave after acidic residues correlates with the presence of the δ-subunit. It is noteworthy that yeast proteasome mutants that have lost this peptidylglutamyl activity map to a single subunit (Hilt *et al.*, 1993) that is closely related in sequence to human delta (DeMartino *et al.*, 1991). Similarly, one yeast mutant that prevents the cleavage of hydrophobic substrates maps to a gene (Heinemeier *et al.*, 1993) that is highly homologous to LMP7. Important findings from Baumeister's (Seemüller *et al.*, 1995) and Huber's (Löwe *et al.*, 1995) laboratories indicate that the active sites of the 20 S proteasome are associated with terminal threonine groups of β-subunits (Chapter 5; Seemüller *et al.*, 1995). In contrast to several other β-subunits, both LMPs and their homologues contain a terminal threonine residue and can be modified by [^{3}H]lactacystin (Craiu *et al.*, 1997a). These observations together suggest that the δ-subunit catalyzes the peptidylglutamyl activity, and that LMP7 is responsible for the chymotrypsinlike activity. The programmed elimination of X and Y subunits after γ-IFN treatment can account for the decrease in cleavages after acidic residues (Gaczynska *et al.*, 1993a, 1994; Aki *et al.*, 1994) and contributes to the enhanced chymotrypsin- and trypsinlike activities of these "immunoproteasomes."

10. IMMUNOLOGICAL SIGNIFICANCE OF THE CHANGES IN PEPTIDASE ACTIVITY

Although the γ-IFN-induced β-subunits do not seem to alter rates of breakdown of Ub-conjugated or other proteins by isolated 20 or 26 S proteasomes, these modifications clearly stimulate cleavage of small peptides after hydrophobic, basic, and branched-chain amino acids and reduce cleavage after acidic residues. Together, these findings predict that during proteolysis, the proteasomes from the γ-IFN-treated cells should generate more peptides having hydrophobic or basic C-termini and fewer peptides having acidic C-termini (Gaczynska *et al.*, 1993a). In fact, the vast majority of naturally processed peptides found associated with class I heterodimers are of this type (Rammensee *et al.*, 1993, 1995; Gaczynska *et al.*, 1993b). Screening of MHC-bound peptides shows that 73% of MHC class I ligands have hydrophobic or aromatic amino acid residues at their C-termini with leucine amounting to one-third of these cases (Niedermann *et al.*, 1996). About 20% of the MHC class I ligands have a charged amino acid at the C-terminus, and the remaining 7% have other C-termini including small neutral amino acids. For example, Fig. 7 shows the analysis of the C-terminal residues of 124 identified T-cell epitopes, with very strong preference for hydrophobic and basic residues in this position. This distribution is clearly statistically significant, because an analysis of the amino acid composition of over 800 extracellular and intracellular proteins shows approximately equal distributions of these classes of amino acids. In contrast to the C-termini, analysis of the types of cleavages that generate the N-terminal residues of these antigenic peptides (e.g., from the sequence of the corresponding antigenic protein) shows that the responsible protease has no similar preference for the cleavage after hydrophobic and basic residues.

This preference for hydrophobic or basic C-terminal residues correlates with the binding properties of MHC class I molecules (Rammensee *et al.*, 1993, 1995). Thus, the changes in peptidase activities induced by γ-IFN would be expected to favor the production of those types of peptides that bind selectively to MHC class I molecules. Moreover, several studies have now shown that oligopeptides with hydrophobic or basic C-terminal residues are also the preferred substrates for uptake into the ER by the TAP transporter (Heemels *et al.*, 1993; Momburg *et al.*, 1994; Ahn *et al.*, 1996b).

The proteasomes therefore seem to be responsible for the critical cleavages that determine the C-termini of the antigenic peptides (Rock *et al.*, 1994; Gaczynska *et al.*, 1993a). This question was also critically addressed by Craiu *et al.* (1997b) who analyzed the class I presentation of different extensions of the ovalbumin-derived peptide epitope SIINFEKL. They compared the presentation of different C- and N-terminal-extended SIINFEKL minigene constructs, expressed from the vaccinia virus T7 system, under conditions where the proteasome was inhibited with lactacystin or β-lactone. Inhibition of the proteasome prevented the class I presentation of peptides with one to five C-terminal-extended residues, whereas presentation of peptides with N-terminal-extending residues was not affected. These findings clearly confirm that the proteasome is responsible

Figure 7. Most peptides presented on MHC class I molecules have hydrophobic or basic C-termini. (Top) Frequency of the C-terminal amino acids of 124 MHC class I-presented antigenic peptides; (center) frequency of amino acids preceding the N-terminus of MHC class I-presented antigenic peptides; (bottom) frequency for the general distribution of amino acids in proteins. The data for the top and center panels were derived from Rammensee *et al.* (1995) and from available protein sequences collected from the EMBL (Heidelberg, Germany) and the National Research Foundation (Bethesda, MD) data bases. The data for the bottom panel were derived and modified from Nakashima and Nishikawa (1994), and encompassed 894 proteins.

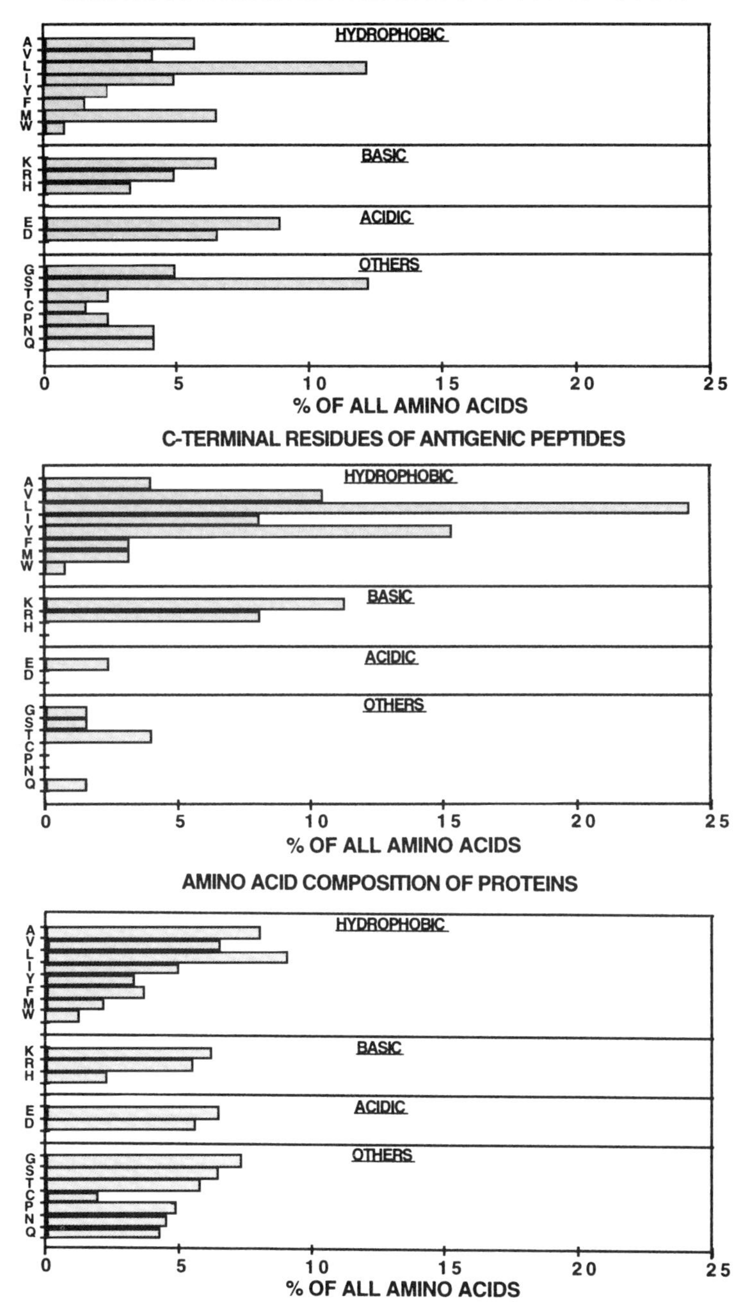
RESIDUES PRECEDING THE N-TERMINUS OF ANTIGENIC PEPTIDES
HYDROPHOBIC
A
V
L
I
Y
F
M
W
BASIC
K
R
H
ACIDIC
E
D
OTHERS
G
S
T
C
P
N
Q
0
5
10
15
20
25
% OF ALL AMINO ACIDS
C-TERMINAL RESIDUES OF ANTIGENIC PEPTIDES
HYDROPHOBIC
A
V
L
I
Y
F
M
W
BASIC
K
R
H
ACIDIC
E
D
OTHERS
G
S
T
C
P
N
Q
0
5
10
15
20
25
% OF ALL AMINO ACIDS
AMINO ACID COMPOSITION OF PROTEINS
HYDROPHOBIC
A
V
L
I
Y
F
M
W
BASIC
K
R
H
ACIDIC
E
D
OTHERS
G
S
T
C
P
N
Q
0
5
10
15
20
25
% OF ALL AMINO ACIDS

for the generation of the C-terminus of the antigenic peptide, but not for its N-terminus (see below).

The γ-IFN-induced changes in peptidase activities do not obviously correlate with the types of proteolytic cleavages that determine the final N-terminus of the antigenic peptides (Gaczynska *et al.*, 1993a) (Fig. 7). However, certain studies suggest that the proteasome may also make, under highly nonphysiological conditions (e.g., SDS activation of the enzyme) and for extended periods of time with multiple rounds of proteasome function on the substrates, the final N-terminus of peptides. Isolated 20 S proteasomes can generate MHC ligands without further need of downstream processing (Dick *et al.*, 1994; Groettrup *et al.*, 1995; Niedermann *et al.*, 1996). But it is still uncertain to what extent these artificial model experiments reflect the behavior of 20 and 26 S proteasomes and whether additional trimming of the N-terminus may occur after the emergence of peptides from the proteasome by amino peptidases either in the cytosol or in the ER (Eisenlohr *et al.*, 1992; Snyder *et al.*, 1994; Elliott *et al.*, 1995). The predominant fate of the great majority of peptides generated by the proteasome is rapid digestion to amino acids by soluble exopeptidases and endopeptidases (Buechler *et al.*, 1994), while a small fraction of these peptides is transported into the ER for antigen presentation. The specialized adaptations in proteasome composition induced by γ-IFN should increase the fractional yield of antigenic peptides and thus enhance the efficiency of antigen presentation. Clearly, the LMPs are not obligatory for antigen presentation (Arnold *et al.*, 1992; Momburg *et al.*, 1992; Yewdell *et al.*, 1994), presumably because proteasomes lacking these subunits still exhibit significant "chymotrypsin"- and "trypsinlike" activities, and still can generate antigenic peptides, although probably to a more limited extent (Gaczynska *et al.*, 1993a).

These findings thus would predict that the mutants lacking LMP2 or LMP7 should have a decreased efficiency, rather than an absolute defect, in antigen processing. This idea was tested by Van Kaer, Tonegawa, and colleagues, who constructed a mouse strain with a specific deletion of LMP2. When studied *in vitro*, peritoneal macrophages and spleen cells from mutant mice exhibited a clear defect in class I presentation of a viral antigen (Van Kaer *et al.*, 1994). Accordingly, proteasomes isolated from these tissues of the mutant mice have altered peptidase activities (Van Kaer *et al.*, 1994) similar to those seen in LMP-deficient lymphoblasts (Gaczynska *et al.*, 1993a). Similar studies of antigen presentation have also been carried out by von Boehmer and colleagues who showed that mutant mice lacking LMP7 also have significantly reduced surface levels of MHC class I molecules and impaired MHC class I responses, presumably related to the altered peptide hydrolytic activities of their proteasomes (Fehling *et al.*, 1994; Stohwasser *et al.*, 1996). Clearly, the γ-IFN-induced changes in proteasome-subunit composition have important consequences for host defense. The generation of LMP2 and LMP7 double knock-out mice and of mice lacking the MECL-1 subunit will allow definitive tests of these important questions.

One important unexpected observation of Van Kaer *et al.* (1994) was that proteasomes isolated from muscle or brain of the LMP2-deficient mice (in contrast to those from liver or spleen) did not differ significantly in peptidase activities from those of wild-type mice. This observation was accounted for by the low levels of LMP2 and LMP7 expression in muscle and brain of normal rats. The amounts of LMP2 and LMP7 in proteasomes isolated from various tissues of the wild-type mice also correlated with the respective organs' content of MHC class I molecules (Van Kaer *et al.*, 1994; Chamberlain *et al.*, 1991). Unlike liver or spleen, muscle and brain of normal rats expressed very low levels of MHC class I molecules and presumably are relatively inactive in class I antigen presentation. In liver and spleen, which are active in this process, the LMP genes appear to be constitutively expressed, presumably because γ-IFN or some other cytokine is functional under normal conditions in these tissues. In any case, these findings do indicate that in normal animals, proteasomes in different organs differ in their subunit composition and functional activities. On infection, proteasome composition and function appear to change further through induction of LMPs in many tissues to facilitate antigen presentation.

11. INVOLVEMENT OF THE PA28 ACTIVATOR IN ANTIGEN PRESENTATION

Changes in the subunit composition are not the only mechanism by which γ-IFN alters the functional properties of the proteasome. Cells contain several complexes that can specifically activate or inhibit the 20 S proteasome, but their precise physiological function is generally unclear. For one, the PA28 activator, there is now strong evidence that it plays an important role in MHC class I antigen presentation. The PA28 activator (also called the *11 S regulator*) was identified in crude cell extracts by its ability to enhance several peptide-hydrolyzing activities of 20 S proteasomes in an ATP-independent manner (Ma *et al.*, 1992; Dubiel *et al.*, 1992). Addition of the PA28 activator complex to purified 20 S proteasome enhances the hydrolysis of small fluorogenic peptide substrates by increasing the V_{max} and decreasing the K_m. However, it does not affect the rate of degradation of large protein substrates to acid-soluble peptides. These observations suggest that PA28 does not influence the initial binding or cleavage of protein substrates, but does affect the subsequent cleavage of oligopeptides to smaller products (see below).

This activator is a 200-kDa complex composed of two subunits, PA28α and PA28β. Cloning of the genes for these proteins demonstrated that they are almost ~ 50% identical and differ slightly in molecular mass, ~ 27 and ~ 28 kDa (Mott *et al.*, 1994; Ahn *et al.*, 1995). They are also homologous to a nuclear protein, K_i,

Because of its homology to PA28α and PA28β, K_i was recently named PA28γ, and it appears to form a homopolymeric complex that can also associate with the 20 S proteasome.

In cell extracts, PA28α and PA28β are present both as free subunits and in a heteromeric complex with a native molecular mass of ~ 200 kDa. Electron microscopic analysis showed that the PA28 complex is composed of either a hexameric or hepatameric ringlike structure containing α- and β-subunits. These complexes form conical rings that cap the 20 S proteasome at either or both ends (Gray *et al.*, 1994). Recently, several groups have used coimmunoprecipitation, cross-linking, and *in vitro* reconstitution experiments to demonstrate equal amounts of α- and β-subunits in the PA28 complex, which suggests a hexameric structure (Song *et al.*, 1996; Kuehn and Dahlmann, 1996; K Tanaka, personal communication). PA28α by itself can generate hexameric complexes, but this form is not able to activate the proteasome as effectively as the heteropolymeric PA28 complex (Song *et al.*, 1996; Kuehn and Dahlmann, 1996). When PA28α genes were transfected and overexpressed, and the 20 and 26 S proteasomes were analyzed by glycerol gradient centrifugation, PA28 complexes were only detected in association with the 20 S particles. No association was found with the 26 S complex or with intermediates in 20 S assembly (Yang *et al.*, 1995). The association between the PA28 and the 20 S proteasome is relatively weak and can be disrupted by low concentrations of monovalent or divalent salts (Akopian and Goldberg, unpublished). Formation of the PA28–proteasome complex requires the C-terminus of PA28α (Ma *et al.*, 1993; Song *et al.*, 1996), and blocking studies using monoclonal antibodies against individual proteasome subunits demonstrated that the HC2 α-subunit is essential for this association (Kania *et al.*, 1996). Interestingly, the function of the PA28 activator itself seems also to be inhibited by phosphorylation. PA28β exists in a phosphorylated form, and Etlinger and co-workers (personal communication) demonstrated that phosphatase treatment increases the activity of the PA28 complex.

The first hint for the involvement of the PA28 activator in antigen presentation came from the finding of Rechsteiner and co-workers (Dubiel *et al.*, 1992) that PA28 corresponded to a known γ-IFN-induced gene, IGUPI-5111 (Honore *et al.*, 1993). Therefore, it is induced by γ-IFN together with other important components of the class I antigen presentation pathway, MHC class I molecules, TAP1/2, MECL-1, and LMPs. In addition, in some cells, PA28 is induced at least by α-IFN and TNF-α (Ahn *et al.*, 1996a). In contrast, γ-IFN does not affect the transcription of the PA28γ gene, and rapidly reduces the level of these proteins (K. Tanaka, personal communication).

The addition of purified or recombinant PA28 activator to 20 S proteasomes enhances the hydrolysis of small fluorogenic substrates, whether or not LMPs are present (Ma *et al.*, 1992; Ustrell *et al.*, 1995b). Moreover, when degradation of a 25mer peptide was studied, and the peptide products analyzed by HPLC and mass spectrometry, qualitative as well as quantitative differences in peptide products were seen with PA28 addition (Groettrup *et al.*, 1995). Thus, one effect of PA28 is to increase the variety of peptides generated from antigenic proteins. Rammensee and colleagues analyzed the products of proteasomal hydrolysis of 19- and 21-residue peptides from the JAK1 kinase and a 25mer from the pp89 immediate early protein of murine cytomegalovirus, all of which contain a defined dominant MHC class I ligand (T. P. Dick *et al.*, 1996). With the PA28 activator present, more antigenic peptides were generated in the size range of MHC class I ligands (8–9mers) as well as longer peptides that might function as precursors for the presented peptides. Tanaka and colleagues obtained similar results by analyzing the generation of the Balb/c leukemia RLo1 cytotoxic T-cell epitope from a longer peptide (personal communication). These effects were attributed by T. P. Dick *et al.* (1996) to coordinated dual cleavages by the PA28–proteasome complex, although other mechanisms could also account for these results.

Although the molecular basis for these changes in proteasome function is unclear, an association with PA28 seems to favor the generation of 8- to 10-residue peptides or potential precursors. These findings are consistent with studies of the effects of PA28 overexpression in mouse fibroblasts that constitutively expressed cytotoxic T-cell antigens from murine cytomegalovirus and influenza virus (Groettrup *et al.*, 1996). PA28 overexpression results in an enhanced presentation of these antigens, which could be explained neither by increased expression of MHC class I molecules nor by faster degradation of the antigenic protein. Clearly, PA28 can enhance the capacity of cells for antigen presentation, although the relative importance of PA28 and modifications of proteasome subunits in promoting this process is still unclear. One fundamental question raised by these findings is whether or not the 20 S particle ever functions by itself in protein breakdown *in vivo*. It is noteworthy that PA28 does not appear to interact with the 26 S complex or to enhance rates of protein breakdown by the 20 S particle, although it can alter the pattern of cleavages made in oligopeptides or proteins (Akopian and Goldberg, unpublished observations) by activated 20 S particles. To explain these observations, Groettrup *et al.* (1996) proposed that the 26 S complex initially degrades antigenic proteins into large peptides, which are then cleaved into the MHC ligands by the 20 S particles together with PA28.

12. INVOLVEMENT OF OTHER CYTOSOLIC PROTEASES IN CLASS I PROCESSING

Although the proteasome is clearly the primary site for degradation of most cell proteins, other cytosolic enzymes hydrolyze further the peptide products generated by these particles into amino acids, which are recycled in protein synthesis or used in intermediary metabolism. Only a few peptides normally escape this catabolic process and are presented on the cell surface on MHC class I molecules. These presented peptides are almost all 8–9 residues in length, apparently because the closed structure of the peptide-binding cleft of the MHC class I molecules does not allow longer peptides to bind. Both 20 and 26 S proteasomes degrade proteins to acid-soluble peptides in a highly processive fashion, i.e., they attack a protein substrate and digest it to a typical group of oligopeptides before attacking another protein molecule (Akopian *et al.*, 1997). Unfortunately, the actual size of the peptides released by mammalian proteasomes has not yet been systematically analyzed. Recent studies by Kisselev *et al.* (1998) demonstrated that the highly symmetrical 20 S proteasomes from the archaeon *Thermoplasma acidophilum* degrade large protein processively to peptide products with a mean size of 7–9 residues, like that of the MHC-presented peptides, but ranging in actual length from 5 to 20 residues. Probably, the products of eukaryotic proteasomes are even longer and more heterogeneous in size, because their active sites are fewer in number, more specific, and not distributed symmetrically (Coux *et al.*, 1996). Therefore, many of the epitope-containing peptides generated by the proteasome are longer than 8–9 residues and there must be a peptide-trimming mechanism in the cell to increase the amount and variety of peptides capable of MHC class I binding.

The recent experiments of Craiu *et al.* (1997b) also indicate that cytosolic exoproteases play a role in this postproteasomal trimming of antigenic peptides. When SIINFEKL was expressed on a minigene construct or microinjected into cells, it was efficiently presented. Peptides with longer C-terminal-extended residues were also presented, but not if the cells were treated with a proteasome inhibitor. By contrast, when extended peptides with up to 25 N-terminal additional residues were microinjected or expressed in cells, they could be processed to the antigenic species without the function of the proteasome. Thus, two different proteolytic steps operate in the generation of the antigenic SIINFEKL peptides; proteasomal cleavages seem to define the C-termini of the antigenic peptides, whereas undefined nonproteasomal exoprotease(s) generate the N-termini. Therefore, it seems likely that a trimming reaction has evolved in the cytosol to increase the yield of peptides of optimal size for MHC class I binding (8–10 amino acids). What proteases are involved in this trimming step is still not clear. Several cytosolic aminopeptidases are possible candidates. Interestingly, mRNA for leucine aminopeptidase is known to be upregulated by γ-IFN (Harris *et al.*, 1992), and leucine aminopeptidase activity is increased severalfold in γ-IFN-treated macrophages and HeLa cells (Beninga, unpublished observations). Because of this coordinated induction with MHC class I molecules, LMPs, MECL-1, TAP, and PA28, leucine aminopeptidase is likely to play a major role in this cytosolic trimming of proteasome-derived antigenic peptides.

In addition to their function in peptide processing, cytosolic peptidases may also reduce antigen processing, as they also rapidly degrade proteasome-generated oligopeptides to amino acids. The half-lives of presented peptides, such as SIIN-FEKL, in cytosolic extracts are very short (Akopian and Goldberg, unpublished observations). One class of enzymes that can catalyze this process are the cytosolic metallo-endoproteases. Buechler *et al.* (1994) demonstrated in yeast that a mutant strain with a defect in the PRD1 gene, which codes for the yscD oligopeptidase, has a decreased capacity to hydrolyze peptides generated during intracellular protein breakdown. Similar metalloproteases (thimet oligopeptidases) are found in mammalian cells and may also be involved in this rapid degradation of peptides generated by the proteasome (Barrett *et al.*, 1995). In fact, Akopian *et al.* (unpublished observations) have shown that such an oligopeptidase is responsible for the degradation of SIINFEKL in cell extracts. Moreover, an inhibitor of this process enhanced the presentation of this epitope, when ovalbumin or SIINFEKL was microinjected or expressed in cells. Future experiments should clarify whether this endoprotease is responsible for the breakdown of many antigenic peptides released by the proteasome, whether other enzymes also contribute to this process, and whether these postproteasomal reactions are also regulated by γ-IFN to favor MHC class I presentation.

13. CONCLUSIONS

Recent studies using selective inhibitors of the proteasome have established that these particles catalyze the degradation of most proteins in mammalian cells, including both the rapid breakdown of abnormal and regulatory proteins and the slower degradation of the bulk of cytosolic proteins. In addition, inhibitors of proteasome function (peptide aldehydes, lactacystin, β-lactone) can prevent the production of antigenic peptides and the appearance of MHC molecules on the cell surface. It has also been established that TS-mutations that block ubiquitin conjugation prevent the generation of an antigenic peptide, and that N-terminal modifications of a protein that lead to its rapid ubiquitination and degradation enhance antigen presentation.

Inhibitors of proteasome function (peptide aldehydes, lactacystin, β-lactone) can prevent the production of antigenic peptides and the appearance of MHC molecules on the cell surface. It has also been established that TS-mutations that block ubiquitin conjugation prevent the generation of an antigenic peptide, and that N-terminal modifications of a protein that lead to its rapid ubiquitination and degradation enhance antigen presentation.

In addition, γ-IFN, which stimulates antigen presentation, induces three new proteasome subunits (LMP2, LMP7, MECL-1) that are incorporated in place of normal homologous subunits. The incorporation of the MHC-encoded subunits,

LMP2 and LMP7, leads to an increased capacity to cleave model peptides after hydrophobic, branched-chain, and basic residues, and decreased cleavage after acidic residues. These LMP subunits are found normally in proteasomes from liver and spleen, where peptidase activities are different than those in proteasomes from normal muscle or brain. Transfection of cells with LMP2 or LMP7 genes causes similar changes in these peptidase activities, whereas transfection with the normal subunits X and Y causes changes opposite those induced by γ-IFN. These modifications should enhance the production by proteasomes of those types of peptides, which are preferentially transported into the ER and are selectively bound to MHC class I molecules. Accordingly, deletions of these LMP2 and LMP7 genes in mice reduce the presentation of viral antigens. γ-IFN also induces the PA28 activator complex, which promotes peptide cleavage by 20 S proteasomes and when transfected enhances antigen presentation. However, many products of the proteasome are longer than the 8- to 10-residue peptides found on MHC class I molecules. Although proteasomal cleavage seems to define the C-termini of antigenic peptides, nonproteasomal exopeptidases can generate the N-termini of the MHC-bound peptides, and this trimming process also is regulated by γ-IFN.

Acknowledgments. This work was supported by research grants from the National Institutes of Health and the Human Frontiers Science Program to A.L.G. and a postdoctoral fellowship from the Deutsche Forschungsgemeinschaft to J.B.

14. REFERENCES

Ahn, J. Y., Tanahashi, N., Akiyama, K.-Y., Hisamatsu, H., Noda, C., Tanaka, K., Chung, C. H., Shibmara, N., Willy, P., Mott, J. D., Slaughter, C. A., and DeMartino, G. N., 1995, Primary structure of two homologous subunits of PA28, a γ-interferon inducible protein activator of the 20S proteasome, *FEBS Lett.* **366:**37–42.

Ahn, K., Erlander, M., Leturcq, D., Peterson, P. A., Frueh, K., and Young, Y., 1996a, In vivo characterisation of the proteasomal regulator PA28, *J. Biol. Chem.* **271:**18237–18242.

Ahn, K., Meyer, T. H., Uebel, S., Sempe, P., Djaballa, H., Yang, Y., Peterson, P. A., Frueh, K., and Tampe, R., 1996b, Molecular mechanism and species specificity of TAP inhibition by herpes simples ICP47, *EMBO J.* **15:**3247–3255.

Aki, M., Tamura, T., Tokunaga, F., Iwanaga, S., Kawamura, Y., Shimbara, N., Kagawa, S., Tanaka, K., and Ichihara, A., 1992, cDNA cloning of rat proteasome subunit RC1, a homologue of RING10 located in the human MHC class II region, *FEBS Lett.* **301:**65–68.

Aki, M., Shimbara, N., Takashina, M., Akiyama, K., Kagawa, S., Tamura, T., Tanahashi, N., Yoshimura, T., Tanaka, K., and Ichihara, A., 1994, Interferon-γ induces different subunit organization and functional diversity of proteasomes, *J. Biochem.* **15:**257–269.

Akiyama, K.-Y., Yokota, K.-Y., Kagawa, S., Shimbara, N., Tamura, T., Akioka, H., Nathwang, H. G., Noda, C., Tanaka, K., and Ichihara, A., 1994a, cDNA cloning and interferon gamma down-regulation of proteasomal subunits X and Y, *Science* **265:**1231–1234.

Akiyama, K., Kagawa, S., Tamura, T., Shimbara, N., Takahashi, M., Kristensen, P., Hendil, K. B., Tanaka, K., and Ichihara, A., 1994b, Replacement of proteasome subunits X and Y by LMP7 and LMP2 induced by interferon-γ for acquirement of the functional diversity responsible for antigen processing, *FEBS Lett.* **343:**85–88.

Akopian, T. N., Kisselev, A. F., and Goldberg, A. L., 1997, Processive degradation of proteins and other catalytic properties of the proteasome from Thermoplasma acidophilum, *J. Biol. Chem.* **272:** 1791–1798.

Arnold, D., Driscoll, J., Androlewicz, M., Hughes, H., Cresswell, P., and Spies, T., 1992, Proteasome subunits encoded in the MHC are not generally required for the processing of peptides bound by MHC class I molecules, *Nature* **360:**171–174.

Bachmair, A., and Varshavsky, A., 1989, The degradation signal in a short-lived protein, *Cell* **56:**1019–1032.

Bachmair, A., Finley, D., and Varshavsky, A., 1986, *In vivo* half-life of a protein is a function of its amino-terminal residue, *Science* **234:**179–186.

Barrett A. J., Brown, M. A., Dondo, P. M., Knight, C. G., McKie, N., Rawlings, N. D., and Serizawa, A., 1995, Thimet oligopeptidase and oligopeptidase M or neurolysin, *Methods Enzymol.* **248:** 529–555.

Belich, M. P., Glynne, R. J., Senger, G., Sheer, D., and Trowsdale, J., 1994, Proteasome components with reciprocal expression to that of the MHC-encoded LMP proteins, *Curr. Biol.* **9:**769–776.

Boes, B., Hengel, H., Ruppert, T., Multhaup, G., Koszinowski, U. H., and Kloetzel, P.-M., 1994, Interferon-γ stimulation modulates the proteolytic activity and cleavage site preferences of 20S mouse proteasomes, *J. Exp. Med.* **179:**901–909.

Braciale, T. J., Morrison, L. A., Sweetser, M. T., Sambrook, J., Gething, M.-J., and Braciale, V. L., 1987, Antigen presentation pathways to class I and class II MHC-restricted T lymphocytes, *Immunol. Rev.* **98:**95–114.

Brown, M. G., Driscoll, J., and Monaco, J. J., 1991, Structural and serological similarity of MHC-linked LMP and proteasome (multicatalytic proteinase) complexes, *Nature* **353:**355–357.

Buechler, M., Tisljar, U., and Wolf, D. H., 1994, Proteinase yscD (oligopeptidase D). Structure, function and relationship of the yeast enzyme with mammalian thimet oligopeptidases (metallo-protease, EP 24.15), *Eur. J. Biochem.* **219:**327–339.

Busch, R., and Mellius, E. D., 1996, Developing and shedding inhibitions: How MHC-class II molecules reach maturity, *Curr. Opin. Immunol.* **8:**51–58.

Chamberlain, J. W., Nolan, J. A., Conrad, P. J., Vasavada, H. A., Vasavada, H. H., Ganguly, S., Janeway, C. A., and Weissman, S. M., 1991, Tissue-specific and cell-surface expression of human major histocompatibility complex class I heavy (HLA-B7) and light (β_2-microglobulin) chain genes in transgenic mice, *Proc. Natl. Acad. Sci. USA* **85:**7690–7694.

Ciechanover, A., 1994, The ubiquitin-proteasome proteolytic pathway, *Cell* **79:**13–21.

Condors, C., Vinitsky, A., Micholud, C., and Orlowski, J., 1994, Evidence that the nature of amino acid residues in the P-3 position directs substrates to distinct catalytic sites of the pituitary multicatalytic proteinase complex (proteasome), *Biochemistry* **33:**6483–6489.

Coux, O., Tanaka, K., and Goldberg, A. L., 1996, Structure and function of the 20S and 26S proteasomes, *Annu. Rev. Biochem.* **65:**801–837.

Cox, J. H., Galardy, P., Bennink, J. R., and Yewdell, J. W., 1995, Presentation of endogenous and exogenous antigens is not affected by inactivation of El ubiquitin-activating enzyme in temperature-sensitive cell lines, *J. Immunol.* **154:**511–519.

Craiu, A., Gaczynska, M., Akopian, T., Gramm, C., Fenteany, G., Goldberg, A. L., and Rock, K. L., 1997a, Lactacystin and clast-lactacystin β-lactone modify multiple proteasome b subunits and inhibit intracellular protein degradation and MHC class I antigen presentation, *J. Biol. Chem.* **272:**13437–13445.

Craiu, A., Akopian, T., Goldberg, A. L., and Rock, V. L., 1997b. Two distinct protocatalytic processes in the generation of a major histocompatibility complex class I-presented peptide, *Proc. Natl. Acad. Sci. USA* **94:**10850–10855.

DeMars, R., Rudersdorf, R., Chang, C., Petersen, J., Strandtmann, J., Korn, N., Sidwell, B., and Orr, H. T., 1985, Mutations that impair a posttranscriptional step in expression of HLA-A and -B antigens, *Proc. Natl. Acad. Sci. USA* **82:**8133–8187.

DeMartino, G. N., Orth, K., McCullough, M. L., Lee, L. W., Munn, T. Z., Moomaw, C. R., Dawson, P. A., and Slaughter, C. A., 1991, The primary structures of four subunits of the human, high-molecular-weight proteinase, macropain (proteasome) are distinct but homologous, *Biochim. Biophys. Acta* **1079:**29–38.

Dice, J., 1987, Molecular determinants of protein half-lives in eukaryotic cells, *FASEB J.* **1:**349–357.

Dick, L. R., Aldrich, C., Jameson, S. C., Moomaw, C. R., Pramanik, B. C., Doyle, K., DeMartino, G. N., Bevan, M. J., Forman, J. M., and Slaughter, C. A., 1994, Proteolytic processing of ovalbumin and β-galactosidase by the proteasome to yield antigenic peptides, *J. Immunol.* **152:**3884–3894.

Dick, L. R., Cruikshank, A. A., Greiner, L., Melandri, F. D., Nunes, S., and Stein, R., 1996, Mechanistic studies on the inactivation of the proteasome by lactacystin, *J. Biol. Chem.* **271:**7273–7276.

Dick, T. P., Ruppert, T., Groettrup, M., Kloetzel, P. M., Kuehn, L., Koszinowski, U. H., Stephanovic, S., Schild, H., and Rammensee, H.-G., 1996, Coordinated dual cleavage induced by the proteasome regulator PA28 lead to dominant MHC-ligands, *Cell* **86:**253–262.

Driscoll, J., Brown, M., Finley, D., and Monaco, J. J., 1993, MHC-linked LMP gene products specifically alter peptidase activities of the proteasome, *Nature* **365:**262–264.

Dubiel, W., Pratt, G., Ferrell, K., and Rechsteiner, M., 1992, Purification of an 11S regulator of the multicatalytic protease, *J. Biol. Chem.* **267:**22369–22377.

Eisenlohr, L. C., Bacik, I., Bennink, J. R., Bernstein, K., and Yewdell, J. W., 1992, Expression of a membrane protease enhances presentation of endogenous antigens to MHC class I-restricted T lymphocytes, *Cell* **71:**963–972.

Elliott, T., Willis, A., Cerundolo, V., and Townsend, A., 1995, Processing of major histocompatibility class I-restricted antigens in the endoplasmic reticulum, *J. Exp. Med.* **181:**1481–1491.

Etlinger, J. D., and Goldberg, A. L., 1977, A soluble ATP-dependent proteolytic system responsible for the degradation of abnormal proteins in reticulocytes, *Proc. Natl. Acad. Sci. USA* **74:**54–58.

Fehling, H. J., Swat, W., Laplace, C., Kuhn, R., Rajewsky, K., Muller, U., and von Boehmer, H., 1994, MHC class I expression in mice lacking the proteasome subunit LMP-7, *Science* **265:**1234–1237.

Fenteany, G., Standaert, R. F., Lane, W. S., Choi, S., Corey, E. J., and Schreiber, S. L., 1995, Inhibition of proteasome activities and subunit-specific amino-terminal threonine modification by lactacystin, *Science* **268:**726–731.

Finley, D., and Chau, V., 1991, Ubiquitination, *Annu. Rev. Cell Biol.* **7:**25–69.

Finley, D., Ciechanover, A., and Varshavsky, A., 1984, Thermolability of ubiquitin-activating enzyme from the mammalian cell cycle mutant ts85, *Cell* **37:**43–58.

Früh, K., Gossen, M., Wang, K., Bujard, H., Peterson, P. A., and Yang, Y., 1994, Displacement of housekeeping proteasome subunits by MHC-encoded LMPs: A newly discovered mechanism for modulating the multicatalytic proteinase complex, *EMBO J.* **13:**3236–3244.

Gaczynska, M., Rock, K. L., and Goldberg, A. L., 1993a, Gamma-interferon and expression of MHC genes regulate peptide hydrolysis by proteasomes, *Nature* **365:**264–267.

Gaczynska, M., Goldberg, A. L., and Rock, K. L., 1993b, Role of proteasomes in antigen presentation, *Enzyme Protein* **47:**354–369.

Gaczynska, M., Rock, K. L., Spies, T., and Goldberg, A. L., 1994, Peptidase activities of proteasomes are differentially regulated by the MHC-encoded genes LMP2 and LMP7, *Proc. Natl. Acad. Sci. USA* **91:**9213–9217.

Goczynska, M., Goldberg, A. L., Tanaka, K., Hendil, K. B., and Rock, K. L., 1996, Proteasomal subunits X and Y alter peptidase activities in opposite ways to the interferon-γ-induced subunits LMP2 and LMP7, *J. Biol. Chem.* **271:**17275–17280.

Germain, R. N., and Margulies, D. H., 1993, The biochemistry and cell biology of antigen processing and presentation, *Annu. Rev. Immunol.* **11:**403–450.

Glynne, R., Powis, S. H., Beck, S., Kelly, A., Kerr, L. A., and Trowsdale, J., 1991, A proteasome-related gene between the two ABC transporter loci in the class II region of the human MHC, *Nature* **353:**357–360.

Goldberg, A. L., 1992, The mechanisms and functions of ATP-dependent proteases in bacterial and animal cells, *Eur. J. Biochem.* **203:**9–23.

Goldberg, A. L., 1995, Functions of the proteasome: The lysis at the end of the tunnel, *Science* **268:** 522–523.

Goldberg, A. L., and Rock, K. L., 1992, Proteolysis, proteasomes and antigen presentation, *Nature* **357:**375–379.

Goldberg, A. L., Gaczynska, M., Grant, E., Michalek, M., and Rock, K. L., 1995, Function of the proteasome in antigen presentation, *Cold Spring Harbor Symp. Quant. Biol.* **60:**479–490.

Gonda, D., Bachmair, K. A., Wunning, I., Tobias, J. W., Lane, W. S., and Varshavsky, A., 1989, University and structure of the N-end rule, *J. Biol. Chem.* **264:**16700–16712.

Grant, E. P., Michalek, M. T., Goldberg, A. L., and Rock, A. L., 1995, The rate of antigen degradation by the ubiquitin-proteasome pathway influences MHC class I presentation, *J. Immunol.* **155:** 3750–3758.

Gray, C. W., Slaughter, C. A., and DeMartino, G. N., 1994, PA28 activator protein forms regulatory caps on proteasome stacked rings, *J. Mol. Biol.* **236:**7–15.

Groettrup, M., Ruppert, T., Kuehn, L., Seeger, M., Standera, M., Koszinowski, U. H., and Kloetzel, P. M., 1995, The interferon-γ-inducible 11S regulator (PA28) and the LMP2/LMP7 subunits govern the peptide production by the 20S proteasome *in vitro*, *J. Biol. Chem.* **270:**23808–23815.

Groettrup, M., Soza, A., Eggers, M., Kuehn, L., Dick, T. P., Schild, H., Rammensee, H.-G., Koszinowski, U. H., and Kloetzel, P. M., 1996, A role for the proteasome regulator PA28a in antigen presentation, *Nature* **381:**166–168.

Gronostajski, R., Goldberg, A. L., and Pardee, A. B., 1984, The role of increased proteolysis in the atrophy and arrest of proliferation in serum-deprived fibroblasts, *J. Cell Physiol.* **121:**189–198.

Gronostajski, R., Pardee, A., and Goldberg, A. L., 1985, The ATP dependence of the degradation of short-lived and long-lived proteins in growing fibroblasts, *J. Biol. Chem.* **260:**3344–3349.

Groth, S., Nguyen, V., and Shastri, N., 1996, Generation of naturally processed peptide/MHC class I complexes is independent of the stability of endogenously synthesized precursors, *J. Immunol.* **157:**1894–1904.

Harding, C. V., France, J., Song, R., Farah, J. M., Chatterjee, S., Iqbal, M., and Siman, R., 1995, Novel dipeptide aldehydes are proteasome inhibitors and block MHC-I antigen-processing pathway, *J. Immunol.* **155:**1767–1775.

Harris, C. A., Hunte, B., Krauss, M. R., Taylor, A., and Epstein, L. B., 1992, Induction of leucine aminopeptidase by interferon-γ, *J. Biol. Chem.* **267:**6865–6869.

Heemels, M. T., Schumacher, T. N. M., Wonigeit, K., and Ploegh, H. L., 1993, Peptide translocation by variants of the transporter associated with antigenic processing, *Science* **262:**2059–2063.

Heinemeier, W., Gruhler, A., Möhrle, V., Mahé, Y., and Wolf, D. H., 1993, PRE2, highly homologous to a human major histocompatibility complex-linked RING10 gene, codes for a yeast proteasome subunit necessary for chymotryptic activity and degradation of ubiquitinated proteins, *J. Biol. Chem.* **268:**5115–5120.

Heinemeier, W., Trondle, N., Albrecht, G., and Wolf, D. H., 1994, PRE5 and PRE6, the last missing genes encoding 20S proteasome subunit from yeast? Indication for a set of 14 different subunits in the eukaryotic proteasome core, *Biochemistry* **33:**12229.

Hilt, W., Enenkel, C., Gruhler, A., Thorsten, S., and Wolf, D. H., 1993, The PRE4 gene codes for a subunit of the yeast proteasome necessary for peptidylglutamyl-peptide-hydrolysing activity, *J. Biol. Chem.* **268:**3479–3486.

Hisamatsu, H., Shimbara, N., Saito, Y., Kirstensen, P., Hendil, K. B., Fujiwara, T., Takahashi, E.-I., Tanahashi, N., Tamura, T., Ichihara, A., and Tanaka, K., 1996, Newly identified pair of proteasomal subunits regulated reciprocally by interferon γ, *J. Exp. Med.* **183:**1807–1816.

Honore, B., Leffers, H., Madsen, P., and Celis, J. E., 1993, Interferon-γ up-regulates a unique set of proteins in human keratinocytes, *Eur. J. Biochem.* **218:**241–430.

Kania, M., DeMartino, G. N., Baumeister, W., and Goldberg, A. L., 1996, The proteasome subunit, C2, contains an important site for binding of the PA28 (11S) activator, *Eur. J. Biochem.* **236:**510–516.

Kelly, A., Powis, S. H., Glynne, R., Radley, E., Beck, S., and Trowsdale, J., 1991, Second proteasome-related gene in the human MHC class II region, *Nature* **353:**667–668.

Kisselev, A. F., Akopian, T. N., and Goldberg, A. L., 1998, Range of sizes of peptide products generated during degradation of different proteins by archael proteasomes, *J. Biol. Chem.* **273:**1982–1989.

Kopp, F., Dahlmann, B., and Hendil, K. B., 1993, Evidence indicating that the human proteasome is a complex dimer, *J. Mol. Biol.* **229:**14.

Kuckelkorn, U., Frenzel, S., Kraft, R., Kostka, S., Groettrup, M., and Kloetzel, P.-M., 1995, Incorporation of major histocompatibility complex-encoded subunits LMP2 and LMP7 changes the quality of the 20S proteasome polypeptide processing products independent of interferon-γ, *Eur. J. Immunol.* **25:**2605–2611.

Kuehn, L., and Dahlmann, B., 1996, Reconstitution of proteasome activator PA28 from isolated subunits: Optimal activity is associated with an α,β-heteromultimer, *FEBS Lett.* **394:**183–186.

Kulka, R. G., Raboy, B., Schuster, R., Parag, H. A., Diamond, G., and Ciechanover, A., 1988, A Chinese hamster cell cycle mutant arrested at G phase has a temperature-sensitive ubiquitin-activating enzyme, El, *J. Biol. Chem.* **263:**15726–15731.

Lowe, J., Stock, D., Jap, B., Zwickl, P., Baumeister, W., and Huber, R., 1995, Crystal structure of the 20S proteasome from the archaeon *T. acidophilum* at 3.4 Å resolution, *Science* **268:**533–539.

Ma, C.-P., Slaughter, C. A., and De Martino, G. N., 1992, Identification, purification and characterization of a protein activator (PA28) of the proteasome (macropain), *J. Biol. Chem.* **267:**10515–10523.

Ma, C.-P., Willy, P. J., Slaughter, C. A., and DeMartino, G. N., 1993, PA28, an activator of the 20S proteasome is inactivated by proteolytic modification at its carboxyl terminus, *J. Biol. Chem.* **268:**22514–22519.

Medina, R., Wing, S. S., and Goldberg, A. L., 1995, Increase in levels of polyubiquitin and proteasome mRNA in skeletal muscle during starvation and denervation atrophy, *Biochem. J.* **307:**639–645.

Michalek, M. T., Grant, E. P., Gramm, C., Goldberg, A. L., and Rock, K. L., 1993, A role for the ubiquitin-dependent proteolytic pathway in MHC class I restricted antigen presentation, *Nature* **363:**552–554.

Michalek, M., Grant, E. P., and Rock, K. L., 1996, Chemical denaturation and modification of ovalbumin alters its dependence on ubiquitin conjugation for class I antigen presentation, *J. Immunol.* **157:**617–624.

Momburg, F., Ortiz-Navarrete, V., Neefjes, J., Goulmy, E., van de Wal, Y., Spits, H., Powis, S. J., Butcher, G. W., Howard, J. C., Walden, P., and Hämmerling, G. J., 1992, Proteasome subunits encoded by the major histocompatibility complex are not essential for antigen presentation, *Nature* **360:**174–177.

Momburg, F., Roelse, J., Howard, J. C., Butcher, G. W., Hämmerling, G. J., and Neefjes, J. J., 1994, Selectivity of MHC-encoded peptide transporters from human, mouse and rat, *Nature* **367:** 648–651.

Monaco, J. J., and McDevitt, H. O., 1986, The LMP antigens: A stable MHC-controlled multisubunit protein complex, *Hum. Immunol.* **15:**416–426.

Moore, M. W., Carbone, F. R., and Bevan, M. J., 1988, Introduction of soluble protein into the class I pathway of antigen processing and presentation, *Cell* **54:**777–785.

Mott, J. D., Pramanik, B. C., Moomaw, C. R., Afendis, S. J., DeMartino, G. N., and Slaughter, C. A., 1994, PA28, an activator of the 20S proteasome, is composed of two nonidentical but homologous subunits, *J. Biol. Chem.* **269:**31466–31471.

Nakashima, H., and Nishikawa, K., 1994, Discrimination of intracellular and extracellular proteins using amino acid composition and residue-pair frequencies, *J. Mol. Biol.* **238:**54–61.

Nandi, D., Jiang, H., and Monaco, J. J., 1996, Identification of MECL-1 (LMP-10) as the third IFN-γ inducible proteasome subunit, *J. Immunol.* **156:**2361–2364.

Niedermann, G., King, G., Butz, S., Birsner, U., Grimm, R., Shabanowitz, J., Hunt, D. F., and Eichmann, K., 1996, The proteolytic fragments generated by vertebrate proteasomes: Structural relationship to major histocompatibility complex class I binding peptides, *Proc. Natl. Acad. Sci. USA* **93:**8572–8577.

Orlowski, M., 1990, The multicatalytic proteinase complex, a major extralysosomal proteolytic system, *Biochemistry* **29:**10289–10297.

Orlowski, M., Cardozo, C., and Michaud, C., 1992, Evidence for the presence of five distinct proteolytic components in the pituitary multicatalytic proteinase complex. Properties of two components cleaving bonds on the carboxyl side of branched chain and small neutral amino acids, *Biochemistry* **32:**1563–1572.

Palombella, V. J., Rando, O. J., Goldberg, A. L., and Maniatis, T., 1994, The ubiquitin-proteasome pathway is required for processing the NF-kB1 precursor protein and the activation of NFkB, *Cell* **78:**773–785.

Rammensee, H. G., Falk, K., and Rotzschke, O., 1993, Peptides naturally presented by MHC class I molecules, *Annu. Rev. Immunol.* **11:**2130–2144.

Rammensee, H. G., Friede, T., and Stevanovic, S., 1995, MHC-ligands and peptide motifs: First listing, *Immunogenetics* **41:**178–228.

Read, M. A., Neish, A. S., Luscinskas, F. W., Palombella, V. J., Maniatis, T., and Collins, T., 1995, The proteasome pathway is required for cytokine induced endothelial-leukocyte adhesion molecule expression, *Immunity* **2:**493–506.

Realini, C., Rogers, S. W., and Rechsteiner, M., 1994a, KEKE motifs. Proposed roles in protein–protein association and presentation of peptides by MHC class I receptors, *FEBS Lett.* **348:** 109–113.

Realini, C., Dubiel, W., Pratt, G., Ferrell, K., and Rechsteiner, M., 1994b, Molecular cloning and expression of a γ-interferon-inducible activator of the multicatalytic protease, *J. Biol. Chem.* **269:**20727–20732.

Rechsteiner, M., Hoffman, L., and Dubiel, W., 1993, The multicatalytic and 26S proteases, *J. Biol. Chem.* **268:**6065–6068.

Rivett, A., 1993, Proteasomes: Multicatalytic proteinase complexes, *Biochem. J.* **291:**1–10.

Robertson, M., 1991, Proteasomes in the pathway, *Nature* **353:**300–301.

Rock, K. L., 1996, A new foreign policy: MHC-class I molecules monitor the outside world, *Immunol. Today* **17:**131–137.

Rock, K. L., Gramm, C., Rothstein, L., Clark, K., Stein, R., Dick, L., Hwang, D., and Goldberg, A. L., 1994, Inhibitors of the proteasome block the degradation of most cell proteins and the generation of peptides presented on MHC-class I molecules, *Cell* **78:**761–771.

Schoenheimer, R., 1942, *Dynamic State of Body Constituents*, Harvard University Press, Cambridge, MA.

Seemüller, E., Lupas, A., Stock, D., Lowe, J., Huber, R., and Baumeister, W., 1995, Proteasome from *Thermoplasma acidophilum*. A threonine protease, *Science* **268:**579–582.

Snyder, H. L., Yewdell, J. W., and Bennink, J. R., 1994, Trimming of antigenic peptides in an early secretory compartment, *J. Exp. Med.* **180:**2389–2394.

Song, X., Mott, J. D., von Kampen, J., Pramanik, B., Tanaka, K., Slaughter, C. A., and DeMartino, G., 1996, A model for the quaternary structure of the proteasome activator PA28, *J. Biol. Chem.* **271:** 26410–26417.

Srivastava, P. K., Udono, H., Blachere, N. E., and Li, Z., 1994, Heat shock proteins transfer peptides during antigen processing and CTL priming, *Immunogenetics* **39:**93–98.

Stohwasser, R., Kuckelkorn, U., Kraft, R., Kostka, S., and Kloetzel, P.-M., 1996, 20S proteasome from LMP7 knock out mice reveals altered proteolytic activities and cleavage site preferences, *FEBS Lett.* **383:**109–113.

Takahashi, K., Cease, K., and Berzofsky, J., 1989, Identification of proteases that process distinct epitopes on the same protein, *J. Immunol.* **142:**2221–2229.

Townsend, A., Bastin, J., Gould, K., Brownlee, G., Andrew, M., Coupar, B., Boyle, D., Chan, S., and Smith, G., 1988, Defective presentation to class I-restricted cytotoxic T lymphocytes in vaccinia-infected cells is overcome by enhanced degradation of antigen, *J. Exp. Med.* **168:**1211–1224.

Ustrell, V., Pratt, G., and Rechsteiner, M., 1995a, Effects of interferon gamma and major histocompatibility complex-encoded subunits on peptidase activities of human multicatalytic proteases, *Proc. Natl. Acad. Sci. USA* **92:**584–588.

Ustrell, V., Realini, C., Pratt, G., and Rechsteiner, M., 1995b, Human lymphoblast and erythrocyte multicatalytic proteases: Differential peptidase activities and responses to the 11S regulator, *FEBS Lett.* **376:**155–158.

Van Kaer, L., Ashton-Rickardt, P. G., Eichelberger, M., Gaczynska, M., Hagashima, K., Rock, K. L., Goldberg, A. L., Doherty, P. C., and Tonegawa, S., 1994, Altered peptidase and viral-specific T cell response in *LMP2* mutant mice, *Immunity* **1:**533–541.

Villanueva, M. S., Fischer, P., Feen, K., and Pamer, E. G., 1996, Efficiency of MHC-class I antigen processing: A quantitative analysis, *Immunity* **2:**479–489.

Wing, S. S., Haas, A. L., and Goldberg, A. L., 1995, Increase in ubiquitin–protein conjugates concomitant with the increase in proteolysis in rat skeletal muscle during starvation in denervation atropy, *Biochem. J.* **307:**631–637.

Yang, Y., Frueh, K., Ahn, K., and Peterson, P. A., 1995, In vivo assembly of the proteasomal complex, implications for antigen processing, *J. Biol. Chem.* **270:**27687–27694.

Yewdell, J., Lapham, C., Bacik, I., Spies, T., and Bennink, J. R., 1994, MHC-encoded proteasome subunits LMP2 and LMP7 are not required for efficient antigen presentation, *J. Immunol.* **152:**1163–1170.

York, I., and Rock, K. L., 1996, Antigen processing and presentation by the class I major histocompatibility complex, *Annu. Rev. Immunol.* **14:**369–396.

Zhou, X., Momburg, F., Liu, T., Motal, U. M. A., Jondal, M., Hammerling, G. J., and Ljunggren, H.-G., 1994, Presentation of viral antigens restricted by H-2Kb, Dd or Kd in proteasome subunits LMP2- and LMP7-deficient cells, *Eur. J. Immunol.* **24:**1836–1868.

Ziegler, H. K., and Unanue, E. R., 1982, Decrease in macrophage antigen catabolism caused by ammonium and chloroquine is associated with inhibition of antigen presentation to T cells, *Proc. Natl. Acad. Sci. USA* **79:**175–182.

CHAPTER 8

The N-End Rule Pathway

A. Varshavsky, C. Byrd, I. V. Davydov, R. J. Dohmen, F. Du, M. Ghislain, M. Gonzalez, S. Grigoryev, E. S. Johnson, N. Johnsson, J. A. Johnston, Y. T. Kwon, F. Lévy, O. Lomovskaya, K. Madura, I. Ota, T. Rümenapf, T. E. Shrader, T. Suzuki, G. Turner, P. R. H. Waller, A. Webster, and Y. Xie

A. Varshavsky, C. Byrd, I. V. Davydov, F. Du, J. A. Johnston, Y. T. Kwon, T. Suzuki, G. Turner, A. Webster, and Y. Xie • Division of Biology, California Institute of Technology, Pasadena, California 91125. **R. J. Dohmen** • Institut für Mikrobiologie, Heinrich-Heine-Universität, D-40225 Dusseldorf, Germany. **M. Ghislain** • Unité de Biochimie Physiologique, Université catholique de Louvain, B-1348 Louvain-la-Neuve, Belgium. **M. Gonzalez** • NICHD, National Institutes of Health, Bethesda, Maryland 20892-2725. **S. Grigoryev** • Department of Biology, University of Massachusetts, Amherst, Massachusetts 01003. **E. S. Johnson** • The Rockefeller University, New York, New York 10021-6399. **N. Johnsson** • Max-Delbrück-Laboratorium, D-50829 Cologne, Germany. **F. Lévy** • Ludwig Institute for Cancer Research, CH-1066 Epalinges, Switzerland. **O. Lomovskaya** • Microcide, Inc., Mountain View, California 94043. **K. Madura** • Department of Biochemistry, UMDNJ–Johnson Medical School, Piscataway, New Jersey 08845. **I. Ota** • Department of Chemistry and Biochemistry, University of Colorado, Boulder, Colorado 80309-0215. **T. Rümenapf** • Institut für Virologie, Fachbereich Veterinärmedizin, 35392 Giessen, Germany. **T. E. Shrader** • Department of Biochemistry, Albert Einstein College of Medicine, Bronx, New York 10461. **P. R. H. Waller** • Testa, Hurwitz & Thibeault, Boston, Massachusetts 02110.

Ubiquitin and the Biology of the Cell, edited by Peters *et al.* Plenum Press, New York, 1998.

1. INTRODUCTION

Among the functions of intracellular proteolysis are the elimination of abnormal proteins, the maintenance of amino acid pools in cells affected by stresses such as starvation, and the generation of protein fragments that act as hormones, antigens, or other effectors. Yet another function of proteolytic pathways is selective destruction of proteins whose concentrations must vary with time and alterations in the state of a cell. Metabolic instability is a property of many regulatory proteins. A short *in vivo* half-life* of a regulator provides a way to generate its spatial gradients and allows for rapid adjustments of its concentration (or subunit composition) through changes in the rate of its synthesis. A protein can also be conditionally unstable: long-lived or short-lived depending on the state of a cell. Conditionally short-lived regulators are often deployed as components of control circuits. One example is cyclins—a family of related proteins whose destruction at specific stages of the cell cycle regulates cell division and growth (Chapter 12; Murray and Hunt, 1993). In addition, many proteins are long-lived as components of larger complexes such as ribosomes and oligomeric proteins but are metabolically unstable as free subunits. The short *in vivo* lifetimes of free subunits allow for a less stringent control over the relative rates of their synthesis, as a subunit produced in excess would not accumulate to a significant level.

Features of proteins that confer metabolic instability are called degradation signals, or degrons (Varshavsky, 1991). The essential component of one degradation signal, the first to be discovered, is a destabilizing N-terminal residue of a protein (Bachmair *et al.*, 1986). This signal is called the N-degron. A set of N-degrons containing different destabilizing residues yields a rule, termed the N-end rule, which relates the *in vivo* half-life of a protein to the identity of its N-terminal residue (Table I and Fig. 1) (reviewed by Varshavsky, 1992, 1996a,b). The N-end rule pathway has been found in all species examined, including the bacterium *Escherichia coli* (Tobias *et al.*, 1991; Shrader *et al.*, 1993), the yeast (fungus) *Saccharomyces cerevisiae* (Bachmair and Varshavsky, 1989), and mammalian cells (Gonda *et al.*, 1989; Lévy *et al.*, 1996) (Fig. 1).

The N-end rule was encountered in experiments that explored the metabolic fate of a fusion between Ub and a reporter protein such as *E. coli* β-galactosidase (βgal) in *S. cerevisiae* (Bachmair *et al.*, 1986). In yeast and other eukaryotes, Ub-X-βgal is cleaved, cotranslationally or nearly so, by Ub-specific processing proteases at the Ub-X-βgal junction. This cleavage takes place regardless of the identity of residue X at the C-terminal side of the cleavage site, proline being the

*The *in vivo* degradation of many short-lived proteins, including the engineered N-end rule substrates, deviates from first-order kinetics. Therefore, the term *half-life*, when it is applied to an entire decay curve, is a useful but often crude approximation. A more rigorous terminology for describing non-exponential decay was proposed by Lévy *et al.* (1996) (Section 7).

Table I
The N-End Rule in *E. coli.* and *S. cerevisiae*[a]

	Half-life of X-βgal			Half-life of X-βgal	
Residue X	*E. coli*	*S. cerevisiae*	Residue X	*E. coli*	*S. cerevisiae*
Arg	2 min	2 min	Asn	>10 hr	3 min
Lys	2 min	3 min	Gln	>10 hr	10 min
Phe	2 min	3 min	Cys	>10 hr	>30 hr
Leu	2 min	3 min	Ala	>10 hr	>30 hr
Trp	2 min	3 min	Ser	>10 hr	>30 hr
Tyr	2 min	10 min	Thr	>10 hr	>30 hr
His	>10 hr	3 min	Gly	>10 hr	>30 hr
Ile	>10 hr	30 min	Val	>10 hr	>30 hr
Asp	>10 hr	3 min	Pro	?	?
Glu	>10 hr	30 min	Met	>10 hr	>30 hr

[a]Approximate *in vivo* half-lives of X-βgal proteins in *E. coli* at 36°C (Tobias *et al.*, 1991) and in *S. cerevisiae* at 30°C (Bachmair and Varshavsky, 1989). The question mark at Pro indicates its uncertain status (see Section 6.8). The degradation of X-βgal test proteins in *E. coli*, *S. cerevisiae*, and mouse cells deviates from the first-order kinetics, in that the metabolic stability of older X-βgal molecules is higher than the metabolic stability of younger ones. The half-lives shown here are the "initial" (partial) half-lives, determined between 0 and 10 min of chase. In the terminology proposed by Lévy *et al.* (Section 7), a "partial" half-life in denoted as $t_{0.5}^{y-z}$, where 0.5 and $y - z$ indicate, respectively, the parameter's half-life aspect and the relevant time interval (from y to z min of chase). For the half-lives shown, this notation becomes $t_{0.5}^{0-10}$.

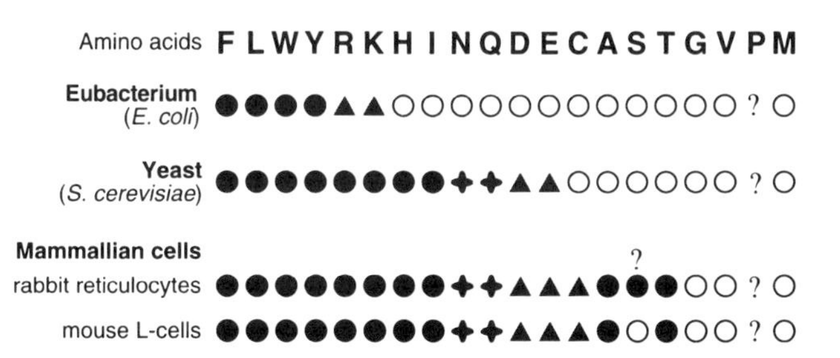

Figure 1. Comparison of eukaryotic and bacterial N-end rules. Open circles denote stabilizing residues. Filled circles denote primary destabilizing (N-d^{p}) residues. Gray triangles and crosses denote, respectively, secondary (N-d^{s}) and tertiary (N-d^{t}) destabilizing residues in the N-end rules of *E. coli* (Tobias *et al.*, 1991), *S. cerevisiae* (Bachmair and Varshavsky, 1989), ATP-supplemented extract from rabbit reticulocytes (Gonda *et al.* 1989), and mouse L-cells (Lévy *et al.*, 1996). A question mark above Ser indicates its uncertain status in the reticulocyte N-end rule (Section 2.2). A question mark at Pro indicates its uncertain status (Section 6.8). Asn, Gln, Asp, Glu, and Cys in the N-end rule of mouse L-cells were directly identified as destabilizing residues, but were assumed to be either tertiary or secondary, by analogy with the same residues in the reticulocyte N-end rule (Lévy *et al.*, 1996). Single-letter abbreviations for amino acids: A, Ala; C, Cys; D, Asp; E, Glu; F, Phe; G, Gly; H, His; I, Ile; K, Lys; L, Leu; M, Met; N, Asn; P, Pro; Q, Gln; R, Arg; S, Ser; T, Thr; V, Val; W, Trp; Y, Tyr.

single exception. By allowing a bypass of the normal N-terminal processing of a newly formed protein, this finding (Fig. 2A) (Bachmair *et al.*, 1986) yielded an *in vivo* method for generating different residues at the N-termini of otherwise identical proteins—a technical advance that led to the N-end rule.

In eukaryotes, the N-degron comprises at least two determinants: a destabilizing N-terminal residue and an internal lysine (or lysines) of a substrate (Bachmair and Varshavsky, 1989; Johnson *et al.*, 1990; Hill *et al.*, 1993; Nishizawa *et al.*, 1993; Dohmen *et al.*, 1994). The Lys residue is the site of formation of a multiUb chain (Chau *et al.*, 1989; Arnason and Ellison, 1994; Spence *et al.*, 1995). Ub is a 76-residue protein whose covalent conjugation to other proteins is involved in a multitude of processes—cell growth and differentiation, signal transduction, apoptosis, DNA repair, transmembrane traffic, and responses to stress, including the immune response. In many of these settings, Ub acts through routes that involve processive degradation of Ub–protein conjugates (Chapter 3; Hershko, 1991; Varshavsky, 1995a).

The binding of an N-end rule substrate by a targeting complex is followed by formation of a substrate-linked multiUb chain (Dohmen *et al.*, 1991). The ubiquitylated* substrate is processively degraded by the 26 S proteasome, an ATP-dependent, multisubunit protease (Chapters 5 and 6). The N-end rule pathway is present in both the cytosol (Bachmair *et al.*, 1986; Lévy *et al.*, 1996) and the nucleus (J. A. Johnston and A. Varshavsky, unpublished data). In this chapter, we summarize the current understanding of the N-end rule.

2. DESIGN, COMPONENTS, AND EVOLUTION OF THE N-END RULE PATHWAY

2.1. Definitions of Terms

The N-end rule: A relation between the metabolic stability of a protein and the identity of its N-terminal residue.

The N-degron: For an active degradation signal to be termed an N-degron, it is necessary and sufficient that it contain a substrate's initial or acquired N-terminal residue whose recognition by the targeting machinery is essential for the activity of this degron.

The pre-N-degron: Features of a protein that are necessary and sufficient, in the context of a given intracellular compartment, for the formation of an N-de-

*To bring ubiquitin-related terms in line with the standard chemical terminology, ubiquitin whose C-terminal (Gly76) carboxyl group is covalently linked to another compound is called the *ubiquityl* moiety, with the derivative terms being *ubiquitylation* and *ubiquitylated*. The abbreviation Ub refers to both free ubiquitin and the ubiquityl moiety.

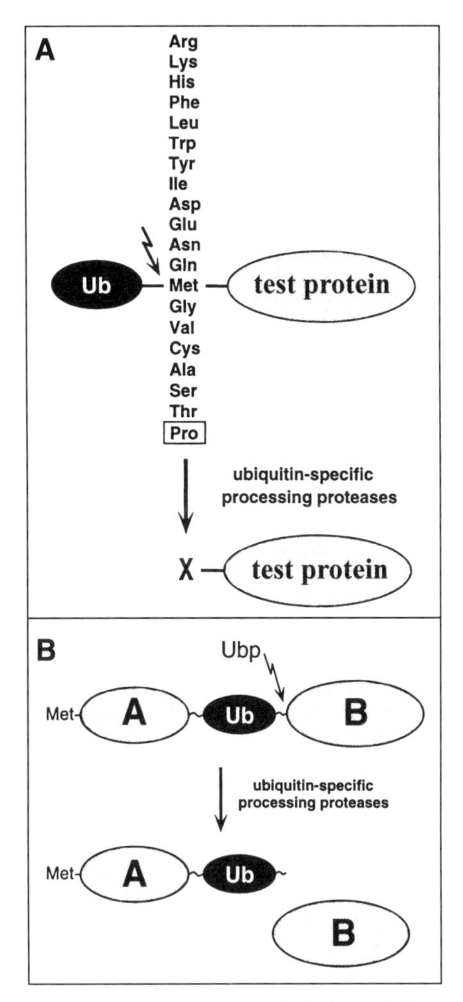

Figure 2. Uses of ubiquitin fusions. (A) The ubiquitin fusion technique. Linear fusions of Ub to other proteins are cleaved at the last residue of Ub both *in vivo* and in cell extracts, making it possible to produce different residues at the N-termini of otherwise identical proteins (Bachmair *et al.*, 1986; Gonda *et al.*, 1989). (B) The UPR (*u*biquitin/*p*rotein/*r*eference) technique (Lévy *et al.*, 1996). A tripartite fusion containing *A*, a reference protein moiety whose C-terminus is linked, via a spacer peptide, to a Ub moiety. The C-terminus of Ub is linked to *B*, a protein of interest. *In vivo*, this tripartite fusion is cleaved, cotranslationally or nearly so, by Ub-specific proteases (UBPs) at the Ub-*B* junction, yielding equimolar amounts of the unmodified protein *B* and the reference protein *A* bearing a C-terminal Ub moiety. If protein *A*-Ub is long-lived, determining the ratio of *A*-Ub to *B* as a function of time or at steady state would yield, respectively, the *in vivo* decay curve or the relative metabolic stability of protein *B*.

gron. For example, a protein may contain a constitutive or conditional recognition site for a processing protease whose cleavage of the protein produces a destabilizing residue at the N-terminus of a cleavage product. If this residue is sterically accessible to the N-end rule pathway, and if other essential determinants of an N-degron are in place, the recognition/cleavage site would function as a determinant of the protein's pre-N-degron. A pre-N-degron of this type is present in engineered Ub fusions whose study led to the finding of the N-end rule (Fig. 2A).

The N-end rule pathway: A set of molecular components that is necessary and sufficient, in the context of a given intracellular compartment, for the recognition and degradation of proteins bearing N-degrons. This "hardware-centric" definition of an N-end rule pathway bypasses semantic problems that arise if, for example, one and the same targeting complex recognizes not only N-degrons but also a degradation signal whose essential determinants do not include the N-terminal residue of a substrate. This definition also encompasses a setting where N-degrons that bear different destabilizing N-terminal residues are recognized by distinct targeting complexes. Neither of these possibilities is entirely hypothetical (Sections 4 and 6.8).

Primary destabilizing residues: Destabilizing activity of these N-terminal residues, denoted N-d^p, requires their physical binding by a protein called N-recognin or E3 (Figs. 3–5). In eukaryotes, the type 1 binding site of N-recognin binds N-terminal Arg, Lys, or His—a set of basic N-d^p residues—whereas the type 2 site binds N-terminal Phe, Leu, Trp, Tyr, or Ile—a set of bulky hydrophobic N-d^p residues (Section 2.2). Accordingly, the N-d^p residues are subdivided into type 1 (N-d^{p1}) and type 2 (N-d^{p2}) residues. The N-d^p residues of *E. coli*—Phe, Leu, Trp, and Tyr—are exclusively type 2 (N-d^{p2}) residues (Section 2.2).

Secondary destabilizing residues: These N-terminal residues, denoted N-d^s, are Arg and Lys in *E. coli*; Asp and Glu in *S. cerevisiae*; and Asp, Glu, and Cys in mammalian cells (Figs. 3–5). In eukaryotes, destabilizing activity of N-d^s residues requires their accessibility to Arg-tRNA-protein transferase (R-transferase). In bacteria such as *E. coli*, destabilizing activity of the N-d^s residues Arg and Lys requires their accessibility to Leu,Phe-tRNA-protein transferase (L/F-transferase) (Figs. 4 and 5C).

Tertiary destabilizing residues: N-terminal Asn and Gln residues, denoted N-d^t. Destabilizing activity of N-d^t residues requires their accessibility to an N-terminal amidohydrolase (Nt-amidase), the presence of the rest of an N-degron, and the presence of the N-end rule pathway (Figs. 3 and 5A,B). The two latter requirements are common to all destabilizing residues.

Stabilizing residues: A stabilizing residue, when it replaces a destabilizing N-terminal residue in the context of active N-degron, renders this N-degron inactive (more precisely, lowers its activity below a predetermined threshold). Gly, Val, and Met are stabilizing residues that are common to all of the known N-end rules (Fig. 1). A stabilizing residue is a "default" residue, in that it is

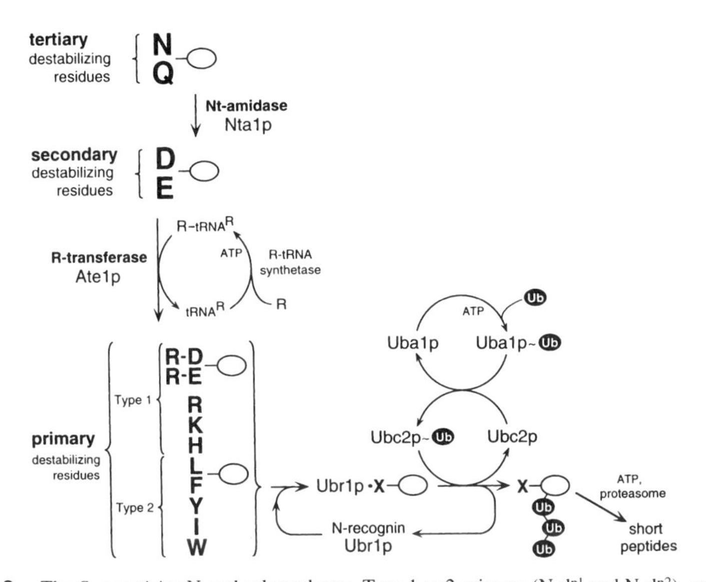

Figure 3. The *S. cerevisiae* N-end rule pathway. Type 1 or 2 primary (N-d^{p1} and N-d^{p2}), secondary (N-d^{s}), and tertiary (N-d^{t}) destabilizing N-terminal residues are defined in Section 2.1. The shaded ovals denote the rest of a protein substrate. The conversion of N-d^{t} residues *N* and *Q* into N-d^{s} residues *D* and *E* is mediated by N-terminal amidohydrolase (Nt-amidase), encoded by *NTA1*. The conjugation of N-d^{p1} residue *R* to N-d^{s} residues *D* and *E* is mediated by Arg-tRNA-protein transferase (R-transferase), encoded by *ATE1*. A complex of N-recognin and the Ub-conjugating (E2) enzyme Ubc2p catalyzes the conjugation of activated Ub, produced by the Ub-activating (El) enzyme Uba1p, to a Lys residue of the substrate, yielding a substrate-linked multiUb chain. Uba1p~Ub and Ubc2p~Ub denote covalent (thioester-mediated) complexes of these enzymes with Ub. A multiubiquitylated substrate is degraded by the 26 S proteasome.

stabilizing because targeting components of an N-end rule pathway do not bind to it (or modify it) efficiently enough even in the presence of other essential determinants of an N-degron. The emerging complexity of N-degrons may result in exceptions to this definition. For example, destabilizing activity of the N-terminal Pro may be strongly influenced by the sequence context immediately downstream of Pro (Section 6.8).

2.2. N-Recognin (E3)

2.2.1. Fungi

In *S. cerevisiae*, N-recognin is a 225-kDa protein (encoded by *UBR1*) that selects potential N-end rule substrates through the binding to their N-d^{p} residues Phe, Leu, Trp, Tyr, Ile, Arg, Lys, or His (Bartel *et al.*, 1990; Baker and Varshavsky,

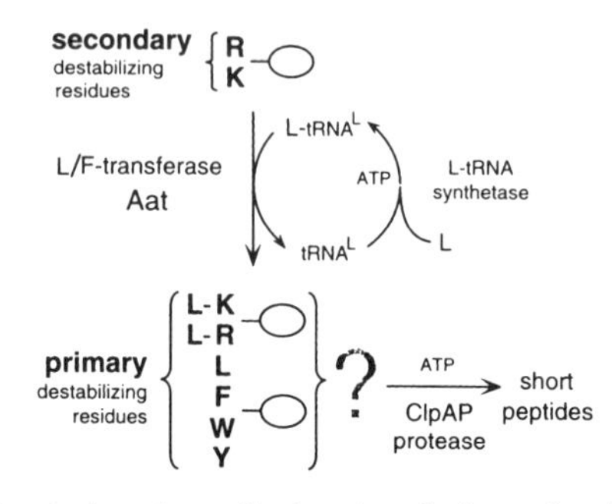

Figure 4. The *E. coli* N-end rule pathway. Conjugation of primary destabilizing ($N-d^p$) residues *L* or *F* to $N-d^s$ residues *R* and *K* is catalyzed by Leu,Phe-tRNA-protein transferase (L/F-transferase), encoded by *aat* (Tobias *et al.*, 1991). *In vivo*, L/F-transferase appears to conjugate mainly, if not exclusively, *L* (Shrader *et al.*, 1993). The degradation of a protein substrate bearing one of the $N-d^p$ residues *F*, *L*, *W*, or *Y* is carried out by the ATP-dependent protease ClpAP, encoded by *clpA* and *clpP*. A question mark denotes an ambiguity about the nature of N-recognin in *E. coli* (Section 2.2).

1991; Madura *et al.*, 1993). Null *ubr1* mutants are unable to degrade the normally short-lived N-end rule substrates, but are only slightly (~ 3%) growth-impaired on standard media (Bartel *et al.*,, 1990). The deduced amino acid sequence of *S. cerevisiae* N-recognin (Ubr1p) is 49% identical to the sequence of a putative N-recognin from another budding yeast, *Kluyveromyces lactis* (P. R. H. Waller and A. Varshavsky, unpublished data).

A physical interaction between N-recognin and an $N-d^p$ residue of a substrate was inferred in particular from the results of *in vitro* binding assays. In one such experiment, the purified, ^{35}S-labeled Val-βgal or Arg-βgal, produced from the corresponding Ub fusions (Fig. 2A) and bearing, respectively, a stabilizing and a primary destabilizing ($N-d^{p1}$) N-terminal residue, were added to an extract from *S. cerevisiae* cells that overexpressed Ubr1p (N-recognin) bearing an epitope tag. Precipitation of Ubr1p from the extract with an antiepitope antibody coimmuno-precipitated Arg-βgal but not the otherwise identical Val-βgal (Bartel, 1990). Similar results were obtained with Leu-DHFR (derived from Ub-Leu-DHFR) (DHFR, mouse dihydrofolate reductase), another engineered N-end rule substrate (Bartel *et al.*, 1990).

N-recognin has at least two substrate-binding sites. The type 1 site is specific for the basic N-terminal residues Arg, Lys, and His. The type 2 site is specific for the bulky hydrophobic N-terminal residues Phe, Leu, Trp, Tyr, and Ile. At present, these sites are defined operationally—through dipeptide-based competition experiments. Specifically, a dipeptide bearing a destabilizing N-terminal residue was found to inhibit the degradation of a test N-end rule substrate if that substrate's

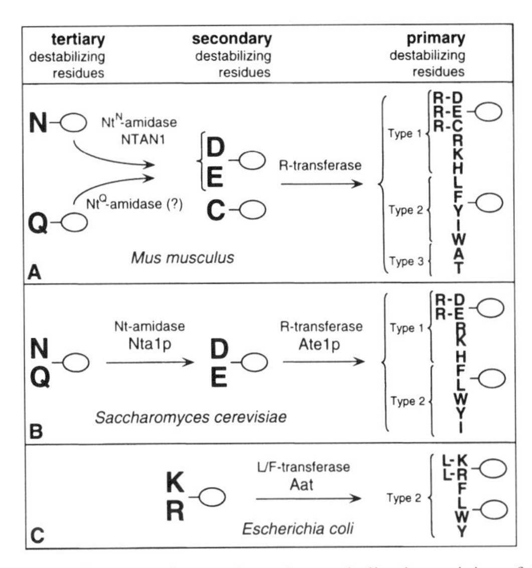

Figure 5. Comparison of enzymatic reactions that underlie the activity of tertiary (N-d[t]) and secondary (N-d[s]) destabilizing residues in different organisms. (A) Mouse (*Mus musculus*) fibroblast-like L-cells and rabbit (*Oryctolagus cuniculus*) reticulocytes (Lévy *et al.*, 1996; Gonda *et al.*, 1989). (B) The yeast *Saccharomyces cerevisiae* (Bachmair and Varshavsky, 1989). (C) The bacterium *Escherichia coli* (Tobias *et al.*, 1991). The *E. coli* N-end rule lacks N-d[t] residues. The postulated mammalian Nt[Q]-amidase (question mark in A) remains to be identified.

N-terminal residue was of the same type (1 or 2) as the dipeptide's N-terminal residue (Reiss *et al.*, 1988; Gonda *et al.*, 1989; Baker and Varshavsky, 1991).

A genetic dissection of the type 1 and type 2 sites in *S. cerevisiae* N-recognin (Ubrlp) has shown that either of these sites can be mutationally inactivated without significantly perturbing the other site. Mutations that selectively inactivate the type 1 or the type 2 site are located within the ~ 50-kDa N-terminal region of the 225-kDa N-recognin (A. Webster, M. Ghislain, and A. Varshavsky, unpublished data).

The concentration of a dipeptide required for ~ 90% inhibition of the N-end rule pathway is ~ 100 μM (Baker and Varshavsky, 1991; Gonda *et al.*, 1989; Reiss *et al.*, 1988), consistent with the K_d (dissociation constant) of N-recognin/dipeptide complex of roughly 10 μM. (The crudeness of this estimate stems in part from the unknown extent of dipeptide degradation in the course of inhibition assays.) That an N-end rule substrate such as Arg-βgal can be specifically co-immunoprecipitated with Ubrlp (N-recognin) from cell extracts (Bartel, 1990) suggests an even lower K_d.

2.2.2. Metazoans*

E3α, the mammalian counterpart of *S. cerevisiae* N-recognin, has been characterized biochemically in extracts from rabbit reticulocytes (Hershko and Ciechanover, 1992). Another N-recognin, termed E3β, which apparently binds to substrates bearing N-terminal Ala and Thr (and possibly also Ser) (Gonda *et al.*, 1989; Lévy *et al.*, 1996), has been described as well (Hershko and Ciechanover, 1992). Ala and Thr are destabilizing in the N-end rule of mouse L-cells *in vivo* (Lévy *et al.*, 1996), whereas Ala, Thr, and Ser are destabilizing in ATP-supplemented extract from rabbit reticulocytes (Gonda *et al.*, 1989) (Fig. 1). N-terminal Ser of many cytosolic and nuclear proteins in cotranslationally acetylated *in vivo* (Arfin and Bradshaw, 1988). N-terminal acetylation of a protein is expected to preclude its binding by N-recognin (Reiss *et al.*,1988; Varshavsky, 1992). It is unknown whether N-terminal Ser of Ser-βgal is acetylated in mouse L-cells. Therefore, it remains to be seen whether the difference in destabilizing activity between N-terminal Ser in reticulocyte extract (Gonda *et al.*, 1989) and in fibroblastlike L-cells (Lévy *et al.*, 1996) (Fig. 1) defines a difference between the N-end rules of these cell types or whether it is an artifact of inefficient acetylation of N-terminal Ser in reticulocyte extract. Ala, Thr, and Ser are stabilizing residues in the yeast and bacterial N-end rules (Fig. 1).

Arg is the strongest destabilizing residue in *S. cerevisiae*, followed closely by Lys, Phe, Leu, Trp, Asp, and Asn (Table I). Tyr and Gln are of intermediate strength, whereas Ile and Glu are the least destabilizing residues in the same settings. However, even the least destabilizing N-terminal residue such as Ile confers a much shorter half-life on a reporter protein than any of the stabilizing residues (Table I).

The N-terminal Pro, in the context of engineered substrates such as X-βgals, does not confer short half-lives on these substrates in either *S. cerevisiae*, mouse L-cells, or ATP-supplemented reticulocyte extract (Gonda *et al.*, 1989; F. Lévy, T. Rümenapf, and A. Varshavsky, unpublished data). By contrast, the N-terminal Pro is a destabilizing residue in the context of c-Mos protein (Nishizawa *et al.*, 1992, 1993) (Section 6.8).

2.2.3. Prokaryotes

All eukaryotes examined have both Ub and the N-end rule pathway. Some, but not all, prokaryotes contain Ub as well (Durner and Boger, 1995; Wolf *et al.*, 1993). The bacterium *E. coli* lacks Ub but does have an N-end rule pathway (Fig.

*The term *metazoans* denotes, strictly speaking, multicellular organisms of the animal kingdom. However, in the absence of a single-word term for multicellular organisms, *metazoans* is used throughout this chapter to refer to the entire set of multicellular organisms.

4) (Tobias *et al.*, 1991). Screens for mutations that inactivate the N-end rule pathway have identified three genes: *clpA*, *clpP*, and *aat* (Tobias *et al.*, 1991; Shrader *et al.*, 1993). The *aat* gene encodes L/F-transferase (Section 2.4). ClpA (81 kDa) and ClpP (21 kDa) form an ~ 750-kDa complex, ClpAP, which exhibits ATP-dependent protease activity *in vitro* (Gottesman and Maurizi, 1992), and is a functional counterpart of the eukaryotic 26 S proteasome in the *E. coli* N-end rule pathway (Fig. 4).

ClpP exhibits a chymotrypsinlike protease activity *in vitro* (Gottesman and Maurizi, 1992). ClpA is the ATP-binding component of ClpAP. *In vitro* studies have shown that ClpA can act as a chaperone in the activation of RepA, the replication initiator encoded by the plasmid P1 (Wickner *et al.*, 1994). Specifically, ClpA monomerizes RepA, thereby allowing it to bind to the plasmid's origin of replication. In addition, ClpA, but not the other chaperones such as DnaJ and DnaK, can target RepA for degradation by the ClpAP protease *in vitro* (Wickner *et al.*, 1994; Hayes and Dice, 1996). The *in vivo* ramifications of these results, and in particular their relevance to the proteolytic function of ClpAP in the *E. coli* N-end rule pathway (Fig. 4) remain to be examined. The *in vitro* assembly of ClpAP protease from purified ClpA and ClpP requires ATP hydrolysis by ClpA (Seol *et al.*, 1995). *In vivo*, ClpP associates not only with ClpA (forming ClpAP protease), but also with the ClpA homologues ClpB or ClpX, forming, respectively, ClpBP or ClpXP proteases (Gottesman *et al.*, 1993; Wawrzynov *et al.*, 1995). In contrast to ClpA, whose mutational elimination stabilizes the normally short-lived N-end rule substrates (Tobias *et al.*, 1991), the elimination of either ClpB or ClpX appears not to perturb the *E. coli* N-end rule pathway (O. Lomovskaya and A. Varshavsky, unpublished data).

Both the initial genetic screen (Tobias *et al.*, 1991), which utilized a βgal-based reporter, and a selection-based screen (O. Lomovskaya and A. Varshavsky, unpublished data) failed to identify an *E. coli* N-recognin. One possibility is that ClpAP protease itself is the N-recognin. Given the structure of the *E. coli* N-end rule (Figs. 1 and 4), an *E. coli* N-recognin is expected to bind N-terminal Phe, Leu, Trp, or Tyr (a subset of eukaryotic N-d^{p2} residues) but not Arg or Lys (a subset of eukaryotic N-d^{p1} residues). In other words, an *E. coli* N-recognin is expected to have a type 2 but not a type 1 substrate-binding site of its eukaryotic counterpart. Whether the ClpA subunit of ClpAP protease functions as N-recognin remains to be determined.

2.3. N-Terminal Amidases

2.3.1. Fungi

The *S. cerevisiae* N-terminal amidohydrolase (Nt-amidase), encoded by *NTA1*, is a 52-kDa enzyme that deamidates Asn or Gln if they are located at the

N-terminus of a polypeptide (Figs. 3 and 5B) (Baker and Varshavsky, 1995; Grigoryev *et al.*, 1995). Null *nta1* mutants are unable to degrade N-end rule substrates that bear N-terminal Asn or Gln, but are unimpaired for growth on standard media. The deduced sequence of Nta1p is not similar to those of the other known amidotransferases, save for the sequence Gly-Ile-Cys-Met that is a part of an 11-residue region conserved among some, but not all, amidotransferases. The conserved cysteine of this sequence is required for the enzymatic activity of Nta1p (Baker and Varshavsky, 1995). The sequence of *S. cerevisiae* Nta1p is 42% identical to that of *K. lactis* Nta1p (S. Grigoryev and A. Varshavsky, unpublished data) but lacks similarities to the sequences of mammalian Nt-amidases (see below).

5′ mapping of *S. cerevisiae NTA1* mRNA detected at least two species, one of which encodes an N-terminally truncated Nta1p (Baker and Varshavsky, 1995). The deduced sequence of 29 residues between Asp4 and Asp34 (of the larger Nta1p species) resembles a mitochondrial translocation signal. Thus, the hypothetical larger Nta1p may be transported into mitochondria, while the smaller Nta1p is expected to reside in the cytosol (Baker and Varshavsky, 1995). Several yeast proteins that are present in both mitochondria and the cytosol are known to be encoded by a single gene, with the mitochondrial form produced through translation from an upstream start codon (Chatton *et al.*, 1988; Ellis *et al.*, 1989; Chiu *et al.*, 1992).

The hypothesis of two Nta1p species remains to be verified directly. This conjecture would account for the currently unexplained presence of Glu (instead of Gln) at the N-termini of at least two mitochondrial proteins in *S. cerevisiae*. Specifically, the gene-deduced amino acid sequences of mitochondrial stress protein Hsp70 and subunit IV of cytochrome *c* oxidase predicted Gln at the proteins' mature N-termini, which result from site-specific cleavages by the mitochondrial signal peptidase. Instead, the isolated proteins were found to bear N-terminal Glu, indicating N-terminal deamidation (Maarse *et al.*, 1984; Scherer *et al.*, 1990). (At present, it is not excluded that the observed N-terminal deamidation of these proteins was an *in vitro* artifact caused by transient access of the normally cytosolic Nta1p to mitochondrial proteins during their isolation.) The putative mitochondrial Nt-amidase may be a component of a mitochondrial N-end rule pathway. In eukaryotes, the N-end rule pathway has been found in both the cytosol (Bachmair *et al.*, 1986) and the nucleus (J. Johnston and A. Varshavsky, unpublished data). It is unknown whether N-end rule pathways operate in mitochondria, chloroplasts, or other compartments distinct from the nucleus and cytosol.

2.3.2. Metazoans

Stewart *et al.*, (1994, 1995) purified a porcine Nt-amidase that deamidates N-terminal Asn (N) but not Gln (Q), and isolated a cDNA that encodes this enzyme. Grigoryev *et al.* (1996) isolated and characterized a cDNA and an ~ 17-kb gene,* termed *Ntan1*, that encodes a mouse homologue of the porcine amidase, termed Nt^N-amidase. ("N" in the superscript and in *Ntan1* refer to the Asn (N) specificity of this Nt-amidase.) *Ntan1* is located in the proximal region of mouse chromosome 16, and contains 10 exons whose length ranges from 54 to 177 bp. This chromosomal region, and a homologous region of human chromosome 16 do not encompass the sites of uncloned human or mouse mutations (Grigoryev *et al.*, 1996; R. A. Bradshaw and S. M. Arfin, personal communication). The ~ 1.4-kb mouse *Ntan1* mRNA is expressed, at varying levels, in all of the tested mouse tissues and cell lines. The deduced amino acid sequence of the 35-kDa mouse Nt^N-amidase is 88% identical to the sequence of its porcine counterpart, but bears no significant similarity to the sequence of the 52-kDa *S. cerevisiae* Nt-amidase (Nta1p). In contrast to mouse Nt^N-amidase, the yeast Nt-amidase can deamidate either N-terminal Asn or Gln (Baker and Varshavsky, 1995; Grigoryev *et al.*, 1996).

Both Asn and Gln are destabilizing residues in the mammalian N-end rule (Gonda *et al.*, 1989; Lévy *et al.*, 1996). Further, both N-terminal Asn and Gln of the test proteins are deamidated in mammalian cell extracts (Gonda *et al.*, 1989; S. Grigoryev and A. Varshavsky, unpublished data). Therefore, there must exist yet another mammalian Nt-amidase (Nt^Q-amidase), which can deamidate N-terminal Gln (Fig. 5A). The apparent absence of *Ntan1* homologues from the mouse genome (those that could be detected by low-stringency Southern hybridization) suggested that the postulated Nt^Q-amidase (Fig. 5A) may be encoded by a differentially spliced mRNA derived from the *Ntan1* gene. Although no differential splicing (or editing) of mouse *Ntan1* pre-mRNA has been detected thus far, GenBank was found to contain partial nucleotide sequences of apparently *Ntan1* human cDNAs one of which lacks exon II of *Ntan1*. In addition, the *Ntan1* intron I, which abuts exon II, contains noncanonical dinucleotide sequences at both 5′ and 3′ splice sites. This circumstantial evidence for the hypothesis of a single gene encoding both Nt^N-amidase and Nt^Q-amidase (Grigoryev *et al.*, 1996) remains to be verified directly.

Physiological substrates of Nt-amidases are unknown in either fungi or metazoans. Part of the difficulty in identifying these substrates stems from the fact that although a number of apparently cytosolic or nuclear proteins bear, at least initially, the N-terminal sequences Met-Asn or Met-Gln (Stewart *et al.*, 1994), the known Met-aminopeptidases are incapable of removing N-terminal Met if it is followed by an Asn or Gln residue. In addition, at least the Met-Asn sequence is often acetylated at the N-terminal Met (Arfin and Bradshaw, 1988). It is unknown

*The names of mouse genes are in italics, first letter uppercase. The names of mouse proteins are in plain font, all uppercase. The names of *S. cerevisiae* proteins are in plain font, first letter uppercase, with a "p" as a suffix. The names of *E. coli* proteins are in plain font, first letter uppercase—otherwise identical to the names of corresponding genes (Stewart, 1995).

whether any of the resulting Ac-Met-Asn-containing N-terminal regions are processed *in vivo* to yield N-terminal Asn. Recently, certain sequences bearing N-terminal Met-Gln have been found to yield N-terminal Gln *in vivo* (Ghislain *et al.*, 1996) (Section 4).

2.4. Aminoacyl-tRNA-Protein Transferases

2.4.1. Fungi and Metazoans

The *S. cerevisiae* Arg-tRNA-protein transferase (R-transferase), encoded by *ATE1*, is a 58-kDa enzyme that utilizes Arg-tRNA to arginylate N-termini of polypeptides (but not free amino acids) that bear Asp or Glu (Fig. 3). Null *ate1* mutants are unable to degrade N-end rule substrates that bear N-terminal Asn, Gln, Asp, or Glu—the $N\text{-}d^t$ and $N\text{-}d^s$ destabilizing residues (Fig. 3) (Balzi *et al.*, 1990). In contrast to *S. cerevisiae*, where only Asp and Glu are $N\text{-}d^s$ residues, in mammals Cys is an $N\text{-}d^s$ residue as well (Cys is a stabilizing residue in yeast and prokaryotes; Table I and Fig. 1). It is unknown whether the arginylation of N-terminal Asp, Glu, and Cys in mammals is catalyzed by an R-transferase whose specificity is broader than that of its yeast counterpart, or whether N-terminal Cys is arginylated by a distinct R-transferase (Fig. 5A, B).

An extract prepared ~ 2 h after a crush injury to the rat sciatic nerve from a segment of the nerve immediately upstream of the crush site was found to conjugate an ~10-fold higher amount of the added [^{3}H]arginine to the N-termini of unidentified endogenous proteins than an otherwise identical extract prepared from the same region of an unperturbed sciatic nerve (Shyne-Athwal *et al.*, 1988). This finding suggested a crush-induced increase in the level of N-end rule substrates and/or a postcrush induction of the N-end rule pathway. No postcrush increase in arginylation was observed with extracts from the rat optic nerve, which does not regenerate after a crush injury, in contrast to the sciatic nerve (Shyne-Athwal *et al.*, 1988).

2.4.2. Prokaryotes

R-transferase appears to be confined to eukaryotes, whereas Leu, Phe-tRNA-protein transferase (L/F-transferase) is present in bacteria such as *E. coli*, but is apparently absent from eukaryotes (Soffer, 1980; Deutsch, 1984). *E. coli* L/F-transferase is a 27-kDa enzyme encoded by the *aat* gene (Shrader *et al.*, 1993; Abramochkin and Shrader, 1995). *In vitro*, a partially purified *E. coli* L/F-transferase (the Aat protein) catalyzes the conjugation of either Leu, Phe, or (at a much lower rate) Met to N-terminal Arg or Lys of a polypeptide substrate (Soffer, 1980) (Figs. 4 and 5C). *In vivo*, L/F-transferase conjugates mainly, if not exclusively, Leu (Shrader *et al.*, 1993). *E. coli* mutants lacking *aat* are unable to degrade N-end rule substrates that bear N-terminal Arg or Lys. These data identified L/F-transferase as a component of the *E. coli* N-end rule pathway (Tobias *et al.*, 1991; Shrader *et al.*, 1993).

E. coli L/F-transferase is half the size of *S. cerevisiae* R-transferase (234 versus 503 residues). Despite also being smaller than the smallest aminoacyl-tRNA synthetase (Burbaum and Schimmel, 1991), L/F-transferase binds specific aminoacyl-tRNAs, recognizes specific N-terminal residues in polypeptide substrates, and catalyzes formation of the peptide bond (Soffer, 1980; Shrader *et al.*, 1993; Abramochkin and Shrader, 1995). There is a functionally suggestive (but statistically weak) similarity between a 24-residue internal region of the *E. coli* Aat L/F-transferase and a region of the RimL N-terminal acetylase (Abramochkin and Shrader, 1995).

E. coli aat is located ~1 kb from *clpA*, which encodes a subunit of the ATP-dependent protease ClpAP—the proteolytic and possibly also a recognition component of the N-end rule pathway (Section 2.2). Despite being functionally linked, the nearly adjacent *aat* and *clpA* are convergently transcribed. The *aat* gene is the last gene of an operon that contains at least two additional ORFs, *cydC* and *cydD*, which encode ABC-type membrane transporters (Poole *et al.*, 1993; Shrader *et al.*, 1993). CydC and CydD are required for the formation of the cytochrome *bd* terminal oxidase—one of the two terminal oxidases that mediate aerobic respiration (Poole *et al.*, 1993). Because the genes of a bacterial operon tend to encode functionally related proteins, we considered the possibility that the *aat*-encoded L/F-transferase may target for degradation an (unknown) N-end rule substrate that controls the activity or expression of the transporters CydC or CydD. However, a precise deletion of *aat* did not render *E. coli* hypersensitive to azide and zinc ions (a trait of cells deficient in the cytochrome *bd* terminal oxidase), making the above possibility unlikely but still not excluded (O. Lomovskaya and A. Varshavsky, unpublished data).

2.5. Ubiquitin-Conjugating Enzymes

The initial interaction between a potential N-end rule substrate and N-recognin is of moderate affinity (Section 2.2), but becomes much stronger if an internal lysine of the substrate is captured by a targeting complex containing N-recognin and a Ub-conjugating (E2) enzyme. This capture initiates a processive synthesis of a lysine-linked multiUb chain by a complex of the E2 enzyme and N-recognin (E3)—a step that decreases the probability of substrate release and leads to the binding of a multiubiquitylated substrate by the 26 S proteasome (Section 5). The E2 enzymes utilize activated Ub, produced by the Ub-activating (E1) enzyme, to catalyze the formation of isopeptide bonds between the C-terminal Gly76 of Ub and ε-amino groups of lysines in acceptor proteins (Chapters 2 and 3).

In at least some Ub-dependent systems (Scheffner *et al.*, 1995), including apparently the N-end rule pathway (V. Chau and A. Varshavsky, unpublished data), a Ub ligase catalyzes the transfer of Ub (which is initially linked to a Cys residue of El enzyme) through a relay of Ub thioesters before conjugating Ub to a Lys residue of a targeted substrate (Section 5.1). In a substrate-linked multiUb chain, the C-terminal glycine of one Ub moiety is conjugated to an internal lysine of the adjacent Ub moiety, resulting in a chain of Ub–Ub conjugates. In the initially characterized multiUb chains, only Lys48 of Ub was found to be joined to another Ub moiety within a chain (Chau *et al.*, 1989). Recently, multiUb chains mediated by Lys63, Lys29, Lys11, or Lys6 have been described as well (Chapter 2; Arnason and Ellison, 1994; Spence *et al.*, 1995; Johnson *et al.*, 1995; Baboshina and Haas, 1996). It is unknown whether these "non-Lys48" chains play a role in the N-end rule pathway.

In *S. cerevisiae*, the ubiquitylation of N-end rule substrates requires an E2 enzyme encoded by the *UBC2* gene—one of 13 E2-encoding genes in this organism (Dohmen *et al.*, 1991). Ubc2p is physically associated with N-recognin (Ubr1p) (Madura *et al.*, 1993). The N-end rule pathway is inactive in a *ubr1*Δ mutant (Section 2.2) and is nearly inactive in a *ubc2*Δ mutant (Fig. 6). However, the overall effect of *ubc2*Δ on cell growth and sporulation is much more severe than that of *ubr1*Δ (Chapters 2 and 3), indicating that the functions of Ubc2p are not confined to the N-end rule pathway. Furthermore, although Ubc2p is clearly the major E2 component of the N-end rule pathway, studies of the function of the N-end rule in peptide import have suggested that Ubc4p, another *S. cerevisiae* E2, can partially compensate for the absence of Ubc2p (Section 6.3) (C. Byrd and A. Varshavsky, unpublished data).

Madura *et al.* (1993) observed that, paradoxically, the degradation of an N-end rule substrate such as Arg-βgal was inhibited more strongly by deletion of the Asp/Glu-rich "tail" of the *S. cerevisiae* Ubc2p than by deletion of the entire *UBC2* gene (Fig. 6). The highly acidic 23-residue tail of the 172-residue Ubc2p is required for a high-affinity interaction between Ubc2p and N-recognin, but is not required for the intrinsic E2 activity of Ubc2p (Madura *et al.*, 1993). A mutant bearing the "tailless" version of Ubc2p is deficient in sporulation and the N-end rule pathway, but is nearly unperturbed in DNA repair and growth rate, in contrast to a congenic *ubc2*Δ mutant (Madura *et al.*, 1993; Jentsch *et al.*, 1987). Given a partial functional overlap between Ubc2p and Ubc4p (see above and Section 6.3), one interpretation of the data in Fig. 6 is that the absence of Ubc2p results in a compensatory overexpression of at least some of the other *S. cerevisiae* E2 enzymes, including Ubc4p, which may substitute, albeit inefficiently, for Ubc2p in the N-recognin-containing targeting complex. By contrast, the tailless Ubc2p retains at least some of the non-N-end rule functions of Ubc2p, resulting in little or no overexpression of the other E2s. This model accounts for the finding that expression of the tailless Ubc2p (in a *ubc2*Δ background) inhibits the N-end rule pathway more strongly than the absence of Ubc2p (Fig. 6). This model is also supported by the observation that the degradation of Ub-Pro-βgal, whose ineffi-

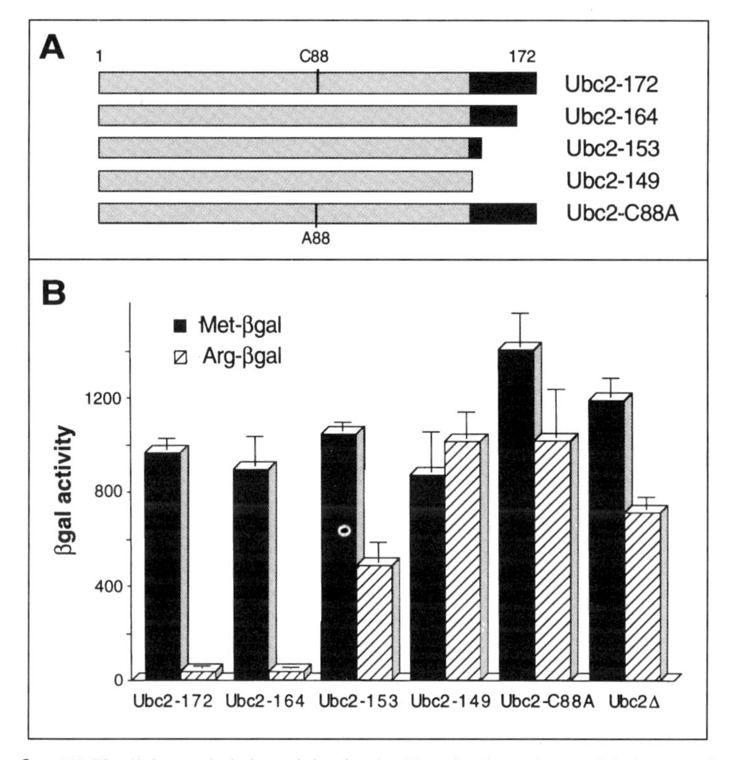

Figure 6. *UBC2* alleles and their activity in the N-end rule pathway (Madura *et al.*, 1993). (A) Schematic representation of the 172-residue Ubc2p, its C-terminally truncated derivatives (Ubc2p-164, Ubc2p-153, and Ubc2p-149), and the enzymatically inactive Ubc2p-C88A, in which the active-site Cys88 has been replaced by Ala (filled portions of the rectangles denote the 23-residue polyacidic tail). (B) Levels of βgal activity in the *S. cerevisiae* strain BBY67 (*ubc2Δ*) carrying plasmids that expressed Ub-Met-βgal (black bars) or Ub-Arg-βgal (striped bars) from the galactose-inducible P_{GAL10} promoter (Bachmair *et al.*, 1986), and also either a low-copy (*CEN*-based) vector alone (denoted as "Ubc2Δ") or vector-derived plasmids that expressed, respectively, Ubc2p-172 (wild-type Ubc2p), Ubc2p-164, Ubc2p-153, Ubc2p-149, and Ubc2p-C88A (Madura *et al.*, 1993). Standard deviations are indicated above the bars.

ciently removed N-terminal Ub moiety is recognized as a degron by a distinct pathway (Johnson *et al.*, 1992, 1995) (Section 4), is more efficient in a *ubc2Δ* mutant than in a congenic *UBC2* strain (Dohmen *et al.*, 1991).

2.6. The N-End Rule as a Witness of Evolution

The hierarchic organization of N-end rules, with their tertiary, secondary, and primary destabilizing residues, is a feature more conserved in evolution than either the Ub dependence of an N-end rule pathway or the identity of enzymatic reactions that mediate the activity of destabilizing residues. In a bacterium such as *E. coli*, which lacks the Ub system, the N-end rule has both N-d^s and N-d^p residues (it lacks N-d^t residues) (Figs. 1, 4, and 5C). The identities of N-d^S residues in *E. coli* (Arg and Lys) are different from those in eukaryotes (Figs. 1 and 5). Bacterial and eukaryotic enzymes that implement the coupling between N-d^s and N-d^p are also different: L/F-transferase in *E. coli* and R-transferase in eukaryotes. Note, however, that bacterial L/F-transferase and eukaryotic R-transferase catalyze reactions of the same type (conjugation of an amino acid to an N-terminal residue of a polypeptide) and utilize the same source of activated amino acid (aminoacyl-tRNA) (Fig. 5).

The hierarchic organization of N-end rules makes possible a "dispersion" of domains that recognize specific destabilizing N-terminal residues among several proteins such as Nt-amidase, R-transferase (L/F-transferase in bacteria), and N-recognin (Fig. 5). Regulatory possibilities of such an arrangement are apparent but physiological substrates of Nt-amidases, R-transferase, and L/F-transferase remain unknown in any organism. It is likely that at least metazoan cells can produce different N-recognins (Section 2.2) and can also regulate either synthesis or activity of Nt-amidase and R-transferase. The resulting changes of the N-end rule may occur in response to alterations in the state of a cell, for example, during cell differentiation. A change of the N-end rule may provide a way to eliminate a set of previously long-lived proteins or to stabilize a set of previously short-lived proteins.

The apparent confinement of R-transferase to eukaryotes and of L/F-transferase to prokaryotes suggests that N-d^s residues were recruited late in the evolution of the N-end rule, after the divergence of prokaryotic and eukaryotic lineages. The lack of sequence similarity between the yeast Nt-amidase and the mammalian Nt^N-amidase, as well as the more narrow specificity of the mammalian enzyme suggest that the N-d^t residues Asn and Gln became a part of the N-end rule much later yet, possibly after the divergence of metazoan and fungal lineages. If so, the N-end rule pathway may be an especially informative witness of evolution: The ancient origins of this proteolytic system, the simplicity and discreteness of changes in the rule books of N-end rules among different species, and the diversity of proteins that either produce or target the N-degron should facilitate phylogenetic deductions—once the components and functions of this pathway become characterized across a broad range of organisms.

2.7. Code versus Hardware

A given N-end rule is defined operationally—for a set of proteins such as X-βgals that differ exclusively by their N-terminal residues. Existing evidence (Varshavsky, 1992) suggests that the ranking aspect of an N-end rule, i.e., an ordering of relative destabilizing activities among 20 fundamental amino acids, is invariant from one protein reporter to another in a given intracellular compartment. (The case of N-end rule substrates bearing N-terminal Pro presents an apparent exception to this conjecture; see Section 6.8.) By contrast, the actual *in vivo* half-lives may differ greatly among *different* proteins bearing one and the same N-terminal residue (Bachmair and Varshavsky, 1989). The cause of these differences is the multicomponent nature of underlying N-degrons. For example, in eukaryotes an N-degron comprises not only a destabilizing N-terminal residue of a protein but also its internal lysine (or lysines), whose quality as a determinant can range from high to nonexistent.

A priori, one and the same N-end rule can be implemented through vastly different assortments of targeting hardware. At one extreme, each destabilizing N-terminal residue may be bound by a distinct N-recognin. Conversely, a single N-recognin may be responsible for the entire rule book of destabilizing residues in a given N-end rule. The actual N-end rule pathways lie between these extremes, and happen to have a hierarchic structure (Figs. 3–5).

3. TARGETING COMPLEX OF THE N-END RULE PATHWAY

The known components of the *S. cerevisiae* N-end rule pathway that mediate steps prior to the proteolysis of a targeted substrate by the 26 S proteasome are Nt-amidase (Nta1p), R-transferase (Ate1p), N-recognin (Ubr1p), a Ub-conjugating (E2) enzyme (Ubc2p), and the Ub-activating (E1) enzyme (Uba1p) (Fig. 3) (Baker and Varshavsky, 1995; Balzi *et al.*, 1990; Bartel *et al.*, 1990; Dohmen *et al.*, 1991; McGrath *et al.*, 1991). In addition to direct evidence for the physical association between N-recognin and Ubc2p (Madura *et al.*, 1993), there is also circumstantial evidence for the existence of a complex between N-recognin, R-transferase, and Nt-amidase (Baker and Varshavsky, 1995). Recently, a high-affinity interaction between Nta1p and Ate1p was demonstrated directly; other data suggest that both Nta1p and Ate1p interact with Ubr1p (M. Ghislain, A. Webster, and A. Varshavsky, unpublished results). In a quaternary Ubc2p-Ubr1p-Nta1p-Ate1p complex suggested by these results, Ate1p and Nta1p interact with each other and Ubr1p (Fig. 7).

Other, perhaps more transient, components of the targeting complex in *S. cerevisiae* are likely to include the 114-kDa Uba1p (E1 enzyme, which must be bound to Ubc2p during the E1→E2 transfer of activated Ub moiety), and also Arg-tRNA synthetase. The latter possibility is suggested by the finding that, in mammals, Arg-tRNA synthetase (whose product, Arg-tRNA, is a cosubstrate of R-transferase) copurifies with R-transferase (Ciechanover *et al.*, 1988; see also Sivaram and Deutsch, 1990). It is also likely that the targeting complex interacts with the 26 S proteasome *in vivo*, for example, during the transfer of a multiubi-

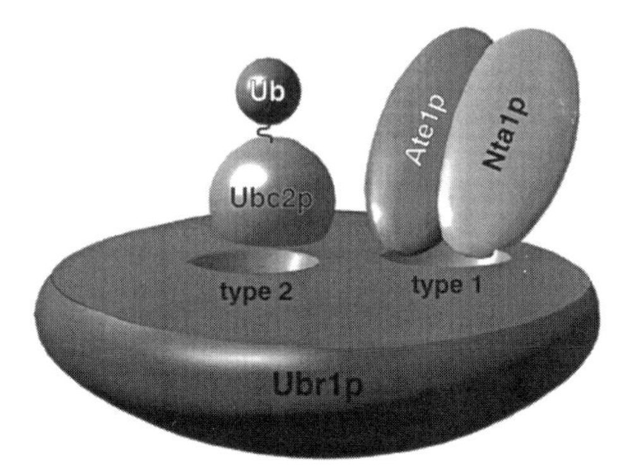

Figure 7. Model of a targeting complex in the *S. cerevisiae* N-end rule pathway. The 20-kDa Ub-conjugating (E2) enzyme Ubc2p is depicted carrying activated Ub linked to Cys88 of Ubc2 through a thioester bond. Both the 52-kDa Nta1p (Nt-amidase) and the 58-kDa Ate1p (R-transferase) bind to Ubr1p in proximity to the type 1 substrate-binding site of Ubr1p. In addition, Nta1p directly interacts with Ate1p (see Sections 2.2 and 3). It is unknown whether any of the four interacting enzymes (Nta1p, Ate1p, Ubc2p, and Ubr1p) is present in the complex as a monomer or a homo-oligomer.

quitylated N-end rule substrate to substrate-binding sites of the proteasome. The proteolytic machine that implements the N-end rule is thus a strikingly diverse assembly of enzymes and binding factors whose total mass is close to that of the large ribosomal subunit. However, even a transient existence of this "meta-complex" is conjectural at present, the alternative possibility being a sequential formation of transient subcomplexes that produce a substrate-linked multiUb chain and relay a substrate toward the 26 S proteasome.

Proteins that function in a multistep metabolic pathway are often physically associated, the examples being replisome, ribosome, and fatty acid synthase. Mechanistic advantages of such assemblies stem from their increased fidelity related to often present editing capabilities, and also from the properties of processivity and channeling that are central to the functioning of multienzyme machines. For example, the association between N-recognin and the Ubc2p Ub-conjugating enzyme endows the latter with the ability to recognize a substrate that bears a destabilizing N-terminal residue, and is also likely to account for the observed processivity of the synthesis of a multiUb chain (Chapter 2).

The effects of overexpressing Nt-amidase and/or R-transferase in *S. cerevisiae* not only suggested the existence of an Nta1p-Ate1p-Ubr1p-Ubc2p complex but also led to the prediction that Nta1p and Ate1p are associated with Ubr1p in proximity to its type 1 substrate-binding site (Fig. 7) (Baker and Varshavsky, 1995). The "proximity" aspect of the postulated complex was invoked to account for the markedly different effects of overexpressed R-transferase on the degradation of N-end rule substrates bearing N-d^{p1} versus N-d^{p2} residues (Baker and

Varshavsky, 1995). A physical proximity of the bound R-transferase to the type 1 site of N-recognin (Fig. 7) is presumed to decrease the accessibility of this site to an N-end rule substrate that bears an N-d^{p1} residue such as Arg and approaches the type 1 binding site of N-recognin from the bulk solvent. By contrast, a substrate that acquired Arg through the arginylation of its N-terminus by the N-recognin-bound R-transferase would be able to reach the (nearby) type 1 binding site of N-recognin directly—without dissociating into the bulk solvent first—a feature known as substrate "channeling" in multistage enzymatic reactions (Ovádi, 1991; Negrutskii and Deutscher, 1991). The mechanics of channeling may involve diffusion of an N-end rule substrate in proximity to surfaces of the targeting complex, analogous to the mechanism of a bifunctional enzyme dihydrofolate reductase-thymidylate synthetase, where the channeling of dihydrofolate apparently results from its movement across the surface of the protein (Knighton *et al.*, 1994).

The expression of genes that encode components of a protein complex is often coregulated to maintain approximately correct stoichiometries of these components. The promoter regions of *NTA1*, *ATE1*, and *UBR1* contain two sequence elements, 11 and 14 bp long, that are common to these promoters but have not been encountered together in other yeast genes (Fig. 8) (Baker and Varshavsky, 1995). This arrangement suggests a coregulated expression of N-recognin (Ubr1p), R-transferase (Ate1p), and Nt-amidase (Nta1p). It is also consistent with the possibility that the N-end rule pathway can be selectively up- or down-regulated under physiological conditions that remain to be identified. The two sequence elements (see Fig. 10) are likely to function as binding sites for regulatory proteins whose combination is specific for genes that encode targeting components of the N-end rule pathway. Recent findings (G. Turner and A. Varshavsky, unpublished data) suggest that the postulated transcriptional regulators act as repressors.

4. THE N-DEGRON AND PRE-N-DEGRON

Nascent proteins contain N-terminal Met (fMet in prokaryotes), which is a stabilizing residue in the known N-end rules (Fig. 1). Thus, the N-degron of an N-end rule substrate must be produced from a pre-N-degron (Section 2.1 and Fig. 9). In an engineered N-end rule substrate, a pre-N-degron contains the N-terminal Ub moiety whose removal by Ub-specific proteases produces the protein's N-degron (Fig. 2A). This design of a pre-N-degron is unlikely to be relevant to physiological N-end rule substrates, because natural Ub fusions (including the precursors of Ub) either contain a stabilizing residue at the Ub–protein junction or bear a mutant Ub moiety that is retained *in vivo* (Özkaynak *et al.*, 1987;

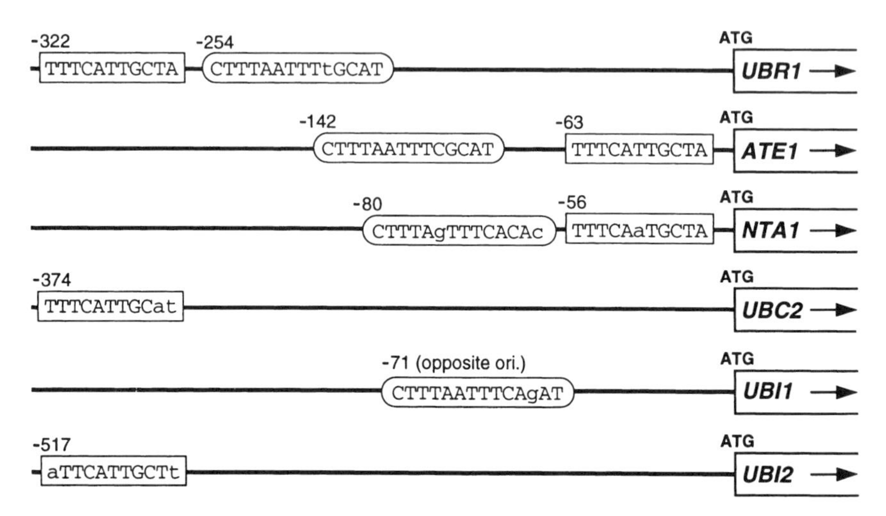

Figure 8. Common sequence motifs in promoters of genes encoding components of the N-end rule pathway. Alignment of the 5′ regions of the *S. cerevisiae* genes *UBR1* (Bartel *et al.*, 1990), *ATE1* (Balzi *et al.*, 1990), and *NTA1* (Baker and Varshavsky, 1995) revealed two distinct regions of similarity: an 11-mer motif 1 (consensus TTTCATTGCTA) and a 14-mer motif 2 (consensus CTTTAATTTCRCAT; R = purine). Mismatches to the consensus are in lowercase. Arrows indicate the direction of transcription. A number of *S. cerevisiae* genes, including at least three that encode known components of the Ub system, contain one of these motifs. Motifs 1 or 2 in the genes of this class, *UBC2*, *HBI1*, and *UBI2* (Dohmen *et al.*, 1991; Özkaynak *et al.*, 1987), are also shown. Motif 2 is present in *UBI1* in the orientation opposite to the one shown. The numbers indicate locations of the motifs relative to either the known (*UBC2*, *UBI1*, *UBI2*) or inferred start codons.

Finley *et al.*, 1989; Watkins *et al.*, 1993). In the latter case, the nonremovable Ub moiety itself may function as a degradation signal targeted by a distinct Ub-dependent proteolytic system, termed the UFD pathway(*U*b *f*usion *d*egradation) (Johnson *et al.*, 1992, 1995).

The known Met-aminopeptidases (Arfin *et al.*, 1995; Li and Chang, 1995; Kendall and Bradshaw, 1992; Chang *et al.*, 1992) remove N-terminal Met if, and only if, the second residue of a protein is stabilizing in the yeast N-end rule (Fig. 1) (Sherman *et al.*, 1985; Huang *et al.*, 1987; Arfin and Bradshaw, 1988; Hirel *et al.*,, 1989; Moerschell *et al.*, 1990). The structural basis of this selectivity is the size of a residue's side chain (Arfin and Bradshaw, 1988). Specifically, the side chains of the residues that are destabilizing in the yeast N-end rule are larger than those of stabilizing residues. The exception is Met—a bulky hydrophobic but stabilizing residue (Fig. 1).

Second-position Ala, Thr, Cys, or Pro, all of which are stabilizing in the yeast and *E. coli* N-end rules (Fig. 1), do not preclude the *in vivo* removal of

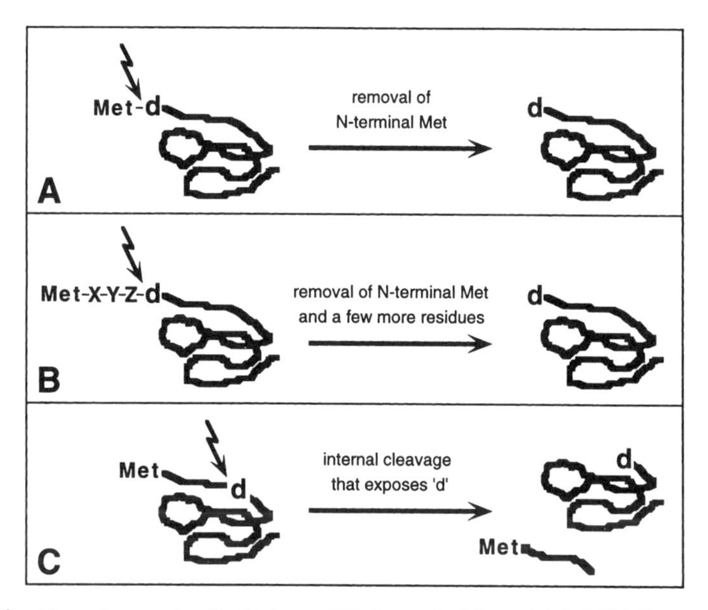

Figure 9. Ways of generating the N-degron. (A) Removal of the protein's initial N-terminal Met. (B) Exoproteolytic and/or endoproteolytic removal of the first several residues. (C) Endoproteolytic cleavage anywhere within a protein that yields a fragment bearing N-terminal "*d*"—a destabilizing residue in a given N-end rule.

N-terminal Met, and therefore can be exposed at the N-termini of proteins that bear the initial Met-Ala, Met-Thr, Met-Cys, or Met-Pro N-termini. Because Ala, Thr, Cys, and Pro (and possibly also Ser in reticulocytes) are destabilizing in the N-end rules of at least some metazoan cells (Fig. 1; Sections 2.2 and 6.8), the already known Met-aminopeptidases are expected to mediate the formation of some physiological N-end rule substrates (Section 6.8).

Endoproteolytic cleavages of viral precursor polyproteins by processing proteases have been found to yield physiological N-end rule substrates (deGroot *et al.*, 1991) (Sections 6.7 and 6.9). However, with a few exceptions (Section 6.9), nonviral cytosolic or nuclear proteins are not synthesized as polyproteins. Thus, a polyprotein-based route to an N-end rule substrate (deGroot *et al.*, 1991) is unlikely to encompass the entire gamut of routes that yield a destabilizing residue at the N-terminus of a protein.

An N-degron might also be produced through a posttranslational conjugation of a destabilizing residue to a protein's *internal* residue, for example, to the ε-amino group of a Lys residue. This possibility is suggested by the following evidence: Whereas only [^{3}H]arginine was shown to be conjugated to the N-termini of acceptor proteins in extracts from mammalian and other metazoan cells, these extracts were also found to catalyze a posttranslational conjugation of several other destabilizing amino acids, in particular Lys and Leu, to internal residues of unidentified acceptor proteins in the extract (Dayal *et al.*, 1990; Shyne-Athwal *et al.*, 1988). The enzymology and functional significance of these tentatively identified and largely unexplored posttranslational modifications are unknown.

Apart from the above possibility, a destabilizing residue at the N-terminus of a physiological N-end rule substrate must be exposed through a proteolytic processing of a pre-N-degron-containing protein at some unspecified distance from the precursor's N-terminus (Fig. 9). Can this distance be just one or a few residues? If so, a short (≤ 10 residues) N-terminal sequence might contain both the recognition motif and the cleavage site(s) for a relevant (unknown) processing protease. Screens for such sequences, carried out in *S. cerevisiae* by Sadis *et al.* (1995) and Ghislain *et al.* (1996), did identify short (≤ 10 residues) N-terminal regions that conferred Ubr1p-dependent metabolic instability on a reporter protein such as βgal or Ura3p. Most of the 65 N-terminal sequences identified by Ghislain *et al.* (1996) were dissimilar to each other. However, certain residues appeared to be excluded from some of the 10 post-Met positions, and in addition, 85% of these deduced sequences bore a destabilizing residue at position 2 (after the deduced N-terminal Met), while the remainder bore Val at this position.

Sadis *et al.* (1995) used a library of 11-residue N-terminal extensions fused to a βgal reporter, and screened for extensions that conferred metabolic instability on βgal. A large fraction of the sequences thus identified targeted βgal for degradation by the N-end rule pathway. None of the sequences identified by Sadis *et al.* (1995) were evidently similar to the sequences identified by Ghislain *et al.* (1996). This lack of sequence similarities may stem in part from differences in the screens' design—randomization immediately after N-terminal Met in the screen by Ghislain *et al.* (1996) but after position 2 in the screen by Sadis *et al.* (1995). As the two screens explored at most 0.00001% of the 10-residue sequence space (Ghislain *et al.*, 1996), it is also possible that the lack of similarities stems from a very large number of 10-residue N-terminal extensions that can produce an N-degron *in vivo*, analogous to a large number of N-terminal sequences that can function as signals for protein translocation across the ER membrane (Kaiser *et al.*, 1987).

Analysis of one N-terminal extension identified by Ghislain *et al.* (1996) has shown that it targets a reporter protein for degradation while retaining its N-terminal Met (M. Gonzalez, F. Lévy, M. Ghislain, and A. Varshavsky, unpublished data). This finding suggests that N-recognin binds not only to N-degrons but also to a degron that consists of an entirely internal sequence motif. By contrast, two other examined extensions were found to be cleaved after N-terminal Met, yielding destabilizing N-terminal residues (Sadis *et al.*, 1995; Ghislain *et al.*, 1996). In sum, we are just beginning to understand the processing reactions that yield a destabilizing N-terminal residue in a nonpolyprotein context.

5. MECHANICS OF THE N-DEGRON

5.1. Stochastic Capture Model

In a stochastic view of the N-degron, each lysine of a potential N-end rule substrate can be assigned a probability of being utilized as a ubiquitylation site. This probability depends on time-averaged spatial position and mobility of the lysine. For some, and often for most of the lysines in an N-end rule substrate, the probability of serving as a ubiquitylation site would be negligible because of the lysine's lack of mobility and/or its distance from a destabilizing N-terminal residue. In this "stochastic capture" model (Fig. 10A,D) (Varshavsky, 1992), the folded conformation of a substrate would be expected to slow down or preclude the search for a Lys residue, unless it is optimally positioned in the folded substrate.

The stochastic capture model was suggested by studies with βgal- and DHFR-based N-end rule substrates (Bachmair and Varshavsky, 1989; Johnson *et al.*, 1990; Dohmen *et al.*, 1994; see also Hill *et al.*, 1993). In addition, this model was supported by a recent analysis of the N-degron in a substrate produced from *S. cerevisiae* triosephosphate isomerase (TIM), a homodimer of 27-kDa subunits whose three-dimensional structure is known (Lolis *et al.*, 1990). In the absence of TIM, yeast cannot survive on glucose-containing media but can grow on media with a nonfermentable carbon source such as glycerol, thereby making possible a colony growth test for approximate ("high" or "low") steady-state levels of TIM. The wild-type N-terminal Ala residue of native TIM is exposed to solvent, whereas the second residue, Arg, is partially buried and forms structure-stabilizing hydrogen bonds and salt bridges with interior residues.

To vary the identity of its N-terminal residue, TIM was expressed as a Ub fusion (Fig. 2A) (R. J. Dohmen and A. Varshavsky, unpublished data). Replacing the stabilizing N-terminal Ala of TIM with a destabilizing residue such as Arg did not yield N-end rule substrates. In striking contrast, a variant of TIM bearing the N-terminal Arg and *also* an altered residue at position 2, either Gly or Cys (instead of wild-type Arg at this position), was rapidly degraded by the N-end rule pathway. To determine whether the identity of a residue at position 2 was the main cause of efficient degradation of Arg-Cys-TIM or Arg-Gly-TIM (in comparison to Arg-Arg-TIM), a tetrapeptide Arg-Thr-Leu-Gln was inserted after the N-terminal Arg in Arg-Arg-TIM, yielding Arg-Arg-Thr-Leu-Gln-Arg-TIM. This substrate, which was identical to the wild-type TIM save for the N-terminal sequence Arg-

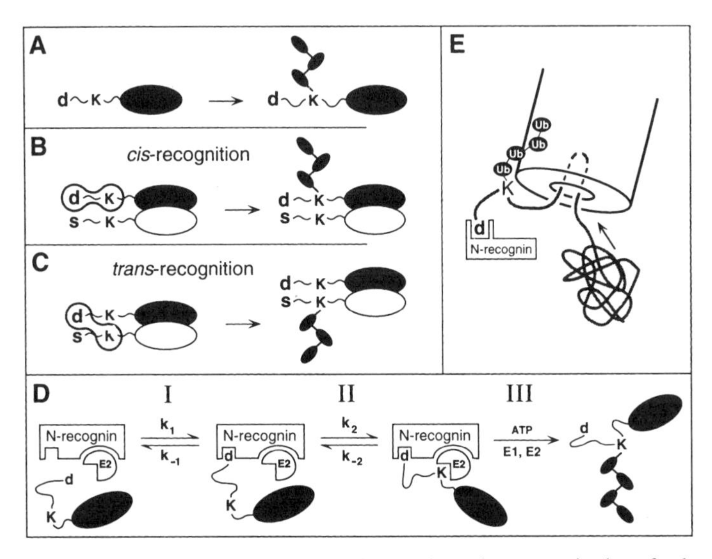

Figure 10. Mechanics of the N-end rule. (A) The two-determinant organization of eukaryotic N-degrons. *d*, a destabilizing N-terminal residue. A chain of black ovals linked to the second-determinant lysine (*K*) denotes a multiUb chain. (B) *Cis* recognition of the N-degron in one subunit of a dimeric protein. The other subunit bears *s*, a stabilizing N-terminal residue. (C) *Trans* recognition, in which the first (*d*) and second (*K*) determinants of the N-degron reside in different subunits of a dimeric protein. (D) A model for the recognition of an N-end rule substrate (Bachmair and Varshavsky, 1989). The reversible binding of N-recognin to a primary destabilizing N-terminal residue (*d*) of a substrate (step I) must be followed by a capture of the second-determinant lysine (*K*) of the substrate by a targeting complex containing a Ub-conjugating (E2) enzyme (step II). It is unknown whether the lysine is captured by E2 (as shown here) or by N-recognin. Ubiquitylation of the substrate commences once the targeting complex is bound to both determinants of the N-degron (step III). This model does not specify, among other things, the details of Ub conjugation (see Section 5.1). (E) The hairpin insertion model (see Section 5.3). A targeted N-end rule substrate bearing a multiUb chain is shown bound to the 26 S proteasome through the chain. The position of a targeting complex containing N-recognin is unknown, and is left unspecified. Only the 20 S core component of the 26 S proteasome is shown. An arrow indicates the direction of net movement of the substrate's polypeptide chain toward active sites in the interior of the proteasome. By analogy with the arrangement of signal sequences during trans-membrane translocation of proteins (Schatz and Dobberstein, 1996), it is proposed that a region of the substrate upstream of its ubiquitylated lysine (*K*) does not move through the proteasome during the substrate's degradation, and may be released intact following a cleavage downstream of the lysine. Variants of this model may also be relevant to the targeting of proteins that bear internal or C-terminal degrons.

Arg-Thr-Leu-Gln (instead of the wild-type Ala residue), was degraded as rapidly as Arg-Cys-TIM, indicating that the steric accessibility of a substrate's destabilizing N-terminal residue (rather than the presence of Cys or Gly at position 2) determines the efficiency of initial targeting by the N-end rule pathway.

A likely interpretation of these results (R. J. Dohmen and A. Varshavsky, unpublished data) is that the type 1 site of N-recognin, which binds $N\text{-}d^{p1}$ residues such as N-terminal Arg, is a deep cleft whose relevant functional groups are at the cleft's bottom. Such a site would require a flexible and exposed N-terminal residue of a substrate for its efficient binding by N-recognin. In this model, both Arg-Arg-Thr-Leu-Gln-Arg-TIM and Arg-Cys-TIM (in contrast to Arg-Arg-TIM) should be bound efficiently by N-recognin, the former because its N-terminal Arg, at the end of a four-residue extension, is exposed well outside the folded TIM globule, and the latter because its non-wild-type Cys at position 2 does not support conformation-stabilizing interactions in this region of TIM; the resulting slower folding and/or a more flexible structure at the TIM's N-terminus should result in a more exposed and mobile N-terminal Arg.

The degradation of Arg-Cys-TIM is inhibited in a *ubc2Δ* mutant, which lacks Ubc2p, the N-end rule-specific Ub-conjugating enzyme (Section 2.5). To determine which of the 21 lysines in a subunit of TIM is its ubiquitylation site, we converted either individual lysines or their clusters into Arg residues, which cannot be ubiquitylated (Hershko and Ciechanover, 1992). With one exception (Lys12, which is known to be required for the enzymatic activity of TIM), all of the thus-modified TIMs had enzymatic activity, as could be inferred from their ability to support cell growth on glucose-containing media. Thus, the multiple Lys→Arg conversions did not result in major alterations of the TIM's conformation. However, none of the changes yielded a significantly longer-lived Arg-Cys-TIM. We concluded that more than one lysine, and apparently many lysines of a newly formed TIM can serve as alternative ubiquitylation sites (R. J. Dohmen and A. Varshavsky, unpublished data).

The initial, βgal-based N-end rule substrates bore a destabilizing N-terminal residue followed by an ~45-residue N-terminal extension joined to a βgal moiety that lacked the first 6 residues of wild-type βgal (Bachmair *et al.*, 1986). The extension, termed e^K [extension (*e*) containing lysines (*K*)], was derived from a C-terminal region of *E. coli* Lac repressor (Johnson *et al.*, 1995). This design of X-βgal fusions—the combination of N-terminally truncated βgal and the apparently flexible, ~45-residue e^K extension—was essential for the rapid *in vivo* degradation of these N-end rule substrates (Bachmair *et al.*, 1986). In particular, the absence of 6 N-terminal βgal residues would be expected to slow down the folding of the nascent βgal moiety and result in a relatively unstable conformation of this region. Indeed, X-βgals that bore the same e^K extension but the intact (nontruncated) βgal moiety were degraded much more slowly than the original X-βgals (F. Lévy and A. Varshavsky, unpublished data).

The e^K extension contained two lysines (Lys15 and Lys17), at least one of which (either one) had to be present for the X-βgal degradation to take place (Bachmair and Varshavsky, 1989). These ubiquitylation-site Lys residues (Chau *et al.*, 1989) were the only lysines within the first ~250 residues of an ~1100-residue X-βgal. The absence of lysines from the first 230 residues of wild-type βgal accounted, most likely, for the nearly all-or-none dependence of X-βgal degradation on the presence of either Lys15 or Lys17 in the e^K extension. Indeed, although similarly designed Ura3p-based substrates (X-Ura3p) were specifically degraded by the N-end rule pathway, their proteolysis was only slightly retarded by the conversion of Lys15/17 into Arg residues, in contrast to the results with X-βgals (F. Lévy and A. Varshavsky, unpublished data). A likely explanation is that the sequence of Ura3p, unlike that of βgal, contains multiple Lys residues close to the N-terminus of Ura3p.

The initial targeting of an engineered N-end rule substrate such as X-βgal results in two alternative outcomes: a processive, Ub-dependent, proteasome-mediated degradation of an ~110-kDa X-βgal to short ($\leq$ 10 residues) peptides, or an apparently single cleavage of X-βgal ~40 residues from its N-terminus that yields a long-lived βgal fragment (Bachmair *et al.*, 1986; Dohmen *et al.*, 1991). This single cleavage is likely to be proteasome-mediated as well, suggesting a stochastic character of the "decision" between two alternative fates of the substrate.

The bipartite design of N-degron (Fig. 10A) is likely to be also characteristic of other Ub-dependent degradation signals—present in a multitude of naturally short-lived proteins that include cyclins (Chapter 12), IκBα (Chapter 10), and c-Jun (Pahl and Baeuerle, 1996; Chapter 9). The first component of these degrons is an internal region of a protein (instead of its N-terminal residue) that is specific for each degradation signal. The second component is an internal lysine (or lysines). A degron may also contain regulatory determinants whose modification (e.g., phosphorylation/dephosphorylation) can modulate the activity of this degron (Nishizawa *et al.*, 1993; Pahl and Baeuerle, 1996).

Recent work (Scheffner *et al.*, 1995) has shown that ubiquitylation of p53, a short-lived mammalian protein that is targeted by a complex containing in particular a recognin (E3) termed E6-AP and a specific E2 enzyme, proceeds through the formation of a Ub-E6-AP thioester. Specifically, the order of Ub transfer in this system is from the Ub-activating (E1) enzyme to the Ub-conjugating (E2) enzyme, from E2 to E6-AP, and from E6-AP to a substrate. The Cys residue of E6-AP that forms a thioester with Ub is located in a region whose sequence is highly conserved among several E6-AP homologues (Scheffner *et al.*, 1995). In another study, the *UBR1*-encoded yeast N-recognin (E3) was overexpressed in *S. cerevisiae* and purified to near homogeneity under gentle conditions (V. Chau and A. Varshavsky, unpublished data). A Ubr1p-based *in vitro* system was set up that contained Ubr1p, a purified *S. cerevisiae* Ubc2p E2 enzyme (isolated from *E. coli* overexpressing Ubc2p), and a purified mammalian El enzyme. When supplemented with ATP, this system produced multiUb chains linked to added substrates, with the specificity of ubiquitylation being that of the N-end rule pathway. For example, a 17-residue peptide bearing a single internal Lys residue and an N-dp N-terminal residue, Arg or Leu, was multiubiquitylated in this system, whereas an otherwise identical peptide bearing a stabilizing N-terminal residue, Val, was not ubiquitylated. Similarly to the finding by Scheffner *et al.* (1995) with E6-AP recognin, we observed a Ubrlp-based Ub thioester in this *in vitro* system (V. Chau and A. Varshavsky, unpublished data). Thus, it is likely that many if not all E3 proteins, in addition to being degron-recognizing components of specific Ub-dependent pathways, are also E2-like enzymes that participate in Ub transesterification relays downstream of the E3-associated E2 enzymes.

5.2. *cis–trans* Recognition and Subunit-Specific Degradation of Oligomeric Proteins

The two determinants of N-degron can be recognized either in *cis* or in *trans* (Fig. 10C,D) (Johnson *et al.*, 1990; F. Lévy and A. Varshavsky, unpublished data). Experiments that revealed the *trans*-recognition have also brought to light a remarkable feature of the N-end rule pathway: Only those subunits of an oligomeric protein that contain the ubiquitylation site (but not necessarily a destabilizing N-terminal residue) are actually degraded (Johnson *et al.*, 1990).

What might be the mechanism of subunit-specific proteolysis? A 'simple' model is suggested by the binding of a substrate-linked multiUb chain to a component of the proteasome (Chapter 6; Varshavsky, 1992) (see also Section 5.3). Specifically, a subunit of an oligomeric substrate bound to the proteasome through a subunit-linked multiUb chain may be the only subunit that undergoes further mechanochemical processing by ATP-dependent, chaperonelike components of the 26 S proteasome. These components mediate the unfolding and translocation steps that cause a movement of the subunit toward active sites in the proteasome's interior, and in the process dissociate this subunit from the rest of the oligomeric substrate. In this mechanism, the initial binding of N-recognin to another subunit—the one that bears a destabilizing N-terminal residue but not the lysine determinant (Fig. 10C)—may be either too transient (lasting, in a "productive" engagement, only long enough for a lysine to be captured on a nearby subunit) or sterically unfavorable for the delivery of this subunit to the interior of the proteasome.

Because other Ub-dependent degradation signals appear to be organized similarly to the N-degron (a "primary" recognition determinant plus an internal lysine or lysines), subunit selectivity is likely to be a general feature of proteolysis by the Ub system (Varshavsky, 1992). Examples of physiologically relevant subunit-selective proteolysis include the degradation of p53 in a complex with the papilloma viral protein E6 (Chapters 3 and 11) and the degradation of cyclin in a complex with a cyclin-dependent kinase (Chapter 12).

Trans-recognition and subunit-specific protein degradation (Johnson *et al.*, 1990) may allow the construction of a new class of dominant negative mutants, in which a long-lived protein could be destabilized by targeting it for degradation in *trans*. Specifically, it may be possible to generate, using the Ub fusion technique (Fig. 2A), the first determinant of the N-degron in a "targeting" polypeptide that is an *in vivo* ligand of the protein of interest and has the additional (if necessary, engineered) property of lacking an efficient ubiquitylation site of the N-degron. In this design, the ubiquitylation-site lysine is provided by the bound protein of interest, which would be targeted for degradation in *trans* (Fig. 10C). Lacking the ubiquitylation site, the targeting ligand would not be destroyed, and therefore should act catalytically rather than stoichiometrically. A short peptide, a peptide mimetic, or even an unrelated compound such as an oligonucleotide or a low-molecular-weight enzyme inhibitor that bears a destabilizing amino acid residue in a stereochemically appropriate context might also function as a targeting ligand in this *trans*-degradation strategy. It has been demonstrated that the N-end rule pathway is capable of *trans*-targeting not only in ATP-supplemented reticulocyte extract (Johnson *et al.*, 1990) but also *in vivo* (F. Lévy and A. Varshavsky, unpublished data).

5.3. The Hairpin Insertion Model and the Function of Multiubiquitin Chain

Formation of a substrate-linked multiUb chain produces an additional binding site (or sites) for components of the proteasome (Chapter 6). The resulting increase in affinity, i.e., a decrease in the rate of dissociation of proteasome–substrate complex, can be used to facilitate proteolysis (Varshavsky, 1992). Suppose that a rate-limiting step that leads, several steps later, to the first proteolytic cleavage of the proteasome-bound substrate is an unfolding (driven by thermal fluctuations) of a relevant region of the substrate. If so, an increase in stability of the proteasome–substrate complex, brought about by the multiUb chain, should facilitate the degradation of substrate, because the longer the allowed "waiting" time, the greater the probability of a required unfolding event. Another (not mutually exclusive) possibility is that a substrate-linked multiUb chain acts as a proximity trap of partially unfolded states of a substrate. This might be achieved through reversible interactions of the chain's Ub moieties with regions of the substrate that undergo local unfolding. A prediction common to both models is that the degradation of a substrate whose conformation poses less of a kinetic impediment to the proteasome should be less dependent on Ub and ubiquitylation than the degradation of an otherwise similar but more stably folded substrate.

How is a proteasome-bound, ubiquitylated protein directed to the interior of the proteasome? This problem is analogous to the one in studies of transmembrane channels for protein translocation (Schatz and Dobberstein, 1996). Could the solutions be similar in these systems, reflecting, perhaps, a common ancestry of translocation channels and proteasomes? The model in Fig. 10E proposes, by analogy with translocation systems, a "hairpin" insertion mechanism for the initiation of proteolysis by the 26 S proteasome. A biased random walk ("thermal ratchet") that is likely to underlie the translocation of proteins across membranes (Schatz and Dobberstein, 1996) may also be responsible for the movement of the substrate's polypeptide chain through the proteasome, with cleavage products diffusing out from the proteasome's distal end and thereby contributing to the net bias in the chain's bidirectional saltations through the proteasome channel.

Two findings indicate that unfolding of a targeted N-end rule substrate is a prerequisite for its degradation by the 26 S proteasome. Methotrexate—a folic acid analogue and high-affinity ligand of DHFR—can inhibit the degradation of an N-end rule substrate such as Arg-DHFR by the N-end rule pathway (Johnston *et al.*, 1995). This result suggests that a critical postubiquitylation step faced by the proteasome includes a "sufficient" conformational perturbation of the proteasome-bound substrate. Further, it was shown that the N-end rule-mediated degradation of a 17-kDa N-terminal fragment of the 70-kDa Sindbis virus polymerase is not precluded by the conversion of all of the fragment's 10 Lys residues into Arg residues, which cannot be ubiquitylated (T. Rümenapf, J. Strauss, and A. Varshavsky, unpublished data). Thus, the ubiquitylation requirement of previously studied N-end rule substrates may be a consequence of their relatively stable conformations. The binding of a largely unfolded substrate (a fragment of Sindbis polymerase is presumed to be such) by the targeting complex of the N-end rule pathway may be sufficient for delivery of the substrate to the proteasome's active sites without assistance by a multiUb chain. In the language of models in Fig. 10D,E, the "waiting" time for a bound and conformationally unstable substrate may be short enough not to require the formation of a dissociation-slowing device such as a multiUb chain.

5.4. The N-End Rule without Ubiquitin

No Ub-like covalent modification of N-end rule substrates has been detected in *E. coli*, in contrast to ubiquitylation of the same substrates in eukaryotes. Moreover, the conversion of ubiquitylation-site lysines of an N-end rule substrate into arginines rendered the substrate long-lived in eukaryotes but did not impair its degradation in *E. coli* (Tobias *et al.*, 1991). Thus, *E. coli* not only lacks a homologue of eukaryotic Ub, but also lacks the requirement for a lysine-specific modification of a substrate. Bacteria may contain proteins whose function in the N-end rule pathway is Ub-like but involves a noncovalent, lysine-independent binding to a targeted substrate. The proposed role of a substrate-linked multiUb chain in "marking" a subunit of a protein for selective destruction leads to another testable conjecture: If a subunit-marking device is absent from the *E. coli* N-end rule pathway, the latter may be incapable of degrading an oligomeric protein "one predetermined subunit at a time."

6. SUBSTRATES AND FUNCTIONS OF THE N-END RULE PATHWAY

The functions of a pathway can be approached experimentally from several perspectives. One can attempt to eliminate the pathway in part or in whole using inhibitors or (often more rigorously) by mutating genes that encode components of the pathway, and then deduce the pathway's functions by analyzing the resulting phenotypes. In the same hope of detecting an informative phenotype, one can also overexpress specific components of the pathway. If a system of interest is nonessential for viability or another distinct trait of wild-type cells, it may be possible to isolate mutants whose viability (or another trait) requires the presence of the pathway. This "synthetic-phenotype" strategy may illuminate connections between the pathway of interest and other pathways. In addition, one can attempt to identify, by any means, physiological substrates of the pathway, and then deduce the pathway's functions through the understanding of the roles these substrates play in a cell. All of these approaches have been employed in the continuing study of N-end rule functions.

6.1. Phenotypes of Mutants that Lack the N-End Rule Pathway

6.1.1. Prokaryotes

clpA$^-$ mutants of *E. coli*, which lack the ClpA subunit of the ClpAP protease (a proteolytic and possibly also a targeting component of the *E. coli* N-end rule pathway; Section 2.2), are unable to degrade N-end rule substrates (Tobias *et al.*, 1991; Shrader *et al.*, 1993). These mutants grow at wild-type rates, appear normal in their resistance to several stresses (Gottesman and Maurizi, 1992), but are less efficient than congenic wild-type cells in degrading proteins that have been induced during carbon starvation (Damerau and St. John, 1993).

6.1.2. Fungi

ubr1Δ mutants of *S. cerevisiae*, which lack N-recognin (Figs. 3 and 7), are unable to degrade N-end rule substrates, but grow at nearly wild-type rates, and are virtually indistinguishable from congenic wild-type cells in their sensitivity to heat and several other stresses (Bartel *et al.*, 1990; Bartel, 1990). Until recently

(Section 6.3), the known consequences of lacking the N-end rule pathway in yeast were a slight (~ 3%) decrease in the rate of growth, a subtle sporulation defect—an increase in the frequency of asci containing fewer than four spores (Bartel *et al.*, 1990)—and an assortment of other poorly understood phenotypes, including a greater resistance to EDTA in the growth medium and an enhanced accumulation of red pigment in *ubr1*Δ cells that lacked the *ADE2* gene (in comparison to congenic *UBR1 ade2* cells) (N. Schnell, B. Bartel, R. J, Dohmen, K. Madura, and A. Varshavsky, unpublished data).

6.1.3. Metazoans

No metazoan mutants in the N-end rule pathway have been identified. The first mouse N-end rule mutants (*ntan1*Δ) are being constructed (Y. T. Kwon and A. Varshavsky, unpublished data).

6.2. The N-End Rule Pathway and Osmoregulation in Yeast

A synthetic lethal screen was used to isolate a *S. cerevisiae* mutant, termed *sln1* (for "*s*ynthetic *l*ethal of *N*-end rule"), whose viability requires the presence of *UBR1* (Ota and Varshavsky, 1992). *SLN1* has been found to encode a eukaryotic homologue of two-component regulators—a large family of proteins previously encountered only in bacteria (Ota and Varshavsky, 1993). The properties of *S. cerevisiae* Sln1p are consistent with it being a sensor component of the osmoregulatory (HOG) pathway—a MAP kinase cascade (Maeda *et al.*, 1994). Sln1p also controls (through a pathway that appears to be distinct from the HOG system) the activity of the essential transcription factor Mcm1p (Yu *et al.*, 1995). Because an otherwise lethal hypomorphic mutation in *SLN1* can be suppressed by the presence of Ubr1p (N-recognin) (Ota and Varshavsky, 1993), it is likely that one or more of the proteins (e.g., kinases) whose activity is downregulated by Sln1p can also be downregulated through their degradation by the N-end rule pathway. The relevant physiological N-end rule substrate(s) remains to be identified.

6.3. The N-End Rule Pathway and Import of Peptides in Yeast

Alagramam *et al.* (1995) found that *ubr1*Δ yeast cells are unable to import di- and tripeptides. Recent results (C. Byrd and A. Varshavsky, unpublished data) indicate that Ubr1p (N-recognin) controls the activity of the peptide transporter Ptr2p, an integral plasma membrane protein, by regulating the synthesis and/or metabolic stability of *PTR2* mRNA. In one model, a transcriptional repressor of *PTR2* is short-lived, being degraded by the N-end rule pathway. Consistent with this mechanism, the control of *PTR2* expression by Ubr1p was found to involve the Ub-conjugating (E2) enzyme Ubc2p, a known component of the N-end rule pathway (Fig. 3). The Ubc4p E2 enzyme can partially compensate for the absence of Ubc2p; a deletion of both *UBC2* and *UBC4* results in cells that do not express Ptr2p and are unable to import peptides, similarly to *ubr1*Δ cells.

An *S. cerevisiae* mutant expressing an allele of *UBR1* that mediated the degradation of substrates bearing an N-d^{p1} residue such as Arg but was inactive with substrates bearing an N-d^{p2} residue such as Leu, was found to be unable to import peptides. By contrast, a "reciprocal" mutant strain, which degrades N-end rule substrates bearing N-d^{p2} but not N-d^{p1} residues, was capable of importing peptides (C. Byrd, A. Webster, and A. Varshavsky, unpublished data).

A screen for mutants that allow a bypass of the requirement for *UBR1* in peptide import identified a gene, *CUP9*, that encodes a homeodomain-containing protein. Cup9p is short-lived; its degradation requires *UBR1* (C. Byrd and A. Varshavsky, unpublished data). Cup9p is likely to be a transcriptional repressor of *PTR2*. Remarkably, an earlier study (Knight *et al.*, 1994) identified *CUP9* as a gene whose inactivation decreases the resistance of *S. cerevisiae* to the toxicity of copper ions in the presence of a nonfermentable carbon source, suggesting a role for Cup9p in detoxification of copper. Although the connection between peptide import and resistance to copper toxicity remains obscure, our findings, taken together with the results of Knight *et al.* (1994), suggest that the N-end rule pathway may be involved in the control of both peptide import and copper homeostasis.

Why was the N-end rule pathway (rather than, for example, another Ub-dependent proteolytic system) "recruited," in the course of evolution, to regulate the import of peptides? A plausible cause is suggested by the ability of N-recognin to bind short peptides bearing N-d^{p} residues (Fig. 7) (Section 2.2). Because more than half of the 20 amino acids are destabilizing in the *S. cerevisiae* N-end rule (Fig. 1 and Table I), a significant fraction of peptides imported into a cell by the Ptr2p transporter would be expected to compete with Cup9p, the short-lived repressor of *PTR2*, for binding to N-recognin. This competition would decrease the rate of Cup9p degradation. The ensuing increase in the level of Cup9p repressor would in turn decrease the level of *PTR2* mRNA, lowering the production of Ptr2p transporter and hence decreasing the rate of peptide import into the cell. The resulting N-recognin-based, "peptide-sensing" negative feedback loop would act to maintain a stable concentration of short peptides in a cell. This potentially useful feature of the system is made possible by the substrate-binding properties of N-recognin.

Apparent metazoan homologues of the yeast Ptr2p transporter have been identified (Paulsen and Skurray, 1994; Steiner *et al.*, 1995), suggesting that the import of peptides in mammalian cells may also be regulated by the N-end rule pathway.

6.4. On a Possible Function of the N-End Rule Pathway in Apoptosis

Until recently, apoptosis ("programmed" cell death) (Vaux and Strasser, 1996) was considered to be an attribute of multicellular but not unicellular organisms. However, given the near identity of cells in a clonally reproducing population of single-cell organisms, selection pressures may favor the emergence of an apoptotic response, for instance to a stress of starvation. By killing a fraction of a quasiclonal cell population, this response may benefit the rest of it. One example is the *mazEF* operon of *E. coli* that encodes a toxin/antitoxin pair (Aizenman *et al.*, 1996). The long-lived MazF protein is toxic; the short-lived MazE binds to MazF and counteracts its toxicity. If the expression of the *mazEF* operon falls below a certain threshold, as can happen during starvation in *E. coli*, the level of antitoxin MazE would decrease more rapidly than the level of toxin MazF, resulting in a starvation-induced programmed cell death (Aizenman *et al.*, 1996). Before the identification of *mazEF* in the *E. coli* chromosome, analogous pairs of genes ("addiction module") have been found in a number of plasmids, where they ensure the plasmids' retention in their hosts. MazE is degraded by the protease ClpAP (Aizenman *et al.*, 1996)—the same protease that degrades N-end rule substrates in *E. coli* (Fig. 4). It is unknown whether MazE contains a pre-N-degron or another degron recognized by ClpAP.

An essential aspect of apoptosis in metazoans may also be controlled by a pair (or pairs) of proteins—"apoptosis modules"—that act similarly to bacterial addiction modules. Further, we suggest that the short-lived component of an apoptosis module may be an N-end rule substrate. One reason for considering this idea is the facility (a single cut) and irreversibility of a process that can convert an initially long-lived antitoxin component of an apoptosis module into a short-lived protein degraded by the N-end rule pathway. Specifically, we propose that the induction of apoptosis by ICE-like proteases (Henkart, 1996) may proceed through the cleavage of an (unknown) antitoxin component of an apoptosis module by an ICE or ICE-like protease. This cleavage, although not necessarily inactivating the antitoxin's function as such, would expose a destabilizing residue at the N-terminus of a cleavage product, rendering it short-lived and thereby releasing a previously inhibited, relatively long-lived toxin. Implicit in this hypothesis is the assumption that a cleavage of antitoxin by an ICE-like protease is, by itself, insufficient (or not immediately sufficient) for the disruption of antitoxin's function, and that processive degradation of a C-terminal cleavage fragment of antitoxin by the N-end rule pathway is a required postcleavage step. It is also possible that an N-end rule substrate produced through a cleavage by an ICE-like protease "high" in the apoptotic cascade may act as a short-lived inhibitor of downstream events—a potential mechanism for downregulating the apoptotic response.

The known targets of ICE-family proteases contain Asp at the P1 position (abutting the site of cleavage) and a small residue, typically Ala or Ser, at the P1′ position—the N-terminus of the C-terminal cleavage fragment. Ala (and possibly also Ser in some settings) is a weakly destabilizing N-d^p residue in the mammalian N-end rule (Section 2.2). However, ICE-family proteases can also cleave peptide bonds whose P1′ position is occupied by Asn—a strongly destabilizing N-d^t residue in the N-end rule (Fig. 5A). Examples of proteins cleaved by ICE-family proteases at the Asp–Asn bond include actin and protein kinase Cδ (Emoto *et al.*, 1995; Kayalar *et al.*, 1996). In addition, the activation of at least one ICE-family protease also involves its proteolytic cleavage at the Asp–Asn bond (Thornberry *et al.*, 1992), suggesting that the enzymatic activation of this or analogous proteases may simultaneously render them short-lived *in vivo*—a possible source of negative control.

In sum, the hypothesis invokes an effector of apoptosis (a "toxin") that is activated when its inhibitor is cleaved by an ICE-like protease, perhaps at the Asp–Asn bond, yielding a short-lived protein degraded by the N-end rule pathway. It is also possible that an N-end rule substrate inhibits apoptosis upstream of at least some ICE-like proteases within the apoptotic cascade. One prediction of these models is that metabolic stabilization of the presumed Asn-bearing (cleaved) inhibitor, for example, through a perturbation of the N-end rule pathway, may inhibit the apoptosis. This and related predictions can be tested in mouse cells that lack the Asn-specific N-terminal amidase (Nt^N-amidase) and/or in cells that lack N-recognin (Fig. 5A).

6.5. The N-End Rule and Clostridial Neurotoxins

Bacteria of the genus *Clostridium* produce tetanus and botulinum neurotoxins that cause paralytic syndromes in susceptible animals (including humans) through the inhibition of neurotransmitter release. The botulinum and tetanus toxins act, respectively, at the peripheral and central synapses. These toxins specifically bind to the presynaptic terminals of neurons and enter the cytosol (Simpson, 1989). Clostridial neurotoxins are zinc endoproteases that cleave specific protein components of the synaptic vesicle docking-fusion complex in the cytosol and thereby block the neurotransmitter release (Schiavo *et al.*, 1992). Among the intracellular proteins whose cleavage by neurotoxins is thought to cause this effect are syntaxin, SNAP-25, and synaptobrevin, also called vesicle-associated membrane protein (VAMP) (Montecucco and Schiavo, 1994; Williamson *et al.*, 1996; Tonello *et al.*, 1996). A complex of these proteins functions in the sequential steps of synaptic vesicle docking, activation, and fusion with the plasma membrane (Söllner *et al.*, 1993).

The experimentally determined P1′ residues of the targets' cleavage sites (i.e., the N-terminal residues of the toxin-produced C-terminal cleavage fragments) vary for different targets and different species of neurotoxins (Montecucco and Schiavo, 1994; Tonello *et al.*, 1996). Nonetheless, all of these P1′ residues are destabilizing in the N-end rule. Each neurotoxin cleaves only one of the three proteins. Whereas syntaxin and SNAP-25 are cleaved by neurotoxins very close to their C-termini, the third protein—the ~14-kDa synaptobrevin (VAMP), which is located on the synaptic vesicle membrane—is cleaved by the tetanus toxin at the Gln76–Phe77 bond, resulting in a cytosol-exposed ~6-kDa fragment that bears N-terminal Phe, a strongly destabilizing N-d^{p2} residue (Figs. 1 and 5A). (The C-terminus of the ~6-kDa fragment is embedded in the synaptic vesicle membrane.) Several species of botulinum toxin also cut synaptobrevin—in the vicinity of its tetanus cleavage site—yielding a C-terminal cleavage fragment whose N-terminal residue is either Phe, Leu, Lys, or Ala, depending on the toxin's subtype (Tonello *et al.*, 1996). With the exception of Ala, all of these N-terminal residues are strongly destabilizing N-d^p residues; Ala is a weakly destabilizing N-d^p residue (Fig. 1 and Section 2.2).

We suggest that the neurotoxic effect of the tetanus toxin and those botulinum toxins that cleave synaptobrevin may be augmented by (perhaps even requires) the degradation of a toxin-generated synaptobrevin fragment by the N-end rule pathway. Analogously to the hypothesis of the N-end rule pathway's involvement in apoptosis (Section 6.4), the present conjecture implies that the toxin-catalyzed, single-site cleavage of synaptobrevin is, by itself, insufficient (or not immediately sufficient) for the disruption of synaptic function, and that the processive degradation of a synaptobrevin fragment by the N-end rule pathway is a relevant post-cleavage step. One prediction of this hypothesis, namely, that a site-specific cleavage of synaptobrevin by a neurotoxin *in vivo* should result in degradation of at least the C-terminal fragment of synaptobrevin, is supported by recent experiments with intact neurons. Using immunoblotting with cell extracts and immunofluorescence microscopy with fixed and permeabilized cells to determine the relative amounts of synaptobrevin before and after the exposure of neurons to the tetanus toxin, Williamson *et al.* (1996) found that synaptobrevin, normally an abundant protein, disappeared from neurons in 16 h or less after treatment with the toxin. By contrast, the large fragment of syntaxin (a product of cleavage by the type C botulinum toxin near the C-terminus of syntaxin) was largely retained in the cells (Williamson *et al.*, 1996).

The second prediction of this hypothesis is that inhibition of the N-end rule pathway should alleviate (or delay) the neurotransmitter-blocking effect of a toxin that acts through the cleavage of synaptobrevin. The results of Williamson *et al.* (1996) are neutral in regard to this prediction, whose direct test must await the development of a sufficiently efficacious N-end rule inhibitor (dipeptides are specific but weak inhibitors; Section 2.2), or the construction of metazoan cell mutants lacking N-recognin. Neurotoxins that cleave syntaxin or SNAP-25 but not synaptobrevin are as toxic as synaptobrevin-cleaving neurotoxins (Tonello *et al.*, 1996). Thus, even if the N-end rule-mediated degradation of a synaptobrevin cleavage fragment does take place, it might be an epiphenomenon, in that the initial cleavage of synaptobrevin by a neurotoxin may be sufficient for the disruption of synaptic vesicle function. Experimental tests with N-end rule inhibitors or mutants should be able to settle this question.

6.6. Gα Subunit of G Protein

Overexpression of the N-end rule pathway was found to inhibit the growth of haploid but not diploid cells (Madura and Varshavsky, 1994). This ploidy-dependent toxicity was traced to the enhanced degradation of Gpa1p, the Gα subunit of the G protein (Blumer and Thorner, 1991) that regulates cell differentiation in response to mating pheromone. The half-life of newly formed Gα at 30°C is ~50 min in wild-type cells, ~10 min in cells overexpressing the N-end rule pathway, and >10 hr in cells lacking the pathway. The degradation of Gα is preceded by its multiubiquitylation (Madura and Varshavsky, 1994). Like other Gα subunits of G proteins, the wild-type Gpa1p of *S. cerevisiae* bears a post-translationally conjugated N-terminal myristoyl moiety, which appears to be retained during the targeting of Gpa1p for degradation. The first 45 residues of the 472-residue Gpa1p can be deleted without a strong alteration in the rate of its Ubr1-dependent degradation. A deletion of the first 88 residues of Gpa1p greatly accelerates its degradation but retains the requirement for Ubr1p (K. Madura, unpublished data). These data suggest that Ubr1p recognizes a feature of the *GPA1*-encoded Gα that is distinct from the N-degron (Section 4). Another, N-degron-based model invokes a *trans*-targeting mechanism (Fig. 10B,C). The metabolic instability of Gpa1p is confined largely to the newly formed Gpa1p molecules. Specifically, if the nascent Gpa1p is pulse-labeled with [^{35}S]methionine in the absence of the N-end rule pathway (in an *S. cerevisiae* strain where Ubr1p is expressed from an inducible promoter), and the pathway is induced 3 h later, the labeled and "chased" Gpa1p is relatively long-lived, in contrast to the newly formed Gpa1p (K. Madura, unpublished data).

The observation that a deletion of the *SST2* gene reduced toxicity of the overexpressed N-end rule pathway (Madura and Varshavsky, 1994) suggested that Sst2p may be required for the degradation of Gpa1p (Gα), perhaps in the context of a *trans*-targeting mechanism (Fig. 10B,C). However, direct measurements of the metabolic stability of Gpa1p in the presence and absence of Sst2p (Dohlman *et al.*, 1995; Druey *et al.*, 1996) found no effect of Sst2p on the *in vivo* degradation of Gpa1p, strongly suggesting that the *SST2* conjecture by Madura and Varshavsky (1994) is incorrect.

Physiological implications of the Ubr1p-dependent degradation of Gα remain to be understood. Because the metabolic stability of Gα is expected to be influenced by its functional state—Gα can be GTP- or GDP-bound, covalently modified, or associated with Gβγ, the pheromone receptor, and other Gα ligands—the degradation of Gα in yeast may function either to augment or to inhibit the cell's responses to a pheromone. A G_s-type Gα is short-lived in mouse cells as well (Levis and Bourne, 1992), consistent with the possibility that Gα subunits of other organisms are also degraded by the N-end rule pathway. The activation of mouse Gα shortens its *in vivo* half-life (Levis and Bourne, 1992), suggesting an adaptation-related function of Gα degradation. Further, Obin *et al.* (1994) described the ATP-dependent degradation of all three subunits of the bovine retinal G protein in reticulocyte extract. (It is unknown whether Gβ and/or Gγ subunits of the *S. cerevisiae* G protein are also metabolically unstable.) Hondermarck *et al.* (1992) reported that differentiation of rat pheochromocytoma PC12 cells is inhibited by dipeptides bearing destabilizing N-terminal residues. [These compounds have been shown to inhibit the N-end rule pathway in *S. cerevisiae* (Baker and Varshavsky, 1991); their efficacy as N-end rule inhibitors in mammalian cells remains to be determined.] Given the findings with Gα (Madura and Varshavsky, 1994), one interpretation of these results (Hondermarck *et al.*, 1992) is that inhibitors of the N-end rule pathway may suppress cell differentiation through a metabolic stabilization of the relevant Gα subunits in PC12 cells.

6.7. Sindbis Virus RNA Polymerase

The Sindbis RNA polymerase, also called nsP4 (*n*on*s*tructural *P*rotein *4*), is produced by an endoproteolytic cleavage of the viral precursor polyprotein nsP1234 (Strauss and Strauss, 1994). The nsP4 protein bears N-terminal Tyr (an N-d^{p2} residue; Figs. 1 and 5A), and is degraded by the N-end rule pathway in reticulocyte extract (deGroot *et al.*, 1991). Tyr is an N-terminal residue of other alphaviral RNA polymerases as well (Strauss and Strauss, 1994), suggesting that these homologues of Sindbis polymerase are also degraded by the N-end rule pathway.

In most, but not all, alphaviruses (plus-stranded RNA viruses) the translation of nsP4 requires readthrough of the UGA stop codon at the end of the nsP123 open reading frame (ORF), resulting in a much lower initial level of the nsP4 protein in comparison to that of nsP123-derived proteins. In addition, nsP4 is a short-lived protein *in vivo*, being degraded, presumably, by the N-end rule pathway (the N-end rule-mediated degradation of nsP4 was demonstrated, thus far, in reticulocyte extract but not in Sindbis-infected cells). Although the bulk of newly formed nsP4 is rapidly degraded, a fraction of nsP4 in infected cells is long-lived, suggesting that the latter subpopulation of nsP4 molecules is protected from degradation within a replication complex that contains both viral and host proteins (Strauss and Strauss, 1994, and references therein). This model may be generally applicable, in that physiological N-end rule substrates, including alphaviral RNA polymerases and Gα subunits of G proteins, are likely to exist in several states that differ by covalent modifications of a substrate and/or its associations with other ligands, and that consequently also differ by the rates at which various forms of a substrate are degraded by the N-end rule pathway.

6.8. c-Mos, a Proto-Oncoprotein

This 39-kDa Ser/Thr-kinase is expressed predominantly in male and female germ cells. Sagata and colleagues identified c-Mos as a physiological substrate of the N-end rule pathway that is targeted for degradation through its N-terminal Pro residue (Nishizawa *et al.*, 1992, 1993). Met–Pro–Ser–Pro, the encoded N-terminal sequence of *Xenopus* c-Mos, is conserved among all vertebrates examined (Nishizawa *et al.*, 1993). Because the N-terminal Met–Pro peptide bond is readily cleaved by the major cytosolic Met-aminopeptidases (Section 4), the initially second-position Pro is expected to appear at the N-terminus of nascent c-Mos cotranslationally or nearly so.

The activity of the Pro-based N-degron in c-Mos is inhibited through the phosphorylation of Ser2 (Ser3 in the c-Mos ORF) (Nishizawa *et al.*, 1992, 1993). During the maturation of *Xenopus* oocytes, c-Mos is phosphorylated partially and reversibly, and therefore remains short-lived. Later—at the time of germinal vesicle breakdown and the arrest of mature oocytes (eggs) at the second meiotic metaphase—c-Mos becomes long-lived, owing to its nearly stoichiometric phosphorylation at Ser2 (Watanabe *et al.*, 1991). Fertilization or mechanical activation of a *Xenopus* egg releases the meiotic arrest through the induced degradation of c-Mos—caused by a nearly complete dephosphorylation of phosphoserine-2 (Nishizawa *et al.*, 1992, 1993). Consistent with this model of the N-degron in c-Mos, the replacement of Ser2 with Asp or Glu (whose negative charge mimics that of the phosphoryl group) rendered c-Mos long-lived, whereas the replacement of Ser2 with Ala yielded a constitutively unstable c-Mos (Nishizawa *et al.*, 1993). Lys33 (Lys34 in the c-Mos ORF) is a major ubiquitylation site of the c-Mos N-degron (Nishizawa *et al.*, 1993).

In contrast to N-terminal Pro in the context of c-Mos, the N-terminal Pro followed by the sequence His-Gly-Ser-... [this is the context of engineered N-end rule substrates such as X-βgal and X-DHFR (Bachmair and Varshavsky, 1989)] did not confer a short half-life on a reporter protein in either yeast or mammalian cells (F. Lévy, T. Rümenapf, and A. Varshavsky, unpublished data). One interpretation of these results is that the N-degron of c-Mos, whose conserved N-terminal sequence is Pro-Ser-Pro-..., has a "degron-enabling" internal determinant additional to, and perhaps specific for, the N-terminal Pro. The nature of the N-recognin (E3) that targets N-terminal Pro in the N-degron of c-Mos is unknown.

The c-Mos N-degron is the first example of an N-degron whose activity is regulated by phosphorylation (Nishizawa *et al.*, 1992, 1993). The relative simplicity of the arrangement involved, and close juxtaposition of the positive (Pro) and the conditionally negative (Ser) element in the N-degron of c-Mos may prove to be of utility in designing phosphorylation-regulated conditional degrons.

6.9. Potential N-End Rule Substrates

Most of the directly sequenced cytosolic and nuclear proteins bear stabilizing residues at their N-termini (Bachmair *et al.*, 1986; Varshavsky *et al.*, 1988). Some of the known exceptions—potential N-end rule substrates—are considered below.

6.9.1. Compartmentalized Proteins Retrotransported to the Cytosol

In contrast to cytosolic and nuclear proteins, the proteins that function in (or pass through) the ER, Golgi, and related compartments often bear destabilizing N-terminal residues—the consequence of cleavage specificity of signal peptidases, which remove signal sequences from proteins translocated into the ER (Bachmair *et al.*, 1986). Thus, one function of the N-end rule pathway might be the degradation of previously compartmentalized proteins that "leak" into the cytosol from compartments such as the ER (Varshavsky *et al.*, 1988). Remarkably, it has been found that at least some compartmentalized proteins can be retrotransported to the cytosol through a route that requires specific ER proteins. US11, the ER-resident transmembrane protein encoded by cytomegalovirus, causes the newly translocated MHC class I heavy chain to be selectively retrotransported back to the cytosol, where the heavy chain is degraded by a proteasome-dependent pathway (Wiertz *et al.*, 1996). Similarly, CPY*, a defective vacuolar carboxypeptidase of *S. cerevisiae*, is retrotransported to the cytosol shortly after entering the ER, and is degraded in the cytosol by a Ub/proteasome-dependent pathway that requires the Ubc7p Ub-conjugating enzyme (Hiller *et al.*, 1996). The expected N-terminal residue of the translocated and processed MHC class I heavy chain is Gly (Klein and O'hUigin, 1993)—a stabilizing residue (Fig. 1). The expected N-terminal residue of the wild-type CPY carboxypeptidase whose signal sequence had been cleaved off is Ile (van Voorst *et al.*, 1996)—an N-d^{p2} destabilizing residue (Fig. 1). Whether the N-end rule pathway plays a role in the degradation of retrotransported proteins remains to be determined.

6.9.2. The λ Phage cII Protein

cII is an essential component of a switch that determines whether λ grows lytically or lysogenizes an infected *E. coli* cell (Ptashne, 1992). The *in vivo* half-life of cII in infected *E. coli* can be 3 min and even shorter. The mature N-terminus of isolated (overexpressed) cII bears Arg (Ho *et al.*, 1986)—the third residue in the cII ORF and an N-d^{s} residue in the *E. coli* N-end rule (Table I, Figs. 1, 4, and 5C) (Tobias *et al.*, 1991). In addition, several mutations that increase the metabolic stability of cII have been mapped to the second position (immediately preceding Arg) in the cII ORF. For example, the *can-1* mutation changes the second encoded residue of cII from Val to Ala (Banuett *et al.*, 1986). The N-terminal residue of the overproduced *can-1* cII is largely Ala, the second (and stabilizing) residue, instead of Arg, the third (and destabilizing) residue at the N-terminus of wild-type cII (Ho *et al.*, 1986). If the metabolic stability of cII were determined largely by its Arg-containing N-degron, cII would be expected to be longer-lived in an *E. coli* mutant such as *clpA*, which lacks the N-end rule pathway (Tobias *et al.*, 1991). If so, *clpA* cells would be lysogenized by λ at a higher frequency than congenic wild-type cells. In a test of this conjecture, no significant increase in the frequency of lysogenization was observed with λ-infected *clpA E. coli* (T. Shrader and A. Varshavsky, unpublished data). The physiological significance of these results remains unclear, in part because cII appears to contain more than one degradation signal (Ho *et al.*, 1986, 1988; Herman *et al.*, 1993; Noble *et al.*, 1993).

6.9.3. The Yeast Copperthionein

S. cerevisiae CUP1 encodes a 61-residue, cysteine-rich, copper-binding cytosolic protein called copperthionein—a member of the metallothionein superfamily. However, the purified Cup1p is a 53-residue protein that lacks the first 8 residues of the encoded sequence and bears N-terminal Gln, an N-d^{t} residue in the N-end rule (Figs. 3 and 5B) (Wright *et al.*,1987). Control experiments strongly suggested that the N-terminal processing of Cup1p (by an unknown protease) occurred *in vivo* rather than during the isolation of Cup1p (Wright *et al.*, 1987). The metabolic stability of Cup1p and the lack of deamidation of its N-terminal Gln suggest that this residue is sterically inaccessible to *S. cerevisiae* Nt-amidase (Nta1p) (Section 2.3), which rapidly deamidates the N-terminal Gln or Asn of engineered N-end rule substrates (Baker and Varshavsky, 1995). Thus, the presence of Gln at the N-terminus of the processed Cup1p may be functionally unrelated to its metabolic stability. However, it is also possible that while the N-terminal Gln of the folded, copper-containing Cup1p is not exposed for recognition by the N-end rule pathway, a copper-deficient, partially unfolded Cup1p might become short-lived specifically as a result of increased flexibility and/or solvent exposure of its Gln-containing N-terminal region. Metabolic destabilization of copperthionein under conditions of copper deficiency may serve to reduce the sequestration of copper ions by this protein. It remains to be determined whether Cup1p can be degraded by the N-end rule pathway under conditions of intracellular copper deficiency, and whether the degradation of Cup1p is relevant to *S. cerevisiae* fitness under these conditions.

6.9.4. The Catalytic Subunit of Calpains

Calpains are a family of intracellular, nonlysosomal, calcium-dependent proteases ubiquitous in metazoans (reviewed by Croall and DeMartino, 1991). A calpain is a heterodimer containing an ~ 80-kDa catalytic subunit and an ~ 30-kDa regulatory subunit. Both the large and small subunits of a calpain are autoproteolytically processed on the activation of calpain by Ca^{2+}. The activation is accompanied (and perhaps caused) by dissociation of the calpain heterodimer, yielding a proteolytically active large subunit (Yoshizawa *et al.*, 1995). The processed (activated) ~ 80-kDa subunit of the μ-calpain (a calpain species activated by micromolar levels of Ca^{2+}) was found to bear N-terminal Leu, an $N\text{-}d^{p2}$ residue in the N-end rule (Figs. 1 and 5A) (Saido *et al.*, 1992). It remains to be determined whether the activation of μ-calpain decreases the metabolic stability of its ~ 80-kDa subunit and whether this decrease is mediated by the N-end rule pathway.

6.9.5. Viral Proteins Produced by Cleavages of Polyprotein Precursors

There are many potential N-end rule substrates other than Sindbis RNA polymerase (Section 6.7) that are derived from viral polyproteins (Dougherty and Semler, 1993). One of them is the integrase of the human immunodeficiency virus (HIV), produced by cleavages within the *gag-pol* precursor polyprotein. The processed integrase bears N-terminal Phe (Dougherty and Semler, 1993), a strongly destabilizing $N\text{-}d^{p2}$ residue in the N-end rule (Fig. 5A). Thus, it is possible that, similarly to the Sindbis virus RNA polymerase, at least a fraction of HIV integrase is short-lived *in vivo*.

6.9.6. Histonelike Eukaryotic Proteins Bearing N-Terminal Arg

Gorovsky and colleagues (Wu *et al.*, 1994) isolated a gene, termed *MLH*, of the ciliated protozoan *Tetrahymena thermophilia*, that encodes a 71-kDa polyprotein. *In vivo* proteolytic processing of the polyprotein yields three histonelike, 20- to 25-kDa proteins termed β, δ, and γ. Direct sequencing of the purified proteins showed that β and γ bore N-terminal Arg, an $N\text{-}d^{p1}$ residue in the yeast and mammalian N-end rules (Figs. 1 and 5A,B), whereas the N-terminus of δ (the first sub-ORF in the polyprotein ORF) was blocked (Wu *et al.*, 1994). The putative cleavage sites in the precursor polyprotein that yield β and γ have the consensus sequence Gly/Ala ↓ Arg-Thr-Lys. β, δ, and γ are located in the transcriptionally inert *Tetrahymena* micronucleus, where they apparently bind to the internucleosomal regions of DNA. Deletion of the *MLH* gene eliminates β, δ, and γ, results in less condensed mitotic micronuclear chromosomes, but is compatible with viability and normal growth of vegetative cells (Shen *et al.*, 1995). It is still unknown whether the *MLH* gene products are required for the mating and meiosis, which involve micronuclei. Although β and γ are apparently long-lived in vegetative *Tetrahymena* cells, accumulating in the micronucleus to a level comparable to that of the core histones, it is possible that at least a subset of β and γ molecules is degraded by the N-end rule pathway under other conditions, for example, during sexual conjugation and meiosis. Nothing is known about the N-end rule pathway in *Tetrahymena* and other protozoans.

7. DETERMINING THE N-END RULE AND PROBING THE KINETICS OF *IN VIVO* PROTEIN DEGRADATION

7.1. The UPR Technique

Determination of an N-end rule entails multiple pulse–chase assays. Its advantage of being direct notwithstanding, a conventional pulse–chase assay is fraught with sources of error. For example, immunoprecipitation yields may vary from sample to sample; the volumes of samples loaded on a gel may vary as well. As a result, pulse–chase data tend to be semiquantitative at best, lacking the means to control these errors. Prompted in part by insufficient reproducibility of conventional pulse–chase assays in studies of the N-end rule, and also by the necessity of estimating the degradation of N-end rule substrates and other short-lived proteins *during* the pulse, we devised a method that can be used to produce equimolar amounts of two or more specific proteins in a cell. In this approach, termed the UPR (*u*biquitin/*p*rotein/*r*eference) technique, a Ub moiety is placed between polypeptides *A* and *B* within a linear fusion (Fig. 2B). *In vivo*, this tripartite fusion is cleaved, cotranslationally or nearly so, at the Ub-*B* junction, yielding equimolar amounts of two separate polypeptides, *A*-Ub and *B* (Lévy *et al.*, 1996). More generally, by positioning *n* Ub moieties within a fusion, one can produce equimolar amounts of $n + 1$ separate polypeptides in a cell. In applications such as pulse–chase analysis, the UPR technique can compensate for the scatter of immunoprecipitation yields, sample volumes, and other sources of sample-to-sample variation. In particular, this method (Fig. 2B) allows a direct comparison of proteins' metabolic stabilities from the pulse data alone.

The utility of UPR was tested by using it to determine the N-end rule in L-cells, a fibroblastlike mouse cell line (Lévy *et al.*, 1996) (Section 2.2). The increased accuracy afforded by the UPR technique underscored insufficiency of the current "half-life" terminology, as the *in vivo* degradation of many proteins deviates from first-order kinetics. (Pseudo-first order, actually, as proteolysis involves both a substrate and a proteolytic pathway, the latter being presumed to be present at a constant concentration.) The non-first-order kinetics of protein degradation stems from the fact that a protein molecule *in vivo* is not a fixed structural entity. For example, the probability of degradation of a nascent, partially unfolded, chaperonin-associated protein should be, in general, different from the probability of degradation of a folded counterpart of this protein at a later time in the same cell. In addition, most proteins undergo covalent modifications and associate with other molecules (including other proteins) in a cell, the fraction of a modified or a complex-associated protein being typically less than unity. Some of these modifications and associations are relevant to the protein's function, whereas the rest are caused by a variety of quasirandom events that include protein damage. Thus, an *in vivo* ensemble of protein molecules encoded by one and the same ORF is heterogeneous structurally and/or conformationally, and therefore may not decay exponentially.

For some short-lived proteins, e.g., *S. cerevisiae* Matα2p, deviations from a first-order decay (at times comparable to a half-life) appear to be small (Hochstrasser and Varshavsky, 1990). However, many other short-lived proteins, including the engineered N-end rule substrates, are degraded *in vivo* with a pronounced nonexponential (non-first-order) kinetics. Lévy *et al.* (1996) suggested a terminology for describing nonexponential decay. The proposed parameters, initial decay (ID) and subsequent decay (SD), can be used to describe specific aspects of any decay curve, including the degradation of a protein *during* the pulse. Lévy *et al.* (1996) also suggested a generalized half-life term, $t^{y-z}_{0.5}$, where 0.5 denotes the parameter's half-life aspect and $y - z$ denotes the relevant time interval (from y to z min of chase).

One advantage of the UPR technique (Fig. 2B) is the possibility of comparing the metabolic stabilities of different proteins using pulse data alone. UPR should also allow the determination of relative metabolic stabilities of proteins *in situ*. In one example of this approach, two nearly identical, separately quantifiable reporter proteins (e.g., fluorescent protein domains) that either lack or bear the N-degron would be coproduced in equimolar amounts using UPR. Determining the steady-state molar ratios of these reporters in cell extracts, cells in culture, or individual cells in tissues of transgenic animals should yield a direct ranking of metabolic stabilities of N-end rule substrates bearing different destabilizing N-terminal residues. Analogous *in vivo* UPR assays should be feasible with other short-lived proteins as well.

7.2. UPR and the Problem of Cotranslational Proteolysis

A nascent and still partially unfolded polypeptide chain emerging from the ribosome may transiently present degradation signals that are buried in a conformationally mature protein. In addition, the overall efficiency of *in vivo* protein folding may be significantly lower than 100%, in that at least some polypeptides may often assume "dead-end" conformations whose further temporal evolution results in a correct conformation either slowly or never (reviewed by Hartl, 1996). These and related considerations suggest that a fraction of normal proteins may be degraded cotranslationally—as nascent chains emerging from the ribosome. It should be possible to address this long-recognized but poorly understood problem using a version of UPR termed the double-Ub technique (G. Turner, F. Lévy, and A. Varshavsky, unpublished data). This assay is based on devising a kinetic competition between the processive (proteasome-mediated) degradation of a nascent polypeptide chain and discrete, site-specific cuts in the same chain by Ub-specific proteases (UBPs), which cleave after the last residue of Ub in a fusion, and appear to do so cotranslationally (Bachmair *et al.*, 1986; Johnsson and Varshavsky, 1994a,b).

Consider the fusion *A*-Ub-*B*-Ub-*C*, where *B* carries a "cotranslational" degron, and *A*-Ub and *C* are the reporter moieties. In the absence of cotranslational proteolysis and premature termination of translation, the molar yield of *A*-Ub must equal that of *C*. However, if the *A*-Ub portion of *A*-Ub-*B*-Ub-*C* is cleaved off by UBPs before the rest of the *A*-Ub-*B*-Ub-*C* chain has been completed, a significant fraction of *B*-Ub-*C* chains may be degraded processively (and cotranslationally). As a result, the molar yield of *A*-Ub would now be higher than that of *C*, the difference being the measure of the extent of cotranslational proteolysis. A powerful feature of this approach is that it detects a transient kinetic effect (cotranslational proteolysis) by measuring a "static" ratio of two quantities: the yield of *A*-Ub and the yield of *C*, in response to changes in the length, sequence, and degron content of *B*-Ub.

Among the aims of these ongoing experiments (G. Turner, F. Lévy, and A. Varshavsky, unpublished data) is a test of the hypothesis that one quality control mechanism in the *in vivo* protein folding operates cotranslationally—through processive degradation of nascent polypeptide chains that fail to fold correctly during, or immediately after, their synthesis. An alternative but not mutually exclusive view of cotranslational proteolysis is that it may exist because the intracellular proteolytic pathways that identify and destroy misfolded and otherwise abnormal proteins do so by recognizing normally buried degrons that become exposed in damaged proteins. In this view, a nascent, still being made, and at most partially folded polypeptide chain is "perceived" by these proteolytic pathways as a damaged protein, and is targeted for degradation in kinetic competition with chaperones that facilitate protein folding. These issues can be addressed using a series of mutant proteins with a range of folding defects as moieties *B*, and measuring the *A*/*C* ratio to determine the extent of cotranslational proteolysis characteristic of a given folding defect.

8. APPLICATIONS OF THE N-DEGRON

The portability and modular organization of N-degrons make possible a variety of applications whose common feature is the conferring of a constitutive or conditional metabolic instability on a protein of interest.

8.1. The N-Degron and Conditional Mutants

A frequent problem with conditional phenotypes is their leakiness, i.e., unacceptably high residual activity of either a temperature-sensitive (*ts*) protein at nonpermissive temperature or a gene of interest in the "off" state of its promoter. Another problem is "phenotypic lag," which often occurs between the imposition of nonpermissive conditions and the emergence of a relevant null phenotype. Phenotypic lag tends to be longer with proteins that are required in catalytic rather than stoichiometric amounts. It is also characteristic of those *ts* mutants in which a nonpermissive temperature inactivates a newly formed *ts* protein but not its mature counterpart.

In one application of the N-end rule to the problem of phenotypic lag, Park *et al.* (1992) and Cormack and Struhl (1992) fused a constitutive N-degron (produced as described in Fig. 2A) to a protein expressed from an inducible promoter. This useful method is constrained by the necessity of using a heterologous promoter and by the constitutively short half-life of a target protein, whose levels may therefore be suboptimal under permissive conditions. An alternative approach is to link the N-degron to a normally long-lived protein in an *S. cerevisiae* strain in which the N-end rule pathway can be induced or repressed (Ghislain *et al.*, 1996; R. J. Dohmen, K. Madura, B. Bartel, and A. Varshavsky, unpublished data). The metabolic stabilities, and hence also the levels of N-degron-bearing proteins in this strain, are either normal or extremely low depending on whether Ubr1p (N-recognin) is absent or present. These *ns* (*N*-end rule-*s*ensitive) conditional mutants can be constructed with any cytosolic or nuclear protein whose function tolerates an N-terminal extension.

Yet another design is a portable, heat-inducible N-degron that is inactive at a low (permissive) temperature but becomes active at a high (nonpermissive) temperature (Dohmen *et al.*, 1994). Linking this degron to proteins of interest yields a new class of *ts* mutants, called *td* (*t*emperature-activated *d*egron). The *td* method (Dohmen *et al.*, 1994) does not require an often unsuccessful search for a *ts* mutation in a gene of interest. If the protein of interest can tolerate N-terminal extensions, the corresponding *td* fusions are functionally unperturbed at permissive temperature. By contrast, low activity of a *ts* protein at permissive temperature is a common problem with conventional *ts* mutants. The *td* method eliminates or reduces the phenotypic lag, because the heat activation of N-degron results in rapid disappearance of a *td* protein. Another advantage of the *td* technique is the possibility of employing two sets of conditions: a *td* protein-expressing strain at permissive versus nonpermissive temperature or, alternatively, the same strain versus a congenic strain lacking the N-end rule pathway, with both strains at nonpermissive temperature. This powerful internal control, provided in the *td* technique by two alternative sets of permissive/nonpermissive conditions (Dohmen *et al.*, 1994), is unavailable with conventional *ts* mutants. The *td* method has recently been employed by several laboratories to produce *td* mutants (Hardy, 1996; Caponigro and Parker, 1995).

8.2. The N-Degron and Conditional Toxins

A major limitation of the current pharmacological strategies stems from the absence of drugs that are specific for two or more independent molecular targets. For reasons discussed elsewhere (Varshavsky, 1995b, 1996a), it is desirable to have a therapeutic agent that requires the presence of two or more predetermined targets in a cell for it to be killed, and that would spare a cell if it lacks even one of these targets. Combining two "conventional" drugs against two different targets in a multidrug regimen would not attain this goal, because the two drugs together would perturb not only cells containing both targets but also cells containing just one of the targets. More generally, it is desirable to have drugs that exhibit a combinatorial selectivity, killing (or otherwise modifying) a cell if, and only if, it *contains* a predetermined set of molecular targets and at the same time *lacks* another predetermined set of molecular targets. Therapeutic agents of this, currently unrealistic, selectivity are likely to be free of side effects—the bane of present-day therapies against diseases such as cancer.

A strategy for designing reagents that are sensitive to the presence or absence of more than one target at the same time has recently been proposed (Varshavsky, 1995b, 1996a). The key feature of new reagents, termed *comtoxins* (*co*dominance-*m*ediated *toxins*), is their ability to utilize codominance, a property characteristic of many signals in proteins, including degrons and nuclear localization signals (NLSs). Codominance refers to the following property of these signals: Each degron (or NLS) in a protein bearing two or more degrons (or NLSs) can target the protein for degradation (or transport to the nucleus) independently of other degrons (or NLSs) in the same protein. The crucial property of a degron-based comtoxin is that its intrinsic toxicity is the same in all cells, whereas its half-life (and, consequently, its steady-state level and overall toxicity) in a cell depends on the cell's protein composition, specifically on the presence of "target" proteins that have been chosen to define the profile of a cell to be eliminated. The target proteins would bind to their ligands in a comtoxin molecule, and either physically obstruct the recognition of degrons or inactivate them catalytically, for example by phosphorylation. These and related ideas are described elsewhere (Varshavsky, 1995b, 1996a).

We are exploring the feasibility of comtoxins using the N-degron as a degradation signal and the cytotoxic A-chains of ricin or diphtheria toxin as effector domains (T. Suzuki, I. V. Davydov, and A. Varshavsky, unpublished data). Our current aim is to determine whether the concept of comtoxins can be implemented in the "easy" setting of a cell culture—without addressing, yet, the delivery problem, the immunogenicity of protein drugs, and related concerns.

9. EPILOGUE

Although many things have been learned about the N-end rule since its discovery 12 years ago, the answers to several key questions are still unknown or glimpsed at best. For example, the detailed mechanics of targeting is not understood. Biochemical dissection of the N-end rule pathway reconstituted *in vitro* from defined (cloned) components will be essential for attaining this goal. Crystallographic-quality structural information about N-recognin and the entire targeting complex will be required as well. The recently emerged possibility that N-recognin may target not only N-degrons but also other degradation signals adds yet another level of complexity, which will have to be addressed.

Genetic screens for proteins degraded by the N-end rule pathway are our best hope for bringing to light physiological N-end rule substrates. It is already clear that at least some of these substrates are conditionally unstable—for example, partitioned between a short-lived free substrate and a long-lived complex of the substrate with other proteins. In addition, for some substrates, the rate-limiting step in their degradation may be a processing (cleavage) event that produces an N-degron from a pre-N-degron. If so, a significant fraction of extant substrate molecules may bear a stabilizing N-terminal residue. Given these obstacles to identifying physiological N-end rule substrates, they are likely to be more numerous than is apparent at the present time.

Acknowledgments. We thank Drs. M. Kozak, P. Bjorkman, and D. Kolakofsky for helpful advice, and the colleagues whose names are cited in the text for their permission to discuss unpublished data. Work in this laboratory is supported by grants to A. Varshavsky from the National Institutes of Health (DK39520 and GM31530).

10. REFERENCES

Abramochkin, G., and Shrader, T. E., 1995, The leucyl/phenylalanyl-tRNA-protein transferase. Overexpression and characterization of substrate recognition, domain structure, and secondary structure, *J. Biol. Chem.* **270:**20621–20628.

Aizenman, E., Engelberg-Kulka, H., and Glaser, G., 1995, An *Escherichia coli* 'addiction module' regulated by 3′,5′-bipyrophosphate: A model for programmed bacterial cell death, *Proc. Natl. Acad. Sci. USA* **93:**6059–6063.

Alagramam, K., Naider, F., and Becker, K. M., 1995, A recognition component of the ubiquitin system is required for peptide transport in *Saccharomyces cerevisiae*, *Mol. Microbiol.* **15:**225–234.

Arfin, S. M., and Bradshaw, R. A., 1988, Cotranslational processing and protein turnover in eukaryotic cells, *Biochemistry* **27:**7979–7984.

Arfin, S. M., Kendall, R. L., Hall, L., Weaver, L. H., Stewart, A. E., Matthews, B. W., and Bradshaw, R. A., 1995, Eukaryotic methionyl aminopeptidases: Two classes of cobalt-dependent enzymes, *Proc. Natl. Acad. Sci. USA* **92:**7714–7718.

Arnason, T. A., and Ellison, M. J., 1994, Stress resistance in *S. cerevisiae* is strongly correlated with assembly of a novel type of multiubiquitin chain, *Mol. Cell. Biol.* **14:**7876–7883.

Baboshina, O. V., and Haas, A. L., 1996, Novel multiubiquitin chain linkages catalyzed by the conjugating enzymes E2EPF and RAD6 are recognized by 26S proteasome subunit 5. *J. Biol. Chem.* **271:**2822–2831.

Bachmair, A., and Varshavsky, A., 1989, The degradation signal in a short-lived protein, *Cell* **56:**1019–1032.

Bachmair, A., Finley, D., and Varshavsky, A., 1986, *In vivo* half-life of a protein is a function of its amino-terminal residue, *Science* **234:**179–186.

Baker, R. T., and Varshavsky, A., 1991, Inhibition of the N-end rule pathway in living cells, *Proc. Natl. Acad. Sci. USA* **88:**1090–1094.

Baker, R. T., and Varshavsky, A., 1995, Yeast N-terminal amidase: A new enzyme and component of the N-end rule pathway, *J. Biol. Chem.* **270:**12065–12074.

Balzi, E., Choder, M., Chen, W., Varshavsky A., and Goffeau, A., 1990, Cloning and functional analysis of the arginyl-tRNA-protein transferase gene *ATE1* of *Saccharomyces cerevisiae*, *J. Biol. Chem.* **265:**7464–7471.

Banuett, F., Hoyt, M. A., MacFarlane, L., Echols, H., and Herskowitz, I., 1986, *hflB*, a new *Escherichia coli* locus regulating lysogeny and the level of bacteriophage lambda cII protein, *J. Mol. Biol.* **187:**213–224.

Bartel, B., 1990, Molecular genetics of the ubiquitin system: The ubiquitin fusion proteins and proteolytic targeting mechanisms, Ph.D. thesis, pp. 104–105, MIT, Cambridge, MA.

Bartel, B., Wünning, I., and Varshavsky, A., 1990, The recognition component of the N-end rule pathway, *EMBO J.* **9:**3179–3189.

Blumer, K. J., and Thorner, J., 1991, Receptor-G protein signaling in yeast, *Annu. Rev. Physiol.* **53:** 37–57.

Burbaum, J. J., and Schimmel, P., 1991, Structural relationships and the classification of aminoacyl-tRNA synthetases, *J. Biol. Chem.* **266:**16965–16968.

Caponigro, G., and Parker, R., 1995, Multiple functions for the polyA-binding protein in mRNA decapping and deadenylation in yeast, *Genes Dev.* **9:**2421–2432.

Chang, Y.-H., Teichert, U., and Smith J.A., 1992, Molecular cloning, sequencing, deletion, and overexpression of a methionine aminopeptidase from *Saccharomyces cerevisiae*, *J. Biol. Chem.* **267:**8007–8011.

Chatton, B. P., Walter, P., Ebel, J.-P., LaCroute, F., and Fasiolo, F., 1988, The yeast *VAS1* gene encodes both mitochondrial and cytoplasmic valyl-tRNA synthetases, *J. Biol. Chem.* **263:**52–57.

Chau, V., Tobias, J. W., Bachmair, A., Marriott, D., Ecker, D., Gonda, D. K., and Varshavsky, A., 1989, A multiubiquitin chain is confined to a specific lysine in a targeted short-lived protein, *Science* **243:**1576–1583.

Chiu, M. I., Mason, T. L., and Fink, G. R., 1992, *HTS1* encodes the cytoplasmic and mitochondrial histidyl-tRNA synthetase of *Saccharomyces cerevisiae:* Mutations alter the specificity of compartmentation, *Genetics* **132:**987–1001.

Ciechanover, A., Ferber, S., Ganoth, D., Elias, S., Hershko, A., and Arfin S., 1988, Purification and characterization of arginyl-tRNA-protein transferase from rabbit reticulocytes, *J. Biol. Chem.* **263:**11155–11167.

Cormack, B. P., and Struhl, K., 1992, The TATA-binding protein is required for transcription by all three RNA polymerases in yeast cells, *Cell* **69:**685–694.

Croall, D. E., and DeMartino, G. E., 1991, Calcium-activated neutral protease (calpain) system: Structure, function and regulation, *Physiol. Rev.* **71:**813–847.

Damerau, K., and St. John, A. C., 1993, Role of Clp protease subunits in degradation of carbon starvation proteins in *Escherichia coli*, *J. Bacteriol.* **175:**53–63.

Dayal, V. K., Chakraborty, G., Sturman, J. A., and Ingoglia, N. A., 1990, The site of amino acid addition to posttranslationally modified proteins of regenerating rat sciatic nerves, *Biochim. Biophys. Acta* **1038:**172–177.

deGroot, R. J., Rümenapf, T., Kuhn, R. J., Strauss, E. G., and Strauss, J. H., 1991, Sindbis virus RNA polymerase is degraded by the N-end rule pathway, *Proc. Natl. Acad. Sci. USA* **88:**8967–8971.

Deutch, C. E., 1984, Aminoacyl-tRNA-protein transferases, *Methods Enzymol.* **106:**198–205.

Dohlman, H. G., Apaniesk, D., Chen, Y., Song, J., and Nusskern, D., 1995, Inhibition of G-protein signaling by dominant gain-of-function mutations in Sst2p, a pheromone desensitization factor, *Mol. Cell. Biol.* **15:**3635–3643.

Dohmen, R. J., Madura, K., Bartel, B., and Varshavsky, A., 1991, The N-end rule is mediated by the Ubc2 (Rad6) ubiquitin-conjugating enzyme, *Proc. Natl. Acad. Sci. USA* **88:**7351–7355.

Dohmen, R. J., Wu, P., and Varshavsky A., 1994, Heat-inducible degron: A method for constructing temperature-sensitive mutants, *Science* **263:**1273–1276.

Dougherty, W. G., and Semler, B. L., 1993, Expression of virus-encoded proteinases: Functional and structural similarities with cellular enzymes, *Microbiol. Rev.* **57:**781–822.

Druey, K. M., Blumer, K. J., Kang, V. H., and Kehrl, J. H., 1996, Inhibition of G-protein-mediated MAP kinase activation by a new mammalian gene family, *Nature* **379:**742–746.

Durner, J., and Boger, P., 1995, Ubiquitin in the prokaryote *Anabaena variabilis*, *J. Biol. Chem.* **270:**3720–3725.

Ellis, S. R., Hopper, A. K., and Martin, N. C., 1989, Amino-terminal extension generated from an upstream AUG codon increases the efficiency of mitochondrial import of yeast N^2,N^2-dimethylguanosine-specific tRNA methyltransferases, *Mol. Cell. Biol.* **9:**1611–1620.

Emoto, Y., Manome, Y., Meinhardt, G., Kisaki, H., Kharbanda, S., Robertson, M., Ghayur, T., Wong, W. W., Kamen, R., Weichselbaum, R., and Kufe, D., 1995, Proteolytic activation of protein kinase Cδ by an ICE-like protease in apoptotic cells, *EMBO J.* **14:**6148–6156.

Finley, D., Bartel, B., and Varshavsky, A., 1989, The tails of ubiquitin precursors are ribosomal proteins whose fusion to ubiquitin facilitates ribosome biogenesis, *Nature* **338:**394–401.

Ghislain, M., Dohmen, R. J., Lévy, F., and Varshavsky, A., 1996, Cdc48p interacts with Ufd3p, a WD-repeat protein required for ubiquitin-dependent proteolysis in *Saccharomyces cerevisiae*, *EMBO J.* **15:**4884–4899.

Gonda, D. K., Bachmair, A., Wünning, I., Tobias, J. W., Lane, W. S., and Varshavsky, A., 1989, Universality and structure of the N-end rule, *J. Biol. Chem.* **264:**16700–16712.

Gottesman, S., and Maurizi, M. R., 1992, Regulation by proteolysis: Energy-dependent proteases and their targets, *Microbiol. Rev.* **56:**592–621.

Gottesman, S., Clark, W. P., de Crecy-Lagard, V., and Maurizi, M. R., 1993, ClpX, an alternative subunit for the ATP-dependent Clp protease of *Escherichia coli*, *J. Biol. Chem.* **268:**22618–22626.

Grigoryev, S., Stewart, A. E., Kwon, Y. T., Arfin, S. M., Bradshaw, R. A., Copeland, N. J., and Varshavsky, A., 1996, A mouse amidase specific for N-terminal asparagine: The gene, the enzyme, and their function in the N-end rule pathway, *J. Biol. Chem.* **271:**28521–28532.

Hardy, C. F. J., 1996, Characterization of an essential Orc2p-associated factor that plays a role in DNA replication, *Mol. Cell. Biol.* **16:**1832–1841.

Hartl, F. U., 1996, Molecular chaperones in cellular protein folding, *Nature* **381:**571–580.

Hayes, S. A., and Dice, J. F., 1996, Roles of molecular chaperones in protein degradation, *J. Cell Biol.* **132:**255–258.

Henkart, P. A., 1996, ICE family proteases: Mediators of all apoptotic cell death? *Immunity* **4:**195–201.

Herman, C., Ogura, T., Tomoyasu, T., Hiraga, S., Akiyama, Y., Ito, K., Thomas, R., D'Ari, R., and Bouloc, P., 1993, Cell growth and λ phage development controlled by the same essential *Escherichia coli* gene, *ftsH/hflB*, *Proc. Natl. Acad. Sci. USA* **90:**10861–10865.

Hershko, A., 1991, The ubiquitin pathway for protein degradation, *Trends Biochem. Sci.* **16:**265–268.

Hershko, A., and Ciechanover, A., 1992, The ubiquitin system for protein degradation, *Annu. Rev. Biochem.* **61:**761–807.

Hill, C. P., Johnston, N. L., and Cohen, R. E., 1993, Crystal structure of a ubiquitin-dependent degradation substrate: A three-disulfide form of lysozyme, *Proc. Natl. Acad. Sci. USA* **90:**4136–4140.

Hiller, M. M., Finger, A., Scheiger, M., and Wolf, D. H., 1996, Endoplasmic reticulum associated degradation of a mutated soluble vacuolar enzyme, carboxypeptidase Y, occurs via the ubiquitin–proteasome pathway, *Science* **273:**1725–1728.

Hirel, P. H., Schmitter, J. M., Dessen, P., Fayat, G., and Blanquet, S., 1989, Extent of N-terminal methionine excision from *Escherichia coli* proteins is governed by the side-chain length of the penultimate amino acid, *Proc. Natl. Acad. Sci. USA* **86:**8247–8251.

Ho, Y. S., Wulff, D., and Rosenberg M., 1986, Protein–nucleic acid interactions involved in transcription activation by the phage λ regulatory protein cII, in *Regulation of Gene Expression* (I. Booth and C. Higgins, eds.), pp. 79–103, Cambridge University Press, London.

Ho, Y. S., Mahoney, M. E., Wulff, D., and Rosenberg M., 1988, Identification of the DNA binding domain of the phage lambda cII transcriptional activator and the direct correlation of cII protein stability with its oligomeric forms, *Genes Dev.* **2:**184–195.

Hochstrasser, M., and Varshavsky, A., 1990, *In vivo* degradation of a transcriptional regulator: The yeast α2 repressor, *Cell* **61:**697–708.

Hondermarck, H., Sy, J., Bradshaw, R. A., and Arfin, S. M., 1992, Dipeptide inhibitors of ubiquitin-mediated protein turnover prevent growth factor-induced neurite outgrowth in rat pheochromocytoma PC12 cells, *Biochem. Biophys. Res. Commun.* **189:**280–288.

Huang, S., Elliott, R. C., Liu. P.-S., Koduri, R. K., Weickmann, J. L., Lee, Je-H., Blair, L. C., Ghosh-Dastidar, P., Bradshaw, R. A., Bryan, K. M., Einarson, B., Kolacz, K. H., and Saito, K., 1987, Specificity of cotranslational amino-terminal processing of proteins in yeast, *Biochemistry* **26:**8242–8246.

Jentsch, S., McGrath, J. P., and Varshavsky, A., 1987, The yeast DNA repair gene *RAD6* encodes a ubiquitin-conjugating enzyme, *Nature* **329:**131–134.

Johnson, E. S., Gonda, D. K., and Varshavsky, A., 1990, *Cis-trans* recognition and subunit-specific degradation of short-lived proteins, *Nature* **346:**287–291.

Johnson, E. S., Bartel, B., Seufert, W., and Varshavsky, A., 1992, Ubiquitin as a degradation signal, *EMBO J.* **11:**497–505.

Johnson, E. S., Ma, P. C. M., Ota, I. M., and Varshavsky, A., 1995, A proteolytic pathway that recognizes ubiquitin as a degradation signal, *J. Biol. Chem.* **270:**17442–17456.

Johnsson, N., and Varshavsky, A., 1994a, Ubiquitin-assisted dissection of protein transport across membranes, *EMBO J.* **13:**2686–2698.

Johnsson, N., and Varshavsky, A., 1994b, Split ubiquitin as a sensor of protein interactions *in vivo*, *Proc. Natl. Acad. Sci. USA* **91:**10340–10344.

Johnston, J. A., Johnson, E. S., Waller, P. R. H., and Varshavsky, A., 1995, Methotrexate inhibits proteolysis of dihydrofolate reductase by the N-end rule pathway, *J. Biol. Chem.* **270:**8172–8178.

Kaiser, C. A., Preuss, D., Grisafi, P., and Botstein, D., 1987, Many random sequences functionally replace the secretion signal sequence of yeast invertase, *Science* **235:**312–317.

Kayalar, C., Örd, T., Testa, M. P., Zhong, L.-T., and Bredesen, D. E., 1996, Cleavage of actin by interleukin 1β-converting enzyme to reverse Dnase I inhibition, *Proc. Natl. Acad. Sci. USA* **93:** 2234–2238.

Kendall, R. L., and Bradshaw, R. A., 1992, Isolation and characterization of the methionine aminopeptidase from porcine liver responsible for the cotranslational processing of proteins, *J. Biol. Chem.* **267:**20667–20673.

Klein, J., and O'hUgin, C., 1993, Composite origin of major histocompatibility complex-genes, *Curr. Opin. Genet. Dev.* **3:**923–930.

Knight, S. A. B., Tamai, K. T., Kosman, D. J., and Thiele, D. J., 1994, Identification and analysis of a *Saccharomyces cerevisiae* copper homeostasis gene encoding a homeodomain protein, *Mol. Cell. Biol.* **14:**7792–7804.

Knighton, D. R., Kan, C. C., Howland, E., Janson, C. A., Hostomska, Z., Welsh, K. M., and Matthews, D. A., 1994, Structure of and kinetic channeling in bifunctional dihydrofolate reductase-thymidylate synthase, *Nature Struct. Biol.* **1:**186–194.

Levis, M. J., and Bourne, H. R., 1992, Activation of the α subunit of G_s in intact cells alters its abundance, rate of degradation, and membrane avidity, *J. Cell Biol.* **119:**1297–1307.

Lévy, F., Johnsson, N., Rümenapf, T., and Varshavsky, A., 1996, Using ubiquitin to follow the metabolic fate of a protein, *Proc. Natl. Acad. Sci. USA* **93:**4907–4912.

Li, X., and Chang, Y. H., 1995, Amino-terminal protein processing in *Saccharomyces cerevisiae* is an essential function that requires two distinct methionine aminopeptidases, *Proc. Natl. Acad. Sci. USA* **92:**12357–12361.

Lolis, E., Alber, T., Davenport, R. C., Rose, D., Hartman, F. C., and Petsko, G. A., 1990, Structure of yeast triosephosphate isomerase at 1.9-Å resolution, *Biochemistry* **29:** 6609–6618.

Maarse, A. C., Van Loon, A. P., Riezman, H., Gregor, I., Schatz, G., and Grivell, L. A., 1984, Subunit IV of yeast cytochrome *c* oxidase: Cloning and nucleotide sequencing of the gene and partial amino acid sequencing of the mature protein, *EMBO J.* **3:**2831–2837.

Madura, K., and Varshavsky, A., 1994, Degradation of Gα by the N-end rule pathway, *Science* **265:** 1454–1458.

Madura, K., Dohmen, R. J., and Varshavsky, A., 1993, N-recognin/Ubc2 interactions in the N-end rule pathway, *J. Biol. Chem.* **268:**12046–12054.

Maeda, T., Wurgler-Murphy, S. M., and Sato, H., 1994, A two-component system that regulates an osmosensing MAP kinase cascade in yeast, *Nature* **369:**242–245.

McGrath, J. P., Jentsch, S., and Varshavsky, A., 1991, *UBA1:* An essential yeast gene encoding ubiquitin-activating enzyme, *EMBO J.* **10:**227–237.

Moerschell, R. P., Hosokawa, Y., Tsunasawa, S., and Sherman, F., 1990, The specificities of yeast methionine aminopeptidase and acetylation of amino-terminal methionine *in vivo*, *J. Biol. Chem.* **265:**19638–19643.

Montecucco, C., and Schiavo, G., 1994, Mechanism of action of tetanus and botulinum neurotoxins, *Mol. Microbiol.* **13:**1–8.

Murray, A., and Hunt, T., 1993, *The Cell Cycle*, pp. 60–62, Freeman, San Francisco.

Negrutskii, B. S., and Deutscher, M. P., 1991, Channeling of aminoacyl-tRNA for protein synthesis *in vivo*, *Proc. Natl. Acad. Sci. USA* **88:**4991–4995.

Nishizawa, M., Okazaki, K., Furuno, N., Watanabe, N., and Sagata, N., 1992, The 'second-codon rule' and autophosphorylation govern the stability and activity of Mos during the meiotic cell cycle in *Xenopus* oocytes, *EMBO J.* **11:**2433–2446.

Nishizawa, M., Furuno, N., Okazaki, K., Tanaka, H., Ogawa, Y., and Sagata, N., 1993, Degradation of Mos by the N-terminal proline-dependent ubiquitin pathway on fertilization of *Xenopus* eggs: Possible significance of natural selection for Pro-2 in Mos, *EMBO J.* **12:**4021–4027.
Noble, J. A., Innis, M. A., Koonin, E. V., Rudd, K. E., Banuett, F., and Herskowitz, I., 1993, The *Escherichia coli hflA* locus encodes a putative GTP-binding protein and two membrane proteins, one of which contains a protease-like domain, *Proc. Natl. Acad. Sci. USA* **90:**10866–10870.
Obin, M., Nowell, T., and Taylor, A., 1994, The photoreceptor G-protein transducin is a substrate for ubiquitin-dependent proteolysis, *Biochem. Biophys. Res. Commun.* **200:**1169–1176.
Ota, I. M., and Varshavsky, A., 1992, A gene encoding a putative tyrosine phosphatase suppresses lethality of an N-end rule-dependent mutant, *Proc. Natl. Acad. Sci. USA* **89:**2355–2359.
Ota, I. M., and Varshavsky, A., 1993, A yeast protein similar to bacterial two-component regulators, *Science* **262:**566–569.
Ovádi, J., 1991, Physiological significance of metabolic channeling, *J. Theor. Biol.* **152:**1–22.
Özkaynak, E., Finley, D., Solomon, M. J., and Varshavsky, A., 1987, The yeast ubiquitin genes: A family of natural gene fusions, *EMBO J.* **6:**1429–1440.
Pahl, H. L., and Baeuerle, P. A., 1996, Control of gene expression by proteolysis, *Curr. Opin. Cell Biol.* **8:**340–347.
Park, E. C., Finley, D., and Szostak, J. W., 1992, A strategy for the generation of conditional mutations by protein destabilization, *Proc. Natl. Acad. Sci. USA* **89:**1249–1252.
Paulsen, I. T., and Skurray, R. A., 1994, The POT family of transport proteins, *Trends Biochem. Sci.* **19:**404–405.
Poole, R. K., Hatch, L., Cleeter, M. W. J., Gibson, F., Cox, G. B., and Wu, G., 1993, Cytochrome *bd* biosynthesis in *Escherichia coli*: The sequences of the *cydC* and *cydD* genes suggest that they encode the components of an ABC membrane transporter, *Mol. Microbiol.* **10:**421–430.
Ptashne, M., 1992, *A Genetic Switch*, Cell Press, Cambridge, MA.
Reiss, Y., Kaim, D., and Hershko, A., 1988, Specificity of binding of N-terminal residues of proteins to ubiquitin-protein ligase. Use of amino acid derivatives to characterize specific binding sites, *J. Biol. Chem.* **263:**2693–2698.
Sadis, S., Atienza, C., and Finley, D., 1995, Synthetic signals for ubiquitin-dependent proteolysis, *Mol. Cell. Biol.* **15:**4086–4095.
Saido, T. C., Nagao, S., Shiramine, M., Tsukaguchi, M., Sorimachi, H., Murofushi, H., Tsuchiya, T., Ito, H., and Suzuki, K., 1992, Autolytic transition of μ-calpain upon activation as resolved by antibodies distinguishing between pre- and post-autolysis forms, *J. Biochem.* **111:**81–86.
Schatz, G., and Dobberstein, B., 1996, Common principles of protein translocation across membranes, *Science* **271:**1519–1526.
Scheffner, M., Nuber, U., and Huibregtse, J. M., 1995, Protein ubiquitination involving an E1-E2-E3 enzyme ubiquitin thioester cascade, *Nature* **373:**81–83.
Scherer, P. E., Krieg, U. C., Hwang, S. T., Vestweber, D., and Schatz, G., 1990, A precursor protein partially translocated into yeast mitochondria is bound to a 70 kd mitochondrial stress protein, *EMBO J.* **9:**4315–4322.
Schiavo, G., Benfenati, F., Poulain, B., Rosetto, O., Polverino de Laureto, P., DasGupta, B. R., and Montecucco, C., 1992, Tetanus and botulinum-B neurotoxins block neurotransmitter release by a proteolytic cleavage of synaptobrevin, *Nature* **359:**832–835.
Seol, J. H., Woo, K. M., Kang, M.-S., Ha, B. B., and Chung, C. H., 1995, Requirement of ATP hydrolysis for assembly of ClpA/ClpP complex, the ATP-dependent protease Ti in *Escherichia coli*, *Biochem. Biophys. Res. Commun.* **217:**41–51.
Shen, X., Yu, L., Wir, J. W., and Gorovsky, M. A., 1995, Linker histones are not essential and affect chromatin condensation *in vivo*, *Cell* **82:**47–56.
Sherman, F., Stewart, J. W., and Tsunasawa, S., 1985, Methionine or not methionine at the beginning of a protein? *BioEssays* **3:**27–31.

Shrader, T. E., Tobias, J. W., and Varshavsky A., 1993, The N-end rule in *Escherichia coli*: Cloning and analysis of the leucyl, phenylalanyl-tRNA-protein transferase gene *aat*, *J. Bacteriol.* **175:**4364–4374.

Shyne-Athwal, S., Chakraborty, G., Gage, E., and Ingoglia, N. A., 1988, Comparison of posttranslational protein modification after crush injury to sciatic and optic nerves of rats, *Exp. Neurol.* **99:** 281–295.

Simpson, L. L., ed., 1989, *Botulinum Neurotoxins and Tetanus Toxin*, Academic Press, San Diego.

Sivaram, P., and Deutscher, M. P., 1990, Existence of two forms of rat liver arginyl-tRNA synthetase suggests channeling of aminoacyl-tRNA for protein synthesis, *Proc. Natl. Acad. Sci. USA* **87:** 3665–3669.

Soffer, R. L., 1980, Biochemistry and biology of aminoacyl-tRNA-protein transferases, in *Transfer RNA: Biological Aspects* (D. Söll, J. Abelson, and P. R. Shimmel, eds.), pp. 493–505, Cold Spring Harbor Laboratory Press, Cold Spring Harbor, NY.

Söllner, T., Bennett, M. K., Whiteheart, S. W., Scheller, R. H., and Rothman, J. E., 1993, A protein assembly–disassemby pathway *in vitro* that may correspond to sequential steps of synaptic vesicle docking, activation, and fusion, *Cell* **75:**409–418.

Spence, J., Sadis, S., Haas, A. L., and Finley, D., 1995, A ubiquitin mutant with specific defects in DNA repair and multiubiquitination, *Mol. Cell. Biol.* **15:**1265–1273.

Steiner, H.-Y., Naider, F., and Becker, J. M., 1995, The PTR family: A new group of peptide transporters, *Mol. Microbiol.* **16:**825–834.

Stewart, A., ed., 1995, *TIG Genetic Nomenclature Guide* (supplement to the March 1995 issue of *Trends in Genetics*), Elsevier Science, Cambridge.

Stewart, A. E., Arfin, S. M., and Bradshaw, R. A., 1994, Protein N-terminal asparagine deamidase: Isolation and characterization of a new enzyme, *J. Biol. Chem.* **269:**23509–23517.

Stewart, A. E., Arfin, S. M., and Bradshaw, R. A., 1995, The sequence of porcine protein N-terminal asparagine amidohydrolase: A new component of the N-end rule pathway, *J. Biol. Chem.* **270:** 25–28.

Strauss, J. H., and Strauss, E. G., 1994, The alphaviruses: Gene expression, replication, and evolution, *Microbiol. Rev.* **58:**491–562.

Thornberry, N. A., Bull, H. G., Calaycay, J. R., Chapman, K. T., Howard, A. D., Kostura, M. J., Miller, D. K., Molineaux, S. M., Weidner, J. R., Aunins, J., Elliston, K. O., Ayala, J. M., Casano, F. J., Chin, J., Ding, G. J.-F., Egger, L. A., Gaffney, E. P., Limjuco, G., Palyha, O. C., Raju, S. M., Rolando, A. M., Salley, J. P., Yamin, T.-T., Lee, T. D., Shively, J. E., MacCross, M., Mumford, R. A., Schmidt, J. A., and Tocci, M. J., 1992, A novel heterodimeric cysteine protease is required for interleukin-1β processing in monocytes, *Nature* **356:**768–774.

Tobias, J. W., Shrader, T. E., Rocap, G., and Varshavsky, A., 1991, The N-end rule in bacteria, *Science* **254:**1374–1377.

Tonello, F., Morante, S., Rossetto, O., Schiavo, G., and Montecucco, C., 1996, Cleavage of specific proteins by neurotoxins *Clostridium*. In *Intracellular Protein Catabolism* (K. Suzuki and J. Bond, eds.), pp. 251–260, Plenum Press, New York.

van Voorst, F., Kielland-Brandt, M. C., and Winther, J. R., 1996, Mutational analysis of the vacuolar sorting signal of procarboxypeptidase Y in yeast shows a low requirement for sequence conservation, *J. Biol. Chem.* **271:**841–846.

Varshavsky, A., 1991, Naming a targeting signal, *Cell* **64:**13–15.

Varshavsky, A., 1992, The N-end rule, *Cell* **69:**725–735.

Varshavsky, A., 1995a, The world of ubiquitin, *Eng. Sci.* **58:**27–36.

Varshavsky, A., 1995b, Codominance and toxins: A path to drugs of nearly unlimited selectivity, *Proc. Natl. Acad. Sci. USA* **92:**3663–3667.

Varshavsky, A., 1996a, The N-end rule, *Cold Spring Harbor Symp. Quant. Biol.* **60:**461–478.

Varshavsky, A., 1996b, The N-end rule: Functions, mysteries, uses, *Proc. Natl. Acad. Sci. USA* **93:** 12142–12149.

Varshavsky, A., Bachmair, A., Finley, D., Gonda, D. K., and Wünning I., 1988, The N-end rule of protein turnover, in *Ubiquitin* (M. Rechsteiner, ed.), pp. 287–324, Plenum Press, New York.

Vaux, D. L. and Strasser, A., 1996, The molecular biology of apoptosis, *Proc. Natl. Acad. Sci. USA* **93:** 2239–2244.

von Heijne, G., 1988, Transcending the impenetrable: How proteins come to terms with membranes, *Biochim. Biophys. Acta* **947:** 307–333.

Watanabe, N., Hunt, T., Ikawa, Y., and Sagata, N., 1991, Independent inactivation of MPF and cytostatic factor (Mos) upon fertilization of *Xenopus* eggs, *Nature* **352:**247–248.

Watkins, J. F., Sung, P., Prakash, L., and Prakash, S., 1993, The *Saccharomyces cerevisiae* DNA repair gene *RAD23* encodes a nuclear protein containing ubiquitin-like domain required for biological function, *Mol. Cell. Biol.* **13:**7757–7765.

Wawrzynov, A., Wojtkowiak, D., Marzalec, J., Banecki, B., Jonsen, M., Graves, B., Georgopoulos, C., and Zylicz, M., 1995, The ClpX heat-shock protein of *Escherichia coli*, the ATP-dependent specificity component of the ClpP-ClpX protease, is a novel molecular chaperone, *EMBO J.* **14:** 1867–1877.

Wickner, S., Gottesman, S., Skowyra, D., Hoskins, J., McKenney, K., and Maurizi, M. R., 1994, A molecular chaperone, ClpA, functions like DnaK and DnaJ, *Proc. Natl. Acad. Sci. USA* **91:**12218–12222.

Wiertz, E. J. H. J., Jones, T. R., Sun, L., Bogyo, M., Geuze, H. J., and Ploegh, H. L., 1996, The human cytomegalovirus US11 gene product dislocates MHC class I heavy chains from the endoplasmic reticulum to the cytosol, *Cell* **84:**769–779.

Williamson, L. C., Halpern, J. L., Montecucco, C., Brown, J. E., and Neale, E. A., 1996, Clostridial neurotoxins and substrate proteolysis in intact neurons, *J. Biol. Chem.* **271:**7694–7699.

Wolf, S., Lottspeich, F., and Baumeister, W., 1993, Ubiquitin found in the archaebacterium *Thermoplasma acidophilum*, *FEBS Lett.* **326:**42–44.

Wright, C. F., McKenny, K., Hamer, D. H., Byrd, J., and Winge, D. R., 1987, Structural and functional studies of the amino-terminus of yeast metallothionein, *J. Biol. Chem.* **262:**12912–12919.

Wu, M., Allis, C. D., Sweet, M. T., Cook, R. G., Thatcher, T. H., and Gorovsky, M. A., 1994, Four distinct and unusual linker proteins in a mitotically dividing nucleus are derived from a 71-kilodalton polyprotein, lack $p34^{cdc2}$ sites, and contain protein kinase A sites, *Mol. Cell. Biol.* **14:** 10–20.

Yoshizawa, T., Sorimachi, H., Tomioka, S., Ishiura, S., and Suzuki, K., 1995, Calpain dissociates into subunits in the presence of calcium ions, *Biochem. Biophys. Res. Commun.* **208:**376–383.

Yu, G., Deschenes, R. J., and Fassler, J. S., 1995, The essential transcription factor, Mcm1, is a downstream target of Sln1, a yeast two-component regulator, *J. Biol. Chem.* **270:**8739–8743.

CHAPTER 9

Ubiquitin-Dependent Degradation of Transcription Regulators

Mark Hochstrasser and Daniel Kornitzer

1. INTRODUCTION

The intracellular level or activity of proteins can be regulated at numerous levels. In most cells, one of the principal mechanisms employed is the regulation of transcription initiation. Transcription regulators are therefore among the main regulatory proteins in the cell. As such, control of their concentration—as well as their activity—has to be tightly regulated. Rapid modulation of the concentration of a protein requires it to be relatively short-lived. It is not surprising, then, that the half-lives of many transcription factors are well below the average for cellular proteins. Half-lives as low as 1 min have been measured for some bacterial transcription factors, e.g., cII (Gottesman *et al.*, 1981) and σ^{32} (Strauss *et al.*, 1987). A number of yeast transcription factors are known to have half-lives of only 3–5 min, e.g., Matα2 (Hochstrasser and Varshavsky, 1990) and Gcn4 (Kornitzer *et al.*, 1994). The mammalian transcription factors c-fos (Curran *et al.*, 1984) and c-myc (Luscher and Eisenman, 1988) have half-lives of ~ 20 min. In eukaryotes, most known cases of rapidly degraded transcription regulators involve the

Mark Hochstrasser • Department of Biochemistry and Molecular Biology, University of Chicago, Chicago, Illinois 60637. **Daniel Kornitzer** • Department of Molecular Microbiology, The B. Rappaport Faculty of Medicine, Technion–Israel Institute of Technology, Haifa 31096, Israel.

Ubiquitin and the Biology of the Cell, edited by Peters *et al.* Plenum Press, New York, 1998.

ubiquitin–proteasome pathway. In this chapter, we will review some of the best-studied examples of degradation of transcription factors by this pathway.

2. EVIDENCE LINKING THE UBIQUITIN PATHWAY TO THE DEGRADATION OF SPECIFIC TRANSCRIPTION FACTORS

A variety of methods have been used to link the ubiquitin–proteasome pathway to the degradation of particular proteins. To demonstrate convincingly that a protein is normally an *in vivo* substrate of this pathway requires multiple lines of genetic, pharmacological, and/or biochemical evidence. Few proteins have been so thoroughly examined. However, the degradation of several dozen proteins has now been connected by some means or another to the ubiquitin–proteasome system, suggesting that this pathway is going to be a very common route of selective intracellular protein breakdown. Thus, prima facie evidence for ubiquitin system involvement in the degradation of a particular protein has generally been borne out when more exhaustive studies were subsequently pursued.

The earliest work on ubiquitin-dependent proteolysis was done in rabbit reticulocyte lysates and components purified from these lysates (e.g., Ciechanover *et al.*, 1978; Hershko *et al.*, 1980, 1984; reviewed in Rechsteiner, 1987). The evidence for the direct involvement of the ubiquitin system from such *in vitro* reconstitution studies is often convincing, but its physiological relevance is left in question, most obviously when the substrates used, e.g., hen egg lysozyme or bovine serum albumin, would not normally be in contact with the cytosolic and nuclear ubiquitin system. Conversely, genetic and pharmacological data generally do not demonstrate that the role of the ubiquitin system in the turnover of a particular protein is direct. In some cases, detailed analysis correlating the degree or pattern of substrate ubiquitination to the rate of substrate degradation can often make a compelling case for direct involvement.

Genetic approaches for testing ubiquitin system involvement have been essentially of two kinds. Most commonly, degradation is compared between wild-type cells and cells carrying mutations in components of the ubiquitin–proteasome pathway. This has been done in yeast as well as in mammalian cell lines. The second genetic approach has been to use dominant-negative ubiquitin mutants. High-level expression of chain-terminating ubiquitin mutants in wild-type cells inhibits degradation of ubiquitin system substrates and, importantly, also causes a relative increase in the level of the monoubiquitinated protein (Hochstrasser *et al.*, 1991; Finley *et al.*, 1994). This approach also works *in vitro* when using either ubiquitin point mutants or methylated ubiquitin in which all primary amino groups have been blocked (Chau *et al.*, 1989; Hershko *et al.*, 1991). For at least some substrates in yeast cells, overproduction of ubiquitin derivatives carrying N-terminal peptides also inhibits their turnover but at a step after ubiquitination (Ellison and Hochstrasser, 1991; Hochstrasser *et al.*, 1991).

The ubiquitin–proteasome pathway can be inhibited pharmacologically *in vivo* by protease inhibitors. Certain peptide aldehydes, which often also inhibit other intracellular proteases such as cathepsins and calpains, can block proteasome activity. An irreversible inhibitor that appears to have higher specificity for the proteasome is the fungal metabolite lactacystin (Fenteany *et al.*, 1995). When radiolabeled lactacystin was incubated with mammalian cell extracts, two 20 S proteasome subunits were the only proteins reported to be labeled. However, because of the abundance of 20 S proteasomes (~ 1% soluble cell protein), the possibility of inhibition of other, less abundant cell proteases cannot be discounted. Inhibitor studies with lactacystin and various peptide aldehydes have been done primarily in mammalian cell lines.

In this section, we review the evidence linking the ubiquitin pathway to the degradation of a number of transcription factors. We will not discuss the degradation of p53 or of NF-κB/IκB, which are each the subject of a separate chapter in this volume.

2.1. *Matα2*

The yeast *Saccharomyces cerevisiae* has two haploid cell types, **a** and α, which can mate to form a third cell type, an **a**/α diploid (Fig. 1). Ultimately, cell identity is determined by the set of transcription regulators encoded by the mating type, or *MAT*, locus. Cryptic (unexpressed) **a** or α information at two other genomic sites can be copied into the *MAT* locus, a process initiated by the HO endonuclease (reviewed in Herskowitz *et al.*, 1992). In HO (homothallic) strains, interconversion between **a** and α cell types may occur within a single cell cycle, which can be less than 90 min. This implies the existence of mechanisms that facilitate rapid dismantling of the preceding cell type-specific regulatory network following a switch and that also limit the action of the proteins involved in the switching process itself. Selective protein degradation may underlie such mechanisms by causing rapid disappearance of the regulatory proteins involved once their synthesis has ceased. Indeed, it is now known that all three transcription factors encoded by the two alleles of the *MAT* locus, α1 and α2 from *MAT*α and **a**1 from *MAT***a**, are extremely short-lived proteins in haploid yeast cells (Hochstrasser and Varshavsky, 1990; P. Johnson and M. Hochstrasser, unpublished data). Much more is known about α2 turnover, so the discussion here will focus largely on this protein.

The ubiquitin-dependence of α2 degradation was first suggested by the finding that α2 is multiubiquitinated *in vivo* (Hochstrasser *et al.*, 1991). Modified forms of α2 are present at very low levels (less than 1% those of nonubiquitinated pulse-labeled α2). Overexpression of a chain-terminating ubiquitin mutant

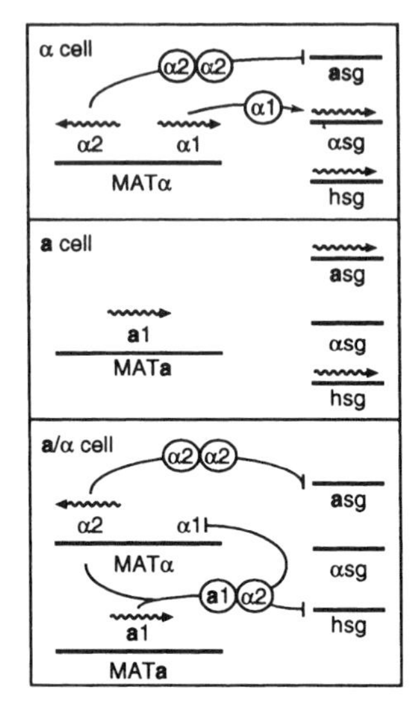

Figure 1. Cell type regulation by products of the *MAT* locus in *S. cerevisiae*. The α2 repressor prevents expression of **a** cell-specific genes (asg) in both α and **a**/α cells, while it also works in conjunction with the **a**1 protein in **a**/α cells to repress α1 and the haploid-specific genes (hsg). α1 is a transcription factor that positively regulates genes required in α (αsg). The **a**1 protein has no known role in **a** cells.

showed a dominant-negative effect, partially inhibiting α2 turnover and causing an increase in the relative level of the monoubiquitinated form of pulse-labeled α2. Moreover, overproduction of a peptide-tagged ubiquitin derivative also partially stabilized α2 (as well as certain artificial substrates of the ubiquitin system). Subsequently, it was shown that four ubiquitin-conjugating (Ubc or E2) enzymes, Ubc4, Ubc5, Ubc6, and Ubc7, are required for rapid intracellular degradation of α2 (Chen *et al.*, 1993). Based on epistasis analysis, these enzymes were inferred to participate in two distinct α2 degradation pathways. Ubc6 and Ubc7 define one of these pathways and appear to physically associate *in vivo*, as measured by a two-hybrid interaction assay. When both pathways were mutationally inactivated, ubiquitination of α2 was significantly reduced (but not eliminated). The Ubc6/Ubc7-containing complex targets the *Deg1* degradation signal (see next section) of α2, a conclusion underscored by the finding that Ubc6 is encoded by *DOA2*, a gene previously implicated in *Deg1*-mediated degradation. These data revealed an unexpected overlap in substrate specificity among diverse ubiquitin-conjugating enzymes and suggested a combinatorial mechanism of substrate selection in which Ubc enzymes partition into multiple ubiquitination complexes.

That the 26 S proteasome is the key enzyme responsible for α2 repressor degradation in yeast cells was demonstrated by extensive genetic analysis. Muta-

Table I
Summary of the Evidence Linking Degradation of Specific Transcription Factors with the Ubiquitin–Proteasome Pathway, and of the *cis*-Acting Signals and *trans*-Acting Proteins Involved in the Degradation[a]

Transcription factor	Ubiquitination shown	Evidence linking degradation to ubiquitin pathway[b]	Ubiquitin pathway enzymes implicated	Degradation signal
p53	Yes	Biochemical; genetic	UbcH5 or E2-F1; E6-E6AP	?
IκB	Yes	Biochemical; pharmacological	26 S proteasome	Ser32, Ser36 phosphorylation; C-terminal PEST motif
MATα2	Yes	Genetic	Ubc4–7; Pre1, Doa3, Pre4, Doa5, Sen3, Cim3, Cim5, Doa4, Ubp14	*Deg1*, *Deg2*
Gcn4	Yes	Genetic	Cdc34, Rad6, Cim5	PEST motif
c-jun	Yes	Biochemical	26 S proteasome	δ domain
c-fos	Yes	Biochemical; genetic	UbcH4 or E2-F1, E3-Fos	C-terminal PEST motif, phosphorylation
myoD	Yes	Biochemical; pharmacological	E2-F1, E3L	?

[a]See text for details and references.
[b]Biochemical, ubiquitination and degradation by the ubiquitin system *in vitro*; genetic: inhibition of degradation in ubiquitin pathway mutants; pharmacological: inhibition of degradation in the presence of specific proteasome inhibitors.

tions in seven different proteins known to be essential components of the proteasome were shown to inhibit degradation of α2 (Table I) (Richter-Ruoff *et al.*, 1994; Chen and Hochstrasser, 1995; DeMarini *et al.*, 1995); these proteins include subunits of the 20 S proteasome core and subunits of the 19 S regulatory complexes. Some of the partial loss-of-function mutations of the proteasome result in 10-fold or greater stabilization of the repressor. Thus, the 26 S proteasome is the major, if not the only, protease that normally degrades α2 in vivo. Two deubiquitinating enzymes, Doa4 and Ubp14, which may function in conjunction with the proteasome *in vivo* (see elsewhere in this volume), are also required for normal rates of α2 proteolysis (Papa and Hochstrasser, 1993).

More recent work provides strong parallels to the data on α2 ubiquitination and degradation. The participation of multiple, distinct Ubc enzymes in the degradation of particular short-lived regulatory proteins may turn out to be the rule rather than the exception. More extensive analysis of Gcn4 turnover (see below) revealed that at least two distinct Ubc proteins are necessary for maximal rates of *in vivo* proteolysis. Where the *in vitro* ubiquitination of a naturally short-lived protein has been investigated in detail, multiple E2 enzymes have generally also been implicated. The best studied are the p53 tumor suppressor and mitotic cyclins (see elsewhere in this volume). However, in these examples, either of at least two different E2s was found to be sufficient for ubiquitination *in vivo*, whereas the *in vivo* data obtained for α2 and Gcn4 indicated that the different E2s were all necessary for maximal levels of degradation. The differences between the inferences drawn from the *in vivo* and *in vitro* data may be related to differences in assay conditions. For instance, some of the *in vitro* analyses have been done at saturating enzyme concentrations, a circumstance that may well not obtain *in vivo*. It is also possible that substrate ubiquitination by different E2s results in different rates of degradation.

The involvement *in vivo* of distinct Ubc enzymes in the same ubiquitination pathway has now also been seen with other substrates in yeast. Both the Ubc6 and the Ubc7 enzymes were shown to be necessary for the degradation of a β-galactosidase derivative bearing a particular synthetic peptide sequence at its N-terminus, and the mutants showed the same epistatic relationship as in the case of α2 turnover (Sadis *et al.*, 1996). The Ubc6 and Ubc7 enzymes were recently shown to participate in the rapid degradation of a temperature-sensitive variant of the Sec61 integral ER membrane protein (Biederer *et al.*, 1996), although the effect of deleting *UBC7* was more severe. Similarly, an aberrant derivative of vacuolar carboxypeptidase Y that fails to get transported beyond the ER is degraded by a ubiquitin- and proteasome-dependent mechanism. This degradation, which appears to occur after the aberrant protein is ejected into the cytosol from the ER, depends on Ubc7 and, to a lesser degree, Ubc6 (Hiller *et al.*, 1996).

2.2. Gcn4

Gcn4 is a yeast transcription factor involved in the activation of amino acid and purine biosynthesis genes (reviewed in Hinnebusch, 1992). The protein forms a stable homodimer via a leucine zipper dimerization domain and binds to a DNA sequence similar to that of the mammalian AP-1 transcription factor (Struhl, 1987). Strong overexpression of Gcn4 slows cell growth. The reason of this effect is unknown, but it may be related to the channeling of cellular resources toward the biosynthesis of unneeded compounds, or to a "squelching" effect (Gill and Ptashne, 1988) on the cellular transcription machinery. Whatever its reason, this overexpression toxicity underscores the requirement for tight regulation of the concentration of the Gcn4 protein. Not surprisingly, then, Gcn4 was found to be normally rapidly degraded.

Under standard growth conditions, the half-life of Gcn4 is 5 min (Kornitzer *et al.*, 1994). Dependence of Gcn4 degradation on the ubiquitin–proteasome pathway was demonstrated by showing a correlation between the level of ubiquitination and the rate of degradation of wild-type Gcn4 versus a more stable Gcn4 mutant (Kornitzer *et al.*, 1994), and by showing that overexpression of a chain-terminating ubiquitin mutant stabilizes Gcn4 (D. Kornitzer and G. R. Fink, unpublished data). Furthermore, two ubiquitin-conjugating enzymes were identified that, when mutated, inhibit Gcn4 degradation: Cdc34/Ubc3 and Rad6/Ubc2. Ubrl, the E3 enzyme that, with Rad6/Ubc2, is responsible for degradation of N-end rule substrates, is not required for Gcn4 degradation, in agreement with the finding that the degradation signal of Gcn4 does not include its N-terminus (see below). A mutant in the proteasomal regulatory subunit Cim5 also inhibited Gcn4 degradation (Kornitzer *et al.*, 1994). Surprisingly, however, mutations in core proteasomal subunits (Prel and Pre2/Doa3), which inhibit degradation of many ubiquitin system substrates, did not affect Gcn4 degradation (D. Kornitzer and G. R. Fink, unpublished data). Presumably, the residual proteolytic activities in these mutants are sufficient for the efficient degradation of Gcn4. These results provide an important cautionary tale for those attempting to conclude that the proteasome is not involved in the degradation of a particular protein based on studies with proteasome mutants. Because the proteasome is essential for viability, only mutants that retain partial proteasomal activity can be used; the degradation of some substrates may be relatively insensitive to partial defects in some proteasome components or to inactivation of only a subset of active sites (Chen and Hochstrasser, 1996).

Gcn4 migrates as three distinct electrophoretic species, probably representing distinct levels of phosphorylation; strikingly, the fastest-migrating species accumulates in a *rad6* mutant, whereas the two slower species accumulate in the *cdc34* mutant, suggesting that differentially phosphorylated forms of the protein are recognized by the two enzymes (Kornitzer *et al.*, 1994). The kinase responsible for Gcn4 phosphorylation is as yet unknown. Other substrates of Cdc34 require phosphorylation by Cdc28 (the yeast homologue of the Cdc2 kinase) prior to degradation: Cln3 (Yaglom *et al.*, 1995), Cln2 (Lanker *et al.*, 1996), and possibly Sicl (Schwob *et al.*, 1994) and Farl (McKinney *et al.*, 1993). In contrast, Cdc28 is not involved in Gcn4 degradation (Kornitzer *et al.*, 1994).

Cdc34 and Rad6 are necessary for efficient Gcn4 degradation *in vivo*, and purified Cdc34 and Rad6 proteins are able to direct the ubiquitination of recombinant Gcn4 *in vitro* (Kornitzer *et al.*, 1994). The *in vitro* reaction, however, is rather inefficient, and only monoubiquitinated substrate molecules are detectable, whereas multiubiquitination is detectable *in vivo*. Other factors are clearly required for efficient Gcn4 ubiquitination. These factors may include both ancillary factors of Cdc34 and Rad6 (ubiquitin ligases?) and the Gcn4 kinase.

2.3. AP-1

This mammalian transcription factor is involved in cell proliferation following mitogenic stimuli (reviewed in Angel and Karin, 1991). The active form of

AP-1 is composed of two subunits, one Jun-type protein and one Fos-type protein, which heterodimerize via a leucine-zipper dimerization domain. Biosynthesis of both subunits is rapidly induced following mitogenic stimuli such as phorbol esters or serum. Deregulated expression of c-Jun and c-Fos can lead to fibroblast transformation, and oncogenic forms of c-Jun and c-Fos (v-Jun and v-Fos, respectively) are found in transforming retroviruses. Both c-Jun and c-Fos have relatively short half-lives (~ 90 and ~ 20 min, respectively), whereas their oncogenic counterparts v-Jun and v-Fos are more stable (Treier *et al.*, 1994; Papavassiliou *et al.*, 1992). This relative stability was suggested to account for at least part of their transforming potential.

2.3.1. c-Jun

Treier *et al.* (1994) found that c-Jun is ubiquitinated *in vivo*; v-Jun displays, along with a lower degradation rate, a lower level of ubiquitination as well, providing correlative evidence that c-Jun is degraded via the ubiquitin pathway. However, a report by Jariel-Encontre *et al.* (1995) shows that c-Jun can be efficiently degraded by purified 26 S proteasomes in the apparent absence of ubiquitin or ubiquitin pathway enzymes. This is not the first report of degradation by the proteasome in the absence of ubiquitination: The unstable enzyme ornithine decarboxylase was also shown to be degraded in such a manner (Murakami *et al.*, 1992). Possibly, c-Jun carries determinants that can be directly recognized by the 26 S proteasome. A caveat in interpreting the *in vitro* c-Jun degradation data is that it is not yet known whether the cell-free degradation of c-Jun purified from *E. coli* follows the same pathway normally used by mammalian cells for endogenously synthesized c-Jun.

2.3.2. c-Fos

This protein has an even shorter half-life than c-Jun, about 20 min, according to some reports (Curran *et al.*, 1984). c-Fos degradation in crude reticulocyte extracts was shown to require the presence of phosphorylated c-Jun (Papavassiliou *et al.*, 1992). Tsurumi *et al.* (1995) demonstrated that the purified 26 S proteasome is able to degrade c-Fos in the presence of ubiquitin system enzymes (purified E1, E2, and partially purified E3 from rabbit reticulocyte extract). Furthermore, they showed that this degradation requires the presence of c-Jun, confirming the results of Papavassiliou *et al.* (1992), and of three distinct kinases, mitogen-activated protein kinase, casein kinase II, and CDC2 kinase. Stancovski *et al.* (1995) demonstrated that c-Fos is degraded *in vivo* via the ubiquitin system by showing that degradation at nonpermissive temperature is impaired in ts20 cells, a mutant cell line that harbors a thermolabile ubiquitin-activating enzyme (Kulka *et al.*, 1988). They further showed that the E2 enzymes involved in p53 degradation, E2-F1 (from rabbit) and UbcH5 (its human homologue), are able to ubiquitinate c-Fos *in vitro*. The E3 enzyme specific for c-Fos was partially purified and shown to be a novel species of E3, called E3-Fos, with a molecular mass of ~ 280 kDa (Stancovski *et al.*, 1995).

Another pathway for the degradation of both c-Jun and c-Fos was suggested by the observation that both proteins are substrates of the calcium-dependent neutral protease (calpain) *in vitro* (Hirai *et al.*, 1991). Furthermore, AP-1 activity was found to correlate with expression of calpastatin, a specific inhibitor of calpains (Hirai *et al.*, 1991), suggesting that calpains have a function in AP-1 degradation *in vivo*. Which of the three pathways of AP-1 degradation—by calpains, ubiquitin–proteasome, or proteasome alone in the case of c-Jun—is responsible for the bulk of the degradation *in vivo* remains an open question. The answer may depend on the cell type or developmental state. A way to differentiate at least between calpain-mediated and proteasome-mediated degradation would be to use the newly developed proteasome-specific inhibitors (Fenteany *et al.*, 1995). In any event, it should be pointed out that calpains are endopeptidases that cleave polypeptides but do not digest them to completion; it therefore cannot be excluded that in some cases at least, the initial cleavage of c-Jun and c-Fos is performed by calpains, followed by further digestion of the cleavage products by the proteasome.

2.4. MyoD

The MyoD gene was isolated as a gene that, when ectopically expressed in fibroblasts, converts them into myoblasts (Davis *et al.*, 1987). MyoD is a transcription factor of the bHLH family (Lassar *et al.*, 1989; Murre *et al.*, 1989) with a half-life of 30–60 min (Thayer *et al.*, 1989). *In vitro*, MyoD can be ubiquitinated by a specific E2 enzyme, E2-F1, and a novel E3 enzyme, E3L (Gonen *et al.*, 1996). In cell culture, addition of proteasome inhibitors leads to inhibition of MyoD degradation and to the accumulation of ubiquitinated forms of the protein, suggesting that the ubiquitin–proteasome system is responsible for degrading MyoD *in vivo* (A. Ciechanover, personal communication).

3. DEGRADATION SIGNALS

How are short-lived transcription factors recognized by the ubiquitin system? Can sequences be identified that are necessary and/or sufficient for proteolytic targeting? The latter question has been addressed in a number of cases by deletion and mutational analyses. In all instances, it appears that discrete structural elements or "degradation signals" target the substrate for rapid turnover. One can distinguish between at least two kinds of determinants in such signals. Certain sequence elements function as recognition sites, i.e., sites on the protein that are recognized by the ubiquitin-conjugation machinery, and other elements provide appropriate sites for ubiquitin ligation, i.e., the specific lysine residues that are linked to ubiquitin. In this section, we review what is known about the degradation signals of some of the better-studied short-lived eukaryotic transcription factors.

3.1. Recognition Sites

For the rapidly degraded transcription factors studied to date, no clear-cut sequence motif analogous, for example, to the cyclin destruction box (see Chapter 12) has been identified. However, the cyclin destruction box is highly degenerate and would not have been recognized were it not for the cloning of many different cyclins from diverse organisms, which allowed the weakly conserved element to be detected. It is likely that recognition sites for ubiquitinating enzyme complexes in rapidly degraded transcription factors will also be degenerate, at least at the primary sequence level.

3.1.1. *MATα2*

The 210-residue α2 repressor appears to contain at least two distinct degradation signals, called *Deg1* and *Deg2*, based on deletion studies with α2-β-galactosidase fusions (Hochstrasser and Varshavsky, 1990). Mutants defective in the degradation of a *Deg1*-βgal fusion protein were subsequently isolated. These *doa* (*d*egradation *o*f *a*lpha2) mutants also fail to degrade the unaltered α2 protein at normal rates, demonstrating that the *Deg1* degradation signal also functions in the context of the intact repressor. No such genetic data are available to support the contention that the *Deg2* signal is important for intact α2 degradation, an important caveat in the interpretation of the *Deg2* deletion data.

The transplantable *Deg1* signal was originally localized to the N-terminal 67 residues of α2 (Hochstrasser and Varshavsky, 1990), but it is now known that short segments within this region can be deleted without interfering with degradation (L. Rakhilina and M. Hochstrasser, unpublished data). Importantly, turnover of these deletion derivatives is still fully dependent on Ubc6 and Ubc7 (see above), as was true of the uninterrupted *Deg1* element (L. Rakhilina and M. Hochstrasser, unpublished data). Thus, the *Deg1* degradation signal is discontinuous. This implies either that the *Deg1* signal comprises a three-dimensional structural feature of the α2 protein formed by noncontiguous sequence elements or that the distinct elements have different functions in proteolysis. Among the segments from the *Deg1* sequence that can be deleted is a so-called PEST sequence (PESTSCORE = 7.43; Rogers *et al.*, 1986) (Fig. 2). PEST sequences are sequences with a high proportion of proline, aspartic and glutamic acid, serine, and

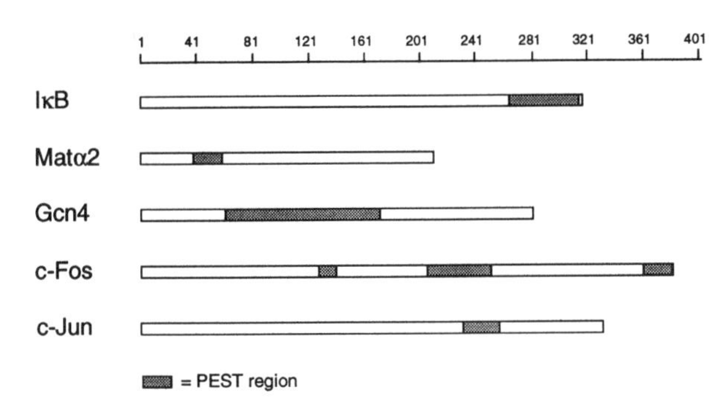

Figure 2. Location of PEST regions in a number of transcription factors. PEST regions are required for the degradation of some—but not all—of the indicated proteins. See text for details.

threonine residues. Such sequences were proposed to serve as recognition sites for cytoplasmic proteases based on the statistical observation that PEST sequences were overrepresented in a sample of rapidly degraded proteins (Rogers *et al.*, 1986). Although there is growing evidence for the importance of such motifs in the degradation of some proteins (see below), the N-terminal PEST element of α2 is neither necessary nor sufficient (Hochstrasser and Varshavsky, 1990) for α2 turnover by the Ubc6/Ubc7-dependent pathway. It is likely that PEST elements function indirectly in degradation, serving principally as sites for phosphorylation, which is necessary for the ubiquitination and degradation of some proteins (see below).

The *Degl* signal in α2 has been subjected to extensive point mutagenesis. Surprisingly, most of the mutations that severely impeded *Degl*-mediated degradation localized to a relative short stretch (~ 16 residues) of the sequence (L. Rakhilina and M. Hochstrasser, unpublished data). When this region was modeled as a helix, the majority of the mutations clustered on the hydrophobic face of the putative amphipathic helix. This helical surface may therefore provide a major recognition site for the Ubc6/Ubc7-containing ubiquitination complex [the helix shows some similarity to the artificial degradation signal described by Sadis *et al.* (1995)]. Strong, albeit indirect support for this idea comes from studies on the regulation of α2 turnover (see below).

3.1.2. IκB

IκB is an inhibitor of the transcription factor NF-κB; binding of IκB to NF-κB results in sequestration of the latter in the cytoplasm (reviewed in Thanos and Maniatis, 1995) (see also Chapter 10). Deletion of the C-terminal 75 (Chen *et al.*, 1995), 61 (Rodriguez *et al.*, 1995), or 41 amino acids (Brown *et al.*, 1995), which contain a PEST-rich sequence (see Fig. 2), stabilizes the protein. Deletion of the N-terminal 54 (Brown *et al.*, 1995) or 36 (Chen *et al.*, 1995) amino acids also stabilizes the protein. This sequence contains phosphorylation sites on two serine residues, at positions 32 and 36. Phosphorylation of these serines was found to be necessary for efficient ubiquitination and subsequent degradation of the protein (Chen *et al.*, 1995; Brown *et al.*, 1995). The targeting signal(s) of IκB thus includes both an N-terminal phosphorylated sequence and a C-terminal PEST sequence.

3.1.3. Gcn4

Deletion analysis established that a PEST-rich region that partially overlaps the Gcn4 activation domain (Fig. 2) is necessary for the degradation of the protein (Kornitzer *et al.*, 1994). However, this region of Gcn4 is not sufficient for degradation because, unlike the whole protein, it cannot by itself destabilize the normally stable βgal protein when fused to it. Point mutations in specific PEST residues in that region of the protein—Ser101, Thr105, Pro106—were also found to partly stabilize Gcn4, corroborating the function of this domain as a degradation signal. Strikingly, however, mutations in other PEST residues in the same region of the protein had no effect. This result suggests that PEST regions are not only defined by a high proportion of PEST residues, but that the specific sequence is important as well.

It has been suggested that an essential attribute of PEST region components of degradation signals is their ability to be phosphorylated (see above). Gcn4 is in fact a phosphoprotein, and a small deletion in the PEST-rich region was found to change the pattern of phosphorylation of Gcn4 (D. Kornitzer and G. R. Fink, unpublished data). However, there is no direct proof yet that the PEST regions themselves are phosphorylated, and it is not known whether Gcn4 phosphorylation plays a role in the degradation of the protein (see below).

3.1.4. c-Jun

A small region between positions 31 and 57 of c-Jun, dubbed the δ domain because it is deleted in v-Jun, was hypothesized by Treier and co-workers to be involved in the degradation of the protein (Treier *et al.*, 1994). Indeed, deletion analysis determined that absence of this region leads to a dramatic reduction in ubiquitination, and a concomitant stabilization of the protein. A slightly larger region, including positions 1–67 of c-Jun, was sufficient by itself to target the ubiquitination and degradation of a normally stable protein, βgal, when fused to it. This larger domain thus represents an independent degradation signal. Surprisingly, however, in the intact c-Jun protein, deletion of the C-terminus beyond position 223 inhibited ubiquitination. Possibly, the C-terminal region contains a second determinant, perhaps the ubiquitination site (see below), which can otherwise be provided by the βgal sequence in the c-Jun(1–67)-βgal fusion protein; alternatively, the C-terminally deleted protein may be conformationally altered in a way that precludes its ubiquitination.

3.1.5. c-Fos

The targeting signal of c-Fos is complex: c-Fos degradation requires a C-terminal PEST-rich sequence as well as dimerization of the protein with phosphorylated c-Jun. c-Fos degradation thus requires a mixture of *cis*- and *trans*-acting determinants. A requirement for dimerization was shown by using a mutant in the leucine zipper of c-Jun, which does not dimerize with c-Fos: c-Fos is not degraded in the presence of this c-Jun mutant (Papavassiliou *et al.*, 1992). Phosphatase treatment of c-Jun prior to binding to c-Fos prevented degradation as well, indicating a role for phosphorylation of c-Jun in its function in c-Fos degradation (Papavassiliou *et al.*, 1992). c-Jun phosphorylation could be mimicked *in vitro* by the simultaneous addition of MAP kinase, casein kinase II, and CDC2 kinase (Tsurumi *et al.*, 1995).

Curran and co-workers noted that v-Fos is more stable than c-Fos (Curran *et al.*, 1984). The difference between the two proteins resides in an out-of-frame deletion that changes the last 48 residues of the sequence to an unrelated sequence of 49 residues, suggesting that at least some of these 48 residues are essential for degradation. Okazaki and Sagata (1995) showed that deleting the 19 C-terminal residues alone (residues 362–380) already greatly increased the steady-state levels of the protein, presumably by reducing its degradation. This C-terminal sequence contains one of three PEST-rich regions of c-Fos (Fig. 2). Phosphorylation of two sites in this region, Ser362 and Ser374, stabilizes the protein, i.e., phosphorylation in this case inactivates the degradation signal. This phosphorylation occurs on transfection of c-Mos, and is apparently mediated by the Mek1/ERK kinase pathway (Okazaki and Sagata, 1995) (see below). Regulation of c-Fos degradation by phosphorylation is reminiscent of the degradation of c-Mos itself, which is inhibited by phosphorylation of the serine residue at position 3 of the protein (Nishizawa *et al.*, 1992, 1993).

3.2. Ubiquitination Sites

Posttranslational ubiquitination of proteins appears to be limited to the ε-amino groups of lysine residues. Physical mapping of *in vivo* ubiquitin ligation sites has only been done for the N-end rule βgal substrates (Chau *et al.*, 1989; see Chapter 8), an artificial substrate of the ubiquitin pathway. On the other hand, mutational analyses have been performed with several substrates to determine which lysines are necessary for substrate degradation, including several transcription factors. Derivatives of *E. coli* βgal that are substrates of the N-end rule pathway are modified by multiubiquitin chains that are largely limited to either of a pair of lysines near the protein N-terminus. Although only assayed by mutation of lysines to nonubiquitinatable arginines, it appears that ubiquitination of some naturally short-lived proteins will not be so tightly constrained.

The yeast α2 repressor can be ubiquitinated at multiple sites, at least some of which are modified by ubiquitin oligomers (Hochstrasser *et al.*, 1991). SDS–polyacrylamide gel electrophoretic separation of α2-containing proteins immunoprecipitated from radiolabeled cell extracts revealed a heterogeneous pattern of ubiquitinated species. Mutation of a specific pair of Lys residues in the linker region between the two globular domains of α2 eliminated a specific monoubiquitinated species. However, the mutant α2 protein continued to be degraded at wild-type rates. This result indicates that some substrate ubiquitination events do not lead to proteolytic targeting *in vivo*.

Whether the ubiquitination of any particular lysine in α2 is necessary for degradation remains an open question. When all 11 lysines in the *Deg1* moiety of a *Deg1*-βgal fusion protein were changed to arginines, the resulting protein continued to be degraded at nearly wild-type rates by a Ubc6/Ubc7-dependent mechanism (L. Rakhilina and M. Hochstrasser, unpublished data). Because ubiquitination of this particular *Deg1*-protein fusion is difficult to detect, it is not yet known if a Lys residue(s) from the βgal sequence provides a ubiquitination site.

In the case of c-Jun, all 17 lysines in the protein were mutated to arginines in clusters of up to 3. No single lysine or cluster of lysines had any significant effect on c-Jun ubiquitination (Treier *et al.*, 1994). Deletion of the C-terminal third of the protein, including 11 of its 17 lysines, however, inhibited ubiquitination. Possibly, this deletion removed the lysines essential for c-Jun ubiquitination; alternatively, of course, lack of ubiquitination of the deletion is a consequence of a disruption of the structure of the protein.

For Gcn4, 11 of the 23 lysines of the protein in closest proximity to the PEST sequence required for degradation were converted to arginines individually or in clusters. No single lysine was necessary for degradation; at least 5 lysines had to be mutated before significant effects on half-life were detected, and dramatic effects were seen only when 9 or more lysines were mutated simultaneously (Kornitzer *et al.*, 1994). The model emerging from these analyses is that any one of a number of lysines can serve as ubiquitin acceptor. In other words, the ubiquitinating lysines do not constitute a part of the targeting signal; rather, after recognition of the targeting signal, the ubiquitination machinery is able to ubiquitinate any lysine that happens to be sterically available.

For IκB, in contrast to the examples presented above, mutating the two N-terminal-most lysines (out of nine), at positions 21 and 22, strongly reduced ubiquitination and degradation (Scherer *et al.*, 1995; Baldi *et al.*, 1996). This double mutation did not affect phosphorylation or activity of the protein. Another example where single or double lysine substitutions were sufficient to reduce ubiquitination and degradation is c-Mos. In c-Mos, substituting the single N-terminal-most lysine (out of nine) with arginine significantly stabilized the protein (Nishizawa *et al.*, 1993). Common to these two substrates, and to N-end rule substrates, is the fact that the target site is located near the N-terminus of the protein. In c-Mos and the N-end rule substrates, the essential lysines are the only ones in close proximity to the N-terminus, at least at the primary sequence level. In IκB, other lysines are found relatively close to the N-terminus, but they are located in a region that is protected from *in vitro* proteolysis by interaction with NF-κB (Jaffray *et al.*, 1995) and might therefore be shielded from the ubiquitination machinery by this interaction as well. The domain containing Lys21 and Lys22, in contrast, is sensitive to *in vitro* proteolysis (Jaffray *et al.*, 1995). These results therefore do not contradict the hypothesis that steric availability rather than a specific sequence context determines the ability of a lysine to be ubiquitinated. Furthermore, this hypothesis would predict that introducing a new lysine at an accessible position in a stable Lys-to-Arg mutant would reconstitute ubiquitination. Indeed, Scherer *et al.* (1995) showed that ubiquitination of an IκB Lys21/22Arg double mutant could be recovered by changing an arginine normally found at position 17 of the protein (i.e., in the protease-sensitive domain) to lysine.

4. REGULATION OF DEGRADATION

Degradation of a number of transcription factors has been shown to be physiologically regulated. These include p53, IκB, c-Jun/c-Fos, Gcn4, and MATα2. A general question regarding regulated degradation is whether the substrate or the degradation system is the target of regulation. As described below, both kinds of regulatory mechanism have been documented.

4.1. p53

The p53 protein is a transcription factor with tumor suppressor activity. p53 is normally relatively unstable, with a half-life of 20–30 min (Reihsaus *et al.*, 1990; Chowdary *et al.*, 1994). Degradation of p53 is increased in cells infected with members of a subclass of papillomaviruses, the high-risk viruses—so called because of their frequent association with cervical cancer (Scheffner *et al.*, 1990). A protein encoded by these viruses, E6, was found to be responsible for the increased degradation of p53 (Scheffner *et al.*, 1990). E6 acts by binding a protein called E6-AP, and this complex binds to p53, thereby promoting the ubiquitination and subsequent degradation of p53 (Scheffner *et al.*, 1993). Thus, the virus recruits the cellular ubiquitin-dependent proteolytic system to inactivate p53 (see Chapter 11).

4.2. IκB

Degradation of IκB is required for activation of the transcription factor NF-κB. Activation of NF-κB can be achieved by a variety of extracellular stimuli, which include cytokines (such as tumor necrosis factor α, interleukin-1β and mitogens (such as phorbol myristate acetate, lipopolysaccharide), among others (reviewed in Thanos and Maniatis, 1995). These compounds activate NF-κB by stimulating degradation of IκB (Alkalay *et al.*, 1995). IκB degradation, as mentioned above, requires phosphorylation on serines 32 and 36; the NF-κB-activating extracellular stimuli trigger a variety of signal transduction pathways, which presumably converge on a kinase(s) that phosphorylates Ser32 and Ser36 of IκB. Such a kinase has been partially purified by Chen *et al.* (1996). Remarkably, activity of this ~ 700-kDa multisubunit kinase requires the ubiquitin system. Presumably, ubiquitination of a component of the kinase is necessary for kinase activity (see Chapter 10).

4.3. c-Fos

Mutation-induced stabilization of the v-Fos protein relative to c-Fos was suggested to play a role in its oncogenic potential. However, little is known about the physiological regulation of c-Fos turnover. According to one report (Miao and Curran, 1994), serum stimulation results in moderate (twofold) extension of the half-life of the protein. c-Fos degradation was shown to require dimerization with phosphorylated c-Jun (see above); however, when the source of the c-Jun protein used in the *in vitro* c-Fos degradation assay is from TPA-stimulated cells, degradation of c-Fos is reduced (Papavassiliou *et al.*, 1992). Thus, it is possible that c-Jun constitutes the target for physiological regulation of c-Fos degradation.

In another report, ectopic expression of the proto-oncogene product c-Mos, a serine/threonine protein kinase, was found to reduce c-Fos degradation (Okazaki and Sagata, 1995), and this stabilization of c-Fos could be attributed to phosphorylation of two serine residues in its C-terminal domain. Deletion of this domain also stabilized c-Fos. The simplest explanation for the effect of the c-Mos-induced phosphorylation is that it inactivates one of the degradation signals of the protein. c-Mos probably does not directly phosphorylate c-Fos; rather, it it thought to activate the Mek1/ERK kinase pathway, which in turn phosphorylates c-Fos. This assumption was supported by the fact that transfection of constitutively active Mek1 stabilizes c-Fos, whereas expression of a protein phosphatase known to inactivate the ERK kinases inhibited stabilization of c-Fos by c-Mos (Okazaki and Sagata, 1995). The ERK kinases are mitogen-activated protein kinases that play a role in entry of quiescent cells into the cell cycle in response to mitogenic stimuli (reviewed in Ruderman, 1993). It is therefore possible that phosphorylation of the C-terminal region of c-Fos by the Mek1/ERK pathway is part of the physiological regulation of c-Fos stability by mitogenic stimuli (Fig. 3).

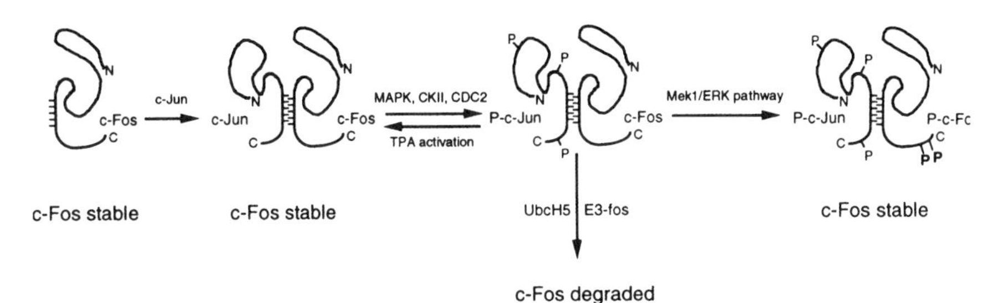

Figure 3. Summary of the known requirements for and inhibitors of c-Fos degradation. Note that the physiological relevance of the indicated factors for c-Fos degradation has not been proven. Note also that the actual phosphorylation sites on c-Jun are unknown.

4.4. Gcn4

Translation of Gcn4 is induced under conditions of starvation for purines or amino acids (reviewed in Hinnebusch, 1992). The mechanism of this regulation centers on modification of the translation initiation factor eIF2 by the kinase Gcn2. The site-specific phosphorylation of eIF2 both lowers the rate of translation initiation of most mRNAs and increases *GCN4* mRNA translation up to 10-fold. In addition to this translational regulation, starvation has the effect of extending the half-life of the Gcn4 protein from 5 min to 20–30 min (Kornitzer *et al.*, 1994). Starvation conditions used included shifting leucine or tryptophan auxotrophs, or a histidine bradytroph to media lacking these amino acids. Shifting a temperature-sensitive aminoacyl-tRNA synthetase mutant to the restrictive temperature also stabilizes Gcn4, suggesting that it is an increase in free tRNA concentration in the cytosol, rather than the decrease in available amino acids, that constitutes the proximal signal for Gcn4 stabilization (Fig. 4). The mechanism by which Gcn4 degradation is inhibited is still unknown, but the fact that it is unaffected by mutations in Gcn2 suggests that it is distinct from the mechanism responsible for Gcn4 translation regulation.

4.5. *MAT*α2

An interesting feature of the α2 repressor is that it functions in two distinct kinds of multimeric complexes (Fig. 1). In α cells, an α2 dimer associates with the general factors Mcm1, Ssn6, and Tup1 to form a repressor complex capable of specific recognition of **a** cell-specific gene operators. In **a**/α diploid cells, this same complex is formed, but, in addition, α2 binds to the **a**1 homeodomain protein forming a heterodimer that associates with Ssn6 and Tup1 to create a repressor complex that specifically recognizes a novel set of DNA sites, those regulating the haploid cell-specific genes. Most of the early work on α2 degradation had been

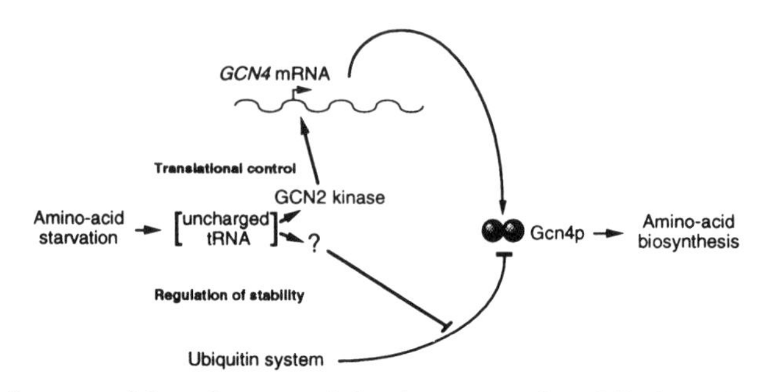

Figure 4. Summary of the pathways regulating the concentration of Gcn4 protein in response to amino acid starvation.

done in α haploid cells, where the protein has a half-life of ~ 4 min. Surprisingly, it was recently found that the α2 protein is far more stable in **a**/α diploid cells (P. Johnson and M. Hochstrasser, unpublished data). The increase in α2 metabolic stability depends on the presence of the **a**1 polypeptide but not on cell ploidy. Genetic experiments indicate that it is the physical association of **a**1 with α2 *per se*, rather than the genetic program engendered by the **a**1/α2-containing complex, that causes the increase in α2 half-life.

Careful analysis of α2 turnover in **a**/α cells revealed biphasic degradation kinetics, with rapid degradation detected early in the chase at a rate approaching that observed in α cells and a much slower second phase in which α2 half-life increases to well over 90 min. This is consistent with the presence of two populations of α2 molecules that do not exchange very rapidly: a short-lived population of complexes containing the α2 homodimer and a long-lived population containing the **a**1/α2 heterodimer. Three distinct helical segments of the α2 protein mediate its association with **a**1. The N-terminal-most helix overlaps the sequence element in the *Deg1* degradation signal that had been implicated in *Deg1*-mediated turnover (see earlier section). A simple way to account for the regulation of α2 stability by **a**1 binding would be to propose that **a**1 interaction with α2 blocks access to the *Deg1* signal (and possibly additional signals) by the ubiquitination machinery. Consistent with this model, mutations in the N-terminal region of α2 that cause defects in **a**1 binding have stronger defects in **a**1-dependent stabilization of α2 than does a mutation in the C-terminal interaction domain.

Remarkably, the **a**1 repressor behaves in an exactly analogous fashion. The protein is very short-lived in **a** cells (half-life <3 min) but is dramatically stabilized in **a**/α cells. Mutations in α2 that interfere with **a**1 binding result in much more rapid **a**1 turnover. Collectively, the data on **a**1 and α2 degradation suggest a simple but elegant means of regulating protein degradation by cell type-specific protein–protein interactions. It would not be surprising to see this kind of control with other transcription factors, many of which form multiple homomeric and heteromeric protein complexes.

Other instances of physiological regulation of transcription factor degradation are found in *E. coli*, where two sigma factors, σ^{32} and σ^{S}, which are normally very unstable, are stabilized by various stress conditions (Strauss *et al.*, 1987; Tilly *et al.*, 1989) or by entry into stationary phase (Muffler *et al.*, 1996), respectively. Although *E. coli* lacks a ubiquitin system, the proteases involved in degradation of these transcription factors do have similarities with the proteasome. The σ^{32} factor is degraded by HflB/FtsH, an apparent metalloproteinase with an AAA-type ATPase domain; this domain is found in the six known 26 S proteasome ATPase regulatory subunits (Tomoyasu *et al.*, 1995). The σ^{S} protein is degraded by ClpPX (Schweder *et al.*, 1996), a protease with structural and functional similarities with the proteasome (Kessel *et al.*, 1995). RssB, an apparent member of a two-component regulator system, was shown to be involved in regulation of σ^{S} degradation (Muffler *et al.*, 1996). It is not yet clear whether the similarities between the degradation systems in prokaryotes and eukaryotes extend beyond the proteases to the degradation signals and regulatory pathways.

5. CONCLUDING REMARKS

A number of transcription factors have been found to be short-lived. Rapid degradation is a hallmark of regulatory proteins both because it enables the rapid modulation of the concentration of the protein following changes in its rate of synthesis and because changes in the rate of degradation can serve as an additional level of control. Therefore, it is probable that with the examination of the stability of additional transcription factors, many will be found to be rapidly degraded.

In this chapter, we concentrated on those transcription factors where good evidence exists for their degradation by the ubiquitin system. A number of other transcription factors have been shown to be rapidly degraded, e.g., c-Myc, c-Myb (Luscher and Eisenman, 1988), and Ftz (Kellerman *et al.*, 1990), but very little or nothing is known about the mechanisms by which they are degraded. It should also be pointed out that even in instances where good evidence exists for involvement of the ubiquitin system, ubiquitin-independent degradation as an alternative mechanism cannot be excluded [as was seen in the case of c-Jun (Jariel-Encontre *et al.*, 1995; Hirai *et al.*, 1991)].

Degradation signals in transcription factors, and in unstable proteins in general, are still incompletely defined. There are probably a number of different classes of such signals. An additional complication is that many degradation signals are likely to depend on secondary and tertiary structural features of the protein, which contribute only indirectly to formation of the signal. Careful dissection of multiple degradation signals will be needed to characterize the protein motifs recognized by the ubiquitination machinery.

PEST sequences may represent one of the classes of degradation signals. However, the exact nature of PEST sequences remains to be clarified; it does not seem likely that a high proportion of PEST residues is enough by itself to confer the specificity necessary for protein–protein interaction. Mutational analyses support the notion of sequence-specific attributes within PEST sequences being necessary for proteolytic targeting (Kornitzer *et al.*, 1994; Yaglom *et al.*, 1995). Presence of phosphorylation sites within PEST sequences might constitute one of the specificity determinants (Yaglom *et al.*, 1995).

Some of the problems specific to degradation of transcription factors have yet to be addressed. For one, where does degradation occur? Transcription factors are synthesized in the cytosol, and act in the nucleus; some transcription factors are known to shuttle in and out of the nucleus. Components of the ubiquitin–proteasome pathway are present in both of these compartments, but the specific enzymes required for degradation of a particular protein might be confined to only one of them. In this case, the substrate protein would be shielded from degradation in the other compartment. Another question, as transcription factors are often bound to DNA and/or are found in large transcription complexes, is whether the factors become more resistant (or more sensitive) in these states to ubiquitination/degradation than when freely floating in the cytoplasm. The first data on this point are emerging for α2 degradation in yeast, but this question needs to be more broadly investigated. The effect of DNA binding on proteolysis has begun to be addressed in the case of p53, where the DNA-bound form of p53 was found to be protected from E6-mediated ubiquitination (Molinari and Milner, 1995). It will be interesting to see whether DNA binding will as a rule protect transcription factors from degradation. Whether the incorporation of such factors into transcription complexes affects their degradation has not been examined in detail. In general, however, it is clear that presence in a complex does not necessarily protect a protein from degradation, as IκB is degraded even though it is bound to NF-κB, and as MATα2-βgal is efficiently degraded even when found in a complex with stable subunits (Hochstrasser and Varshavsky, 1990).

6. REFERENCES

Alkalay, I., Yaron, A., Hatzubai, A., Orian, A., Ciechanover, A., and Ben-Neriah, Y., 1995, Stimulation-dependent IκBα phosphorylation marks the NF-κB inhibitor for degradation via the ubiquitin-proteasome pathway, *Proc. Natl. Acad. Sci. USA* **92:**10599–10603.

Angel, P., and Karin, M., 1991, The role of Jun, Fos and the AP-1 complex in cell proliferation and transformation, *Biochim. Biophys. Acta* **1072:**129–157.

Baldi, L., Brown, K., Franzoso, G., and Siebenlist, U., 1996, Critical role for lysines 21 and 22 in signal-induced, ubiquitin-mediated proteolysis of IκB-α, *J. Biol. Chem* **271:**376–379.

Biederer, T., Volkwein, C., and Sommer, T., 1996, Degradation of subunits of the Sec61p complex, an integral component of the ER membrane, by the ubiquitin-proteasome pathway, *EMBO J.* **15:** 2069–2076.

Brown, K., Gerstberger, S., Carlson, L., Franzoso, G., and Siebenlist, U., 1995, Control of IκB-α proteolysis by site-specific, signal-induced phosphorylation, *Science* **267:**1485–1488.

Chau, V., Tobias, J. W., Bachmair, A., Marriott, D., Ecker, D. J., Gonda, D. K., and Varshavsky, A., 1989, A multiubiquitin chain is confined to specific lysine in a targeted short-lived protein, *Science* **243:**1576–1583.

Chen, P., and Hochstrasser, M., 1995, Biogenesis, structure and function of the yeast 20S proteasome, *EMBO J.* **14:**2620–2630.

Chen, P., and Hochstrasser, M., 1996, Autocatalytic subunit processing couples active site formation in the 20S proteasome to completion of assembly, *Cell* **86:**961–972.

Chen, P., Johnson, P., Sommer, T., Jentsch, S., and Hochstrasser, M., 1993, Multiple ubiquitin-conjugating enzymes participate in the in vivo degradation of the yeast Matα2 repressor, *Cell* **74:** 357–369.

Chen, Z., Hagler, J., Palombella, V. J., Melandri, F., Scherer, D., Ballard, D., and Maniatis, T., 1995, Signal-induced site-specific phosphorylation targets IkBα to the ubiquitin-proteasome pathway, *Genes Dev.* **9:**1586–1597.

Chen, Z. J., Parent, L., and Maniatis, T., 1996, Site-specific phosphorylation of IκBα by a novel ubiquitination-dependent protein kinase activity, *Cell* **84:**853–862.

Chowdary, D. R., Dermody, J. J., Jha, K. J., and Ozer, H. L., 1994, Accumulation of p53 in a mutant cell line defective in the ubiquitin pathway, *Mol. Cell. Biol.* **14:**1997–2003.

Ciechanover, A., Hod, Y., and Hershko, A., 1978, A heat-stable polypeptide component of an ATP-dependent proteolytic system from reticulocytes, *Biochem. Biophys. Res. Commun.* **81:**1100–1105.

Curran, T., Miller, A. D., Zokas, L., and Verma, I. M., 1984, Viral and cellular *fos* proteins: A comparative analysis, *Cell* **36:**259–268.

Davis, R. L., Weintraub, H., and Lassar, A. B., 1987, Expression of a single transfected cDNA converts fibroblasts to myoblasts, *Cell* **51:**987–1000.

DeMarini, D. J., Papa, F. R., Swaminathan, S., Ursic, D., Rasmussen, T. P., Culbertson, M. R., and Hochstrasser, M., 1995, The yeast *SEN3* gene encodes a regulatory subunit of the 26S proteasome complex required for ubiquitin-dependent protein degradation in vivo, *Mol. Cell. Biol.* **15:**6311–6321.

Ellison, M. J., and Hochstrasser, M., 1991, Epitope-tagged ubiquitin. A new probe for analyzing ubiquitin function, *J. Biol. Chem.* **266:**21150–21157.

Fenteany, G., Standaert, R. F., Lane, W. S., Choi, S., Corey, E. J., and Schreiber, S. L., 1995, Inhibition of proteasome activities and subunit-specific amino-terminal threonine modification by lactacystin, *Science* **268:**726–731.

Finley, D., Sadis, S., Monia, B. P., Boucher, P., Ecker, D., Crooke, S. T., and Chau, V., 1994, Inhibition of proteolysis and cell cycle progression in a multiubiquitination-deficient yeast mutant, *Mol. Cell. Biol.* **14:**5501–5509.

Gill, G., and Ptashne, M., 1988, Negative effect of the transcriptional activator GAL4, *Nature* **334:** 721–724.

Gonen, H., Stancovski, I., Shkedy, D., Hadari, T., Bercovich, B., Bengal, E., Mesilati, S., Abu-Hatoum, O., Schwartz, A. L., and Ciechanover, A., 1996, Isolation, characterization, and partial purification of a novel ubiquitin-protein ligase, E3, *J. Biol. Chem.* **271:**302–310.

Gottesman, S., Gottesman, M. E., Shaw, J. E., and Pearson, M. L., 1981, Protein degradation in *E. coli*: The *lon* mutation and bacteriophage λN and cII protein stability, *Cell* **24:**225–235.

Hershko, A., Ciechanover, A., Heller, H., Haas, A. L., and Rose, I. A., 1980, Proposed role of ATP in protein breakdown: Conjugation of proteins with multiple chains of the polypeptide of ATP-dependent proteolysis, *Proc. Natl. Acad. Sci. USA* **77:**1365–1368.

Hershko, A., Leshinsky, E., Ganoth, D., and Heller, H., 1984, ATP-dependent degradation of ubiquitin–protein conjugates, *Proc. Natl. Acad. Sci. USA* **81:**1619–1623.

Hershko, A., Ganoth, D., Pehrson, J., Palazzo, R. E., and Cohen, L. H., 1991, Methylated ubiquitin inhibits cyclin degradation in clam embryo extracts, *J. Biol. Chem.* **266:**16376–16379.

Herskowitz, I., Rine, J., and Strathern, J., 1992, Mating-type determination and mating-type interconversion in *Saccharomyces cerevisiae*, in *The Molecular and Cellular Biology of the Yeast Saccharomyces cerevisiae*, Vol. 2 (E. W. Jones, J. R. Pringle, and J. R. Broach, eds.), Cold Spring Harbor Laboratory Press, Cold Spring Harbor, NY. pp. 583–656.

Hiller, M. M., Finger, A., Schweiger, M., and Wolf, D. H., 1996, ER degradation of a misfolded luminal protein by the cytosolic ubiquitin–proteasome pathway, *Science* **273:**1725–1728.

Hinnebusch, A. G., 1992, General and pathway-specific regulatory mechanisms controlling the synthesis of amino acid biosynthetic enzymes in *Saccharomyces cerevisiae*, in *The Molecular and Cellular Biology of the Yeast Saccharomyces*, Vol. 2 (E. W. Jones, J. R. Pringle, and J. R. Broach, eds.), Cold Spring Harbor Laboratory Press, Cold Spring Harbor, NY. pp. 319–414.

Hirai, S., Kawasaki, H., Yaniv, M., and Suzuki, K., 1991, Degradation of transcription factors, c-Jun and c-Fos, by calpain, *FEBS Lett.* **287:**57–61.

Hochstrasser, M., and Varshavsky, A., 1990, In vivo degradation of a transcriptional regulator: The yeast α2 repressor, *Cell* **61:**697–708.

Hochstrasser, M., Ellison, M. J., Chau, V., and Varshavsky, A., 1991, The short-lived MATα2 transcriptional regulator is ubiquitinated in vivo, *Proc. Natl. Acad. Sci. USA* **88:**4606–4610.

Jaffray, E., Wood, K. M., and Hay, R. T., 1995, Domain organization of IκBα and sites of interaction with NF-κB p65, *Mol. Cell. Biol.* **15:**2166–2172.

Jariel-Encontre, I., Pariat, M., Martin, F., Carillo, S., Salvat, C., and Piechaczyk, M., 1995, Ubiquitinylation is not an absolute requirement for degradation of c-Jun protein by the 26S proteasome, *J. Biol. Chem.* **270:**11623–11627.

Kellerman, K. A., Mattson, D. M., and Duncan, I., 1990, Mutations affecting the stability of the *fushi tarazu* protein of *Drosophila*, *Genes Dev.* **4:**1936–1950.

Kessel, M., Maurizi, M. R., Kim, B., Kocsis, E., Trus, B. L., Singh, S. K., and Steven, A. C., 1995, Homology in structural organization between E. coli ClpAP protease and the eukaryotic 26 S proteasome, *J. Mol. Biol.* **250:**587–594.

Kornitzer, D., Raboy, B., Kulka, R. G., and Fink, G. R., 1994, Regulated degradation of the transcription factor Gcn4, *EMBO J.* **13:**6021–6030.

Kulka, R. G., Raboy, B., Schuster, R., Parag, H. A., Diamond, G., Ciechanover, A., and Marcus, M., 1988, A Chinese hamster cell cycle mutant arrested at G2 phase has a temperature-sensitive ubiquitin-activating enzyme El, *J. Biol. Chem.* **263:**15726–15731.

Lanker, S., Valdivieso, M. H., and Wittenberg, C., 1996, Rapid degradation of the G1 cyclin Cln2 induced by CDK-dependent phosphorylation, *Science* **271:**1597–1601.

Lassar, A. B., Buskin, J. N., Lockshon, D., Davis, R. L., Apone, S., Hauschka, S. D., and Weintraub, H., 1989, MyoD is a sequence-specific DNA binding protein requiring a region of myc homology to bind to the muscle creatine kinase enhancer, *Cell* **58:**823–831.

Luscher, B., and Eisenman, R. N., 1988, c-myc and c-myb protein degradation: Effect of metabolic inhibitors and heat shock, *Mol. Cell. Biol.* **8:**2504–2512.

McKinney, J. D., Chang, F., Heintz, N., and Cross, F. R., 1993, Negative regulation of *FAR1* at the start of the yeast cell cycle, *Genes Dev.* **7:**833–843.

Miao, G. G., and Curran, T., 1994, Cell transformation by *c-fos* requires an extended period of expression and is independent of the cell cycle, *Mol. Cell. Biol.* **14:**4295–4310.

Molinari, M., and Milner, J., 1995, p53 in complex with DNA is resistant to ubiquitin-dependent proteolysis in the presence of HPV-16 E6, *Oncogene* **10:**1849–1854.

Muffler, A., Fischer, D., Altuvia, S., Storz, G., and Hengge-Aronis, R., 1996, The response regulator RssB controls stability of the σ^S subunit of RNA polymerase in *Escherichia coli*, *EMBO J.* **15:**1333–1339.

Murakami, Y., Matsufuji, S., Kameji, T., Hayashi, S., Igarashi, K., Tamura, T., Tanaka, K., and Ichihara, A., 1992, Ornithine decarboxylase is degraded by the 26S proteasome without ubiquitination, *Nature* **360:**597–599.

Murre, C., McCaw, P. S., and Baltimore, D., 1989, A new DNA binding and dimerization motif in immunoglobulin enhancer binding, *daughterless*, *MyoD* and *myc* proteins, *Cell* **56:**777–783.

Nishizawa, M., Okazaki, K., Furuno, N., Watanabe, N., and Sagata, N., 1992, The 'second-codon rule' and autophosphorylation govern the stability and activity of Mos during the meiotic cell cycle in *Xenopus* oocytes, *EMBO J.* **11:**2433–2446.

Nishizawa, M., Furuno, N., Okazaki, K., Tanaka, H., Ogawa, Y., and Sagata, N., 1993, Degradation of Mos by the N-terminal proline (Pro2)-dependent ubiquitin pathway on fertilization of *Xenopus* eggs: Possible significance of natural selection for Pro2 in Mos, *EMBO J.* **12:**4021–4027.

Okazaki, K., and Sagata, N., 1995, The Mos/MAP kinase pathway stabilizes c-Fos by phosphorylation and augments its transforming activity in NIH 3T3 cells, *EMBO J.* **14:**5048–5059.

Papa, F., and Hochstrasser, M., 1993, The yeast *DOA4* gene encodes a deubiquitinating enzyme related to a product of the human *tre-2* oncogene, *Nature* **366:**313–319.

Papavassiliou, A. G., Treier, M., Chavrier, C., and Bohmann, D., 1992, Targeted degradation of c-Fos, but not v-Fos, by a phosphorylation-dependent signal on c-Jun, *Science* **258:**1941–1944.

Rechsteiner, M., 1987, Ubiquitin-mediated pathways for intracellular proteolysis, *Annu. Rev. Cell Biol.* **3:**1–30.

Reihsaus, E., Kohler, M., Kraiss, S., Oren, M., and Montenarh, M., 1990, Regulation of the level of the oncoprotein p53 in non-transformed and transformed cells, *Oncogene* **5:**137–145.

Richter-Ruoff, B., Wolf, D. H., and Hochstrasser, M., 1994, Degradation of the yeast MATα2 transcriptional regulator is mediated by the proteasome, *FEBS Lett.* **354:**50–52.

Rodriguez, M. S., Michalopoulos, I., Arenzana-Seisdedos, F., and Hay, R. T., 1995, Inducible degradation of IκBα in vitro and in vivo requires the acidic C-terminal domain of the protein, *Mol. Cell. Biol.* **15:**2413–2419.

Rogers, S., Wells, R., and Rechsteiner, M., 1986, Amino acid sequences common to rapidly degraded proteins: The PEST hypothesis, *Science* **234:**364–368.

Ruderman, J. V., 1993, MAP kinase and the activation of quiescent cells, *Curr. Opin. Cell Biol.* **5:** 207–213.

Sadis, S., Atienza, C. J., and Finley, D., 1995, Synthetic signals for ubiquitin-dependent proteolysis, *Mol. Cell. Biol.* **15:**2086–4094.

Scheffner, M., Werness, B. A., Huibregtse, J. M., Levine, A. J., and Howley, P. M., 1990, The E6 oncoprotein encoded by human papillomavirus types 16 and 18 promotes the degradation of p53, *Cell* **63:**1129–1136.

Scheffner, M., Huibregtse, J. M., Vierstra, R. D., and Howley, P. M., 1993, The HPV-16 E6 and E6-AP complex functions as a ubiquitin-protein ligase in the ubiquitination of p53, *Cell* **75:**495–505.

Scherer, D. C., Brockman, J. A., Chen, Z., Maniatis, T., and Ballard, D. W., 1995, Signal-induced degradation of IκBα requires site-specific ubiquitination, *Proc. Natl. Acad. Sci. USA* **92:**11259–11263.

Schweder, T., Lee, K.-H., Lomovskaya, O., and Matin, A., 1996, Regulation of *Escherichia coli* starvation sigma factor (σS) by ClpPX protease, *J. Bacteriol.* **178:**470–476.

Schwob, E., Böhm, T., Mendenhall, M. D., and Nasmyth, K., 1994, The B-type cyclin kinase inhibitor p40-SIC1 controls the G1 to S transition in S. cerevisiae, *Cell* **79:**233–244.

Stancovski, I., Gonen, H., Orian, A., Schwartz, A. L., and Ciechanover, A., 1995, Degradation of the proto-oncogene product c-Fos by the ubiquitin proteolytic system in vivo and in vitro: Identification and characterization of the conjugating enzymes, *Mol. Cell. Biol.* **15:**7106–7116.

Strauss, D. B., Walter, W. A., and Gross, C. A., 1987, The heat shock response of *E. coli* is regulated by changes in the concentration of sigma 32, *Nature* **329:**348–351.

Struhl, K., 1987, The DNA-binding domains of the jun oncoprotein and the yeast GCN4 transcriptional activator protein are functionally homologous, *Cell* **50:**841–846.

Thanos, D., and Maniatis, T., 1995, NF-κB: A lesson in family values, *Cell* **80:**529–532.
Thayer, M. J., Tapscott, S. J., Davis, R. L., Wright, W. E., Lassar, A. B., and Weintraub, H., 1989, Positive autoregulation of the myogenic determination gene MyoD1, *Cell* **58:**241–248.
Tilly, K., Spence, J., and Georgopoulos, C., 1989, Modulation of stability of the Escherichia coli heat shock regulatory factor sigma, *J. Bacteriol.* **171:**1585–1589.
Tomoyasu, T., Gamer, J., Bukau, B., Kanemori, M., Mori, H., Rutman, A. J., Oppenheim, A. B., Yura, T., Yamanaka, K., Niki, H., Hiraga, S., and Ogura, T., 1995, Escherichia coli FtsH is a membrane-bound, ATP-dependent protease which degrades the heat-shock transcription factor σ^{32}, *EMBO J.* **14:**2551–2560.
Treier, M., Staszewski, L. M., and Bohmann, D., 1994, Ubiquitin-dependent c-jun degradation in vivo is mediated by the δ domain, *Cell* **78:**787–798.
Tsurumi, C., Ishida, N., Tamura, T., Kakizuka, A., Nishida, E., Okumura, E., Kishimoto, T., Inagaki, M., Okazaki, K., Sagata, N., Ichihara, A., and Tanaka, K., 1995, Degradation of c-Fos by the 26S proteasome is accelerated by c-Jun and multiple protein kinases, *Mol. Cell. Biol.* **15:**5682–5687.
Yaglom, J., Linskens, H. K., Sadis, S., Rubin, D. M., Futcher, B., and Finley, D., 1995, p34-Cdc28-mediated control of Cln3 degradation, *Mol. Cell. Biol.* **15:**731–741.

CHAPTER 10

Role of the Ubiquitin–Proteasome Pathway in NF-κB Activation

Zhijian J. Chen and Tom Maniatis

1. INTRODUCTION

The transcription factor NF-κB (nuclear factor κB) is a heterodimeric protein that plays a pivotal role in immune and inflammatory responses (for recent reviews see Verma *et al.*, 1995; Finco and Baldwin, 1995; Thanos and Maniatis, 1995; Siebenlist *et al.*, 1994; Baeuerle and Henkel, 1994). Both subunits of NF-κB (p50 and p65) are members of the Rel family of transcriptional activator proteins, but they differ in their modes of synthesis. The p65 subunit of NF-κB is produced as a mature protein, whereas the p50 subunit is generated by the proteolytic processing of a 105-kDa precursor protein (p105). The N-termini of p105 and p65 are homologous; both contain the Rel homology region, which includes protein dimerization and DNA binding domains, and a nuclear localization signal (Fig. 1). In addition, the p65, but not the p50, subunit contains a transcriptional activation domain. The C-terminus of p105 is characterized by the presence of a sequence motif known as ankyrin repeats, and a PEST domain (Ghosh *et al.*, 1990). Ankyrin repeats constitute a protein interaction domain (Gilligan and Bennett, 1993), and the PEST domain is associated with rapidly degraded proteins (Rechsteiner and

Zhijian J. Chen • ProScript, Inc., Cambridge, Massachusetts 02139. **Tom Maniatis** • Department of Molecular and Cellular Biology, Harvard University, Cambridge, Massachusetts 02138.

Ubiquitin and the Biology of the Cell, edited by Peters *et al.* Plenum Press, New York, 1998.

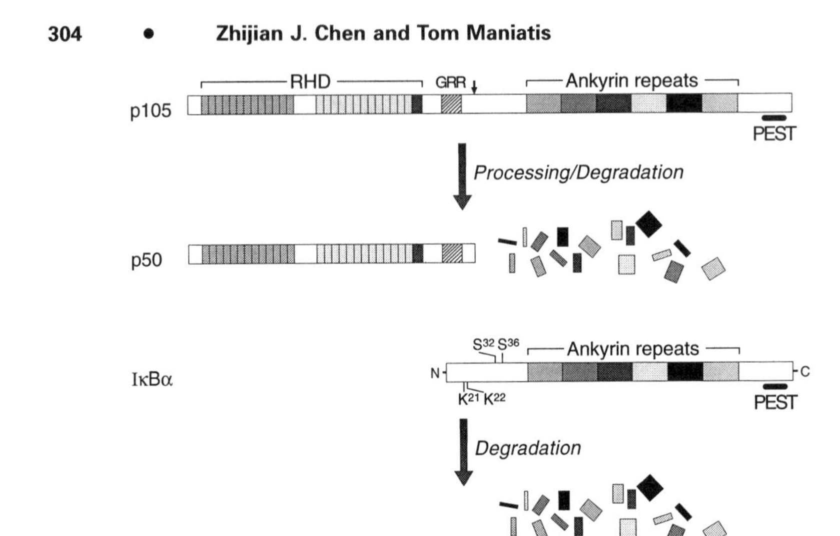

Figure 1. (Top) Processing of p105 to p50. The p105 protein is comprised of a highly conserved 300-amino-acid region known as the Rel-homology domain (RHD) at its N-terminus and ankyrin repeats at its C-terminus. The RHD is composed of two subdomains (dark hatched area and light hatched area). Although only the C-terminal domain of RHD supports protein dimerization, the entire RHD contributes to DNA binding. A nuclear localization sequence is also present within RHD (small black box). In the middle of the molecule is a glycine-rich region (GRR), which directs the cleavage of p105 to generate the p50 subunit of NF-κB. Following cleavage, the C-terminus of p105 is rapidly degraded. The ubiquitin–proteasome pathway is required for the processing of p105 to p50. (Bottom) Degradation of IκBα. IκBα has a tripartite structure with an N-terminal region, a central region containing six ankyrin repeats, and a C-terminal region enriched in PEST sequences. The N-terminus of IκBα contains two highly conserved serine residues, S32 and S36, which are phosphorylated on stimulation of cells. Phosphorylation at these two sites targets IκBα for ubiquitination at two specific lysine residues, K21 and K22. Following ubiquitination, IκBα is rapidly degraded by the 26 S proteasome.

Rogers, 1996). Proteolytic processing of p105 involves the degradation of the C-terminus containing the ankyrin repeats and PEST domain, leaving the 50-kDa N-terminal Rel homology region intact (Fig. 1).

NF-κB was originally thought to be a B-cell-specific transcription factor (Sen and Baltimore, 1986), but was subsequently shown to be present in virtually all cell types, sequestered in the cytoplasm either as a p50/p65 heterodimer in a complex with a member of the inhibitor κB (IκB) family of inhibitor proteins (Baeuerle and Baltimore, 1988) or as a p105/p65 heterodimer (Naumann *et al.*, 1993; Rice *et al.*, 1992) (Fig. 2). The most thoroughly characterized member of the IκB family is IκBα, which forms a stable complex with NF-κB, thus blocking its nuclear localization signal. As shown in Fig. 1, IκBα contains ankyrin repeats and a PEST domain and is therefore strikingly similar to the C-terminus of p105. In fact, in certain cell types the C-terminus of p105 is expressed as a separate NF-κB

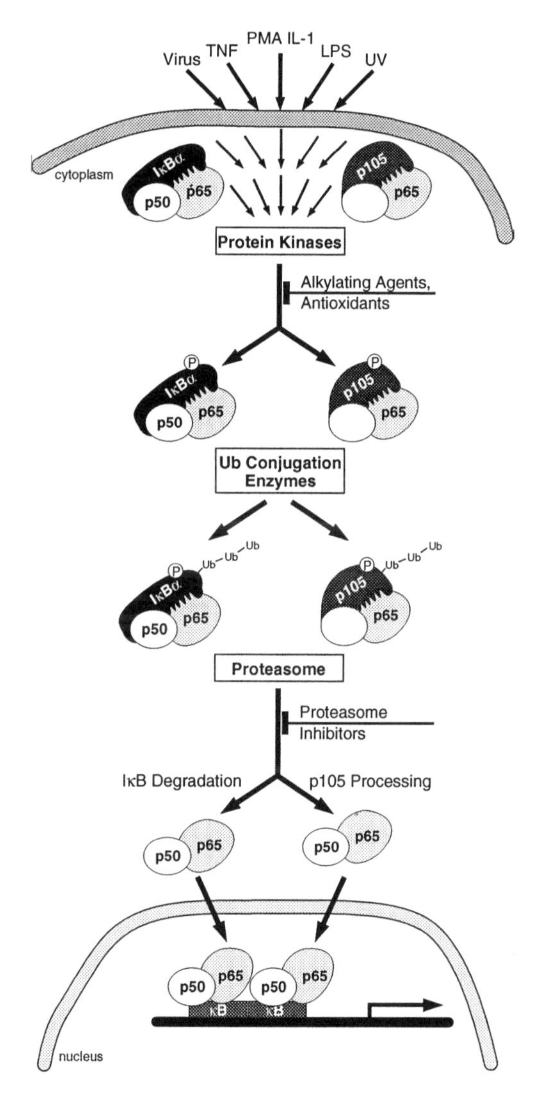

Figure 2. NF-κB activation pathway. In quiescent cells, NF-κB is sequestered in the cytoplasm either as a ternary complex of p50/p65 bound to IκBα or as a binary complex of p105 and p65. After stimulation with various NF-κB stimuli, IκBα is rapidly phosphorylated by protein kinase(s). This signal-induced phosphorylation leads to site-specific ubiquitination of IκBα and its subsequent degradation by the 26 S proteasome. The degradation of IκBα releases NF-κB, which translocates to the nucleus to turn on the transcription of its target genes. Agents that prevent the phosphorylation (e.g., alkylating agents and antioxidants) or degradation of IκBα (e.g., proteasome inhibitors) have been shown to block NF-κB activation. A similar pathway for signal-induced processing of p105 to p50 may also exist, and this may contribute to the activation of NF-κB. However, direct evidence for this possibility has not been reported.

inhibitor protein known as IκBγ (Grumont and Gerondakis, 1994; Inoue *et al.*, 1992). Other Rel family members are capable of forming heterodimers with p105, p50, or p65 and with themselves, and other members of the IκB family of inhibitors can bind to Rel family heterodimers (reviewed by Siebenlist *et al.*, 1994). In this chapter we focus entirely on NF-κB (p50/p65) and IκBα, as their behavior reflects principles that are thought to be shared by the members of both families of proteins.

The quiescent forms of NF-κB can be rapidly activated by a remarkably large number of extracellular signals, including a number of cytokines such as tumor necrosis factor α (TNF-α) and interleukin 1 (IL-1), processes such as antigen-dependent T-cell activation, viral or bacterial infection, and stresses such as UV irradiation and reactive oxygen (Fig. 2). The endpoint of all of these signaling pathways is the phosphorylation of p105 and IκB, and the processing of the former and degradation of the latter. Although the relationship between p105 phosphorylation and processing has not been clearly demonstrated, phosphorylation of IκBα is a prerequisite for its rapid degradation. In any case, the consequence of p105 processing and IκBα degradation is the nuclear translocation of NF-κB and the transcriptional activation of genes that are targeted by this transcription factor.

During the past few years, considerable progress has been made in understanding the mechanisms involved in p105 processing and the signal-dependent degradation of IκBα. The most unexpected finding was that the ubiquitin–proteasome pathway is required for both processes. Evidence supporting this finding is the subject of this review.

2. ROLE OF THE UBIQUITIN–PROTEASOME PATHWAY IN p105 PROCESSING

The first hint that the ubiquitin–proteasome pathway might be involved in the activation of NF-κB was provided by studies of p105 processing (Fan and Maniatis, 1991). *In vivo* pulse–chase experiments established that p50 was generated by proteolytic processing of p105, and deletion studies revealed that the C-terminal portion of p105 including part of the ankyrin repeats is dispensable for this processing. To investigate the mechanism of p105 processing, an *in vitro* system was established. Truncated p105 (p60) was produced by *in vitro* translation in wheat germ extracts. When the labeled protein was incubated in HeLa S100 extracts (supernatant of 100,000*g* spin), mature p50 was produced. Surprisingly, this *in vitro* processing required ATP, suggesting that p105 might be processed by the ATP-dependent ubiquitin–proteasome pathway (Fan and Maniatis, 1991).

In vitro reconstitution experiments definitively demonstrated that ubiquitin and the purified 26 S proteasome are required for p105 processing (Palombella *et al.*, 1994). When the truncated p105 precursor (p60) was incubated in extracts depleted of proteasomes, a ladder of bands corresponding to multiubiquitinated proteins could be observed by gel electrophoresis. The number and intensity of these bands increased on the addition of excess ubiquitin. Two types of experiments provided evidence that the proteasome is required for p105 processing *in vivo* (Palombella *et al.*, 1994). First, peptide-aldehyde inhibitors shown to be selective for the proteasome (Rock *et al.*, 1994) blocked p105 processing. This conclusion was recently confirmed by the observation that lactacystin, a highly specific proteasome inhibitor that acts by covalently binding to a subunit of the proteasome (Fenteany *et al.*, 1995), also inhibits p105 processing *in vitro* and *in vivo* (J. Hagler *et al.*, unpublished). Second, when full-length p105 was expressed in yeast, functional p50 was produced. Most importantly, this processing was blocked in yeast carrying mutations in the proteasome (Palombella *et al.*, 1994) or in other genes involved in the ubiquitin–proteasome pathway (C. Sears *et al.*, unpublished).

In an effort to identify activities required to ubiquitinate p105 *in vitro*, Orian *et al.* (1995) showed that the ubiquitin carrier protein E2-F1 or UbcH5 can reconstitute the p105 ubiquitination reaction *in vitro*, but another E2 (E2-14K) was not active in this assay. These studies also identified a novel ubiquitin-protein ligase (E3) capable of reconstituting the *in vitro* p105 ubiquitination activity. This 320-kDa protein is distinct from E6-AP, the E3 required for the ubiquitination of p53 *in vitro* (Scheffner *et al.*, 1993).

Recently, significant progress has been made in identifying the amino acid sequences in p105 required for its processing. Lin and Ghosh (1996) showed that a glycine-rich region (GRR) located near the C-terminus of p50 is both necessary and sufficient for p105 processing *in vivo*. Specifically, they showed that deletion of the GRR prevents p105 processing. Furthermore, when the GRR was inserted into a heterologous protein, it directed a cleavage downstream of the insertion site. Interestingly, the fate of the C-terminal fragment depends on the protein into which the GRR is inserted. In the case of p105, the C-terminus is degraded so rapidly that it cannot be detected. By contrast, when the C-terminus of p105 was replaced by IκBα, both p50 and IκBα could be detected following processing *in vivo*. Although the reasons for these differences are not understood, the fact that the C-terminus can be detected in at least one case clearly indicates that GRR-dependent cleavage can be uncoupled from the degradation of the C-terminus under some circumstances. Thus, it is possible that p105 processing is a two-step mechanism: GRR-dependent cleavage followed by proteolysis of the C-terminus. It is important to note, however, that p105 processing *in vitro* does not occur in the absence of ubiquitin or the 26 S proteasome, and cleavage is not observed *in vivo* in the presence of inhibitors of the proteasome (Palombella *et al.*, 1994). A Ub–proteasome-dependent endoproteolytic cleavage is clearly unprecedented and difficult to reconcile with the structure of the proteasome where the active sites of the proteases are buried within the narrow channel passing through the core of the proteasome. Clearly, more work is necessary to explain this interesting phenomenon.

Another unresolved issue is the regulation of p105 processing. As previously mentioned, p105 is phosphorylated in response to inducers of NF-κB (Naumann and Scheidereit, 1994; Mellits *et al.*, 1993), and there is some evidence that this results in an increase in the rate of p105 processing (Donald *et al.*, 1995; Mercurio *et al.*, 1993; Mellits *et al.*, 1993). However, significant levels of p50 are produced in the absence of NF-κB induction, and the kinetics of the increase in p105 processing is slow relative to the signal-dependent degradation of IκBα (Donald *et al.*, 1995; Mellits *et al.*, 1993). In addition, p105 is processed in extracts derived from uninduced mammalian cells, and is processed constitutively in yeast (Palombella *et al.*, 1994). In Jurkat T cells the rate of p105 processing is very low, and the increase in p50 production on induction with phorbol 12-myristate 13-acetate (PMA) requires *de novo* protein synthesis (Harhaj *et al.*, 1996). In addition, excess Rel proteins appear to decrease the level of p105 processing. Thus, it is possible that much of the p50 that accumulates in cells is generated by p105 processing prior to its association with a Rel partner. An additional twist was provided by the observation that lipopolysaccharide (LPS) induction of a monocytic cell line in the presence of cycloheximide resulted in the degradation of p105 without an increase in p50 (Harhaj *et al.*, 1996). On the basis of these observations the authors proposed that p105 processing occurs constitutively with newly synthesized protein, and that induction results in the degradation, not the processing, of preexisting p105. However, additional studies will be required to establish the validity and generality of this hypothesis.

Recently, Fujimoto *et al.* (1995) demonstrated that p105 can be phosphorylated *in vitro* with a cyclin-dependent kinase (CDK), and that the sites of phosphorylation map to the C-terminus of p105. Single amino acid substitutions at serine residues 894 and 908 prevented this phosphorylation *in vitro* and *in vivo*, and led to an increase of the steady-state levels of p105 *in vivo*. It is not clear whether this increase is related to a decrease in the turnover or processing of p105. For example, it is possible that phosphorylation of the C-terminus of p105 simply increases its rate of turnover without affecting its processing in the absence of signaling. Moreover, the kinase responsible for p105 phosphorylation *in vivo* has not been identified, nor have the sites of signal-dependent phosphorylation been mapped.

At present, the function of p105 processing remains a mystery. One hint is provided by the observation that, unlike p65, p50 homodimers are present in the nucleus prior to induction, and these molecules appear to function as transcriptional repressors rather than activators (Kang *et al.*, 1992). Thus, p105 processing may be a mechanism for the constitutive production of p50 homodimers, which may repress NF-κB-dependent gene expression in the absence of inducer. However, it was recently suggested that p105 processing may be involved in the activation of NF-κB in some cell types (Baldassarre *et al.*, 1995). In this study, it was found that p105 was complexed with p65 in the nucleus of Epstein–Barr virus-immortalized B cells. When these cells were treated with mitomycin C, NF-κB was activated as a result of p105 processing in the nucleus. The steady-state protein levels of IκBα and p105 in the cytoplasm were not affected by mitomycin C treatment. Therefore, the nuclear processing of p105 could directly contribute to NF-κB activation in response to some signaling molecules. Whatever its function, the production of a p105-like molecule is conserved in evolution, as indicated by the recent identification of a p105 homologue in *Drosophila* (Dushay *et al.*, 1996). This protein, called Relish, contains a Rel homology region at its N-terminus, and ankyrin repeats and a PEST domain at its C-terminus. It is not known whether this protein is processed to generate a p50-like protein.

3. SIGNAL-DEPENDENT PHOSPHORYLATION AND DEGRADATION OF IκBα

Compared with p105 processing, a much clearer relationship has been established between the signal-dependent phosphorylation of IκBα and its inactivation. The first suggestion that phosphorylation may be involved in the activation of NF-κB was provided by the observation that the latent DNA binding activity of NF-κB could be activated *in vitro* by purified protein kinases such as protein kinase C, protein kinase A, and a heme-regulated kinase that phosphorylates the translation initiation factor eIF-2 (Ghosh and Baltimore, 1990; Shirakawa and Mizel, 1989). The target of phosphorylation was shown to be IκB, and the phosphorylated IκB could no longer inhibit the DNA binding activity of NF-κB. Based on this observation, it was proposed that NF-κB is activated via phosphorylation and dissociation of IκB. This model was supported by the observation that IκBα is indeed phosphorylated *in vivo* following stimulation with agents that activate NF-κB (Beg *et al.*, 1993; Cordle *et al.*, 1993; Mellits *et al.*, 1993). Subsequently, however, several groups showed that IκBα was rapidly degraded following stimulation with NF-κB inducers, and the kinetics of IκBα degradation correlated with the activation of NF-κB (Beg *et al.*, 1993; Brown *et al.*, 1993; Cordle *et al.*, 1993; Henkel *et al.*, 1993; Mellits *et al.*, 1993). To accommodate this new finding, the original phosphorylation/dissociation model was modified. In this model, phosphorylation of IκBα triggers its dissociation from NF-κB, and the free IκBα is then rapidly degraded. This proposal was supported by the observation that free IκBα is much more labile than IκBα bound to NF-κB (Scott *et al.*, 1993; Sun *et al.*, 1993; Rice and Ernst, 1993).

According to this model, phosphorylation of IκBα is solely responsible for the activation of NF-κB, and IκBα degradation is a secondary event following its phosphorylation and dissociation. If this were correct, one would expect that inhibition of IκBα degradation would not affect the activation of NF-κB. However, serine protease inhibitors [e.g., tosyl-Phe-chloromethylketone (TPCK) or *N*α-*p*-tosyl-L-lysine-chloromethylketone (TLCK)], which inhibit the degradation of IκBα, were shown to block NF-κB activation (Chiao *et al.*, 1994; Mellits *et al.*, 1993; Henkel *et al.*, 1993). Unfortunately, interpretation of these results was complicated by the finding that the serine protease inhibitors used in those studies actually prevent the inducible phosphorylation of IκBα rather than its degradation (Alkalay *et al.*, 1995a; Finco *et al.*, 1994; Miyamoto *et al.*, 1994). This inhibition by the serine protease inhibitors could be either a nonspecific effect or a result of inhibiting an unidentified upstream protease required for the activation of a kinase. In any case, there was no direct evidence that proteolysis of IκBα was required for the activation of NF-κB.

The role of IκBα degradation in the activation of NF-κB was established by the use of a different class of protease inhibitors, namely, the peptide aldehyde inhibitors of proteasome (Palombella *et al.*, 1994; Traenckner *et al.*, 1994). Unlike TPCK or other alkylating agents, the proteasome inhibitors did not block the inducible phosphorylation of IκBα. On the contrary, these inhibitors blocked the degradation of IκBα and led to the accumulation of a hyperphosphorylated form of IκBα. More importantly, the proteasome inhibitors prevented the activation of NF-κB by a variety of inducing agents. This inhibitory effect on NF-κB is readily explained by the data showing that phosphorylated IκBα remains associated with NF-κB (Brockman *et al.*, 1995; Brown *et al.*, 1995; DiDonato *et al.*, 1995; Lin *et al.*, 1995; Palombella *et al.*, 1994; Traenckner *et al.*, 1994; Miyamoto *et al.*, 1994). Therefore, IκBα degradation is an obligatory step in the activation of NF-κB.

The peptide aldehyde inhibitors of the proteasome also inhibit other proteases such as calpains. Although the rank order potency of these inhibitors toward the proteasome correlates very well with their inhibitory effect on IκBα degradation, more specific proteasome inhibitors were required to definitively rule out the possibility that other proteases might be involved in IκBα degradation. Recently, two classes of highly specific proteasome inhibitors became available: lactacystin (Fenteany *et al.*, 1995) and peptide boronics (Adams and Stein, 1996). Both inhibitors, which have no inhibitory effects on any other proteases tested so far, potently inhibit NF-κB activation via stabilization of the phosphorylated form of IκBα (Z. Chen *et al.*, manuscript in preparation; J. Hagler *et al.*, manuscript in preparation). These inhibitors also specifically inhibit NF-κB-dependent expression of the human interleukin-2 gene without interfering with the function of other transcription factors such as Fos/Jun, NFAT, or Oct-1 (Z. Chen *et al.*, manuscript in preparation). These studies with all three classes of proteasome inhibitors definitively establish that the proteasome mediates the signal-dependent degradation of IκBα.

Although the inducible phosphorylation of IκBα is not sufficient for NF-κB activation, it is required. When the signaling pathway leading to the phosphorylation of IκBα is blocked by agents such as the antioxidant pyrrolidinedithiocarbamate (PDTC) or alkylating agents (e.g., TPCK), NF-κB activation is prevented (Mellits *et al.*, 1993; Henkel *et al.*, 1993). Even more compelling evidence supporting the importance of inducible phosphorylation was provided by recent studies of the N-terminal phosphorylation sites of IκBα. When two conserved serine residues, S32 and S36, were mutated to alanine or glycine, the signal-induced phosphorylation of IκBα was abolished (Brockman *et al.*, 1995; Brown *et al.*, 1995; see also DiDonato *et al.*, 1996; Sun *et al.*, 1996; Traenckner *et al.*, 1995; Whiteside *et al.*, 1995). That S32 and S36 are indeed the inducible phosphorylation sites has been confirmed by direct two-dimensional peptide mapping (DiDonato *et al.*, 1996). Importantly, the phosphorylation-defective mutants could not be degraded following stimulation, and they function as dominant-negative mutants in preventing the activation of NF-κB by multiple stimuli (Sun *et al.*, 1996; Brockman *et al.*, 1995; Brown *et al.*, 1995; Traenckner *et al.*, 1995). Therefore, signal-induced phosphorylation of IκBα is a prerequisite for its degradation.

Although phosphorylation of IκBα is necessary for its degradation, it is not sufficient (Sun *et al.*, 1996; Aoki *et al.*, 1996; Brown *et al.*, 1995; Rodriguez *et al.*, 1995; Whiteside *et al.*, 1995). The IκBα molecule can be divided into three domains: The N-terminal domain, the ankyrin repeats, and C-terminal PEST domain (Fig. 2). As discussed above, deletion of the N-terminus or amino acid substitutions at positions S32 and S36 prevent phosphorylation of IκBα and block its signal-dependent degradation. These N-terminal mutations do not prevent the binding of IκBα to NF-κB. In contrast, mutations in the ankyrin repeats abolish the binding of IκBα to NF-κB. As to the function of the C-terminal PEST domain of IκBα, two studies found that deletion of this region had no effect on its ability to associate with NF-κB, nor did it prevent the signal-dependent degradation of IκBα (Sun *et al.*, 1996; Aoki *et al.*, 1996). In other studies this region was found to be required for signal-dependent degradation (Brown *et al.*, 1995; Whiteside *et al.*, 1995; Rodriguez *et al.*, 1995).

In the first two studies, human IκBα mutants were examined in human cells, whereas in two of the latter studies, human IκBα mutants were studied in mouse cells (Brown *et al.*, 1995; Whiteside *et al.*, 1995). The study by Rodriguez *et al.* (1995) used the IκBα mutant lacking the sixth ankyrin repeat, which is necessary for association with NF-κB. Thus, the discrepant results may be explained by differences in the strength of interaction between IκB mutants and NF-κB in different studies. Assuming that the results with the human assay systems are correct, the PEST domain does not appear to be involved in the signal-dependent degradation of IκBα. Rather, this region may determine the half-life of IκBα in unstimulated cells (Sun *et al.*, 1996; Aoki *et al.*, 1996). When a short region that includes the sixth ankyrin repeat is deleted, IκBα no longer associates with NF-κB but is still phosphorylated in a signal-dependent manner (Sun *et al.*, 1996; Aoki *et al.*, 1996). However, this mutant protein is not degraded in a signal-dependent manner. Thus, the binding of IκBα to NF-κB is not required for signal-dependent phosphorylation, but may be required for additional steps in the pathway to degradation.

The signal-induced degradation of IκBα is very rapid, usually with a half-life of less than 10 min following stimulation. Concomitant with the degradation of IκBα, NF-κB is activated. The activation of NF-κB leads to the resynthesis of IκBα as a result of the presence of NF-κB binding sites in the promoter of IκBα (Chiao *et al.*, 1994; Sun *et al.*, 1993). The newly synthesized IκBα not only prevents nuclear translocation of NF-κB, but also enter the nucleus to remove active NF-κB from DNA and terminate the transcription of NF-κB-responsive genes (Arenzana-Seisdedos *et al.*, 1995). The NF-κB/IκB complex is then either transported back to the cytoplasm or degraded within the nucleus. This autoregulatory mechanism ensures the transient activation of NF-κB.

4. SIGNAL-DEPENDENT UBIQUITINATION OF IκBα

The observation that proteasome inhibitors stabilize the phosphorylated form of IκBα suggested but did not prove that phosphorylated IκBα is a target for degradation by the ubiquitin–proteasome pathway. The first direct evidence that ubiquitination is required for IκBα degradation was recently reported (Chen *et al.*, 1995; Alkalay *et al.*, 1995b). When Jurkat T cells were treated with TNF-α or the phosphatase inhibitor calyculin A in the presence of proteasome inhibitors, a ladder of high-molecular-weight bands corresponding to multiubiquitinated IκBα could be observed (Chen *et al.*, 1995). Using HeLa cell cytoplasmic extracts supplemented with the phosphatase inhibitor okadaic acid, which is known to stimulate NF-κB *in vivo*, phosphorylation and ubiquitination of IκBα were observed *in vitro*. When a panel of IκBα mutants was examined in this *in vitro* assay, it was found that mutations that abolish the phosphorylation and degradation of IκBα *in vivo* also prevent ubiquitination *in vitro* (Chen *et al.*, 1995). Alkalay *et al.* (1995b) reported that the signal-induced degradation of IκBα is diminished in cells where a thermolabile E1 is inactivated at nonpermissive temperature. More recently, Roff *et al.* (1996) and DiDonato *et al.* (1996) showed that the S32A/S36A mutant of IκBα is defective in ubiquitination *in vivo*.

Following *in vitro* ubiquitination, IκBα remains associated with NF-κB, and ubiquitinated IκBα is selectively degraded by the 26 S proteasome (Chen *et al.*, 1995). Whether ubiquitinated IκBα is associated with NF-κB *in vivo* is less clear, as ubiquitinated IκBα in stimulated cells cannot be coprecipitated using antibodies against NF-κB (Roff *et al.*, 1996). It is possible that ubiquitination *per se* does not dissociate IκBα from NF-κB. Rather, a protein with chaperonin activity could release ubiquitinated IκBα from NF-κB *in vivo*.

Studies of the sites of ubiquitination on IκBα have identified two highly conserved lysine residues, K21 and K22, that are required for signal-induced degradation of IκBα *in vivo* (Scherer *et al.*, 1995; see also Baldi *et al.*, 1996; DiDonato *et al.*, 1996; Rodriguez *et al.*, 1996). This double mutant is competent for phosphorylation, but its multiubiquitination is markedly reduced, presumably because of the lack of ubiquitination sites. Single point mutations on K21 or K22 do not impair the degradation of ubiquitination of the mutant protein, suggesting that either lysine residue can serve as a ubiquitination site or, alternatively, the nearby lysine residues can substitute for one another to accept a multiubiquitin chain. It has been difficult to locate a specific ubiquitination site in several other substrates, apparently because the ubiquitination machinery may utilize alternative lysines when the principal ubiquitination sites are disrupted (Ciechanover, 1994). In the case of IκBα, site-specific ubiquitination may result from the binding of NF-κB to IκBα, which "protects" distal regions of IκBα and exposes the N-terminal region so that K21 and K22 are sterically accessible to the ubiquitination machinery.

5. ENZYMES INVOLVED IN THE PHOSPHORYLATION AND UBIQUITINATION OF IκBα

5.1. Ubiquitination Enzymes

Biochemical fractionation efforts have begun to reveal the enzymes involved in the ubiquitination of IκBα. Although the E3 responsible for the signal-induced ubiquitination of IκBα remains to be identified, the E2 that ubiquitinates IκBα *in vitro* has been shown to be a member of the UBC4/UBC5 family (Chen *et al.*, 1996; Alkalay *et al.*, 1995b). This family of E2s has been shown to be involved in the ubiquitination of a variety of proteins including p53, cyclins, p105, and the yeast transcription factor Matα2. Deletion of the genes encoding these E2s in yeast results in defective turnover of many short-lived and abnormal proteins, and the mutant cells are unable to cope with stress responses. Given the increasingly large number of E2s in the UBC4/5 family (Nuber *et al.*, 1996), it is necessary to identify the specific E2 in this family that catalyzes the ubiquitination of IκBα. In addition, it is important to determine whether these E2s are indeed involved in the ubiquitination of IκBα *in vivo*.

5.2. A Ubiquitination-Dependent IκB Kinase

Mutants of IκBα lacking serines 32 and 36 are resistant to induced phosphorylation by a variety of stimuli, suggesting that different signal transduction pathways converge on a specific kinase or kinases. The identity of IκB kinase has been a subject of intense investigation since the discovery of IκB. Several serine/threonine kinases, including protein kinase C (Ghosh and Baltimore, 1990), protein kinase A (Shirakawa and Mizel, 1989), and casein kinase II (McElhinny *et al.*, 1996; Barroga *et al.*, 1995; Kuno *et al.*, 1995), have been shown to phosphorylate IκBα *in vitro*. However, none of these kinases phosphorylate IκBα at serines 32 and 36, which is a major criterion for a bona fide IκBα kinase. Various kinases have also been implicated in the regulation of NF-κB *in vivo*, such as protein kinase C (Diaz-Meco *et al.*, 1994), double-stranded RNA-dependent protein kinase (Kumar *et al.*, 1994), ceramide-dependent protein kinase (Schutze *et al.*, 1992), tyrosine kinases (Devary *et al.*, 1993), Raf (Finco and Baldwin, 1993; Li and Sedivy, 1993), and pelle (Shelton and Wasserman, 1993). However, these kinases probably function at various steps upstream of IκBα phosphorylation, for none of these kinases have been shown to directly phosphorylate IκBα (or the *Drosophila* IκB homologue Cactus) at relevant sites.

Recently, a protein kinase capable of phosphorylating IκBα at serines 32 and 36 was identified (Chen *et al.*, 1996). Surprisingly, the activity of this large (~ 700 kDa), multisubunit kinase complex requires the ubiquitin-activating enzyme (E1), a specific E2 of the UBC4/UBC5 family, and ubiquitin. Kinetic studies suggest that this IκBα kinase is activated by a prior ubiquitination event (Fig. 3). Ubiquitination, not Ub-dependent degradation, activates the kinase, as proteolytic activity is not required for IκBα phosphorylation in the partially purified system. Furthermore, proteasome inhibitors do not inhibit the phosphorylation of IκBα *in*

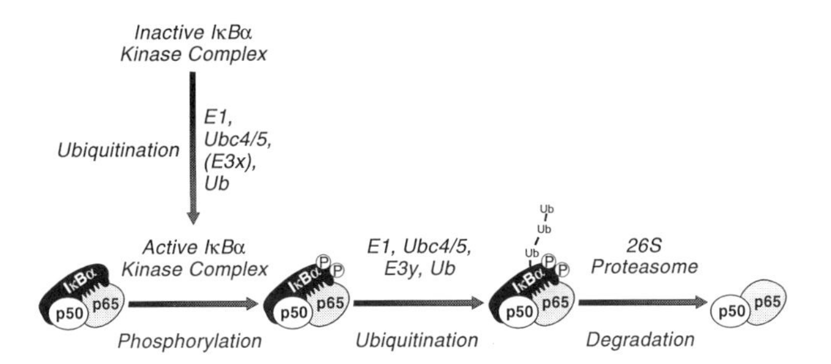

Figure 3. Ubiquitin-dependent phosphorylation and degradation of IκBα. The large IκBα kinase complex is activated by a ubiquitination event involving E1, and E2 of the UBC4/UBC5 family, ubiquitin, and an E3 that may be part of the kinase complex (E3x). The activated kinase phosphorylates IκBα at serines 32 and 36, leading to its ubiquitination at lysines 21 and 22. This second ubiquitination step also involves an E2 of the UBC4/UBC5 family, but the E3 in this case (E3y) may be distinct from E3x, as the IκBα kinase could catalyze the phosphorylation but not ubiquitination of IκBα in the presence of E1, UBC4, and ubiquitin. The ubiquitinated IκBα remains associated with NF-κB (p50/p65) but is selectively degraded by the 26 S proteasome. Therefore, there is a two-step ubiquitination requirement that eventually leads to the activation of NF-κB.

vivo or *in vitro*. Therefore, ubiquitination directly regulates the enzymatic activities of the IκBα kinase complex. Ubiquitination could induce a conformational change in the kinase complex to activate the kinase activity. Alternatively, ubiquitination may inactivate an inhibitor of IκBα kinase, leading to the activation of the kinase activity. The answers to these questions await identification of the ubiquitinated subunit on the IκBα kinase complex.

Although the same E2, UBC4/5, is involved in both phosphorylation and ubiquitination of IκBα, the E3s involved in these two processes appear to be distinct, as partially purified IκBα kinase is able to phosphorylate but unable to ubiquitinate IκBα in the presence of El, UBC4/5, and ubiquitin. It is unlikely that ubiquitination of IκBα kinase is regulated at the E2 level, as all known E2s have constitutive enzymatic activity. On the other hand, it is possible that the E3 component of the IκBα kinase complex is regulated. One precedent of regulated E3 activity is provided by cyclosome, a 20 S complex that is involved in the ubiquitination of mitotic cyclins (see Chapter 12).

The discovery of a ubiquitination-dependent IκBα kinase adds another layer of complexity to the NF-κB signal transduction pathway. Although much is known about the fate of IκBα following its phosphorylation, little is known about the connection between the phosphorylation of IκBα and the upstream signaling events. For example, activation of NF-κB by TNF-α originates from the binding of TNF-α to its cognate receptor, which elicits a cascade of events including the recruitment of the TRADD, TRAF2, and RIP proteins to a signaling complex (Hsu *et al.*, 1996), generation of the second messenger ceramide which activates ceramide-dependent kinase (Schutze *et al.*, 1992), activation of the JNK pathway (Meyer *et al.*, 1996; Hirano *et al.*, 1996), and production of reactive oxygen intermediates (Schmidt *et al.*, 1995; Schreck *et al.*, 1991). The exact sequence of these events and their relationship to the ubiquitination of IκBα kinase remain to be determined. One possibility is that a component of IκBα kinase is phosphorylated by one of the upstream kinases (e.g., JNK), and this phosphorylation step leads to its ubiquitination and subsequent activation of the IκB kinase activity. Consistent with this possibility, conditions that allow detection of IκBα kinase activity in HeLa cell extracts also result in elevated levels of JNK activity (F. Lee, personal communication).

6. BASAL PHOSPHORYLATION AND DEGRADATION OF IκBα

Besides inducible phosphorylation, IκBα also undergoes basal phosphorylation independent of stimulatory signals. Basal phosphorylation of IκBα occurs within the terminal PEST domain of IκBα, and is carried out by casein kinase II *in vitro* (McElhinny *et al.*, 1996; Barroga *et al.*, 1995; Kuno *et al.*, 1995). Basal phosphorylation is not required for induced phosphorylation or degradation of IκBα, but it may enhance basal degradation of free IκBα (Schwarz *et al.*, 1996). In unstimulated cells, IκBα is a labile protein that is continuously turning over, and free IκBα is much more unstable than IκBα bound to NF-κB. This signal-independent degradation of free IκBα is important to ensure that essentially all IκBα is complexed to NF-κB to maintain its inactive state in unstimulated cells. The degradation of free IκBα may also be carried out by the ubiquitin–proteasome pathway, as it can be ubiquitinated in an *in vitro* system consisting of E1, UBC4, and the human papillomavirus type 16 E6 associated protein (E6-AP; Z. Chen, unpublished data). Interestingly, when IκBα is complexed to a Rel protein, its ubiquitination is suppressed, consistent with the *in vivo* observation that bound IκBα is more stable than the free form.

7. SIGNAL-INDUCED DEGRADATION OF OTHER IκBα PROTEINS

7.1. IκBβ

Like IκBα, IκBβ is also degraded on stimulation of cells by some agents such as LPS, interleukin 1β, and the Tax protein of the type 1 human T-cell leukemia virus (HTLV-1; McKinsey *et al.*, 1996; DiDonato *et al.*, 1996; Thompson *et al.*, 1995). The degradation of IκBβ is distinct from that of IκBα in three respects: (1) The kinetics of IκBβ degradation is much slower; (2) only a small subset of stimulatory agents causes IκBβ degradation; and (3) there is no re-synthesis of IκBβ following its degradation, probably because of the lack of NF-κB-responsive sites in the promoter of IκBβ. Based on these observations, it was postulated that the degradation of IκBβ may account for the persistent activation of NF-κB by some stimuli such as LPS and Tax.

The N-terminus of IκBβ contains two conserved serine residues (S19 and S23) that correspond to the serine residues present in IκBα. Mutation of these two serines abolishes the signal-induced degradation of IκBβ (McKinsey *et al.*, 1996; DiDonato *et al.*, 1996), suggesting that induced phosphorylation of IκBβ is important for its degradation. The degradation of IκBβ appears to be carried out by the ubiquitin–proteasome pathway, as IκBβ is stablized by the specific proteasome inhibitor lactacystin (McKinsey *et al.*, 1996). However, ubiquitination of IκBβ has yet to be directly demonstrated. In addition, it remains to be seen whether the IκBα kinase also phosphorylates IκBβ in a ubiquitination-dependent manner.

7.2. Cactus

Another IκB protein that undergoes signal-induced degradation is the *Drosophila* IκB homologue Cactus (Belvin *et al.*, 1995). The *Drosophila* NF-κB homologue Dorsal is sequestered in the cytoplasm in early embryos by virtue of its association with Cactus. A signal transduction pathway consisting of 12 known genes is selectively activated in the ventral region of the embryo leading to the localized nuclear translocation of Dorsal. This in turn activates genes required for normal dorsal–ventral patterning of the early embryo (Geisler *et al.*, 1992). A number of parallels exist between IκB and Cactus. First, activation of the dorsal–ventral signaling pathway leads to the rapid degradation of Cactus (Belvin *et al.*, 1995). Second, signal-dependent degradation of Cactus requires the N-terminus and ankyrin repeat region, but not the C-terminal PEST domain. Mutants in the latter region affect the turnover of Cactus in the absence of signal, but have no effect on signal-dependent degradation. Third, mutants that stabilize Cactus block dorsal–ventral signaling (Belvin *et al.*, 1995). The only potential difference between the two systems is that Cactus does not appear to require Dorsal for its signal-dependent degradation (Belvin *et al.*, 1995). However, it is possible that Cactus associates with a yet to be identified *Drosophila* Rel family member, in the absence of Dorsal. Additional studies will be required to determine whether signal-dependent degradation of Cactus requires the ubiquitin–proteasome pathway, and whether this degradation depends on association of a Rel family dimer.

8 CONCLUSIONS AND PERSPECTIVES

The ubiquitin–proteasome pathway is known to be involved in many biological processes such as the stress response, cell cycle control, DNA repair, and receptor endocytosis. Studies of NF-κB regulation clearly demonstrate that this pathway also actively participates in signal transduction and transcriptional regulation. Ubiquitination is involved in at least three important steps in NF-κB activation: the processing of p105 to p50, the activation of IκB kinase, and the degradation of IκBα (and possibly other IκB proteins). The processing of p105 provides the first example of limited proteolysis by the proteasome, whereas the ubiquitination-dependent IκB kinase reveals a novel regulatory function of ubiquitination without involving proteolysis. Owing to rapid progress in the past few years, we can now record two critical steps that lead to the activation of NF-κB: Signal-induced phosphorylation of IκBα at serines 32 and 36 targets this inhibitor for ubiquitination at lysines 21 and 22; and the ubiquitinated IκBα is rapidly degraded by the 26 S proteasome, thus allowing NF-κB to enter the nucleus to turn on its target genes.

However, much remains to be learned about the mechanism that governs the phosphorylation, ubiquitination, and degradation of IκB proteins. How is IκBα kinase regulated by the upstream signaling events? What is the mechanism by which ubiquitination activates the IκBα kinase activity? What is the E3 responsible for the signal-dependent ubiquitination of IκBα? What is the structural basis for the specificity of the IκBα E3 that allows it to distinguish between phosphorylated and nonphosphorylated IκBα? Finally, how is the phosphorylation of IκBα coupled to its ubiquitination and subsequent degradation? In the areas of p105 processing, there are also some outstanding issues. What is the mechanism for the limited proteolysis of p105 by the proteasome? In addition to the constitutive processing of p105 to p50, is there a signal-dependent processing that contributes to NF-κB activation? What is the endoprotease that recognizes the glycine-rich region (GRR)? How is the GRR-directed cleavage coupled to the subsequent degradation by the proteasome? The next few years will witness an exciting period when the answers to these questions are uncovered and, as always, surprises are not unexpected.

9. REFERENCES

Adams, J., and Stein, R., 1996, Novel inhibitors of the proteasome and their therapeutic use in inflammation, *Ann. Rep. Med. Chem.*, in press.

Alkalay, I., Yaron, A., Hatzubai, A., Jung, S., Avraham, A., Gerlitz, O., Pashut-Lavon, I., and Ben-Neriah, Y., 1995a, In vivo stimulation of IκB phosphorylation is not sufficient to activate NF-κB, *Mol. Cell. Biol.* **15:**1294–1301.

Alkalay, I., Yaron, A., Hatzubai, A., Orian, A., Ciechanover, A., and Ben-Neriah, Y., 1995b, Stimulation-dependent IκBα phosphorylation marks the NF-κB inhibitor for degradation via the ubiquitin-proteasome pathway, *Proc. Natl. Acad. Sci. USA* **92:**10599–10603.

Aoki, T., Sano, Y., Yamamoto, T., and Inoue, J.-I., 1996, The ankyrin repeats but not the PEST-like sequences are required for signal-dependent degradation of IκBα, *Oncogene* **12:**1159–1164.

Arenzana-Seisdedos, F., Thompson, J., Rodriguez, M. S., Bachelerie, F., Thomas, D., and Hay, R. T., 1995, Inducible nuclear expression of newly synthesized IκBα negatively regulates DNA-binding and transcriptional activities of NF-κB, *Mol. Cell. Biol.* **15:**2689–2696.

Baeuerle, P. A., and Baltimore, D., 1988, Activation of DNA-binding activity in an apparently cytoplasmic precursor of the NF-κB transcription factor, *Cell* **53:**211–217.

Baeuerle, P. A., and Henkel, T., 1994, Function and activation of NF-κB in the immune system, *Annu. Rev. Immunol.* **12:**141–179.

Baldassarre, F., Mallardo, M., Mezza, E., Scala, G., and Quinto, I., 1995, Regulation of NF-κB through the nuclear processing of p105 (NF-κB1) in Epstein–Barr virus-immortalized B cell lines, *J. Biol. Chem.* **270:**31244–31248.

Baldi, L., Brown, K., Franzoso, G., and Siebenlist, U., 1996, Critical role for lysines 21 and 22 in signal-induced, ubiquitin-mediated proteolysis of IκBα, *J. Biol. Chem.* **271:**376–379.

Barroga, C. F., Stevenson, J. K., Schwarz, E. M., and Verma, I. M., 1995, Constitutive phosphorylation of IκBα by casein kinase II, *Proc. Natl. Acad. Sci. USA* **92:**7637–7641.

Beg, A. A., Finco, T. S., Nantermet, P. V., and Baldwin, A. S., Jr., 1993, Tumor necrosis factor and interleukin-1 lead to phosphorylation and loss of IκB-α: A mechanism for NF-κB activation, *Mol. Cell. Biol.* **13:**3301–3310.

Belvin, M. P., Jin, Y., and Anderson, K. V., 1995, Cactus protein degradation mediates *Drosophila* dorsal–ventral signaling, *Genes Dev.* **9:**783–793.

Brockman, J. A., Scherer, D. C., McKinsey, T. A., Hall, S. M., Qi, X., Lee, W. Y., and Ballard, D. W., 1995, Coupling of a signal response domain in IκBα to multiple pathways for NF-κB activation, *Mol. Cell. Biol.* **15:**2809–2818.

Brown, K., Park, S., Kanno, T., Franzoso, G., and Siebenlist, U., 1993, Mutual regulation of the transcriptional activator NF-κB and its inhibitor, IκBα, *Proc. Natl. Acad. Sci. USA* **90:**2532–2536.

Brown, K., Gerstberger, S., Carlson, L., Franzoso, G., and Siebenlist, U., 1995, Control of IκBα proteolysis by site-specific, signal-induced phosphorylation, *Science* **267:**1485–1491.

Chen, Z. J., Hagler, J., Palombella, V. J., Melandri, F., Scherer, D., Ballard, D., and Maniatis, T., 1995, Signal-induced site-specific phosphorylation targets IκBα to the ubiquitin-proteasome pathway, *Genes Dev.* **9:**1586–1597.

Chen, Z. J., Parent, L., and Maniatis, T., 1996, Site-specific phosphorylation of IκBα by a novel ubiquitination-dependent protein kinase activity, *Cell* **84:**853–862.

Chiao, P. J., Miyamoto, S., and Verma, I. M., 1994, Autoregulation of IκBα activity, *Proc. Natl. Acad. Sci. USA* **91:**22–32.

Ciechanover, A., 1994, The ubiquitin-proteasome proteolytic pathway, *Cell* **79:**13–21.

Cordle, S. R., Donald, R., Read, M. A., and Hawiger, J., 1993, Lipopolysaccharide induces phosphorylation of MAD-3 and activation of c-rel and related NF-κB proteins in human monocytic THP-1 cells, *J. Biol. Chem.* **268:**11803–11810.

Devary, Y., Rosette, C., DiDonato, J. A., and Karin, M., 1993, NF-κB activation by ultraviolet light not dependent on a nuclear signal, *Science* **261:**1442–1445.

Diaz-Meco, M. T., Dominguez, L. S., Dent, P., Lozano, J., Municio, M. M., Berra, E., Hay, R. T., Sturgill, T. W., and Moscat, J., 1994, ζPKC induces phosphorylation and inactivation of IκB-α in vitro, *EMBO J.* **13:**2842–2848.

DiDonato, J. A., Mercurio, F., and Karin, M., 1995, Phosphorylation of IκBα precedes but is not sufficient for its dissociation from NF-κB, *Mol. Cell. Biol.* **15:**1302–1311.

DiDonato, J. A., Mercurio, F., Rosette, C., Wu-Li, J., Suyang, H., Ghosh, S., and Karin, M., 1996, Mapping of the inducible IκB phosphorylation sites that signal its ubiquitination and degradation, *Mol. Cell. Biol.* **16:**1295–1304.

Donald R., Ballard, D. W., and Hawiger, J., 1995, Proteolytic processing of NF-κB/IκB in human monocytes, *J. Biol. Chem.* **270:**9–12.

Dushay, M. S., Asling, B., and Hultmark, D., 1996, Origins of immunity: *Relish*, a compound Rel-like gene in the antibacterial defense of *Drosophila*, *Proc. Natl. Acad. Sci. USA*, **93:**10343–10347.

Fan, C.-M., and Maniatis, T., 1991, Generation of p50 subunit of NF-κB by processing of p105 through an ATP-dependent pathway, *Nature* **354:**395–398.

Fenteany, G., Standaert, R. F., Lane, W. S., Choi, S., Corey, E. J., and Schreiber, S. L., 1995, Inhibition of proteasome activities and subunit-specific amino-terminal threonine modification by lactacystin, *Science* **268:**726–731.

Finco, T. S., and Baldwin, A. S., Jr., 1993, κB site-dependent induction of gene expression by diverse inducers of nuclear factor κB requires Raf-1, *J. Biol. Chem.* **268:**17676–17679.

Finco, T. S., and Baldwin, A. S., Jr., 1995, Mechanistic aspects of NF-κB regulation: The emerging role of phosphorylation and proteolysis, *Immunity* **3:**263–272.

Finco, T. S., Beg, A. A., and Baldwin, A. S., Jr., 1994, Inducible phosphorylation of IκBα is not sufficient for its dissociation from NF-κB and is inhibited by protease inhibitors, *Proc. Natl. Acad. Sci. USA* **91:**11884–11888.

Fujimoto, K., Yasuda, H., Sato, Y., and Yamamoto, K., 1995, A role for phosphorylation in the proteolytic processing of human NF-κB1 precursor, *Gene* **165:**183–189.

Geisler, R., Bergmann, A., Hiromi, Y., and Nusslein-Volhard, C., 1992, *Cactus*, a gene involved in dorsoventral pattern formation of Drosophila, is related to the IκB gene family of vertebrates, *Cell* **71:**613–621.

Ghosh, S., and Baltimore, D., 1990, Activation in vitro of NF-κB by phosphorylation of its inhibitor IκB, *Nature* **344:**678–682.

Ghosh, S., Gifford, A. M., Riviere, L. R., Tempst, P., Nolan, G. P., and Baltimore, D., 1990, Cloning of the p50 DNA binding subunit of NF-κB: Homology to rel and dorsal, *Cell* **62:**1019–1029.

Gilligan, D., and Bennett, V., 1993, The spectrin-based membrane skeleton and micron-scale organization of the plasmid membrane, *Annu. Rev. Cell Biol.* **9**:27–66.

Grumont, R. J., and Gerondakis, S., 1994, Alternative splicing of RNA transcripts encoded by the murine p105 NF-κB gene generates IκBα isoforms with different inhibitory activities, *Proc. Natl. Acad. Sci. USA* **91**:4367–4371.

Harhaj, E. W., Maggirwar, S. B., and Sun, S. C., 1996, Inhibition of p105 processing by NF-κB proteins in transiently transfected cells, *Oncogene* **12**:2385–2392.

Henkel, T., Machleidt, T., Alkalay, I., Kronke, M., Ben-Neriah, Y., and Baeuerle, P. A., 1993, Rapid proteolysis of IκB-α is necessary in the activation of transcription factor NF-κB, *Nature* **365**: 182–185.

Hirano, M., Osada, S.-I., Aoki, T., Hirai, S.-I., Hosaka, M., Inoue, J.-I., and Ohno, S., 1996, MEK kinase is involved in tumor necrosis factor α-induced NF-κB activation and degradation of IκBα, *J. Biol. Chem.* **271**:13234–13238.

Hsu, H., Huang, J., Shu, H.-B., Baichwal, V., and Goeddel, D., 1996, TNF-dependent recruitment of the protein kinase RIP to the TNF receptor-1 signaling complex, *Immunity* **4**:387–396.

Inoue, J., Kerr, L. D., Kakizuka, A., and Verma, I. M., 1992, IκBγ, a 70 kD protein identical to the C-terminal half of p110 NF-κB: A new member of the IκB family, *Cell* **68**:1109–1120.

Kang, S.-M., Tran, A.-C., Grilli, M., and Lenardo, M. J., 1992, NF-κB subunit regulation in non-transformed CD4+ T lymphocytes, *Science* **256**:1452–1455.

Kumar, A., Hague, J., Lacoste, J., Hiscott, J., and Williams, B. R. G., 1994, Double-stranded RNA-dependent protein kinase activates transcription factor NF-κB by phosphorylating IκB, *Proc. Natl. Acad. Sci. USA* **91**:6288–6297.

Kuno, K., Ishikawa, Y., Ernst, M. K., Ogata, M., Rice, N. R., Mukaida, N., and Matsushima, K., 1995, Identification of an IκBα-associated protein kinase in a human monocytic cell line and determination of its phosphorylation sites on IκBα, *J. Biol. Chem.* **270**:27914–27919.

Li, S., and Sedivy, J. M., 1993, Raf-1 protein kinase activates the NF-κB transcription factor by dissociating the cytoplasmic NF-κB–IκB complex, *Proc. Natl. Acad. Sci. USA* **90**:9247–9251.

Lin, L., and Ghosh, S., 1996, A glycine-rich region of NF-κB p105 functions as a processing signal for the generation of the p50 subunit, *Mol. Cell. Biol.* **16**:2248–2254.

Lin, Y.-C., Brown, K., and Siebenlist, U., 1995, Activation of NF-κB requires proteolysis of the inhibitor IκB-α: Signal-induced phosphorylation of IκB-α alone does not release active NF-κB, *Proc. Natl. Acad. Sci. USA* **92**:552–556.

McElhinny, J. A., Trushin, S. A., Bren, G., Chester, N., and Paya, C., 1996, Casein kinase II phosphorylates IκBα at S-283, S-289, and T-291 and is required for its degradation, *Mol. Cell. Biol.* **16**:899–906.

McKinsey, T. A., Brockman, J. A., Scherer, D. C., Al-Murrani, S. A., Green, P. L., and Ballard, D. W., 1996, Inactivation of IκBβ by the Tax protein of human T-cell leukemia virus type 1: A potential mechanism for constitutive induction of NF-κB, *Mol. Cell. Biol.* **16**:2083–2090.

Mellits, K. H., Hay, R. T., and Goodbourn, S., 1993, Proteolytic degradation of MAD3 (IκBα) and enhanced processing of the NF-κB precursor p105 are obligatory steps in the activation of NF-κB, *Nucleic Acids Res.* **21**:5059–5066.

Mercurio, F., DiDonato, J. A., Rosette, C., and Karin, M., 1993, p105 and p98 precursor proteins play an active role in NF-κB-mediated signal transduction, *Genes Dev.* **7**:705–718.

Meyer, C. F., Wang, X., Chang, C., Templeton, D., and Tan, T.-H., 1996, Interaction between c-Rel and the mitogen-activated protein kinase kinase kinase 1 signaling cascade in mediating κB enhancer activation, *J. Biol. Chem.* **271**:8971–8976.

Miyamoto, S., Maki, M., Schmitt, M. J., Hatanaka, M., and Verma, I. M., 1994, Tumor necrosis factor α-induced phosphorylation of IκBα is a signal for its degradation but not dissociation from NF-κB, *Proc. Natl. Acad. Sci. USA* **91**:12740–12744.

Naumann, M., and Scheidereit, C., 1994, Activation of NF-κB in vivo is regulated by multiple phosphorylations, *EMBO J.* **13**:4597–4607.

Naumann, M., Wulczyn, F. G., and Scheidereit, C., 1993, The NF-κB precursor p105 and the proto-oncogene product Bcl-3 are IκB molecules and control nuclear translocation of NF-κB, *EMBO J.* **12:**213–222.

Nuber, U., Schwarz, S., Kaiser, P., Schneider, R., and Scheffner, M., 1996, Cloning of human ubiquitin-conjugating enzymes UbcH6 and UbcH7 (E2-F1) and characterization of their interaction with E6AP and RSP5, *J. Biol. Chem.* **271:**2795–2800.

Orian, A., Whiteside, S., Israel, A., Stancovski, I., Schwartz, A. L., and Ciechanover, A., 1995, Ubiquitin-mediated processing of NF-κB transcriptional activator precursor p105, *J. Biol. Chem.* **270:**21707–21714.

Palombella, V. J., Rando, O. J., Goldberg, A. L., and Maniatis, T., 1994, The ubiquitin-proteasome pathway is required for processing the NF-κB1 precursor protein and the activation of NF-κB, *Cell* **78:**773–785.

Rechsteiner, M., and Rogers, S. W., 1996, PEST sequences and regulation by proteolysis, *Trends Biochem. Sci.* **247:**267–271.

Rice, N. R., and Ernst, M. K., 1993, *In vivo* control of NF-κB activation by IκBα, *EMBO J.* **12:**4685–4695.

Rice, N. R., Mackichan, M. L., and Israel, A., 1992, The precursor of NF-κB p50 has IκB-like functions, *Cell* **71:**243–253.

Rock, K. L., Gramm, C., Rothstein, L., Clark, K., Stein, R., Dick, L., Hwang, D., and Goldberg, A. L., 1994, Inhibitors of the proteasome block degradation of most cell proteins and the generation of peptides presented on MHC class I molecules, *Cell* **78:**761–771.

Rodriguez, M. S., Michalopoulos, I., Arenzana-Seisdedos, F., and Hay, R. T., 1995, Inducible degradation of IκBα in vitro and in vivo requires the acidic C-terminal domain of the protein, *Mol. Cell. Biol.* **15:**2413–2419.

Rodriguez, M. S., Wright, J., Thompson, J., Thomas, D., Baleuz, F., Virelizier, J. L., Hay, R. T., and Arenzana-Seisdedos, F., 1996, Identification of lysine residues required for signal-induced ubiquitination and degradation of IκBα in vivo, *Oncogene* **12:**2425–2435.

Roff, M., Thompson, J., Rodriguez, M. S., Jacque, J. M., Baleux, F., Arenzana-Seisdedos, F., and Hay, R. T., 1996, Role of IκBα ubiquitination in signal-induced activation of NF-κB in vivo, *J. Biol. Chem.* **271:**7844–7850.

Scheffner, M., Huibregtse, J. M., Vierstra, R. D., and Howley, P. M., 1993, The HPV-16 E6 and E6AP complex functions as a ubiquitin-protein ligase in the ubiquitination of p53, *Cell* **75:**495–505.

Scherer, D. C., Brockman, J. A., Chen, Z. J., Maniatis, T., and Ballard, D., 1995, Signal-induced degradation of IκBα requires site-specific ubiquitination, *Proc. Natl. Acad. Sci. USA* **92:**11259–11263.

Schmidt, K. N., Traenckner, E. B.-M., Meier, B., and Baeuerle, P. A., 1995, Induction of oxidative stress by okadaic acid is required for activation of transcription factor NF-κB, *J. Biol. Chem.* **270:** 27136–27142.

Schreck, R., Rieber, P., and Baeuerle, P. A., 1991, Reactive oxygen intermediates as apparently widely used messengers in the activation of the NF-κB transcription factor and HIV-1, *EMBO J.* **10:** 2247–2258.

Schutze, S., Pothoff, K., Machleidt, T., Bercovic, D., Wiegmann, K., and Kronke, M., 1992, TNF activates NF-κB by phosphatidylcholine-specific phospholipase C-induced "acidic" sphingomyelin breakdown, *Cell* **71:**765–776.

Schwarz, E. M., Van Antwerp, D., and Verma, I. M., 1996, Constitutive phosphorylation of IκBα by casein kinase II occurs preferentially at serine 293: Requirement for degradation of free IκBα, *Mol. Cell. Biol.* **16:**3554–3559.

Scott, M. L., Fujita, T., Liou, H. C., Nolan, G. P., and Baltimore, D., 1993, The p65 subunit of NF-κB regulates IκB by two distinct mechanisms, *Genes Dev.* **7:**1266–1276.

Sen, R., and Baltimore, D., 1986, Multiple nuclear factors interact with the immunoglobulin enhancer sequences, *Cell* **46:**705–716.

Shelton, C. A., and Wasserman, S. A., 1993, pelle encodes a protein kinase required to establish dorsoventral polarity in the Drosophila embryo, *Cell* **72:**515–525.

Shirakawa, F., and Mizel, S. B., 1989, In vitro activation and nuclear translocation of NF-κB catalyzed by cyclic AMP-dependent protein kinase and protein kinase C, *Mol. Cell. Biol.* **9:**2424–2430.

Siebenlist, U., Franzoso, G., and Brown, K., 1994, Structure, regulation and function of NF-κB, *Annu. Rev. Cell Biol.* **10:**405–455.

Sun, S.-C., Ganchi, P. A., Ballard, D. W., and Greene, W. C., 1993, NF-κB controls expression of inhibitor IκBα: Evidence for an inducible autoregulatory pathway, *Science* **259:**1912–1915.

Sun, S.-C., Elwood, J., and Greene, W. C., 1996, Both amino- and carboxyl-terminal sequences within IκBα regulate its inducible degradation, *Mol. Cell. Biol.* **16:**1058–1065.

Thanos, D., and Maniatis, T., 1995, NF-κB: A lesson in family values, *Cell* **80:**529–532.

Thompson, J. E., Phillips, R. J., Erdjument-Bromage, H., Tempst, P., and Ghosh, S., 1995, IκB-β regulates the persistent response in a biphasic activation of NF-κB, *Cell* **80:**573–582.

Traenckner, E. B.-M., Wilk, S., and Baeuerle, P. A., 1994, A proteasome inhibitor prevents activation of NF-κB and stabilizes a newly phosphorylated form of IκBα that is still bound to NF-κB, *EMBO J.* **13:**5433–5441.

Traenckner, E. B.-M., Pahl, H. L., Henkel, T., Schmidt, K. N., Wilk, S., and Baeuerle, P. A., 1995, Phosphorylation of human IκBα on serines 32 and 36 controls IκBα proteolysis and NF-κB activation in response to diverse stimuli, *EMBO J.* **14:**2876–2883.

Verma, I. M., Stevenson, J. K., Schwarz, E. M., Van Antwerp, D., and Miyamoto, S., 1995, Rel/NF-κB/IκB family: Intimate tales of association and dissociation, *Genes Dev.* **9:**2723–2735.

Whiteside, S. T., Ernst, M. K., LeBail, O., Laurent-Winter, C., Rice, N., and Israel, A., 1995, N- and C-terminal sequences control degradation of MAD3/IκBα in response to inducers of NF-κB activity, *Mol. Cell. Biol.* **15:**5339–5345.

CHAPTER 11

Ubiquitination of the p53 Tumor Suppressor

Jon M. Huibregtse, Carl G. Maki, and Peter M. Howley

p53 was described in 1979 as a protein that coimmunoprecipitated with simian virus 40 (SV40) large T antigen (Lane and Crawford, 1979; Linzer and Levine, 1979). Nearly 10 years later it was discovered that p53 is a tumor suppressor and that mutations in the p53 gene are very frequently associated with many types of human cancers (Levine *et al.*, 1991; Vogelstein and Kinzler, 1992). As our understanding of the biochemical functions and regulation of this protein continues to unfold, the activity most clearly related to its function as a tumor suppressor is its ability to bind DNA in a sequence-specific manner and activate transcription (Funk *et al.*, 1992; Kern *et al.*, 1992; Pietenpol *et al.*, 1994). This activity increases in response to UV and gamma irradiation, related at least in part to an elevation in p53 protein levels (Maltzman and Czyzyk, 1984; Kastan *et al.*, 1991). The effect of increasing p53 activity is to stop cell proliferation, either by pausing the cell cycle or by inducing apoptotic cell death (Lane, 1992). Some of the genes affected by p53 at the transcriptional level have been identified, including p21, mdm-2, and bax (El-Deiry *et al.*, 1993; Juven *et al.*, 1993; Han *et al.*, 1996). A current model is that p53 functions in a pathway that monitors the "status" of the genome,

Jon M. Huibregtse • Department of Molecular Biology and Biochemistry, Rutgers University, Piscataway, New Jersey 08855. **Carl G. Maki and Peter M. Howley** • Department of Pathology, Harvard Medical School, Boston, Massachusetts 02115.

Ubiquitin and the Biology of the Cell, edited by Peters *et al.* Plenum Press, New York, 1998.

preventing the accumulation of somatic mutations either by signaling a cell cycle arrest, allowing the cell time to repair the damage, or by eliminating the damaged cell by inducing apoptosis. Many excellent reviews have covered recent advances in the p53 field in detail (Vogelstein and Kinzler, 1992; Friend, 1994; Ko and Prives, 1996).

This chapter focuses on the role of the ubiquitin system in regulating p53 activities. This will deal primarily with the degradation of p53 induced by the human papillomavirus (HPV) E6 protein, as well as on evidence that p53 is normally subject to ubiquitin-mediated degradation.

1. E6-DEPENDENT UBIQUITINATION OF p53

1.1. The E6 Oncoprotein

One of the clearest examples of a human virus being etiologically linked to cancer is that of HPVs and cancers of the anogenital region, especially cervical cancer (zur Hausen and Schneider, 1987; Bosch *et al.*, 1995). There are over 70 known papillomavirus types (distinguished based on DNA sequence divergence), and more than 35 types infect the anogenital tract and are sexually transmitted. HPV DNA representing a subset of these types is found in approximately 93% of cervical cancers and in their precursor lesions (intraepithelial neoplasias), and the remaining types are associated almost exclusively with benign lesions (DeVilliers, 1989; Chan *et al.*, 1995). This has led to the categorization of the genital HPVs into a "high risk" or cancer-associated group (including HPV16 and 18) and a "low risk" or non-cancer-associated group (including HPV6 and 11).

Only two viral genes, the E6 and E7 genes, are generally expressed in cervical carcinomas and in derived cell lines, suggesting that these genes play a role in carcinogenesis. The viral genome consists of approximately 8000 base pairs of closed circular double-stranded DNA, with all of the major open reading frames (ORFs) derived from one strand of the genome. The major *cis*-regulatory elements for viral transcription are located within the long control region (LCR), and the E6 and E7 genes are the first ORFs downstream of the LCR. When HPV DNA is found in cervical carcinomas, it is frequently integrated into the host genome, with integration occurring just downstream of the E7 gene. Thus, LCR-dependent transcription drives expression of only two viral genes, E6 and E7. Continued expression of E6 and E7 has been shown to be necessary for the growth of cervical carcinoma cells in culture. Antisense RNA experiments in the C4-I cervical carcinoma cell line showed that E6 and E7 expression are necessary for the continued growth and the tumorigenicity of these cells (von Knebel-Doeberitz *et al.*, 1988). Expression of E6 and E7 in the SW756 cervical carcinoma cell line has been shown to be specifically repressed by dexamethasone (von Knebel-Doeberitz *et al.*, 1991). Consequently, dexamethasone causes growth inhibition and decreases the tumorigenicity of these cells in nude mice.

The development of a cell culture assay that reflected the high and low risk classification of the virus types provided further evidence for the role of E6 and E7 in immortalization of the natural host cell type, genital keratinocytes (Hawley-Nelson *et al.*, 1989; Münger *et al.*, 1989). The immortalized cell lines derived from primary genital keratinocytes are not tumorigenic in nude mice, indicating that other genetic events are also selected for in the course of carcinogenesis and that E6 and E7 activities, by themselves, are not sufficient for transformation.

The HPV16 E6 and E7 proteins are both small proteins (160 and 98 amino acids, respectively, in HPV16), and both contain a metal (presumably zinc) binding domain (Barbosa *et al.*, 1989) consisting of CXXC-X_{29}-CXXC. E6 proteins contain two of these motifs and E7 proteins contain one. Figure 1 aligns sequences of two high-risk (types 16 and 18) and two low-risk (types 6 and 11) HPV E6 proteins. The overall identity among the high-risk HPV E6 proteins is approximately 49% (67% similarity), whereas the identity between the high- and low-risk HPV E6 proteins is approximately 33% (50% similarity, comparing HPV16 and 11 E6). The most dissimilar regions between the high- and low-risk groups are in the first 30 amino acids, before the first CXXC, and the extreme C-terminus, after the last CXXC. Efforts to determine the cellular pathways involved in immortalization by the E6 and E7 genes led to the p53 and pRB (retinoblastoma) tumor suppressors, respectively, proteins that had been previously shown to be targets of the transforming proteins of other DNA tumor viruses. Adenovirus E1A protein, SV40 large T antigen, and the HPV E7 proteins, through conserved protein sequence elements, bind to the underphosphorylated form of pRB and inactivate its G1 cell cycle inhibitory function (DeCaprio *et al.*, 1988; Whyte *et al.*, 1988; Dyson *et al.*, 1992). The high-risk HPV E7 proteins interact with pRB with higher affinity than the low-risk E7 proteins, suggesting

```
HPV16:   MFQDPQERPRKLPQLCTELQTTIHDIILECVYCKQQLLRREVYDFAFRDLCIVYRDGNPYAVCDKCLKFYSKISEYR-
HPV18: MARFEDPTRRPYKLPDLCTELNTSLQDIEITCVYCKTVLELTEVFEFAFKDLFVVYRDSIPHAACHKCIDFYSRIRELR-
HPV11:  MESKDASTSATSIDQLCKTFNLSLHTLQIQCVFCRNALTTAEIYAYAYKNLKVVWRDNFPFAACACCLELQGKINQYR-
HPV6:   MESANASTSATTIDQLCKTFNLSMHTLQINCVFCKNALTTAEIYSYSYKHLKVLFRGGYPYAACACCLEFHGKINQYR-

HPV16:  HYCYSLYGTTLEQQYNKPLCDLLIRCINCQKPLCPEEKQRHLDKKQRFHNIRGRWTGRCMSCC------RSSRTRRETQL
HPV18:  HYSDSVYGDTLEKLTNTGLYNLLIRCLRCQKPLNPAEKLRHLNEKRRFHNIAGHYRGQCHSCCNRARQERLQR-RRETQV
HPV11:  HFNYAAYAPTVEEETNEDILKVLIRCYLCHKPLCEIEKLKHILGKARFIKLNNQWKGRCLHCWTTCMEDLLP
HPV6:   HFNYAAYAPTVEEETNEDILKVLIRCYLCHKPLCEVEKVKHILTKARFIKLNCTWKGRCLHCWITCMEDMLP
```

Figure 1. Alignment of the complete amino acid sequence of two high-risk HPV E6 proteins, types 16 and 18, with two low-risk HPV E6 proteins, types 11 and 6. The putative metal binding domains are indicated with brackets above the sequence, with the CXXC motifs shaded.

that this interaction is an important determinant of immortalizing potential. In parallel with this finding, it was shown that the E6 proteins of the high-risk HPVs, but not of the low-risk HPVs, coimmunoprecipitated with p53 *in vitro* (Werness *et al.*, 1990). There is now abundant evidence that the high-risk E6 proteins, SV40 TAg, and the large adenovirus E1B proteins all abrogate the transcriptional transactivation activity of wild-type p53 *in vivo* (Mietz *et al.*, 1992; Yew and Berk, 1992).

E6 inactivates p53 by a mechanism that is distinct from that of T antigen and the E1B proteins. In SV40 T-immortalized cells the half-life of p53 is extended and the steady-state level of p53 is greatly elevated. Ad5 E1B has a similar effect on p53 stability and overall concentration. In contrast, the level of p53 in HPV-containing cervical carcinoma cell lines and HPV-immortalized keratinocytes is generally lower than that seen in primary cells and the half-life is decreased. p53 in normal cultured human keratinocytes has a half-life of approximately 1–4 hr, whereas in HPV-immortalized keratinocytes the half-life is 15–30 min (Hubbert *et al.*, 1992). A biochemical basis for this observation was suggested by *in vitro* experiments showing that E6 could stimulate the ubiquitination and degradation of p53 when both proteins were translated in rabbit reticulocyte lysate (Scheffner *et al.*, 1990). A model therefore developed wherein the function of the high-risk HPV E6 proteins in carcinogenesis is to promote cellular proliferation through targeted degradation of the p53 tumor suppressor.

Further evidence that HPV16 E6 inactivates p53 *in vivo* is that E6-expressing cells containing wild-type p53 generally do not undergo cell cycle arrest in response to either irradiation or DNA-damaging drugs (Kessis *et al.*, 1993). This correlates with p53 protein levels, which do not increase after irradiation of E6-expressing cells, presumably because E6 stimulates the rapid degradation of p53. On the other hand, there is also evidence that E6 is not functionally equivalent to an inactivation of p53 by somatic mutation. p53 is detectable in most cervical carcinoma cell lines, albeit at a low level (Scheffner *et al.*, 1991). This remaining p53 is transcriptionally active and it appears that the E6-mediated inactivation can be overcome by high doses of radiation or genotoxic drugs (Butz *et al.*, 1995).

Additional circumstantial evidence suggesting that E6 inactivates p53 comes from studies that have looked at the status of p53 alleles in cervical carcinoma cell lines. In HPV-containing cell lines, both p53 alleles are generally wild-type by sequence analysis, whereas in the relatively small number of HPV-negative cervical carcinoma lines, both p53 alleles have been found to be disrupted by mutation (Crook *et al.*, 1991; Scheffner *et al.*, 1991). This suggests that inactivation of p53 is selected for in virtually all cervical cancers, but that it can occur either by expression of E6 or by p53 mutation. Exceptions to this pattern have been reported (i.e., p53 mutations in HPV-positive cells), which may indicate that in some cases, possibly in metastases of HPV-containing carcinomas, p53 mutation may provide an additional selective growth advantage (Crook and Vousden, 1992).

1.2. E6-AP and the Ubiquitination of p53

The initial demonstration of a stable interaction between high-risk HPV (types 16 and 18) E6 proteins and p53 was performed by coimmunoprecipitation at 4°C, with both proteins being translated *in vitro* in rabbit reticulocyte lysate (Werness *et al.*, 1990). It was later found that incubation of HPV16 or 18 E6 with p53 at room temperature leads to the multiubiquitination and degradation of p53 via the action of enzymes endogenous to the reticulocyte lysate (Scheffner *et al.*, 1990). A protein present in the reticulocyte lysate, now known as E6-AP (*E6*-Associated *P*rotein), is essential for formation of the initial E6/p53 complex (Huibregtse *et al.*, 1991). This 100-kDa protein was originally detected because HPV16 E6 protein produced by *in vitro* translation in a wheat germ extract system, which does not contain E6-AP, did not interact with bacterially produced p53 (GST-p53). The addition of a small amount of reticulocyte lysate to wheat germ extract-translated E6 protein restores its ability to interact with p53. Human E6-AP was purified based on its ability to stimulate the binding of E6 to p53 and its cDNA was cloned (Huibregtse *et al.*, 1993a). A functional E6-AP homologue has been detected in every mammalian cell type examined so far, and the rat and mouse E6-AP proteins are over 95% identical to human E6-AP (J. M. Huibregtse, unpublished). E6-AP activity has not been detected in any nonmammalian cell type, including yeast, plants, or insect cells, although over 30 eukaryotic cDNAs from both mammalian and nonmammalian organisms have been identified that are related to E6-AP. These comprise a family of E6-AP-related proteins, defined by conservation of a C-terminal domain of approximately 350 amino acids (Huibregtse *et al.*, 1995). This domain is referred to as the hect domain (*h*omologous to *E*6-AP *c*arboxy *t*erminus), and as discussed below, is critical in the function of E6-AP.

The association of E6 and E6-AP precedes the formation of a ternary complex with p53. Neither E6 nor E6-AP, by itself, forms a stable complex with p53. An initial structure–function analysis of E6-AP mapped three functional domains: (1) the E6 binding domain, (2) the minimal region required for association with p53, and (3) the minimal region required for ubiquitination of p53 (Huibregtse *et al.*, 1993a). An 18-amino-acid region within the central portion of E6-AP was found to be necessary and sufficient for binding HPV16 E6. The minimal region of E6-AP required to form a stable complex with p53 spans approximately 500 amino acids and includes, as predicted, the E6 binding domain. Finally, determinants within the C-terminal 84 amino acids of E6-AP, although dispensable for ternary complex formation, are required for ubiquitination of p53. These results are summarized in Fig. 2. It is still not known which protein, E6 or E6-AP, makes physical contact with p53 in the ternary complex. E6 might activate E6-AP for p53 binding, or vice versa, or perhaps both proteins together form a binding surface for p53. Mutations in both E6 and E6-AP have been identified that do not

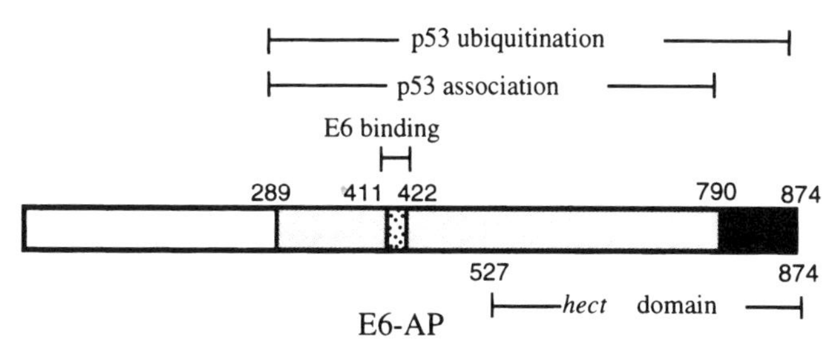

Figure 2. Structure–function of E6-AP. The minimal region of E6-AP necessary for E6 binding is a 12-amino-acid peptide from amino acids 411 to 422. The minimum region necessary for stable p53 association spans a region of approximately 500 amino acids, from residues 289 to 790. The region from 790 to the end of the protein is required in addition to the p53 binding domain for ubiquitination of p53 (Huibregtse *et al.*, 1993b). The approximately 350-amino-acid region of similarity to the family of E6-AP-related proteins, the hect domain, is also shown.

affect formation of the E6/E6-AP complex, yet abrogate formation of a complex with p53 (Mietz *et al.*, 1992; Huibregtse *et al.*, 1993a).

The low-risk HPV E6 proteins (HPV6 and 11) do not detectably interact with E6-AP, which according to the above model for complex formation, would account for why they do not associate with or lead to the ubiquitination of p53. It has recently been reported, however, that bovine papillomavirus (BPV-1) E6 protein, a major transforming protein of this virus, also associates with human E6-AP (Chen *et al.*, 1995). Interestingly, there is no evidence that BPV-1 E6 inactivates p53, suggesting that this E6 protein, like the high-risk HPV E6 proteins, may exert its function through the ubiquitin proteolysis system, but that the target of BPV E6-AP may be a protein other than p53.

Complete *in vitro* reconstitution of E6/E6-AP-dependent ubiquitination of p53 revealed that, in addition to ATP and ubiquitin, only two other proteins were required: the E1 ubiquitin-activating enzyme and an E2 ubiquitin-conjugating enzyme (Scheffner *et al.*, 1993, 1994). Based on the definition of E3 ubiquitin-protein ligases as activities in addition to E1 and E2 enzymes required for ubiquitination of a particular substrate, these results implied that the E6/E6-AP complex was an E3 activity. In addition, E6-AP was shown to have E3 activity in the absence of the HPV E6 protein in that it was able to stimulate the ubiquitination of other as yet unidentified cellular proteins (Scheffner *et al.*, 1993).

A mechanism proposed for the E3 protein encoded by the *S. cerevisiae UBR1* gene was that it acts as a "docking protein," binding an activated E2 protein at one site and a specific substrate at another (Dohmen *et al.*, 1991), orienting them so that the E2 can catalyze the ubiquitination reaction. A stable interaction between the E6/E6-AP complex, or either protein by itself, with an E2 protein, has so far not been demonstrated. In the course of looking for such an interaction, however, it was observed that E6-AP forms a ubiquitin thioester in an E1- and E2-dependent manner (Scheffner *et al.*, 1995). This suggested an alternative model for protein ubiquitination in which the E3 protein is part of a "thioester cascade." This model is shown in Fig. 3, with the suggestion of perhaps a fourth class of protein involved in protein ubiquitination, represented by the HPV16 E6 protein. This fourth class of proteins might be defined as auxiliary proteins that aid in directing the substrate specificity of the E3 proteins. The formation of the E6-AP ubiquitin thioester is not dependent on E6, nor does E6 influence the efficiency of formation of the E6-AP thioester, consistent with the idea that E6 plays a role in directing substrate specificity of E6-AP.

Characterization of the E6-AP gene structure and E6-AP cDNAs has shown that at least four distinct mRNAs are generated by differential splicing (Yamamoto *et al.*, 1997). The mRNAs are predicted to encode three protein isoforms that utilize different initiating methionines and differ from each other only at their extreme N-termini, by at most 23 amino acids. Functional differences between the isoforms have so far not been elucidated, although all three isoforms are capable of catalyzing E6-dependent ubiquitination of p53 *in vitro* (J. M. Huibregtse and P. M. Howley, unpublished data).

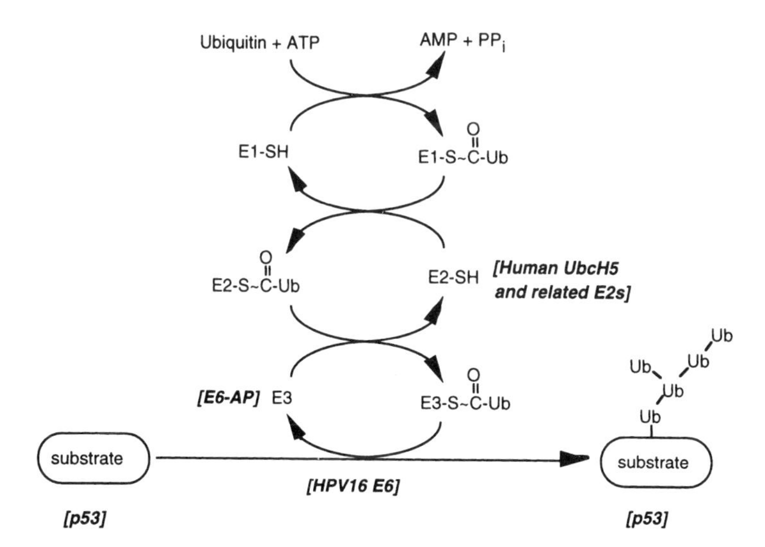

Figure 3. Thioester cascade model for ubiquitination of p53 (Scheffner *et al.*, 1995). Following formation of an E1-ubiquitin thioester, the E1 transfers the ubiquitin to the active site of an E2 ubiquitin-conjugating protein (human UbcH5 or related family member; Nuber *et al.*, 1996). The E2 then transfers the ubiquitin to the active-site cysteine of E6-AP, which then catalyzes the multiubiquitination of p53. HPV16 E6 appears to function as an auxiliary factor to E6-AP, directing the E6-AP/p53 association.

Based on physical mapping, the human E6-AP gene (chromosome 15q11-q13, also referred to as *UBE3A*) was a candidate for the affected gene in Angelman syndrome (AS) (Nakao *et al.*, 1994), a disorder characterized by moderate to severe mental retardation, absence of speech, abnormal gait, and inappropriate laughter. AS is one of the best-studied cases of a genetic disorder in which imprinting plays a role: 70% of cases involve maternal deletions at chromosome 15q11-q13, 2% involve paternal uniparental disomy (UPD) at this region, and 2–3% result from "imprinting mutations" that alter the methylation pattern at this region. The remaining 25% of the cases show biparental inheritance without methylation abnormalities or deletions. Mutations resulting in truncated E6-AP protein products were recently identified in this latter class of patients, strongly suggesting that E6-AP is the affected gene in at least some cases of AS (Kishino *et al.*, 1997; Matsuura *et al.*, 1997). Interestingly, earlier results suggested that E6-AP was not the AS gene as its expression was not exclusive to the maternal allele in cell types examined (Nakao *et al.*, 1994). This, together with the recent findings, suggests that AS might be caused by a deficiency in a specific maternally expressed isoform, perhaps generated through differential splicing. Alternatively, maternal-specific expression of E6-AP might be restricted to a particular cell type or developmental stage. Future work will hopefully demonstrate the stage at which E6-AP activity is critical, as well as the proteins that are targeted by E6-AP, in neurological development.

1.3. The E6-AP-Related Family of E3 Proteins

As mentioned earlier, the C-terminal 84 amino acids of E6-AP contain determinants that, although dispensable for E6/p53 ternary complex formation, are required for ubiquitination of p53 (Huibregtse *et al.*, 1993a). Mutation of the only cysteine residue within this region (to an alanine or serine) resulted in a protein that was unable to form a ubiquitin thioester and was unable to ubiquitinate p53, although these mutations did not affect the ability of the protein to bind to p53 in an E6-dependent manner. This strongly suggested that this cysteine residue, 33 amino acids from the C-terminus of E6-AP, is the site of ubiquitin-thioester formation. This cysteine residue and several surrounding amino acids are conserved among all of the hect domain-containing proteins. Several of the hect domain-containing proteins, including yeast Rsp5, a rat 100-kDa protein, and *Schizosaccharomyces pombe* Pub1, have now been shown to also form ubiquitin thioesters with similar requirements to E6-AP (Huibregtse *et al.*, 1995; Nefsky and Beach, 1996), suggesting that these proteins represent a family of E3 ubiquitin-protein ligases.

Evidence that E6-AP plays a role *in vivo* in the E6-mediated degradation of p53 comes from studies using mutant forms of E6-AP that function in a dominant-negative manner to inhibit p53 degradation. Transient expression in HPV-positive cervical carcinoma HeLa cells of certain E6-AP mutants, including the active-site Cys-to-Ala mutant, results in a stabilization and prolonged half-life of p53 (A. Talis, J. M. Huibregtse, and P. M. Howley, manuscript in preparation). The catalytically inactive E6-AP protein, which is still capable of forming a stable complex with E6 and p53, is presumably titrating E6 and p53 away from the endogenous wild-type E6-AP, resulting in the stabilization of p53.

An alignment of the complete hect domain of E6-AP with the homologous regions of three other hect E3s is shown in Fig. 4. The overall similarity between any two of the hect domains of these proteins is approximately 40%, with approximately 25% amino acid identity. Fig. 5 is a partial list of known ORFs that appear to encode hect E3s, along with the amino acid sequence immediately surrounding the active-site cysteine residue. There are a minimum of 16 human hect E3s, many of which contain closely related homologues in other mammalian species, and exactly 5 in the yeast *Saccharomyces cerevisiae*, where the complete genome has been sequenced. Hect domain proteins have also been found in *Arabidopsis thaliana*, *C. elegans*, and *Drosophila*. In cases where the complete cDNA sequences are known, the encoded proteins range in size from 85 kDa to over 600 kDa, the latter being the second largest human protein characterized to date (Rosa *et al.*, 1996). In every case the active-site cysteine is between 32 and 36 amino acids from the C-terminus of the protein.

Understanding the mechanisms underlying the function of the hect E3 proteins will be an important step toward understanding how substrate specificity of the ubiquitin system is determined and controlled. A model proposed for hect E3 function, based largely on E6-dependent ubiquitination of p53 and the pattern of similarity among the proteins, is that these proteins consist of two broad domains: an N-terminal substrate recognition domain and a C-terminal catalytic domain that ubiquitinates bound substrates. The model is shown schematically in Fig. 6.

The first recombinant E2 enzyme to be shown to function with E6-AP was *Arabidopsis thaliana* (*At*) UBC8 (Scheffner *et al.*, 1993). The experiment was suggested by the observation that a human E2 activity that could function in E6/E6-AP-dependent ubiquitination of p53 displayed chromatographic properties similar to those of *At* UBC8 (Girod and Vierstra, 1993). Specifically, both UBC8 and the human E2 activities were retained by cation-exchange columns but not by anion-exchange columns. At UBC8 belongs to a subfamily of E2s that includes *S. cerevisiae UBC4* and 5 and at least five other proteins from *Arabidopsis* (UBC8–12) (Girod *et al.*, 1993). This family of E2 proteins has been shown to play a major role in protein turnover in both yeast and *Arabidopsis* (Seufert and Jentsch, 1990; Girod and Vierstra, 1993). Five human E2 cDNAs (UbcH5A, 5B, 5C, 6, and 7) have now been cloned that encode proteins capable of supporting E6-AP ubiquitin-thioester formation (Scheffner *et al.*, 1994; Jensen *et al.*, 1995; Nuber *et al.*, 1996). UbcH5A, 5B, 5C, and 6 belong to the UBC4/5 family, but UbcH7 does not, indicating that distinct subfamilies of E2 proteins are capable of activating E6-AP.

```
E6-AP :  RRDHIIDDALVRLEMIAMENPADLKKQLYVEF-----EGEQGVDEGGVSK
D25215:  RRNNLVGDALRELSIHSDI---DLKKPLKVIF-----DGEEAVDAGGVTK
Rsp5  :  RRKNIFEDAYQEIMRQTPE---DLKKRLMIKF-----DGEEGLDYGGVSR
D13635:  RRNYIYEDAYDKLSPENEP---DLKKRIRVHLLNAHGLDEAGIDGGGIFR

E6-AP:   EFFQLVVEEIFNPDIGMFTYDESTKLFWFNPSSFETEGQ----FTLIGIV
D25215:  EFFLLLLKELLNPIYGMFTYYQDSNLLWFSDTCFVEHNW----FHLIGIT
Rsp5 :   EFFFLLSHEMFNPFYCLFEYSAYDNYTIQINPNSGINPEHLNYFKFIGRV
D13635:  EFLNELLKSGFNPNQGFFKTTNEGLLYPNPAAQMLVGDSFARHYYFLARM

E6-AP:   LGLAIYNNCILDVHFPMVVYRKLMGKKGLFVD-LGDSHPVLYQSLKDLLE
D25215:  CGLAIYNSTVVDLHFPLALYKKLLNVKPGLED-LKELSPTEGRSLQELLD
Rsp5 :   VGLGVFHRRFLDAFFVGALYKMMLRKKVVLQD----MEGVDAEVYNSLNW
D13635:  LGKALYENMLVELPFAGFFLSKLLGTSADVDIHHLASLDPEVYKNLLFLK

E6-AP:   YVGNVEDDMMITFQISQTNLFGNPMMYDLKENGDKIPITNENRKEFVNLY
D25215:  YPGEDVEETFCLNFTICRESYGVIEQKKLIPGGDNVTVCKDNRQEFVDAY
Rsp5 :   MLENSIDGVLDLTFSADDERFGEVVTVDLKPDGRNIEVTDGNKKEYVELY
D13635:  SYEDDVE-ELGLNFTVVNNDLGEAQVVELKFGGKDIPVTSANRIAYIHLV

E6-AP:   SDYILNKSVEKQFKAFRRGFHMVTNESPLKYLFRPEEIELLICGSRNLDF
D25215:  VNYVFQISVHEWYTAFSSGFLKV-CGGKVLELFQPSELRAMMVGNSNYNW
Rsp5 :   TQWRIVDRVQEQFKAFMDGFNEL-IPEDLVTVFDERELELLIGGIAEID-
D13635:  ADYRLNRQIRQHCLAFRQGLANV-VSLEWLRMFDQQEIQVLISGAQVPIS

E6-AP:   QALEETTEYD---GGYTRDSVLAKNGPDTERLPTSHTCFNVLLLPEYSSK
D25215:  EELEETAIYK---GDYSATHPTQSTASGEEYLPVAHTCYNLLDLPKYSSK
Rsp5  :  --IEDWKKHTDYRG-YQESDEV-EKAGEVQQLPKSHTCFNRVDLPQYVDY
D13635:  --LEDLKSFTNYSGGYSADHPVHNGGSDLERLPTASTCMNLLKLPEFYDE

E6-AP:   EKLKERLLKAITYAKGFGML
D25215:  EILSARLTQALDNYEGFSLA
Rsp5 :   DSMKQKLTLAVEETIGFGQE
D13635:  TLLRSKLLYAIECAAGFELS
```

Figure 4. Alignment of the hect domain of E6-AP with that of two human hect E3s (D25215 and D13635) and Rsp5, a yeast hect E3. The domain is considered to begin at an RR sequence found in most of the hect domains, although in some cases the homology extends farther in the N-terminal direction. Only the most highly conserved residues (identical or closely related residues within all four proteins) are indicated in bold and the active-site cysteine is underlined. In each case the last amino acid shown is the C-terminal residue of the protein.

Database designation	Common name	Sequence	Mass
Human and other mammalian			
gb|L07557|HUMAPE6ONC	E6-AP	LPTSHTCFNVLLLPEY	100 kD
gb|U50842|RNU50842	ratp100	LPTANTCISRLYVPLY	100 kD
dbj|D85414|MUSNEDD4N	mouseNEDD4	LPRAHTCFNRLDLPPY	≥110 kD
dbj|D25215|HUMORFKA		LPVAHTCYNLLDLPKY	118 kD
dbj|D13635|HUMRSC1083		LPTASTCMNLLKLPEF	124 kD
dbj|D28476|HUMKG1C		LPSVMTCVNYLKLPDY	220 kD
gb|U50078|HSU50078	p619	LPTSQTCFFQLRLPPY	619 kD
gb|U08214|RSU08214	UREB1	LPSAHTCFNQLDLPAY	?
gb|N57610|N57610		LPTSSTCINMLKLPEY	?
gb|W45913|W45913		LPRSHTCFNRLDLPPY	?
gb|T62800|T62800		LPVSHTCFNLLDLPKY	?
emb|F07060|HSC1TH111		LPESYTCFFLLKLPRY	?
gb|T74302|T74302		CPELHTCFNRLXLPPY	?
gb|R18042|R18042		LPTAHTCFNQLCLPTY	?
gb|T97237|T97237|		XPTSITCHNILSLPKY	?
C. elegans			
Z34800		FPRAHTCFNRLQLPSY	?
dbj|D27057|CELK002BXR		LPVAQTCFNLLDLPNI	?
dbj|D33580|CELK033A8R		LPTASTCFNLLKLPNY	?
emb|Z54284|CED2085		LPTSATCMNMLRIPKY	?
gb|U58755|CELC34D4		LPPGNTCVHYLKLPEY	?
dbj|C11864|C11864		LPRSHTCFNRLDLPPY	?
S. cerevisiae			
gb|S53418|S53418	*RSP5*	LPKSHTCFNRVDLPQY	92 kD
emb|Z49536|SCYJR036C		LPLAHTCFNEICLWNY	103 kD
emb|Z72663|SCYGL141W		LPTASTCVNLLKLPDY	105 kD
emb|Z28010|SCYKL010C	*UFD4*	LPSVMTCANYLKLPKY	167 kD
dbj|D63905|YSCTOM1	*TOM1*	LPSSHTCFNQLNLPPY	374 kD
Schizo. pombe			
emb|Y07592|SPPUB1	Pub1	LPKAHTCFNRLDLPPY	87 kD
Arabidopsis thaliana			
gb|R65295|R65295		LPSAHTCFNQLDLPEY	?
gb|T88393|T88393		LPSVMTCANYLKLPPY	?
Drosophila			
gb|L14644|DROHYDISC	Hyperplastic discs	LPTANTCISRLYIPLY	318 kD
emb|Z50361|DM145E2S		LPTASTCTNLLKLPPF	?

Figure 5. A partial catalog of hect domain-containing proteins. The database designation is given in each case, along with the common name, if any. For E6-AP (Huibregtse *et al.*, 1995), rat p100 (Müller *et al.*, 1992), NEDD4 (Kumar *et al.*, 1992), and UreB1 (Gu *et al.*,, 1994), apparent homologues have been found in both human and rodent species. Other reports have described *S. pombe* Publ (Nefsky and Beach, 1996). *Drosophila* Hyperplastic discs protein (Mansfield *et al.*, 1994), human p619 (Rosa *et al.*, 1996), and yeast UFD4 (Johnson *et al.*, 1995). Sixteen amino acids in the vicinity of the active-site cysteine are shown for each protein sequence, with the highly conserved residues indicated in bold. The predicted molecular masses are indicated for those proteins for which the complete ORF is thought to have been identified. Additional apparent hect E3s are present in sequence databases, but only those for which sequence is available that encompasses the active-site cysteine are shown.

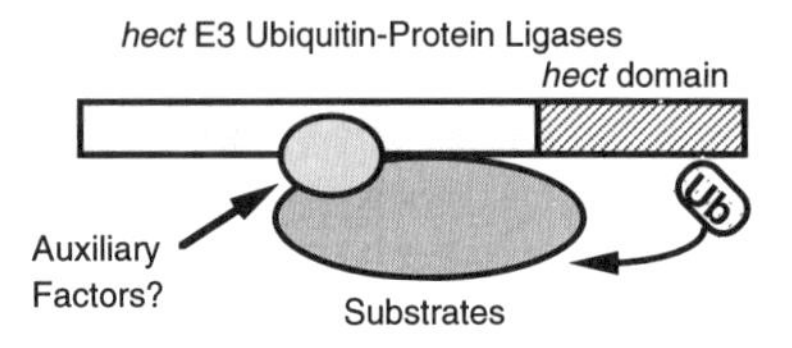

Figure 6. Model for hect E3 function. Based on characterization of E6-AP and the pattern of similarity among the E6-AP-related proteins, the hect E3s appear to consist of two broad domains: an N-terminal substrate recognition domain, which stably associates with substrates, and a C-terminal catalytic domain, the hect domain, that ubiquitinates bound substrates. A ubiquitin thioester at the active-site cysteine is an obligatory intermediate in this reaction. Auxiliary factors, perhaps analogous to HPV16 E6, might in some cases function in directing the association with substrates.

UbcH7 is identical to E2-F1, which was implicated in E6-independent ubiquitination of p53 *in vitro* (Ciechanover *et al.*, 1994). It is not yet known if the particular E2 protein that activates E6-AP influences the substrate specificity of E6-AP, although there is precedent for redundancy of E2 function in targeting a single substrate: Four distinct E2 genes have been shown to affect MATα2 degradation in yeast (Chen *et al.*, 1993).

1.4. Structure–Function Analyses of E6 and p53

Although the initial report of the E6/p53 interaction suggested that only the high-risk HPV E6 proteins could associate with p53 (Werness *et al.*, 1990), Crook *et al.* (1991) found that low-risk (HPV6) E6 proteins could also associate with p53, albeit with a reduced affinity. They also proposed that both the low- and high-risk E6 proteins associated with p53, but that only the high-risk E6 proteins promoted the ubiquitination of p53. A p53 binding domain was localized to the C-terminal half of the E6 proteins, whereas a "ubiquitination" domain was mapped to the N-terminal half of HPV16 E6. Contrary to these findings, Foster *et al.* (1994) as well as our own labs (J. M. Huibregtse and P. M. Howley, unpublished data) failed to reveal separable functional domains within the high-risk HPV E6 proteins.

The recent study by Li and Coffino (1996), which demonstrated two modes of E6 binding to p53, may explain some of these discrepancies. Consistent with the original description (Werness *et al.*, 1990), they found that E6 binding that led to p53 ubiquitination was specific for the high-risk E6 proteins and was dependent on E6-AP. The interaction was sensitive to mutations throughout the E6 molecule and the binding was to the core DNA binding domain of p53 (discussed below). Both the high- and low-risk E6 proteins, however, could participate in a second mode of p53 binding, which was independent of E6-AP and did not lead to p53 ubiquitination. This binding was mediated by the C-terminal domain of the E6 proteins and was directed to the C-terminus of p53. Furthermore, this result was only obtained with GST-E6 fusion proteins produced in *E. coli*. It remains unclear if this second type of low-risk HPV E6/p53 interaction has any biological relevance, although all reports appear to be in agreement that this E6-AP-independent E6/p53 interaction does not lead to the ubiquitination of p53.

Most cancer-associated mutations of p53 are localized to certain "hot spots" within the core DNA binding domain of p53 (Levine *et al.*, 1991; Friend, 1994). Most of these mutations reduce or eliminate the ability of E6 to bind and ubiquitinate p53 *in vitro* (Scheffner *et al.*, 1992b). The only mutations within the DNA binding domain that were found to not affect the ability of E6 to target p53 were mutations at amino acid 273 (R to either L or C). These results suggested that the core DNA binding domain of p53 is critical for E6-mediated p53 ubiquitination. This is consistent with another report showing that targeting of p53 by E6 was preferential for the wild-type form of p53, as analyzed by a conformation-dependent epitope (Medcalf and Milner, 1993). Li and Coffino showed directly that the core DNA binding domain of p53 was recognized by E6/E6-AP *in vitro* using a series of N- and C-terminally truncated p53 mutants (Li and Coffino, 1996). p53 binding to DNA and E6/E6-AP appear to be mutually exclusive (Lechner and Laimins, 1994; Molinari and Milner, 1995).

p53 is phosphorylated at multiple sites outside of the core DNA binding domain, within both the N-terminal transactivation domain and the c-terminal domain, which contains the oligomerization domain of p53. No function has been unambiguously associated with these phosphorylation events, although phosphorylation at C-terminal sites by casein kinase II or protein kinase C can stimulate DNA binding *in vitro* (Hupp *et al.*, 1992; Takenaka *et al.*, 1995). Phosphorylation does not appear to play a role in E6-mediated ubiquitination of p53. Unphosphorylated GST-p53 synthesized in bacteria is a substrate of E6/E6-AP (Scheffner *et al.*, 1993) and the region containing the phosphorylated residues of p53 can be deleted without affecting E6/E6-AP recognition (Li and Coffino, 1996). The latter result also suggests that E6/E6-AP can recognize both monomeric and oligomeric forms of p53 *in vitro*, as an earlier study had also indicated (Medcalf and Milner, 1993).

With respect to the lysine(s) of p53 that are the site of ubiquitination, mutagenesis of all of the p53 lysines demonstrated that no single lysine residue is critical for E6-dependent ubiquitination *in vitro* (Crook *et al.*, 1996). A triple mutation of lysines 381, 382, and 386 abrogated ubiquitination *in vitro*, although the same mutation did not affect the ability of E6 to target p53 for degradation *in vivo*. This implies that E6-AP has a certain degree of flexibility in catalyzing ubiquitination of a bound substrate, yet at the same time, we know that E6-AP shows a high degree of specificity in the initial binding to p53. These apparent contradictions might be explained by an unfolding of the substrate following the initial binding event, which might involve other protein factors.

Interestingly, E6 can still target p53 for ubiquitination when it is in complex with SV40 T antigen, even though both appear to recognize the core DNA binding domain of p53 (Scheffner *et al.*, 1990, 1992b). The binding of the cellular mdm2 protein, which recognizes the N-terminal domain of p53, also does not interfere with E6/E6-AP-mediated ubiquitination of p53 (Martson *et al.*, 1994).

2. E6-INDEPENDENT UBIQUITINATION OF p53

There is evidence to suggest that p53, even in the absence of HPV E6 protein, is a target of the ubiquitin system. p53 has a short half-life in normal cells, with reports varying between 20 min and 4 h (Linzer and Levine, 1979; Hubbert *et al.*, 1992), and it is stabilized in a mouse cell line containing a temperature-sensitive mutation within the El ubiquitin-activating enzyme (Chowdary *et al.*, 1994). Highly specific peptide aldehyde inhibitors of the 26 S proteasome, as well as reagents that reduce ATP production, also lead to a stabilization of p53 (Gronostajski *et al.*, 1984; Maki *et al.*, 1996). Ubiquitin–p53 conjugates have been detected in both γ-irradiated and untreated cells, suggesting that ubiquitin-dependent proteolysis plays a role in the normal turnover of p53 (Maki *et al.*, 1996). In addition, p53 can be ubiquitinated in an *in vitro* system, albeit less efficiently than in the presence of HPV16 E6 (Ciechanover *et al.*, 1991, 1994). A role for the ubiquitin system in regulating p53 activities is suggested by the observation that several types of DNA-damaging agents lead to a dramatic increase in p53 protein levels (Maltzman and Czyzyk, 1984; Kastan *et al.*, 1991). Stabilization of p53 protein accounts for at least part of the increase, suggesting that DNA-damaging agents may somehow signal a repression of ubiquitin-mediated degradation of p53. It is not known whether E6-AP is involved in the E6-independent ubiquitination of p53, either *in vitro* or *in vivo*.

These results raise the question of why the HPV E6 protein should promote the degradation of a protein that is apparently already being turned over quite rapidly. The first consideration is that the half-life of p53 in normal cultured genital keratinocytes, the natural host cell for HPV16, is approximately 4 hr, and expression of HPV16 E6 decreases the half-life to 15–30 min (Hubbert *et al.*, 1992). This difference is perhaps sufficient to account for the biological effects of HPV16 E6. A more compelling consideration, however, is the fact that expression of the HPV16 E7 protein alone, as with expression of Ad E1A alone, actually leads to an increase in p53 levels and activities (Debbas and White, 1993; Lowe and Ruley, 1993; Demers *et al.*, 1994). It has been proposed that this is related to a mechanism by which the cell senses that the cell cycle progression signal, elicited by the E7 and E1A oncoproteins via the pRB/E2-F pathway, is inappropriate (White, 1994). The response is an activation of p53, which signals the cell to arrest in G1 or undergo apoptosis. Therefore, one can perhaps think of the high-risk HPV E6 proteins as not increasing the degradation of an already rapidly turned-over protein, but rather as preventing the accumulation of p53, and subsequent cell cycle arrest, that would otherwise occur under conditions of E7 expression alone. This is likely to be critical in the normal life cycle of the virus, as viral DNA replication is dependent on host cell S phase-specific DNA replication machinery.

3. REMAINING QUESTIONS

Many questions remain concerning the biochemical mechanism of how E6 and E6-AP function in the ubiquitination of p53. With respect to the E6 proteins, it is not clear if their interaction with E6-AP serves to redirect the activity of E6-AP from its normal substrates to p53, or if E6 is simply enhancing the normal E6-AP-dependent ubiquitination of p53. Experiments with HPV16 E6–E7 fusion proteins suggest that E6 might function in redirecting E6-AP substrate specificity (Scheffner *et al.*, 1992b). This study found that a fusion protein consisting of the HPV16 E6 and E7 proteins can stimulate the ubiquitination of pRB. Although these experiments were done before characterization of E6-AP, in retrospect, it appears that the fusion protein is able to bind to E6-AP through E6 sequences and to pRB through E7 sequences. With E6-AP thus in close proximity to pRB, it catalyzes the multiubiquitination of pRB. One can therefore think of E6 as having binding determinants for both E6-AP and p53, bringing the two proteins in proximity for ubiquitination of E6-AP. The caveat to this model is that, as mentioned above, it is not known whether E6 or E6-AP makes direct contact with p53 in the complex. It is equally likely that E6 alters E6-AP structure in such a way that it is able to directly bind to and then ubiquitinate p53.

Recent experiments have suggested that the E6 proteins might also activate the ubiquitin-protein ligase activity of E6-AP. HPV16 E6 has been shown to cause self-ubiquitination of E6-AP *in vitro* (S. Beaudenon, J. M. Huibregtse, and P. M. Howley, manuscript in preparation). This reaction is dependent on the active-site cysteine of E6-AP as well as an intact E6 binding domain on E6-AP, and appears to be related to the intramolecular transfer of ubiquitin from the active-site cysteine to lysine residues within the N-terminal region of E6-AP. Experiments in mammalian cells have shown that E6 expression also leads to ubiquitin-mediated degradation of E6-AP *in vivo*, suggesting that one of the functions of E6 may actually be to degrade E6-AP. This might explain the fact that, as discussed above, E6 expression does not lead to a complete loss of wild-type p53 activity. "Left-over" p53 might therefore reflect a balance between E6-stimulated degradation of p53 and degradation of E6-AP itself. Alternatively, this phenomenon, regardless of its biological significance, may simply reflect an activation of the catalytic activity of E6-AP such that any nearby lysine residues, even if they are on E6-AP itself, are subject to ubiquitination.

What are the functions of the low-risk HPV E6 proteins? The lack of a biological assay for the low-risk E6 proteins has hampered progress on this point. So far the high- and low-risk HPV E6 proteins have not been shown to have any biochemical function in common other than metal ion binding. One intriguing possibility is that the low-risk E6 proteins might interact with a different hect E3 protein and influence the turnover of proteins related to their particular replication cycle.

Are there other targets of the high-risk HPV E6/E6-AP complex, and if so, are they related to HPV-associated cellular immortalization? Experiments hinting at this have involved analysis of a particular mutated HPV16 E6 protein, SAT_{8-10} (amino acids 8–10, RPR, changed to SAT, the amino acids found at the analogous positions of several low-risk HPV E6 proteins). The SAT_{8-10} protein is able to interact with E6-AP-like wild-type HPV16 E6, yet it is inactive in associating with or stimulating the ubiquitination of p53 (J. M. Huibregtse, unpublished; cited in Mietz *et al.*, 1992). Interestingly, both wild-type HPV16 E6 and the SAT_{8-10} mutant were found to stimulate telomerase activity in genital keratinocytes (Klingelhutz *et al.*, 1996). Elevated telomerase activity is a phenomenon often associated with cellular immortalization and is believed to be necessary in preventing the gradual shortening of chromosome ends that otherwise occurs with continual cell division. The fact that both HPV16 E6 and the SAT_{8-10} mutant stimulate telomerase activity suggests that the E6 proteins may function in a p53-independent immortalization pathway that may yet be dependent on E6-AP.

What are the natural substrates of E6-AP? Is there a cellular homologue of the HPV E6 proteins that functions in the ubiquitination of p53? What is the mechanism underlying the stabilization of p53 in response to DNA-damaging agents? These are just some of the questions that remain unanswered at this point. We should also not lose sight of what other viral oncoproteins might still tell us about p53 regulation. Even though p53 was first identified through its interaction with SV40 T antigen, we still do not know the mechanism underlying the stabilization of p53 in T antigen-expressing cells. In the case of the large adenovirus E1B proteins, the Ad5 protein binds and stabilizes p53, perhaps similarly to T antigen (Sarnow *et al.*, 1982), although the Ad12 protein stabilizes p53 apparently without physically interacting with it (van den Heuvel *et al.*, 1993), indicating that it is interfering with p53 function and turnover through another mechanism. A complete understanding of the mechanisms underlying HPV E6-dependent turnover as well as stabilization of p53 in response to other viral oncoproteins is likely to aid in understanding the stabilization of p53 that accompanies DNA damage. This in turn is likely to be critical in being able to activate and control the tumor suppressive function of p53, a goal that has broad clinical applications.

4. REFERENCES

Barbosa, M. S., Lowy, D. R., and Schiller, J. T., 1989, Papillomavirus polypeptides E6 and E7 are zinc-binding proteins, *J. Virol.* **63:**1404–1407.

Bosch, F. X., Manos, M. M., Muñoz, N., Sherman, M., Jansen, A. M., Peto, J., Schiffman, M. H., Moreno, V., Kurman, R., and Shah, K. V., 1995, Prevalence of human papillomavirus in cervical cancer: A worldwide perspective, *J. Natl. Cancer Inst.* **87:**796–802.

Butz, K., Shahabeddin, L., Geisen, C., Spitkovsky, D., Ullman, A., and Hoppe-Seyler, F., 1995, Functional p53 protein in human papillomavirus-positive cancer cells, *Oncogene* **10:**927–936.

Chan, S.-Y., Delius, H., Halpern, A. L., and Bernard, H.-U., 1995, Analysis of genomic sequences of 95 papillomavirus types: Uniting typing, phylogeny, and taxonomy, *J. Virol.* **69:**3074–3083.

Chen, J. C., Reid, C. E., Band, V., and Androphy, E. J., 1995, Interaction of papillomavirus E6 oncoproteins with a putative calcium-binding protein, *Science* **269:**529–531.

Chen, P., Johnson, P., Sommer, T., Jentsch, S., and Hochstrasser, M., 1993, Multiple ubiquitin-conjugating enzymes participate in the in vivo degradation of the yeast MATα2 repressor, *Cell* **74:**357–389.

Chowdary, D. R., Dermody, J. J., Jha, K. K., and Ozer, H. L., 1994, Accumulation of p53 in a mutant cell line defective in the ubiquitin pathway, *Mol. Cell. Biol.* **14:**1997–2003.

Ciechanover, A., DiGiuseppe, J. A., Bercovich, B., Orlan, A., Richter, J. D., Schwartz, A. L., and Brodeur, G. M., 1991, Degradation of nuclear oncoproteins by the ubiquitin proteolysis system *in vitro*, *Proc. Natl. Acad. Sci. USA* **88:**139–143.

Ciechanover, A., Shkedy, D., Oren, M., and Bercovich, B., 1994, Degradation of the tumor suppressor protein p53 by the ubiquitin-mediated proteolytic system requires a novel species of ubiquitin-carrier protein, E2, *J. Biol. Chem.* **269:**9582–9589.

Crook, T., and Vousden, K. H., 1992, Properties of p53 mutations detected in primary and secondary cervical cancers suggest mechanisms of metastasis and involvement of environmental carcinogens, *EMBO J.* **11:** 3935–3940.

Crook, T., Wrede, D., and Vousden, K. H., 1991, p53 point mutation in HPV negative human cervical carcinoma cell lines, *Oncogene* **6:**873–875.

Crook, T., Ludwig, R. L., Marston, N. J., Willkomm, D., and Vousden, K. H., 1996, Sensitivity of p53 lysine mutants to ubiquitin-directed degradation targeted by human papillomavirus E6, *Virology* **217:**285–292.

Debbas, M., and White, E., 1993, Wild-type p53 mediates apoptosis by E1A, which is inhibited by E1B, *Genes Dev.* **7:**546–554.

DeCaprio, J. A., Ludlow, J. W., Figge, J., Shew, J.-Y., Huang, C.-M., Lee, W.-H., Marsilio, E., Paucha, E., and Livingston, D. M., 1988, SV40 large tumor antigen forms a specific complex with the product of the retinoblastoma susceptibility gene, *Cell* **54:**275–283.

Demers, G. W., Foster, S. A., Halbert, C. L., and Galloway, D. A., 1994, Growth arrest by induction of p53 in DNA damaged keratinocytes is bypassed by human papillomavirus 16 E7, *Proc. Natl. Acad. Sci. USA* **91:**4382–4386.

DeVilliers, E. M., 1989, Heterogeneity of the human papillomavirus group, *J. Virol.* **63:**4898–4903.

Dohmen, R. J., Madura, K., Bartel, B., and Varshavsky, A., 1991, The N-end rule is mediated by the UBC2(RAD6) ubiquitin-conjugating enzyme, *Proc. Natl. Acad. Sci. USA* **88:**7351–7355.

Dyson, N., Guida, P., Münger, K., and Harlow, E., 1992, Homologous sequences in adenovirus E1A and human papillomavirus E7 proteins mediate interaction with the same set of cellular proteins, *J. Virol.* **66:**6893–6902.

El-Deiry, W. S., Tokino, T., Velculescu, V. E., Levy, D. B., Parsons, R., Trent, J. M., Lin, D., Mercer, W. E., Kinzler, K. W., and Vogelstein, B., 1993, WAF1, a potential mediator of p53 tumor suppression, *Cell* **75:**817–825.

Foster, S. A., Demers, G. W., Etscheid, B. G., and Galloway, D. A., 1994, The ability of human papillomavirus E6 proteins to target p53 for degradation in vivo correlates with their ability to abrogate actinomycin D-induced growth arrest, *J. Virol.* **68:**5698–5705.
Friend, S., 1994, p53: A glimpse at the puppet behind the shadow play, *Science* **265:**334–335.
Funk, W. D., Pak, D. T., Karas, R. H., Wright, W. E., and Shay, J. W., 1992, A transcriptionally active DNA-binding site for human p53 protein complexes, *Mol. Cell. Biol.* **12:**2866–2871.
Girod, P. A., and Vierstra, R. D., 1993, A major ubiquitin conjugation system in wheat germ extracts involves a 15-kDa ubiquitin-conjugating enzyme (E2) homologous to the yeast UBC4/UBC5 gene products, *J. Biol. Chem.* **268:**955–960.
Girod, P.-A., Carpenter, T. P., van Nocker, S., Sullivan, M. L., and Vierstra, R. D., 1993, Homologs of the essential ubiquitin conjugating enzymes UBC1, 4, and 5 in yeast are encoded by a multigene family in *Arabidopsis thaliana, Plant J.* **3:**545–552.
Gronostajski, R. M., Goldberg, A. L., and Pardee, A. B., 1984, Energy requirement for degradation of tumor associated protein p53, *Mol. Cell. Biol.* **4:**442–448.
Gu, J., Ren, K., Dubner, R., and Iadarola, M. J., 1994, Cloning of a DNA binding protein that is a tyrosine kinase substrate and recognizes an upstream initiator-like sequence in the prodynorphin promoter, *Mol. Brain Res.* **24:**77–88.
Han, J., Sabbatini, P., Perez, D., Rao, L., Modha, D., and White, E., 1996, The E1B 19K protein blocks apoptosis by interacting with and inhibiting the p53-inducible and death-promoting Bax protein, *Genes Dev.* **10:**461–477.
Hawley-Nelson, P., Vousden, K. H., Hubbert, N. L., Lowy, D. R., and Schiller, J. T., 1989, HPV16 E6 and E7 proteins cooperate to immortalize human foreskin keratinocytes, *EMBO J.* **8:**3905–3910.
Hubbert, N. L., Sedman, S. A., and Schiller, J. T., 1992, Human papillomavirus type 16 E6 increases the degradation rate of p53 in human keratinocytes *J. Virol.* **66:**6237–6241.
Huibregtse, J. M., Scheffner, M., and Howley, P. M., 1991, A cellular protein mediates association of p53 with the E6 oncoprotein of human papillomavirus types 16 or 18, *EMBO J.* **10:**4129–4135.
Huibregtse, J. M., Scheffner, M., and Howley, P. M., 1993a, Cloning and expression of the cDNA for E6-AP, a protein that mediates the interaction of the human papillomavirus E6 oncoprotein with p53, *Mol. Cell. Biol.* **13:**775–784.
Huibregtse, J. M., Scheffner, M., and Howley, P. M., 1993b, Localization of the E6-AP regions that direct human papillomavirus E6 binding, association with p53, and ubiquitination of associated proteins, *Mol. Cell. Biol.* **13:**4918–4927.
Huibregtse, J. M., Scheffner, M., Beaudenon, S., and Howley, P. M., 1995, A family of proteins structurally and functionally related to the E6-AP ubiquitin-protein ligase, *Proc. Natl. Acad. Sci. USA* **92:**2563–2567.
Hupp, T, R., Meek, D. W., Midgley, C. A., and Lane, D. P., 1992, Regulation of the specific DNA binding function of p53, *Cell* **71:**875–886.
Jensen, J. P., Bates, P. W., Yang, M., Vierstra, R. D., and Weissman, A. M., 1995, Identification of a family of closely related human ubiquitin conjugating enzymes, *J. Biol. Chem.* **270:**30408–30413.
Johnson, E. S., Ma, P. C. M., Ota, I, M., and Varshavsky, A., 1995, A proteolytic pathway that recognizes ubiquitin as a degradation signal, *J. Biol. Chem.* **270:**17442–17456.
Juven, T., Barak, Y., Zauberman, A., George, D. L., and Oren, M., 1993, Wild type p53 can mediate sequence-specific transactivation of an internal promoter within the mdm2 gene, *Oncogene* **8:** 3411–3416.
Kastan, M. B., Onyerkwere, O., Sidransky, D., Vogelstein, B., and Craig, R. W., 1991, Participation of p53 protein in the cellular response to DNA damage, *Cancer Res.* **53:**6304–6311.
Kern, S. E., Pietenpol, J. A., Thiagalingam, S., Seymour, A., Kinzler, K. W., and Vogelstein, B., 1992, Oncogenic forms of p53 inhibit p53-regulated gene expression, *Science* **256:**827–830.
Kessis, T. D., Slebos, R. J., Nelson, W. G., Kastan, M. B., Plunkett, B. S., Han, S. M., Lorincz, A. T.,

Hedrick, L., and Cho, K. R., 1993, Human papillomavirus 16 E6 expression disrupts the p53-mediated cellular response to DNA damage, *Proc. Natl. Acad. Sci. USA* **90:**3988–3992.

Kishino, T., Lalande, M., and Wagstaff, J., 1997, *UBE3*/E6-AP mutations cause Angelman syndrome, *Nature Genet.* **15:**70–73.

Klingelhutz, A. J., Foster, S. A., and McDougall, J. K., 1996, Telomerase activation by the E6 gene product of human papillomavirus type 16, *Nature* **380:**79–82.

Ko, L. J., and Prives, C., 1996, p53: Puzzle and paradigm, *Genes Dev.* **10:**1054–1072.

Kumar, S., Tomooka, Y., and Noda, M., 1992, Identification of a set of genes with developmentally down-regulated expression in the mouse brain, *Biochem. Biophys. Res. Commun.* **185:**1155–1161.

Lane, D. P., 1992, p53, guardian of the genome, *Nature* **358:**15–16.

Lane, D. P., and Crawford, L. V., 1979, T antigen is bound to a host protein in SV40-transformed cells, *Nature* **278:**261–263.

Lechner, M. S., and Laimins, L. A., 1994, Inhibition of p53 DNA binding by human papillomavirus E6 proteins, *J. Virol.* **68:**4262–4273.

Levine, A, J., Momand, J., and Finlay, C. A., 1991, The p53 tumour suppressor gene, *Nature* **351:** 453–456.

Li, X., and Coffino, P., 1996, High-risk human papillomavirus E6 protein has two distinct binding sites within p53, of which only one determines degradation, *J. Virol.* **70:**509–516.

Linzer, D. I. H., and Levine, A. J., 1979, Characterization of a 54K dalton cellular SV40 tumor antigen present in SV40-transformed cells and uninfected embryonal carcinoma cells, *Cell* **17:**43–52.

Lowe, S. W., and Ruley, H. E., 1993, Stabilization of the p53 tumor suppressor is induced by adenovirus 5 ElA and accompanies apoptosis, *Genes Dev.* **7:**535–545.

Maki, C. G., Huibregtse, J. M., and Howley, P. M., 1996, In vivo ubiquitination and proteasome-mediated degradation of p53, *Cancer Res,* **56:**2649–2654.

Maltzman, W., and Czyzyk, L., 1984, UV irradiation stimulates levels of p53 cellular tumor antigen in nontransformed mouse cells, *Mol. Cell. Biol.* **4:**1689–1694.

Mansfield, E., Hersperger, E., Biggs, J., and Shearn, A., 1994, Genetic and molecular analysis of *hyperplastic discs*, a gene whose product is required for regulation of cell proliferation in *Drosophila melanogaster* imaginal discs and germ wells, *Dev. Biol.* **165:**507–526.

Martson, N. J., Crook, T., and Vousden, K. H., 1994, Interaction of p53 with MDM2 is independent of E6 and does not mediate wild type transformation suppressor function, *Oncogene* **9:**2707–2716.

Matsuura, T., Sutcliffe, J. S., Fang, P., Galjaard, R.-J., Jiang, Y., Benton, C. S., Rommens, J. M., and Beaudet, A. L., 1997, *De novo* truncating mutations in E6-AP ubiquitin-protein ligase gene (*UBE3A*) in Angelman syndrome, *Nature Genet.* **15:**74–77.

Medcalf, E. A., and Milner, J., 1993, Targeting and degradation of p53 by E6 of human papillomavirus type 16 is preferential for the 1620+ p53 conformation, *Oncogene* **8:**2847–2851.

Mietz, J. A., Unger, T., Huibregtse, J. M., and Howley, P. M., 1992, The transcriptional transactivation function of wild-type p53 is inhibited by SV40 large T-antigen and by HPV-16 E6 oncoprotein, *EMBO J.* **11:**5013–5020.

Molinari, M., and Milner, J., 1995, p53 in complex with DNA is resistant to ubiquitin-dependent proteolysis in the presence of HPV-16 E6, *Oncogene* **10:**1849–1854.

Müller, D., Rehbein, M., Baumeister, H., and Richter, D., 1992, Molecular characterization of a novel rat protein structurally related to poly(A) binding proteins and the 70K protein of the U1 small nuclear ribonuclear particle (snRNP), *Nucleic Acids Res.* **20:**1471–1475.

Münger, K., Phelps, W. C., Bubb, V., Howley, P. M., and Schlegel, R., 1989, The E6 and E7 genes of the human papillomavirus type 16 together are necessary and sufficient for transformation of primary human keratinocytes, *J. Virol.* **63:**4417–4421.

Nakao, M., Sutcliffe, J. S., Durtschi, B., Mutirangura, A., Ledbetter. D. H., and Beaudet, A., 1994, Imprinting analysis of three genes in the Prader-Willi/Angelman region: SNRNP, E6-associated protein, and PAR-2 (D15S225E), *Hum. Mol. Genet.* **3:**309–315.

Nefsky, B., and Beach, D., 1996, Publ acts as an E6-AP-like protein ubiquitin ligase in the degradation of cdc25, *EMBO J.* **15:**1301–1312.

Nuber, U., Schwarz, S., Kaiser, P., Schneider, R., and Scheffner, M., 1996, Cloning of human ubiquitin-conjugating enzymes UbcH6 and UbcH7 (E2-F1) and characterization of their interaction with E6-AP and RSP5, *J. Biol. Chem.* **271:**2795–2800.

Pietenpol, J. A., Tokino, T., Thiagalingam, S., El-Deiry, W. S., Kinzler, K. W., and Vogelstein, B., 1994, Sequence-specific transcriptional activation is essential for growth suppression by p53, *Proc. Natl. Acad. Sci. USA* **91:**1998–2002.

Rosa, J. L., Casaroli-Marano, R. P., Buckler, A. J., Vilaro, S., and Barbacid, M., 1996, p619, a giant protein related to the chromosome condensation regulator RCC1, stimulates guanine nucleotide exchange in ARF1 and Rab proteins, *EMBO J.* **15:**4262–4273.

Sarnow, P., Ho, Y. S., Williams, J., and Levine, A. J., 1982, Adenovirus E1b-58kd tumor antigen and SV40 large tumor antigen are physically associated with the same 54kd cellular protein in transformed cells, *Cell* **28:**387–394.

Scheffner, M., Werness, B. A., Huibregtse, J. M., Levine, A, J., and Howley, P. M., 1990, The E6 oncoprotein encoded by human papillomavirus types 16 and 18 promotes the degradation of p53, *Cell* **63:**1129–1136.

Scheffner, M., Münger, K., Byrne, J. C., and Howley, P. M., 1991, The state of the p53 and retinoblastoma genes in human cervical carcinoma cell lines, *Proc. Natl. Acad. Sci. USA* **88:**5523–5527.

Scheffner, M., Münger, K., Huibregtse, J. M., and Howley, P. M., 1992a, Targeted degradation of the retinoblastoma protein by human papillomavirus E7-E6 fusion proteins, *EMBO J.* **11:**2425–2431.

Scheffner, M., Takahashi, T., Huibregtse, J. M., Minna, J. D., and Howley, P. M., 1992b, Interaction of the human papillomavirus type 16 oncoprotein with wild-type and mutant p53 proteins, *J. Virol.* **66:**5100–5105.

Scheffner, M., Huibregtse, J. M., Vierstra, R. D., and Howley, P. M., 1993, The HPV-16 E6 and E6-AP complex functions as a ubiquitin-protein ligase in the ubiquitination of p53, *Cell* **75:**495–505.

Scheffner, M., Huibregtse, J. M., and Howley, P. M., 1994, Identification of a human ubiquitin-conjugating enzyme that mediates the E6-AP-dependent ubiquitination of p53, *Proc. Natl. Acad. Sci. USA* **91:**8797–8801.

Scheffner, M., Nuber, U., and Huibregtse, J. M., 1995, Protein ubiquitination involving an E1–E2–E3 enzyme thioester cascade, *Nature* **373:**81–83.

Seufert, W., and Jentsch, S., 1990, Ubiquitin conjugating enzymes UBC4 and UBC5 mediate selective degradation of short-lived and abnormal proteins, *EMBO J.* **9:**543–550.

Takenaka, I., Morin, F., Seizinger, B. R., and Kley, N., 1995, Regulation of the sequence-specific DNA binding function of p53 by protein kinase C and protein phosphatases, *J. Biol. Chem.* **270:**5405–5411.

van den Heuvel, S. J., van Laar, T., The, I., and van der Eb, A. J., 1993, Large E1B proteins of adenovirus types 5 and 12 have different effects on p53 and distinct roles in cell transformation, *J. Virol.* **67:**5226–5234.

Vogelstein, B., and Kinzler, K. W., 1992, p53 function and dysfunction, *Cell* **70:**523–526.

von Knebel-Doeberitz, M., Oltersdorf, T., Schwarz, E., and Gissmann, L., 1988, Correlation to modify human papillomavirus early gene expression with altered growth properties in C4-I cervical carcinoma cells, *Cancer Res.* **48:**3780–3785.

von Knebel-Doeberitz, M., Bauknecht, T., Bartsch, D., and zur Hausen, H., 1991, Influence of chromosomal integration on glucocorticoid-regulated transcription of growth-stimulating papillomavirus genes E6 and E7 in cervical carcinoma cells, *Proc. Natl. Acad. Sci. USA* **88:**1411–1415.

Werness, B. A., Levine, A. J., and Howley, P. M., 1990, Association of human papillomavirus types 16 and 18 E6 proteins with p53, *Science* **248:**76–79.

White, E., 1994, p53, guardian of Rb, *Nature* **371:**21–22.

Whyte, P., Buchkovich, K. J., Horowitz, J. M., Friend, S. H., Raybuck, M., Weinberg, R. A., and

Harlow, E., 1988, Association between an oncogene and an antioncogene: The adenovirus E1a proteins bind to the retinoblastoma gene product, *Nature* **334:**124–129.

Yamamoto, Y., Huibregtse, J. M., and Howley, P. M., 1997, The human E6-AP gene (*UBE3A*) encodes three potential protein isoforms generated by differential splicing, *Genomics,* **41:**263–266.

Yew, P. R., and Berk, A, J., 1992, Inhibition of p53 transactivation required for transformation by adenovirus early 1B protein, *Nature* **357:**82–85.

zur Hausen, H., and Schneider, A., 1987, The role of papillomaviruses in human anogenital cancers, in *The Papovaviridae* (N. Salzman and P. M. Howley, eds.), pp. 245–263, Plenum Press, New York.

CHAPTER 12

Cell Cycle Control by Ubiquitin-Dependent Proteolysis

Jan-Michael Peters, Randall W. King, and Raymond J. Deshaies

1. INTRODUCTION

The prime directive of the dividing cell is to pass on two identical copies of its genetic blueprint. This requires accurate replication of DNA and accurate segregation of replicated chromosomes. The cell cycle control machinery regulates progression from one phase of the chromosome cycle to the next, ensuring that DNA replication (S phase) alternates with chromosome segregation (mitosis) to maintain a constant ploidy. In most cells, growth is coordinated with division to maintain a constant average cell size; the length of the gap phases that occur before (G1) and after (G2) S phase must therefore be regulated. In contrast, early embryonic division occurs in the absence of cell growth, so that the cell cycle consists solely of rapidly alternating S and M phases. Despite these distinct physiologies, a common biochemical machinery regulates cell division in all eukaryotic cells.

Jan-Michael Peters and Randall W. King • Department of Cell Biology, Harvard Medical School, Boston, Massachusetts 02115. **Raymond J. Deshaies** • Division of Biology, California Institute of Technology, Pasadena, California 91125. *Present address of J.-M.P.*: Research Institute of Molecular Pathology, A-1030 Vienna, Austria.

Ubiquitin and the Biology of the Cell, edited by Peters *et al.* Plenum Press, New York, 1998.

Transitions from one stage of the cell cycle to the next are often associated with the activation or inactivation of protein kinases known as the cyclin-dependent kinases (CDKs; reviewed in Morgan, 1995; King *et al.*, 1994; Nasmyth, 1993). Although the cell cycle of fungi is controlled by only a single CDK (called Cdc28p in budding yeast and Cdc2p in fission yeast), a family of divergent cyclins, which are essential activators of the kinase, affords functional diversification. In budding yeast, the cyclins Cln1p, Cln2p, and Cln3p are expressed early in the cell cycle and promote events that are unique to G1 phase. Clb5p and Clb6p are expressed later and promote S phase, whereas the mitotic cyclins Clb1p through Clb4p drive entry into mitosis. Although the family of CDKs has expanded in higher eukaryotes, they appear to regulate the same events as their fungal relatives. The CDKs of higher eukaryotes form complexes with functional homologues of the budding yeast cyclins, with cyclins D and E functioning during G1, cyclins E and A during S phase, and cyclins A and B during mitosis.

This chapter reviews the important role that ubiquitin-mediated proteolysis plays in regulating cellular division (for related reviews see King *et al.*, 1996a; Deshaies, 1995a). The proteolysis of cell cycle regulators is intimately coupled to protein phosphorylation catalyzed by CDKs: Phosphorylation plays an important role in regulating ubiquitin-mediated proteolysis during the cell cycle; conversely, degradation of cyclins and CDK inhibitors is essential for the proper regulation of CDKs. Although many of the components involved in regulating cell cycle progression have been identified, we have only begun to catch a glimpse of how the complex interplay of protein phosphorylation and protein degradation produces orderly progression through the cell cycle.

In the first half of this chapter, we discuss a ubiquitin-dependent proteolytic pathway, the *CDC34* pathway, that regulates the transition from G1 to S phase. In the second half of the review, we describe a second ubiquitin-dependent proteolytic system, the APC pathway, that triggers anaphase and the exit from mitosis by initiating the destruction of proteins that regulate sister chromatid cohesion and mitotic cyclins, respectively (Fig. 1).

2. THE ROLE OF UBIQUITIN-DEPENDENT PROTEOLYSIS IN THE G1/S TRANSITION

In most eukaryotic cells, division is made contingent on cell growth that occurs during G1. Cells become committed to a round of cell division only after they have reached a point in G1 called START in budding yeast or the Restriction Point in animal cells. A yeast or animal cell that is deprived of essential nutrients or growth factors while it is in early G1 phase will not divide and instead will enter a period of quiescence (G0). However, once the cell has passed START, it will

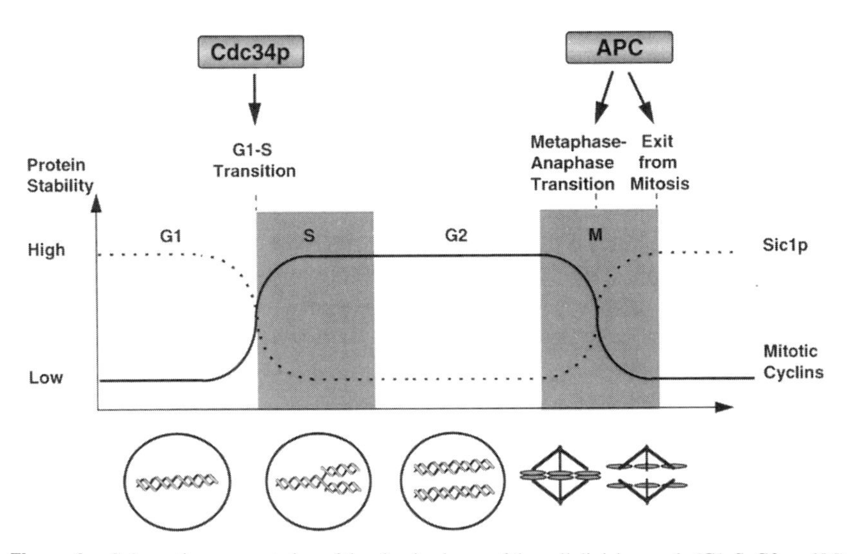

Figure 1. Schematic representation of the classic phases of the cell division cycle (G1, S, G2, and M) and the chromosome cycle (bottom panel), indicating when the Cdc34p- and APC-dependent ubiquitination pathways control important transitions during the cell cycle. The relative stability of Sic1p, a key substrate of the Cdc34p pathway, and of mitotic cyclins, the most prominent substrates of the APC pathway, are also indicated. Sic1p is a protein that has so far only been found in budding yeast, but functionally related proteins are thought to play a similar role in the G1/S transition in other eukaryotes. Although the stability of Sic1p and mitotic cyclins appears complementary, their proteolysis seems to be regulated by principally different mechanisms (for details see text and Figs. 2 and 5). It is also important to note that the restabilization of mitotic cyclins at the G1–S transition does not depend on the *CDC34* pathway and can occur, although reversibly, in cdc4^{ts} cells (Schwob *et al.*, 1996; Amon *et al.*, 1994). Note that the metaphase–anaphase transition, the "point of no return" in the process of chromosome segregation, does not coincide with any of the classical transitions from one cell cycle phase to the next but instead divides mitosis into two separate phases.

divide regardless of the concentration of nutrients and growth factors in the growth medium.

Whereas both passage through START and the subsequent transition from G1 to S phase require the activity of CDKs complexed with G1-specific cyclins, these two events are temporally separable and are regulated distinctly. In budding yeast, different genes are required for START (e.g., *CDC28*, G1 cyclins, *CDC37*) and the G1/S transition (*CDC4*, *CDC34*, *CDC53 and SKP1*). At START, budding yeast cells assemble active G1 cyclin/CDK complexes, accumulate DNA replication proteins, and inactive S phase-inducing cyclin/CDK complexes, and extinguish pathways characteristic of the pre-START G1 phase (e.g., cyclin B proteolysis). Later, at the G1/S transition, a further series of changes occurs: Latent S phase-inducing cyclin/CDK complexes are activated, a G1-specific pattern of gene expression yields to a pattern characteristic of S phase, and pathways restricted to post-START G1 phase, such as the initiation of bud growth, are shut off. Recent evidence indicates that the conversion from a post-START G1 state to an S phase state at the G1/S transition is regulated by the phosphorylation of the CDK inhibitor Sic1p, which renders it susceptible to ubiquitination by the *CDC34* pathway and subsequent destruction.

2.1. Physiology and Biochemistry of the *CDC34* Pathway

2.1.1. Background

An important role for ubiquitin-mediated proteolysis in cell cycle control was suggested by the observation that the *CDC34* gene of *Saccharomyces cerevisiae* encodes a ubiquitin-conjugating enzyme (Goebl *et al.*, 1988). On incubation at the restrictive temperature, strains harboring $cdc34^{ts}$ mutations accumulate high levels of G1 cyclins (Deshaies *et al.*, 1995; Yaglom *et al.*, 1995), produce new daughter buds, and complete spindle pole body duplication. Nevertheless, $cdc34^{ts}$ cells fail to initiate DNA replication and remain arrested in a G1-like state (Goebl *et al.*, 1988). This constellation of phenotypes indicates that *CDC34* function is dispensible for the execution of START, but is required for cells to transit from G1 to S phase. Besides *CDC34*, genetic analyses have revealed that *CDC4*, *CDC53* (Mathias *et al.*, 1996), *SKP1* (Bai *et al.*, 1996), and G1 cyclins (*CLN1–3*) (Schneider *et al.*, 1996) are also required for cells to pass from G1 into S phase (Table 1). Genetic and biochemical studies summarized below indicate that the products of

Table I
Genes Implicated in the *CDC34* Pathway

Gene	Implicated in proteolysis *in vivo* (protein)	Implicated in ubiquitination *in vitro* (protein)	Gene essential?	References
CDC4	Yes (Sic1p, Cdc6p)	Yes (Sic1p)	Yes	Piatti *et al.* (1996), Schwob *et al.* (1994)
CDC34	Yes (Cln3p)	Yes (Sic1p)	Yes	Deshaies *et al.* (1995), Yaglom *et al.* (1995), Schwob *et al.* (1994)
CDC53	Yes (Cln2p, Sic1p)	Yes (Sic1p)	Yes	Willems *et al.* (1996), Mathias *et al.* (1996)
SKP1	Yes (Cln2p, Sic1p)	Yes (Sic1p)	Yes	Bai *et al.* (1996)
GRR1	Yes (Cln2p)	Not tested	No	Barral *et al.* (1995)
SCM4	Not tested	Not tested	No	Smith *et al.* (1992)
UBS1	Not tested	Not tested	No	Prendergast *et al.* (1996)
CLN1–3	Yes (Sic1p)	Yes (Sic1p)	Yes	Schneider *et al.* (1996)

these genes act in the same pathway and that Cdc34p, Cdc4p, Cdc53p, and Skplp are part of a multisubunit ubiquitin-conjugating complex. Thus, for simplicity, we will hereafter refer to *CDC34*, *CDC4*, *CDC53*, *SKP1*, and their products collectively as the *CDC34* pathway.

2.1.2. Physiological Role of the *CDC34* Pathway in G1 Cells

How do *CDC4*, *CDC34*, *CDC53*, and *SKP1* promote the transition from G1 to S phase? An elegant study by Schwob *et al.* (1994) indicated that the *CDC34* pathway is required for the activation of a set of cyclin B/Cdc28p complexes that trigger the initiation of DNA synthesis. A budding yeast strain deleted for all six known cyclin B genes (*CLB1–CLB6*) and sustained by expression of *CLB5* from the regulatable *GAL1* promoter arrests cell division at the same point as *CDC34* pathway *ts* mutants when expression of *CLB5* is extinguished. This observation suggested that the *CDC34* pathway might be required for the generation of active Clb/Cdc28p complexes. Indeed, *CDC34* pathway *ts* mutants accumulate inactive Clb5p/Cdc28p complexes, and extracts from these mutants contain high levels of a Clb5p/Cdc28p inhibitor. A candidate for this inhibitor was Siclp/Sdb25p (for simplicity, we will use "Siclp" throughout the remainder of this review), which was previously identified as both a tightly bound inhibitor of the Cdc28p protein kinase (Nugroho and Mendenhall, 1994; Mendenhall, 1993), and a multicopy suppressor of the lethal anaphase arrest of *dbf2*^{ts} mutants (Donovan *et al.*, 1994). Remarkably, *cdc34*^{ts} *sic1*Δ, *cdc4*^{ts} *sic1*Δ, *cdc53*^{ts} *sic1*Δ, and *skp1*^{ts} *sic1*Δ double mutants do not accumulate Clb5p/Cdc28p inhibitory activity, and are able to replicate DNA at the nonpermissive temperature (Bai *et al.*, 1996; Schwob *et al.*, 1994). Furthermore, whereas Siclp disappears as wild-type cells transit from G1 to S phase, Siclp stably accumulates in *CDC34* pathway *ts* mutants. These observations led to the hypothesis that the *CDC34* pathway specifies the ubiquitin-dependent destruction of Siclp, which otherwise restrains Clb5p/Cdc28p complexes from triggering the initiation of DNA synthesis (Schwob *et al.*, 1994). This hypothesis has recently been confirmed by the observation that expression of a nondegradable form of Siclp blocks cell cycle progression at the G1/S boundary (Verma *et al.*, 1997a).

Although its role in promoting the destruction of Siclp accounts for the cell cycle arrest of *CDC34* pathway mutants, the *CDC34* pathway may also contribute to the regulation of G1/S by limiting the accumulation of G1 cyclins. The G1 cyclins Cln2p and Cln3p are substrates of the *CDC34* pathway (Willems *et al.*, 1996; Deshaies *et al.*, 1995; Yaglom *et al.*, 1995), and *cis*-acting mutations that stabilize Cln2p and Cln3p shorten G1 and perturb the regulation of START (Lanker *et al.*, 1996; Yaglom *et al.*, 1995; Cross and Blake, 1993; Tyers *et al.*, 1992).

2.1.3. Biochemical Functions of *CDC34* Pathway Components

How does the *CDC34* pathway promote the destruction of Sic1p at the G1/S transition? Because *CDC34* encodes a ubiquitin-conjugating enzyme, the most parsimonious model is that Cdc34p ubiquitinates Sic1p, thereby targeting it for degradation by the 26 S proteasome. Although the molecular function of Cdc34p can be inferred from its sequence, the sequences of Cdc4p, Cdc53p, and Skp1p are less informative. *CDC4* encodes an 86-kDa protein that contains a recently described Skp1p interaction motif named the "F box," 7–8 copies of the WD-40 repeat, and unique N- and C-terminal domains (Bai *et al.*, 1996; Peterson *et al.*, 1984). Characterization of *cdc4* mutant alleles suggests that all four domains of *CDC4* are essential (Bai *et al.*, 1996; Mathias *et al.*, 1996; Johnson, 1991). *CDC53* encodes a 94-kDa protein that is related to at least five genes of unknown function in a variety of organisms including humans and nematodes (Kipreos *et al.*, 1996; Mathias *et al.*, 1996). *SKP1* encodes a 23-kDa protein (Bai *et al.*, 1996) that is 48% identical to a human protein that was originally identified via its association with SKP2/cyclin A/CDK2 complexes (Zhang *et al.*, 1995). Human *SKP1* can complement a yeast *skp1*^{ts} mutant (Bai *et al.*, 1996).

Although the functions of Cdc4p, Cdc53p, and Skp1p are unclear, genetic and biochemical data suggest that they collaborate with Cdc34p to effect multiubiquitination of unstable proteins. All double combinations of *cdc4*^{ts}, *cdc34*^{ts}, and *cdc53*^{ts} mutations are inviable at the normally permissive temperature, and overexpression of *CDC53* or *SKP1* suppresses *cdc4*^{ts}, and overexpression of *CDC4* suppresses *skp1*^{ts} and *cdc53*^{ts} mutants (Bai *et al.*, 1996; Mathias *et al.*, 1996). As is often the case, such genetic interactions are indicative of physical interactions between these polypeptides: Cdc4p and Cdc53p are associated with Cdc34p in yeast lysates (Mathias *et al.*, 1996), and Cdc53p–Cdc34p, Cdc53p–Skp1p, and Skp1p–Cdc4p binary interactions can be demonstrated with recombinant proteins (Feldman *et al.*, 1997; Skowyra *et al.*, 1997; Bai *et al.*, 1996; Willems *et al.*, 1996).

Besides interacting with Cdc4p and Cdc34p, Cdc53p has also been identified as a component of affinity-purified Cln2p/Cdc28p complexes (Willems *et al.*, 1996). *CDC53* is required for rapid Cln2p destruction, and Cdc53p associates with phosphorylated, but not unphosphorylated Cln2p complexed with Cdc28p, suggesting that Cdc53p targets phosphorylated Cln2p for Cdc34p-dependent ubiquitination and destruction. No direct reconstitution of Cdc53p interaction with phosphorylated Cln2p was reported, however, leaving open the possibility that Cdc53p and phospho-Cln2p interact through the agency of an unidentified protein.

Recently, reconstitution of Sic1p ubiquitination *in vitro* has provided direct evidence that Cdc34p, Cdc4p, Cdc53p, and Skp1p define a novel ubiquitination pathway. Sic1p is extensively multiubiquitinated on incubation in wild-type, but not *cdc4*^{ts} or Cdc34p-depleted extracts, and addition of recombinant Cdc34p or Cdc4p rescues the defect of the corresponding mutant extract (Verma *et al.*, 1997b). Furthermore, efficient ubiquitination of Sic1p can be reconstituted by mixing together purified E1 enzyme, ubiquitin, Cln2p/Cdc28p (see Note Added in Proof), Cdc34p, Cdc4p, Cdc53p, and Skp1p. These data indicate that the genetically identified members of the Cdc34p pathway define a set of components that is both necessary and sufficient to catalyze Sic1p ubiquitination. This *in vitro* system should now permit a systematic dissection of the functions of the enigmatic Cdc4p, Cdc53p, and Skp1p proteins.

Because Cdc4p, Cdc53p, and Skp1p assist Cdc34p in the ubiquitination of substrates, these proteins appear to fulfill the operational definition of an E3 enzyme, a component that is required for substrate ubiquitination in the presence of a ubiquitin-charged E2 (Hershko *et al.*, 1983). Intriguingly, there is no apparent homology between Cdc4p/Cdc53p/Skp1p and Ubr1p (see Chapter 8) or the E6-AP family (see Chapters 3 and 11) of E3s. Thus, Sic1p ubiquitination may proceed by a mechanism different from that described for the E6-AP-mediated ubiquitination of p53, and Cdc4p, Cdc53p, Skp1p may be charter members of a novel class of E3 enzymes.

At least five lines of evidence suggest that Cdc4p, Cdc53p, and Skp1p do not always function in lockstep to mediate the ubiquitination of Cdc34p pathway substrates. First, besides its role in Sic1p destruction, Skp1p also serves as an essential subunit of the centromere-binding CBF3 complex (Connelly and Hieter, 1996; Stemmann and Lechner, 1996). Second, whereas *CDC4*, *CDC53*, and *CDC34* function are required for the destruction of Sic1p, *CDC4* is not required for the *CDC34*- and *CDC53*-dependent destruction of Cln2p (R. J. Deshaies, unpublished data). Third, although Cln2p proteolysis is severely reduced in *grr1Δ* cells (Barral *et al.*, 1995), Grr1p is unlikely to be directly involved in Sic1p proteolysis because *grr1Δ* cells are viable and exhibit a shortened G1 phase. Fourth, Cdc4p apparently does not associate with the pool of Cdc53p that binds to phosphorylated Cln2p (Willems *et al.*, 1996). Fifth, mutant alleles of *SKP1* have reciprocal effects on proteolysis; *skp1-11*^{ts} stabilizes Sic1p but not Cln2p, whereas *skp1-12*^{ts} stabilizes Cln2p but not Sic1p (Bai *et al.*, 1996). These observations can be reconciled by assuming that the individual subunits of the putative Cdc4p/Cdc53p/Skp1p complex are in dynamic equilibrium in living cells. For example, by substituting Grr1p for Cdc4p, the substrate specificity of the complex could readily be changed from Sic1p to Cln2p. Such rearrangements could be mediated by the Skp1p-binding F box domains of Cdc4p and Grr1p (Bai *et al.*, 1996). Likewise, Skp1p might tether Cdc53p and its partners to the kinetochore, thereby accounting for the Cdc34p-dependent ubiquitination of the kinetochore protein Ndc10p (Yoon and Carbon, 1995). The shorthand used so far in this review, "*CDC34* pathway," is thus probably an oversimplification of the boundaries of Cdc34p, Cdc4p, Cdc53p, and Skp1p function.

2.2. Regulation of *CDC34*-Dependent Ubiquitination

2.2.1. CDK-Dependent Phosphorylation Triggers Destruction of Sic1p and G1 Cyclins

Accumulating evidence indicates that substrate phosphorylation triggers Cdc34p-dependent ubiquitination (Fig. 2). A causal link between substrate phosphorylation and ubiquitin-dependent destruction via the Cdc34p pathway was originally suggested by the observations that (1) Cln2p and Cln3p are partially stabilized by *cdc28*^{ts} mutations; (2) destruction of β-galactosidase-Cln3p chimeras requires Cdc28p-dependent phosphorylation of a CDK consensus site in the destabilization domain of Cln3p; and (3) Cdc34p-dependent ubiquitination of Cln2p *in vitro* requires Cdc28p activity (Deshaies *et al.*, 1995; Yaglom *et al.*, 1995). Lanker *et al.* (1996) strengthened this hypothesis by demonstrating that a mutated version of Cln2p lacking seven consensus Cdc28p phosphorylation sites (Cln2p-4T3S) is dramatically stabilized *in vivo*. Surprisingly, Cln2p-4T3S is

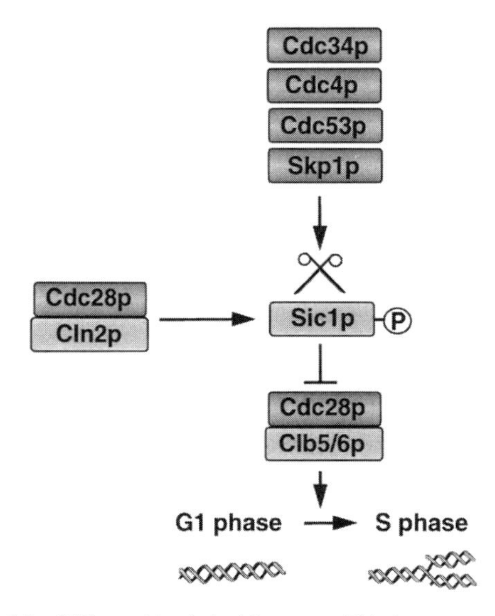

Figure 2. A model of the G1/S transition in budding yeast. This figure summarizes data presented in this review on the mechanism and regulation of Sic1p destruction: In G1 cells Sic1p binds to and inhibits Cdc28p-Clb kinase complexes and thus prevents the initiation of DNA replication. This inhibition is released once Sic1p becomes phosphorylated by Cdc28p-Cln2p kinase, which allows the ubiquitination and subsequent destruction of Sic1p by the *CDC34* pathway. It is unknown if Cdc4p, Cdc53p, Skp1p, and Cdc34p exist as a stable particle. Although the putative Cdc4p/Cdc53p/Skp1p complex is shown to be interacting directly with phosphate groups on Sic1p, it is not yet known how phosphorylation targets Sic1p for ubiquitination (see Section 2.3.2 and Fig. 3).

substantially more stable than either a deletion mutant of Cln2p (Cln2p-Δxs) that is unable to bind Cdc28p or wild-type Cln2p expressed in *cdc28*^{ts} cells (Lanker *et al.*, 1996; Deshaies *et al.*, 1995). In a double mutant analysis, the Δxs mutation is epistatic to the 4T3S allele, implying that there exist two pathways for Cln2p destruction: The first targets Cln2p that is bound to and phosphorylated by Cdc28p, whereas the second preferentially degrades free Cln2p, and does not depend on Cdc28p activity. It is important to note that proteolysis of phosphorylated Cln2p persists in *cdc34*^{ts} cells (albeit at a rate about fourfold slower than that observed in wild type) (Willems *et al.*, 1996; Deshaies *et al.*, 1995). This residual destruction of Cln2p may be related to either the leakiness of the *cdc34*^{ts} mutant, the activity of the phosphorylation-independent Cln2p destruction pathway mentioned above, or the presence of yet a third pathway for Cln2p proteolysis. Regardless of the underlying cause, the complex interplay of destruction pathways may confound analyses of ubiquitin-dependent proteolysis (Salama *et al.*, 1994).

Recently, Blondel and Mann (1996) published a study of Cln1p and Cln2p degradation in various mutant strains. The major findings in this paper are as follows. Both Clb-deficient and *ubc9* mutants accumulate stable Cln1/2p. Furthermore, the stabilization of Cln2p in *cdc34* mutants (Bai *et al.*, 1996; Willems *et al.*, 1996, Deshaies *et al.*, 1995) is suggested to be an artifact of cell cycle arrest at the G1/S boundary, as Cln1/2p are unstable in *cdc34 sic1*Δ mutants. These data suggest that Clbs antagonize Cln accumulation both by inhibiting *CLN* transcription (Amon *et al.*, 1993) and by stimulating Cln proteolysis, although it is unclear if Clb-dependent proteolysis of Cln1/2p is important for proper cell cycle regulation. Whereas the data in this paper appear to be at odds with the suggestion that Cln2p is a Cdc34p substrate (Bai *et al.*, 1996; Willems *et al.*, 1996; Deshaies *et al.*, 1995), several points deserve comment. First, it is unclear if the polyclonal antibody used in this report detects primarily Cln1p or Cln2p; little has been published regarding the degradation of Cln1p, and it is unclear if it behaves as a Cdc34p substrate. Second, it is difficult to exclude the possibility that deletion of *SIC1* derepresses a normally cryptic Cln1/2p degradation pathway. Third, it will be interesting to see if stabilization of Cln2p in *cdc53* and *skp1* mutants is *SIC1*-dependent, as these genes have been reported to be required for Cln2p proteolysis at cell cycle positions other than the G1/S boundary (Bai *et al.*, 1996; Willems *et al.*, 1996).

Several lines of evidence suggest that Cdc34p-dependent destruction of Sic1p is triggered by the Cdc28p-dependent phosphorylation of Sic1p at the G1/S transition. *In vivo* analyses have revealed that Sic1p destruction is prevented in cells arrested at the *cdc28*^{ts} block (Schwob *et al.*, 1994), is substantially delayed in a *cln1*Δ*cln2*Δ mutant (E. Schwob, personal communication), and requires continuous expression of G1 cyclins on reversal of the *cdc34*^{ts} block (Schneider *et al.*, 1996). In fact, switching on Sic1p destruction probably is, along with the polarization of membrane assembly, a key function of G1 cyclins since deletion of

SIC1 advances the timing of S phase entry in a *cln1Δcln2Δ* mutant (Dirick *et al.*, 1995) and suppresses the lethal cell division arrest of a *cln1Δcln2Δcln3Δ* mutant (Schneider *et al.*, 1996; Tyers, 1996). What exactly are the G1 cyclins doing to activate Sic1p destruction? By analogy to Cln2p, one might imagine that G1 cyclin/CDK complexes phosphorylate Sic1p, rendering it competent for ubiquitination by the *CDC34* pathway. Alternatively (or in addition), Cln2p/Cdc28p might activate components of the *CDC34* pathway or quench inhibitors that oppose the *CDC34* pathway. Evaluation of Sic1p ubiquitination reactions reconstituted *in vitro* revealed that G1 cyclin activity is required for the assembly of multiubiquitin chains on Sic1p, and Sic1p is an excellent substrate for Cln2p/Cdc28p protein kinase complexes (Verma *et al.*, 1997b). Moreover, purified phospho-Sic1p is ubiquitinated by the *CDC34* pathway in the absence of Cln2p/Cdc28p activity, indicating that direct phosphorylation of Sic1p by Cln2p/Cdc28p is both necessary and sufficient to trigger its ubiquitination (Verma *et al.*, 1997a). The hypothesis that phosphorylation triggers Sic1p destruction is further supported by the analysis of nonphosphorylatable mutants of Sic1p (Verma *et al.*, 1997a). Mutation of CDK phosphorylation sites in Sic1p both reduces Sic1p ubiquitination *in vitro* and stabilizes Sic1p *in vivo*. Furthermore, expression of nonphosphorylatable Sic1p *in vivo* inhibits progression into S phase, strongly suggesting that phosphorylation-dependent destruction of Sic1p is a prerequisite for the initiation of DNA replication.

2.2.2. Are Cdc34p, Cdc4p, Cdc53p, and Skp1p Regulated?

Little is known about the regulation of the activity, abundance, localization, or posttranslational modification state of *CDC34* pathway components *in vivo*. Cdc34p is phosphorylated by an unknown kinase on serine residues, but the role of this modification and its cell cycle timing have not been reported (Goebl *et al.*, 1994). Although the phosphorylation status of Cdc4p, Cdc53p, and Skp1p have not been directly evaluated, Cdc53p does not serve as a substrate for the tightly bound Cln2p/Cdc28p protein kinase (Willems *et al.*, 1996). Both Cdc34p and Cdc53p form thiol-insensitive conjugates with ubiquitin *in vivo*, but the significance of these modifications is unclear (Willems *et al.*, 1996; Goebl *et al.*, 1994).

The experiments summarized in the previous section are consistent with the notion that destruction of *CDC34* pathway substrates is controlled not by regulating the activity of the *CDC34* pathway enzymes, but by modulating the phosphorylation of the substrate. According to this view, substrates submit themselves, on their phosphorylation by regulated protein kinases, to be disposed of via the action of a constitutively active *CDC34* pathway. The *CDC34* pathway is active at least throughout the period of CDK activity *in vivo*: *CDC34* pathway (and CDK) substrates Cln2p (Willems *et al.*, 1996), Sic1p (Bai *et al.*, 1996), and Cdc6p (Piatti *et al.*, 1996) are unstable in cells arrested during S phase or mitosis. An important caveat is that it is not yet clear whether phosphorylation of *CDC34* pathway substrates by CDK complexes is sufficient to trigger their ubiquitination *in vivo*; certain forms of regulation (i.e., compartmental localization) may not be preserved in the *in vitro* ubiquitination system.

2.3. Recognition of Substrates by the *CDC34* Pathway

2.3.1. Other Substrates of Cdc34p

Besides Siclp, Cln2p, and Cln3p, several other candidate substrates of the Cdc34p pathway have been identified (Table II). The transcription factor Gcn4p (Kornitzer *et al.*, 1994) and the G1 cyclin/Cdc28p inhibitor Farlp (Henchoz *et al.*, 1997; McKinney *et al.*, 1993) accumulate in *cdc34^{ts}* cells, and the multiubiquitination of both proteins in yeast extract is stimulated by Cdc4p and Cdc34p (Y. Chi and R. J. Deshaies, unpublished data). Likewise, the normally unstable Cdc6p is dramatically stabilized in nocodazole-arrested *cdc4^{ts}* cells (Piatti *et al.*, 1996), and serves as a substrate for Cdc4p/Cdc34p-stimulated ubiquitination *in vitro* (Y. Chi and R. J. Deshaies, unpublished data). The Cbf2p subunit of the CBF3 kinetochore-binding complex may also be a Cdc34p substrate; ubiquitinated forms of Cbf2p are detected in wild-type but not *cdc34^{ts}* cells at the nonpermissive temperature. Cbf2p is not extensively multiubiquitinated, and the formation of short ubiquitin chains on Cbf2p does not require *CDC4* (Yoon and Carbon, 1995). Perhaps ubiquitination of Cbf2p influences some aspect of its function other than

Table II
Known and Suspected Substrates of the *CDC34* Ubiquitination Pathway

Protein	*CDC34*-dependent instability *in vivo*?	Cdc34p-dependent ubiquitination *in vitro*?	References
Siclp	Yes	Yes	Schwob *et al.* (1994)
Clnlp	Not tested	Yes	Y. Chi and R. J. Deshaies (unpublished data)
Cln2p	Yes	Yes	Deshaies *et al.* (1995)
Cln3p	Yes	Not tested	Yaglom *et al.* (1995)
Farlp	Not tested	Yes	McKinney and Cross (1995), Y. Chi and R. J. Deshaies (unpublished data)
Gcn4p	Yes	Yes	Kornitzer *et al.* (1994), Y. Chi and R. J. Deshaies (unpublished data)
Cdc6p	Yes	Yes	Piatti *et al.* (1996), Y. Chi and R. J. Deshaies (unpublished data)
Cbf2p	No	Yes	Yoon and Carbon (1995)

its stability. It will be interesting to determine if Cbf2p is targeted for ubiquitination by its CBF3 partner Skplp (Connelly and Hieter, 1996; Stemmann and Lechner, 1996). Lastly, whereas *cdc34*^{ts} *sic1*Δ double mutants fail to arrest at the G1/S transition, they remain temperature-sensitive for growth and arrest in G2/M phase, suggesting that the accumulation of a Cdc34p substrate other than Siclp can block cell division (Schwob *et al.*, 1994).

2.3.2. Signals Involved in Substrate Recognition

Sequence comparisons have failed to identify conserved destruction signals in known *CDC34* pathway substrates, but most Cdc34p substrates contain multiple segments rich in proline, glutamate/aspartate, serine, and threonine (PEST sequences), which are commonly found in unstable proteins (Rogers *et al.*, 1986). It is important to note that PEST elements are related by composition, not by primary sequence similarity. Mutational analyses have shown that PEST segments of Cln2p, Cln3p, and Gcn4p can be important for maximal rates of proteolysis, but individual elements are neither essential nor sufficient for degradation (Yaglom *et al.*, 1995; Kornitzer *et al.*, 1994; Salama *et al.*, 1994). Surprisingly, ~ 40% of the proteins predicted from the yeast genome sequence contain high-scoring PEST regions (D. Mathog and R. J. Deshaies, unpublished data). Taken together with the mutational data, this observation suggests that the PEST algorithm is an imperfect prognosticator of a bona fide destruction-targeting sequence. An accurate description of the true destruction signal embedded within Cdc34p substrates that contain active PEST regions will probably require detailed analyses of the interaction between these substrates and components of the *CDC34* pathway. It is interesting to speculate that the *CDC34* pathway may be the physiological route of destruction for rapidly degraded proteins that contain destabilizing PEST domains.

Besides PEST sequences, a second feature shared by all *CDC34* pathway substrates is that phosphorylation appears to play an important role in their ubiquitin-mediated destruction. All candidate Cdc34p substrates except for Gcn4p are phosphorylated in a *CDC28*-dependent manner prior to either their ubiquitination or degradation, and mutation of Cdc28p consensus phosphorylation sites or inactivation of Cdc28p itself stabilizes Cln2p (Lanker *et al.*, 1996; Deshaies *et al.*, 1995), Cln3p (Yaglom *et al.*, 1995), Siclp (see Section 2.2.1), and Farlp (Henchoz *et al.*, 1997). In this instance, Gcn4p may be the exception that proves the rule: *in vitro* studies suggest that Cdc34p/Cdc4p-dependent ubiquitination of Gcn4p requires a Gcn4p kinase that is distinct from Cdc28p (Y. Chi and R. J. Deshaies, unpublished data). An important question is how phosphorylation serves to target substrates to the *CDC34* pathway. Two possible mechanisms are presented in Fig. 3. Phosphorylation may create an epitope that is directly recognized by the

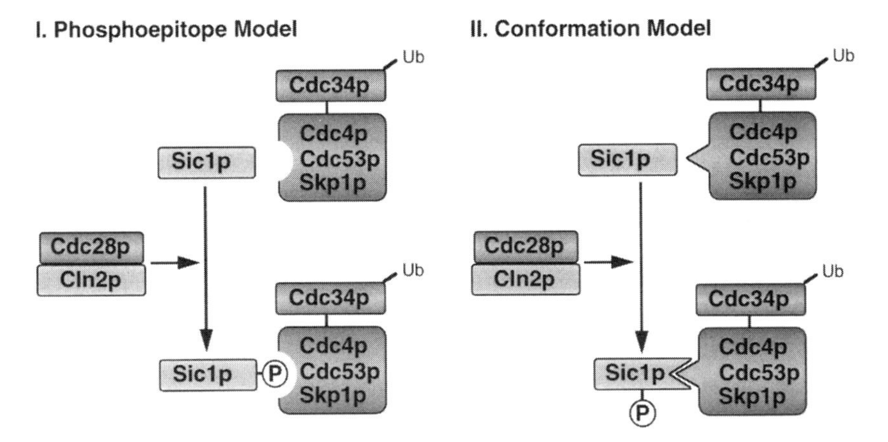

Figure 3. Two models showing how phosphorylation renders Sic1p a target for ubiquitination by the Cdc34p pathway (for details see text).

ubiquitination machinery, much like tyrosine phosphorylation creates an epitope recognized by SH2 domains. Alternatively, phosphorylation may perturb the conformation of protosubstrates, revealing a cryptic peptide-based ubiquitination signal.

As PEST elements are often enriched in the S/T P dipeptide that serves as a minimal Cdc28p phosphorylation site, the essence of PEST-mediated instability may be the propensity of PEST sequences to serve as a prominent target for phosphorylation (Yaglom *et al.*, 1995). The actual destruction signal encrypted within PEST elements is probably more complex, however, because many proteins are phosphorylated in eukaryotic cells, and it is improbable that all CDK substrates are rendered unstable by phosphorylation. Rather, it seems likely that either a protein must be phosphorylated on multiple residues, or a phosphorylated residue must be juxtaposed with a specific peptide "cosignal" to allow for effective substrate recognition by the *CDC34* pathway. Similar context effects have been observed for the recognition of phosphorylated tyrosine residues by SH2 domains.

2.4. Conservation of *CDC34* Pathway Function

2.4.1. Vertebrate *CDC34*

A vertebrate *CDC34* homologue was isolated in a screen for human genes that can rescue a budding yeast checkpoint mutant that fails to arrest cell division in the presence of unreplicated DNA (Plon *et al.*, 1993). Human *CDC34* suppresses the thermosensitive growth of *cdc34*^{ts} strains, implying that yeast and human Cdc34p perform similar functions. One of the targets of human Cdc34p may be the CDK inhibitor p27, which participates in the regulation of the Restriction Point (Coats *et al.*, 1996). p27 is stable in quiescent cells but is unstable in proliferating cells (Pagano *et al.*, 1995). Furthermore, the destruction of p27 in both reticulocyte and fibroblast cell extracts is inhibited by dominant-negative human Cdc34p (Pagano *et al.*, 1995). Like p27, the myogenic regulator MyoD is also stabilized by dominant-negative Cdc34p (Song *et al.*, 1996). A Cdc34p-like protein has also been identified in *Xenopus* eggs, and *Xenopus* egg extracts immunodepleted of Cdc34p and affiliated polypeptides fail to sustain cyclin E/Cdk2-dependent replication of double-stranded sperm DNA (Yew and Kirschner, 1997). The picture emerging so far is consistent with the notion that vertebrate Cdc34p, like budding yeast Cdc34p, counteracts the activity of an inhibitor(s) of DNA replication by targeting it for ubiquitin-mediated destruction.

2.4.2. The *cullin* Family

CDC53 is also conserved among eukaryotes. The *lin-19* gene of *C. elegans* (renamed *cul-1*, for *cullin*) encodes a protein 30% identical to *CDC53* (Kipreos *et al.*, 1996). Comparison of *cul-1* and *CDC53* sequences with those present in expressed sequence tag databases revealed that there are at least six *cullins* in humans, five in *C. elegans*, three in budding yeast, and one in fission yeast (Kipreos *et al.*, 1996). All five of the *C. elegans cul* genes encode proteins of similar length (~ 740–800 amino acids), and all pairwise combinations with CUL1 exhibit 24–28% sequence identity. Cdc53p and CUL1 probably possess similar biochemical activities, as human *CUL1* complements the thermosensitive phenotype of a *cdc53*^{ts} mutant (S. Lyapina and R. J. Deshaies, unpublished data). In contrast to the cell division arrest of *cdc53*^{ts} mutants in yeast, *cul-1* mutants exhibit hyperplasia of all larval tissues, suggesting that wild-type CUL1 normally restrains cell division (Kipreos *et al.*, 1996). A possible explanation for this discrepancy is that CUL-1 may be responsible only for the degradation of positive cell cycle regulators, such as G1 cyclins, whereas other cullin family members may be important for the degradation of a Sic1p-like protein. It will be interesting to see whether any members of the *cullin* family behave as tumor suppressor genes in humans. *Cullins* may also play a role in mitotic degradation events, as a subunit of the anaphase-promoting complex has recently been found to be a novel member of this protein family (see Section 3.1.2b; J.-M. Peters, H. Yu, R. W. King, and M. Kirschner, unpublished data; W. Zachariae and K. Nasmyth, personal communication).

2.5. Links between Ubiquitin-Dependent Proteolysis and the Control of G1 Phase in Vertebrate Cells

The classical experiments of Pardee and co-workers led to the hypothesis that the accumulation of an unstable polypeptide(s) positively regulated the commitment to cell division during Gl phase (Pardee, 1987). Subsequent molecular analyses have revealed that both positive (cyclins D and E, c-Myc, E2F) and negative (p27, p53) regulators of Gls progression are unstable (Clurman *et al.*, 1996; Hateboer *et al.*, 1996; Hofmann *et al.*, 1996; Won and Reed, 1996; Pagano *et al.*, 1995; Matsushime *et al.*, 1991; Ramsay *et al.*, 1984; Reich and Levine, 1984). With the exceptions of p27 and the virally targeted destruction of p53 in cells infected by human papillomavirus (for reviews, see chapter 11 and Deshaies, 1995b), the identity of the pathways by which these proteins are degraded in normal cells remains unknown. Nevertheless, cyclin E (Clurman *et al.*, 1996; Won and Reed, 1996) shares an important characteristic of Cdc34p substrates inasmuch as it is stabilized by mutation of a CDK consensus phosphorylation site. The same may be true for p53 (Lin and Desiderio, 1993).

Progression through Gl phase may also be influenced by proteins that disassemble multiubiquitin chains, because two genes that encode deubiquitinating enzymes—*tre-2* and *DUB1*—have been linked to cell proliferation (Matoskova *et al.*, 1996a,b; Zhu *et al.*, 1996; Papa and Hochstrasser, 1993; Nakamura *et al.*, 1992; see also Chapter 4). However, detailed insight into the substrates and physiological roles of both *DUB-1* and *tre-2* awaits future investigations.

3. UBIQUITIN-DEPENDENT PROTEOLYSIS IN MITOSIS

Mitosis is the process by which the cell divides its duplicated DNA and other cellular contents. Whereas the early events of mitosis such as chromosome condensation and spindle assembly are thought to be initiated by the activation of CDC2 and other mitotis-specific protein kinases, later events such as the separation of sister chromatids require the activation of a ubiquitin-dependent proteolytic system. The first substrates of this system that were identified are the mitotic cyclins, whose destruction inactivates CDC2 and allows exit from mitosis. Recent studies have revealed that the degradation of other substrates regulates the metaphase–anaphase transition and the disassembly of the mitotic spindle (Fig. 4). An important question is how the CDK-dependent entry into mitosis is coordinated with the exit from mitosis, which is driven by proteolysis. The cell cycle machinery must allow sufficient time for the chromosomes to condense and align on the spindle before the proteolytic system that initiates sister chromatid

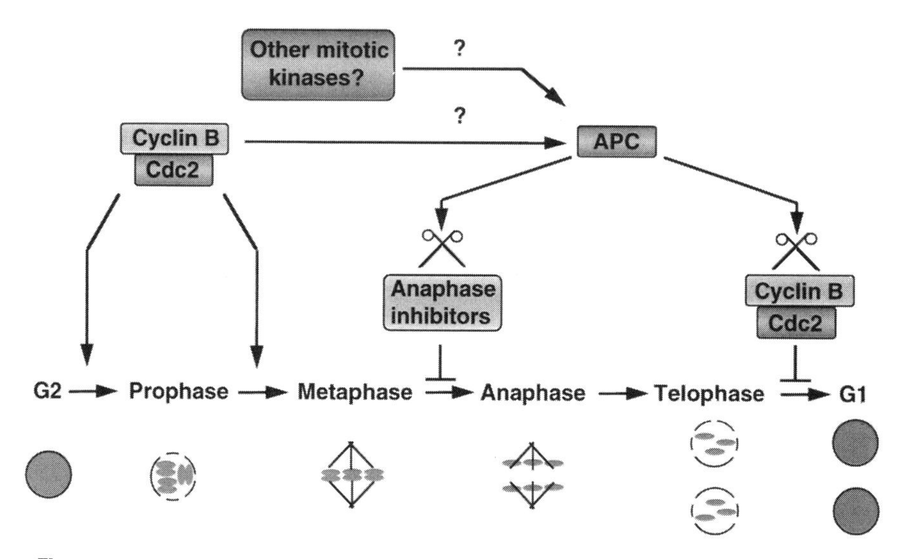

Figure 4. Mitotic functions of the APC pathway. Schematic representation of the morphological phase of mitosis, indicating how APC-mediated ubiquitination regulates progression through mitosis. Degradation of anaphase inhibitors such as budding yeast Pds1p and fission yeast Cut2 is required for the initiation of chromosome segregation, whereas the proteolysis of mitotic cyclins such as cyclin B is essential for exit from mitosis (for references see text).

separation is activated. Although the central players of the mitosis-specific proteolytic pathway have recently been identified, little is known about how these components are switched on and off during the cell cycle.

3.1. Mitotic Cyclin Degradation

In rapidly dividing marine embryos, mitotic cyclins accumulate steadily throughout interphase and are rapidly degraded at anaphase (Evans *et al.*, 1983). This intriguing observation provided an early clue that proteolysis may regulate cell division. This hypothesis was confirmed by two important findings: first, that maturation-promoting factor (MPF), a kinase activity capable of inducing entry into mitosis, is composed of a heterodimer of cyclin B and a homologue of fission yeast Cdc2p (Labbé *et al.*, 1989; Arion *et al.*, 1988; Dunphy *et al.*, 1988); and second, that degradation of cyclin B is necessary for inactivation of MPF and exit from mitosis (Murray *et al.*, 1989). The latter finding has subsequently been confirmed in many organisms, emphasizing the universal importance of cyclin degradation in regulating exit from mitosis (Yamano *et al.*, 1996; Luo *et al.*, 1994; Rimmington *et al.*, 1994; Surana *et al.*, 1993; Gallant and Nigg, 1992; Ghiara *et al.*, 1991).

Higher eukaryotes contain two subclasses of mitotic cyclins, called A and B, which share the capacity to activate CDC2; cyclin A is also able to activate the related kinase CDK2, and plays a role in regulating S phase. Although the mitotic destruction of cyclins A and B in clam embryos appears to be mediated by the same pathway (Sudakin *et al.*, 1995), their proteolysis has been reported to differ in two respects: First, cyclin A is degraded in advance of cyclin B; second, B-type cyclins are preferentially stabilized by an arrest mechanism, called the *spindle assembly checkpoint*, that prevents the separation of sister chromatids if they are not properly attached to the mitotic spindle (Edgar *et al.*, 1994; Minshull *et al.*, 1994; Hunt *et al.*, 1992; Whitfield *et al.*, 1990). Why the destruction of cyclins A and B differs in these respects and whether these differences are important for the orderly progression through mitosis are presently unknown.

Although degradation of mitotic cyclins can be observed in most dividing somatic cells and early embryos, the syncytial nuclear divisions of the early *Drosophila* embryo provide an interesting exception. In this system, bulk fluctuations in cyclin levels are only observed once cellularization occurs (Edgar *et al.*, 1994). A similar situation has been reported for the syncytial slime mold *Physarum polycephalum* (Cho and Sauer, 1994). This suggests that widespread cyclin degradation and kinase inactivation may be necessary only for cell cycles that involve cytokinesis. In syncytial cycles, however, local degradation of cyclin may suffice to allow nuclear division.

Several features of cyclin proteolysis are worth noting. First, cyclin degradation appears relatively specific, as the majority of labeled proteins in marine embryos remains stable through mitosis (Evans *et al.*, 1983). Second, in early clam embryos, cyclin was found to be unstable during only a brief period at the end of mitosis (Hunt *et al.*, 1992), suggesting that the activity of the cyclin degradation system is regulated during the cell cycle. Subsequent studies have indicated that the cyclin degradation system remains active from late mitosis through G1 in cells whose cell cycle contains G1 and G2 phases such as budding yeast (Amon *et al.*, 1994) and somatic mammalian cells (Brandeis and Hunt, 1996).

The identification of nondegradable mutants of cyclin B was crucial to the biochemical analysis of the cyclin degradation system, whose fleeting activity made it difficult to study. Deletion of the N-terminal 90 amino acids of cyclin B results in a protein (cyclin B $\Delta 90$) that is stable in mitosis, but retains the capacity to activate CDC2 (Murray *et al.*, 1989). *Xenopus* egg extracts containing cyclin B $\Delta 90$ remain arrested in a mitotic state with a constitutively activated cyclin degradation system. Using these mitotically arrested extracts, Glotzer *et al.* (1991) found that the N-terminus of sea urchin cyclin B is sufficient to target a heterologous protein (protein A) for mitosis-specific degradation. Labeling of cyclin B–protein A chimeras to high specific radioactivity revealed the presence of ubiquitinated intermediates during the course of degradation, providing the first evidence that cyclin B is degraded by the ubiquitin pathway. Ubiquitinated intermediates are 10-fold more abundant in mitotic extracts than in interphase extracts, where cyclins are stable, suggesting that regulated ubiquitination can account for the cell cycle dependence of cyclin proteolysis. Subsequent experiments in clam extracts demonstrated that methylated ubiquitin, an inhibitor of polyubiquitin chain formation, inhibits cyclin degradation, confirming that mitotic cyclin proteolysis is ubiquitin-dependent (Hershko *et al.*, 1991).

3.1.1. Signals that Target Mitotic Cyclins for Destruction

The N-termini of mitotic cyclins share little overall sequence similarity, with the exception of a 9-amino-acid motif that has been called the *destruction box* (D box; Glotzer *et al.*, 1991; Table III). Mutation of conserved positions of the D box is sufficient to prevent both ubiquitination and proteolysis of cyclins A and B, suggesting that it serves as a ubiquitination signal (King *et al.*, 1996b; Sudakin *et al.*, 1995; Kobayashi *et al.*, 1992; Lorca *et al.*, 1992a; Glotzer *et al.*, 1991). Although a cyclin B N-terminal fragment as short as 27 amino acids can target a heterologous protein for degradation, physiological rates of degradation (half-life less than 5 min) appear to require the entire N-terminus (King *et al.*, 1996b). This may indicate that the N-terminus is necessary for the proper presentation and recognition of the D box, which is presumed to be recognized by a component of the anaphase-promoting complex (see below). There does not appear to be a strict requirement for a particular lysine residue to serve as a ubiquitin acceptor site, as cyclin B can be ubiquitinated at multiple lysine residues in a manner sufficient to support degradation at physiological rates (King *et al.*, 1996b).

Although the N-terminal fragments of *Arbacia* cyclin B and *Xenopus* cyclin B1 are rapidly degraded in mitotic *Xenopus* extracts, the N-terminal fragments of other cyclins, such as *Xenopus* cyclin A1 or cyclin B2, are not (Stewart *et al.*, 1994; van der Velden and Lohka, 1993). This may reflect the fact that these cyclins need to be bound to CDC2 to be degraded (Stewart *et al.*, 1994; van der Velden and Lohka, 1994). Surprisingly, the D box of cyclin A is not sufficient to substitute for that of cyclin B (King *et al.*, 1996b; Klotzbücher *et al.*, 1996), although the cyclin B D box can target A-type cyclins for destruction (Klotzbücher *et al.*, 1996). Although A- and B-type cyclins are thought to be ubiquitinated by the same components (Sudakin *et al.*, 1995), structural differences between A- and B-type D boxes may contribute to the distinct physiological properties of cyclin A and cyclin B degradation discussed earlier.

Recently, D boxes have also been found to be important for the mitosis-specific ubiquitination and degradation of several noncyclin proteins (see below). A comparison of D boxes in cyclins and noncyclins shows that only residues 1 (arginine), 4 (isoleucine, leucine, or phenylalanine), and 9 (usually asparagine) are conserved.

Table III
Destruction Box Sequences[a]

Protein	Position	Sequence
B-type cyclins		
S. pombe Cdc13p	59	RHALDDVSN
S. cerevisiae Clb2p	25	RLALNNVTN
A. nidulans NIME	52	RAALGDVSN
A. thaliana Cyc1	38	RQVLGDIGN
D. melanogaster Cyc B	37	RAALGDLQN
A. punctulata Cyc B	42	RAALGNISN
S. solidissima Cyc B	40	RNTLGDIGN
X. laevis Cyc B1	36	RTALGDIGN
X. laevis Cyc B2	30	RAALGEIGN
G. gallus Cyc B2	32	RAVLEEIGN
G. gallus Cyc B3	51	RSAFGDITN
H. sapiens Cyc B1	42	RTALGDIGN
A-type cyclins		
D. melanogaster Cyc A	160	RSILGVIQS
S. solidissima Cyc A	33	RAALGVITN
X. laevis Cyc A1	41	RTVLGVIGDN
X. laevis Cyc A2	26	RTVLGVLQEN
G. gallus Cyc A	93	RAALGTVGE
M. musculus Cyc A1	37	RTVLGVLTEN
H. sapiens Cyc A	47	RAALAVLKSGN
Other APC substrates		
S. cerevisiae Pds1p	85	RLPLAAKDN
S. cerevisiae Ase1p	760	RQLFPIPLN
X. laevis Geminin	33	RRTLKVIQP

[a]Amino acid sequences are given in one-letter code. The position of the first shown amino acid residue is indicated. For a detailed discussion see King *et al.* (1996b).

3.1.2. Components Involved in Mitotic Cyclin Degradation

Both biochemical and genetic approaches have played an important role in the identification of components involved in mitotic cyclin degradation. The clam system was the first to be successfully fractionated into several distinct activities required for ubiquitination of cyclin B N-terminal fusion proteins *in vitro* (Hershko *et al.*, 1994). Cyclin ubiquitination in clam extracts requires two fractions in addition to ubiquitin and E1: an E2 activity, called E2-C, and a putative E3 activity that associates with particulate material and is active only when derived from mitotic extracts. On solubilization, this E3-containing component, called the *cyclosome*, sediments as a large particle on glycerol gradients (Sudakin *et al.*, 1995).

Fractionation of *Xenopus* extracts yielded similar, although not identical results (King *et al.*, 1995). In this system, all components are soluble, and two distinct E2 activities can sustain D box-dependent cyclin ubiquitination *in vitro*. As in the clam system, both E2 activities are constitutively active through the cell cycle, and a large protein complex, containing the putative E3 activity, is active only when purified from mitotic extracts. The purified complex contains eight distinct proteins (Peters *et al.*, 1996; King *et al.*, 1995), several of which are homologous to proteins required for cyclin proteolysis in budding yeast (Zachariae *et al.*, 1996; Irniger *et al.*, 1995). Earlier genetic studies in different fungi had shown that these conserved proteins are essential for the transition from metaphase to anaphase (see below). The E3 complex purified from *Xenopus* and characterized in yeast was therefore called the *anaphase-promoting complex* (APC).

3.1.2a. Ubiquitin-Conjugating Enzymes. One of the E2s involved in cyclin ubiquitination in *Xenopus* extracts is a homologue of the budding yeast Ubc4/5p subfamily of E2 enzymes, which has been implicated in the ubiquitination of many different proteins (King *et al.*, 1995). Mutation of Ubc4p family members in yeast has not been reported to cause a cell cycle arrest (Seufert and Jentsch, 1990), but deletion of *UBC4* is synthetically lethal in yeast strains carrying a mutated form of the APC subunit Cdc23p (Irniger *et al.*, 1995), suggesting that Ubc4p and APC function in the same biochemical pathway.

Purification and cloning of the second E2 activity from *Xenopus* and clam extracts (called UBCx and E2-C, respectively) has revealed that they are orthologues (Aristarkhov *et al.*, 1996; Yu *et al.*, 1996). Dominant-negative versions of the human relative, designated UBC-H10, arrest cells at metaphase, consistent with a role of this E2 in the APC pathway (Townsley *et al.*, 1997). This E2 appears to be a novel enzyme, most closely related to human UBC2. In *Xenopus*, either UBC4 or UBCx is sufficient to support D box-dependent cyclin ubiquitination in the presence of El and purified APC. UBCx is active at slightly lower concentrations than UBC4, and at saturating concentration can convert twice as much substrate into ubiquitin conjugates. However, the conjugates formed are of lower average molecular mass than those formed in the presence of UBC4, suggesting that the reaction may be less processive (Yu *et al.*, 1996).

UBCx and E2-C appear to have homologues in budding and fission yeast, although the role of the yeast enzymes in cyclin degradation has not been directly tested. Mutation of the *S. pombe* gene UbcP4 leads to a metaphase arrest that can be rescued by overexpression of the APC subunit Cut9p, suggesting that this enzyme functions in the APC pathway (Osaka *et al.*, 1997).

In budding yeast, mutation of yet another E2-related enzyme, Ubc9p, stabilizes the B-type cyclins Clb2p and Clb5p (Seufert *et al.*, 1995). However, unlike Clb2p, whose ubiquitination and degradation is APC-dependent (see below),

Clb5p stability is not cell cycle regulated (Seufert *et al.*, 1995), does not require a D box, and is not affected in *cdc16* and *cdc23* mutants (Irniger *et al.*, 1995), suggesting that its degradation is not mediated by APC. In *Xenopus* extracts a homologue of Ubc9p is not required for APC-dependent cyclin B ubiquitination (King *et al.*, 1995). and mutation of *UBC9* does not interfere with Clb2p ubiquitination in yeast extracts *in vitro* (Zachariae and Nasmyth, 1996). Recently, vertebrate UBC9 has been to found to assemble with two other proteins, RanGAP1 and RanBP2, into a complex that is involved in regulating nuclear transport of karyophilic proteins (Mahajan *et al.*, 1997; Saitoh *et al.*, 1997). The stabilization of Clb2p in *ubc9* mutants could therefore be related to a defect in the nuclear transport machinery, which would prevent the nuclear uptake of proteins that are required for progression into mitosis or for mitotic cyclin proteolysis in the yeast nucleus. Interestingly, the targeting of the UBC9–RanGAP1–RanBP2 complex to the nuclear pore complex requires modification with SUMO-1, a ubiquitin related protein (Mahajan *et al.*, 1997; see Chapter 2 for more details). UBC9 can form a thioester with SUMO-1 (M. Scheffner, personal communication), explaining the previous observation that *Xenopus* UBC9 is unable to form thioesters with conventional ubiquitin (R. W. King, J.-M. Peters, and M. W. Kirschner, unpublished results).

3.1.2b. The APC/Cyclosome. Several genes have been identified in yeasts and filamentous fungi that are required for progression through the metaphase–anaphase transition. A subset of these proteins contain the tetratricopeptide repeat (TPR) motif, a repeated 34-amino-acid sequence that is thought to mediate protein–protein interactions (Hirano *et al.*, 1990). Consistent with this hypothesis, the TPR-containing proteins Cdc27p, Cdc16p, and Cdc23p are part of a macromolecular complex in budding yeast (Lamb *et al.*, 1994). Mutations in homologous proteins in fission yeast (Cut9 and Nuc2; Samejima and Yanagida, 1994; Hirano *et al.*, 1988) and *Aspergillus* (BIMA; O'Donnell *et al.*, 1991) also cause defects in anaphase, emphasizing the universal role of these proteins in regulating progression through mitosis. Furthermore, injection of antibodies raised against a human homologue of CDC27 into mammalian tissue culture cells results in a metaphase arrest (Tugendreich *et al.*, 1995).

A genetic screen in budding yeast identified three genes required for mitotic cyclin proteolysis: *CDC16*, *CDC23*, and *CSE1* (Irniger *et al.*, 1995). However, it was unclear whether the corresponding proteins regulate or directly catalyze mitotic cyclin proteolysis. Utilizing antibodies raised against human CDC27 and CDC16, it was found that the large E3 complex in *Xenopus* extracts cofractionates with both proteins, and that immunopurified CDC27 complexes are sufficient to support cyclin B ubiquitination in the presence of recombinant UBC4 and E1 (King *et al.*, 1995). Purification and peptide sequencing of the components of the E3 complex has revealed that it consists of at least eight tightly associated subunits

(Table IV), including *Xenopus* homologues of budding yeast Cdc27p, Cdc16p, and Cdc23p (Peters *et al.*, 1996). Four of the remaining proteins have previously not been identified, whereas the final subunit is a homologue of the *Aspergillus* protein BIME (Peters *et al.*, 1996), which is required for anaphase progression (Osmani *et al.*, 1988). BIME was originally proposed to be a transmembrane protein (Engle *et al.*, 1990), but in *Xenopus* extracts BIME is only detectable in association with APC, which is soluble (Peters *et al.*, 1996). Extension of the screen for genes required for cyclin proteolysis in *S. cerevisiae* has demonstrated that *CDC26*, and a yeast homologue of *BIME*, called *APC1*, are also required for Clb2p ubiquitination (Zachariae *et al.*, 1996). A BIME-related protein in fission yeast is also required for mitotic cyclin ubiquitination and is a component of APC in that organism (Yamashita *et al.*, 1996). The budding yeast APC has also been affinity-purified and shown to contain several subunits in addition to Cdc16p, Cdc23p, Cdc27p, and the homologue of BIME. Three of these are homologous to *Xenopus* APC2, APC4, and APC5 (W. Zachariae, M. Mann, and K. Nasmyth, personal communication).

Neither Cse1p, a protein required for accurate chromosome segregation in budding yeast (Xiao *et al.*, 1993), nor its vertebrate relative CAS (Scherf *et al.*, 1996; Brinkmann *et al.*, 1995) are tightly associated with APC in yeast or *Xenopus* (Zachariae *et al.*, 1996; Peters *et al.*, 1996), and the role of this protein in cyclin degradation is presently unclear: Whereas yeast extracts from *cse1* mutants are deficient in Clb2p ubiquitination (Zachariae and Nasmyth, 1996), cyclin B ubiquitination can be reconstituted from purified *Xenopus* enzymes in the absence of *Xenopus* CAS (Peters *et al.*, 1996). Another protein that may play a role in mitotic cyclin ubiquitination is Suc1, a small CDK-binding protein in fission yeast that has homologues in budding yeast (Cks1p), mammals (CKS), and *Xenopus* (p9). In the absence of Suc1, fission yeast cells arrest in mitosis with high levels of Cdc2 kinase activity (Basi and Draetta, 1995; Moreno *et al.*, 1989) and *Xenopus* extracts immunodepleted with p9 antibodies are unable to degrade cyclin B (Patra and Dunphy, 1996). The observation that *Xenopus* p9 is required for cyclin B degradation in calcium-treated cytostatic factor (CSF) extracts, in which APC has already been activated (see Section 3.4.2), suggests that p9 may play a direct role in cyclin ubiquitination, perhaps by facilitating recognition of cyclin B–CDC2 complexes by APC. This view is supported by the recent observation that the mitotic form of the clam cyclosome can bind to fission yeast Suc1 *in vitro* (Sudakin *et al.*, 1997).

Given the large number of components involved in cyclin ubiquitination, it is perhaps not surprising that the reaction mechanism remains obscure. Two possible models of how APC may ubiquitinate substrates are shown in Fig. 5. Although APC meets the functional criteria for a ubiquitin-protein ligase or E3, as it is sufficient for ubiquitination in the presence of E1 and E2, none of the APC subunits identified to date bear sequence resemblance to either one of the two well-defined E3 enzymes Ubr1p and E6-AP (see Chapters 8 and 11). However,

Table IV
Subunits of Xenopus APC Purified from Interphase (i) and Mitotic (m) Extract and Their Fungal Homologues

Subunit	M_r ($\times 10^3$) i	M_r ($\times 10^3$) m	Homologues *S. cerevisiae*	Homologues *S. pombe*	Homologues *A. nidulans*	References
APC1	210	220	Apc1p	Cut4	BIME	Peters *et al.* (1996), Yamashita *et al.* (1996), Zachariae *et al.* (1996), Starborg *et al.* (1994), Engle *et al.* (1990)
APC2	112	112	Apc2p	—	—	Peters *et al.* (1996), H. Yu, J.-M. Peters, R. W. King, and M. Kirschner (unpublished data), W. Zachariae, M. Mann, and K. Nasmyth (personal communication)
APC3	100	130	Cdc27p	Nuc2	BIMA	Peters *et al.* (1996), Zachariae *et al.* (1996), King *et al.* (1995), Lamb *et al.* (1994), O'Donnell *et al.* (1991), Hirano *et al.* (1988)
APC4	100	100	Apc4p	—	—	Peters *et al.* (1996), H. Yu, J.-M. Peters, R. W. King, and M. Kirschner (unpublished data), W. Zachariae, M. Mann, K. Nasmyth (personal communication)
APC5	82	82	Apc5p	—	—	Peters *et al.* (1996), H. Yu, J.-M. Peters, R. W. King, and M. Kirschner (unpublished data), W. Zachariae, M. Mann, and K. Nasmyth (personal communication)
APC6	75	78	Cdc16p	Cut9	—	Peters *et al.* (1996), Zachariae *et al.* (1996), Irniger *et al.* (1995), King *et al.* (1995), Lamb *et al.* (1994), Samejima and Yanagida (1994)
APC7	69	69	Cdc23p	—	—	Peters *et al.* (1996), H. Yu, J.-M. Peters, R. W. King, and M. Kirschner (unpublished data)
APC8	66	69	—	—	—	Peters *et al.* (1996), Zachariae *et al.* (1996), Irniger *et al.* (1995), Lamb *et al.* (1994), Sikorski *et al.* (1990)
—			Cdc26p	—	—	Zachariae *et al.* (1996)

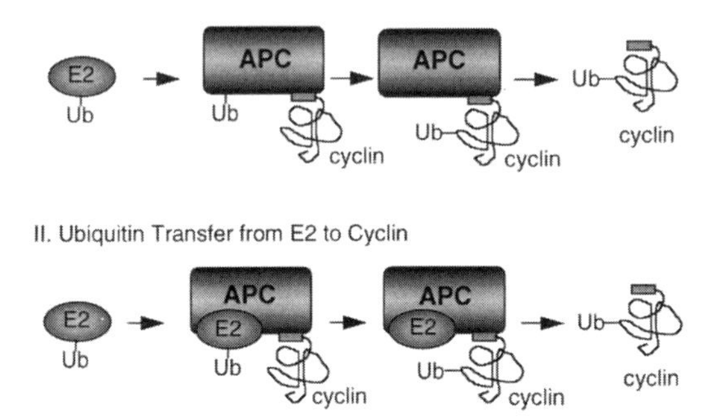

Figure 5. Two models of how APC could mediate cyclin ubiquitination. For a detailed discussion see Yu *et al.* (1996).

both human APC2 and its budding yeast homologue show limited but significant sequence similarity with members of the cullin family which includes Cdc53p, a component of the *CDC34* pathway (J.-M. Peters, H. Yu, R. W. King and M. W. Kirschner, unpublished data; W. Zachariae and K. Nasmyth, personal communication). It is therefore conceivable that the APC and the *CDC34* pathways utilize similar substrate recognition or ubiquitination mechanisms.

The formation of ubiquitin thioester intermediates, characteristic for members of the E6-AP family of E3s and also observed with Ubr1p (see Chapters 3 and 8), has not been detected with APC (King *et al.*, 1995). This suggests that APC may not serve as the final ubiquitin donor for the reaction, but rather that APC may serve to bind both the E2 and the substrate. Substrate specificity is presumed to be mediated by a component of the APC rather than by the E2, although it has not yet been determined whether the substrate binds the APC or the E2 directly.

3.1.2c. The 26 S Proteasome. Several genetic experiments have suggested that the 26 S proteasome plays an important role in mitotic cyclin proteolysis and exit from mitosis. Mutation of proteasome subunits in both budding and fission yeast results in a G2/M arrest that is similar to the arrest point of APC mutants (Gordon *et al.*, 1996, 1993; Ghislain *et al.*, 1993), and budding yeast proteasome mutants accumulate Clb2p (Ghislain *et al.*, 1993). It has also been reported that budding yeast proteasome mutants can be rescued by deletion of the *CLB2* gene (Friedman and Snyder, 1994) and are growth-arrested by overexpression of Clb2p (Richter-Ruoff and Wolf, 1993) or Clb5p (Seufert *et al.*, 1995). Although there is no direct biochemical experiment demonstrating that ubiquitinated mitotic cyclins are degraded by the 26 S proteasome, cyclin B proteolysis is inhibited in cells (Sherwood *et al.*, 1993) or extracts (J.-M. Peters and M. W.

Kirschner, unpublished data) treated with proteasome inhibitors. The 26 S proteasome itself does not appear to be regulated during mitosis, because model substrates such as ubiquitinated lysozyme (Mahaffey *et al.*, 1993) and purified cyclin–ubiquitin conjugates (J.-M. Peters, M. Glotzer and M. W. Kirschner, unpublished results) are rapidly degraded in both interphase and mitotic *Xenopus* extracts.

3.2. Chromosome Segregation

In most cell types, cyclin B is degraded at about the time that sister chromatids separate, suggesting that its proteolysis might be functionally related to anaphase. Furthermore, cyclin B is stabilized in two physiological circumstances in which sister chromatid segregation is also prevented: the spindle assembly checkpoint, and the natural arrest of vertebrate eggs at metaphase of the second meiotic division (see below). The finding that nondegradable mutants of cyclin B block exit from mitosis in *Xenopus* extracts (Murray *et al.*, 1989) and budding yeast (Ghiara *et al.*, 1991) suggested that inactivation of CDC2-cyclin B kinase might trigger chromosome segregation. However, it was found that both frog egg extracts and yeast cells containing nondegradable cyclin B arrest in telophase rather than metaphase (Holloway *et al.*, 1993; Surana *et al.*, 1993). The reconstitution of sister chromatid segregation in *Xenopus* egg extracts (Shamu and Murray, 1992) revealed that even though nondegradable cyclin B does not interefere with anaphase, chromosome segregation is inhibited by high concentrations of a D box-containing fragment of cyclin B, which presumably acts as a competitive inhibitor of APC-mediated ubiquitination reactions (Holloway *et al.*, 1993). The same result was obtained in extracts supplemented with methylated ubiquitin and in extracts lacking endogenous full-length cyclins, suggesting that D box-containing proteins other than cyclin B must be degraded by the APC pathway to initiate anaphase. The recent finding that components necessary for cyclin degradation are also required for anaphase in yeast and mammalian cells further suggests that the two processes are mediated by the same APC-dependent proteolytic pathway (Irniger *et al.*, 1995; Tugendreich *et al.*, 1995).

The work of Holloway *et al.* (1993) predicted the existence of an anaphase inhibitor that would be degraded through D box-dependent, and hence APC-mediated, proteolysis. Recently, two such candidates have been identified in budding and fission yeast. In budding yeast, the *PDS1* gene was identified in a screen for mutants that undergo *p*recocious *d*issociation of *s*ister chromatids in the presence of microtubule inhibitors (Yamamoto *et al.*, 1996a,b). Although the null mutant is viable, *pds 1* mutants have a high rate of chromosome loss, suggesting that Pds1p is important for anaphase fidelity. Although it is not clear how Pds1p regulates anaphase, the protein has all of the hallmarks of a substrate of the APC pathway: Pds1p is degraded during anaphase, and is unstable when expressed during G1, as is the case for mitotic cyclins; Pds1p contains a sequence element similar to the cyclin D box that is essential for its degradation *in vivo*, and Pds1p is stabilized in either *cdc16* or *cdc23* mutants (Cohen-Fix *et al.*, 1996). Biochemical studies support the contention that Pds1p is a substrate of APC: yeast Pds1p is rapidly degraded in a mitosis-specific and D box-dependent fashion in *Xenopus* egg extracts, and Pds1p is ubiquitinated *in vitro* by purified *Xenopus* APC (Cohen-Fix *et al.*, 1996). Intriguingly, whereas *cdc23* or *cdc16* mutants arrest prior to anaphase, about 50% of *pds1 cdc23* or *pds1 cdc16* double mutants progress through anaphase, suggesting that failure to degrade Pds1p partially accounts for the metaphase arrest of *cdc23* and *cdc16* mutants (Yamamoto *et al.*, 1996b). This hypothesis is strongly supported by the observation that transient expression of a Pds1p mutant lacking the D box results in a metaphase arrest (Cohen-Fix *et al.*, 1996).

Another putative anaphase inhibitor has been identified in fission yeast (Funabiki *et al.*, 1996). Degradation of Cut2, which contains two D box-related sequences, appears essential for anaphase, as overexpression of a nondegradable deletion mutant of Cut2 blocks sister chromatid separation (this does not demonstrate, however, that destruction of physiological levels of Cut2 is required for anaphase). The levels of wild-type Cut2 drop only about twofold during anaphase, suggesting that only a subpopulation of Cut2 may have to be degraded for the initiation of sister chromatiol separation. Like other *cut* (*c*ells *u*ntimely *t*orn) mutants, *cut2* mutants initiate septum formation despite being unable to separate chromosomes. Cut2 accumulates in cells containing a mutated version of Cut9, a homologue of budding yeast Cdc16p, suggesting that Cut2 is a substrate of APC.

How Pds1p and Cut2 regulate anaphase is unknown. Holloway *et al.* (1993) proposed that a chromosomal protein might shield sister chromatids from the activity of topoisomerase II until the proteolytic machinery is activated at the metaphase–anaphase transition. Alternatively, the key substrate could be a soluble protein that regulates sister chromatid segregation. Pds1p and Cut2 may have distinct functions, as they share little sequence similarity except for the presence of an N-terminal D box. Moreover, their intracellular distribution differs, with Cut2 localized on the mitotic spindle, and Pds1p distributed diffusely throughout the nucleus. Lastly, whereas Cut2 is essential, cells deleted for *PDS1* are viable but sick. Null mutants of *cut2* arrest at metaphase, indicating that Cut2 may independently promote and inhibit anaphase.

Additional proteins that may regulate the metaphase–anaphase transition have been identified through the analysis of the *Drosophila* mutants *pimples* (*pim*; Stratmann and Lehner, 1996) and *three rows* (*thr*; D'Andrea *et al.*, 1993; Philp *et al.*, 1993). Homozygous *pim* and *thr* mutants fail to separate sister chromatids once the maternal store of wild-type protein has been exhausted after embryonic cell cycle 14. As DNA replication continues in these embryos, chromosomes accumulate with up to four pairs of unseparated sister chromatids held together at the centromeric region. PIM and THR cannot be detected by immunofluorescence microscopy in cells that have undergone anaphase, suggesting that both proteins are either redistributed or degraded at the metaphase–anaphase transition. This resembles the loss of cyclin staining observed during anaphase of *Drosophila* embryos and therefore suggests that PIM and THR may be targets of APC. It has been proposed that PIM and THR function as activators of anaphase and that their mitotic degradation may be required to prevent premature chromosome segregation in the subsequent cell cycle (Stratmann and Lehner, 1996). The phenotype of null *cut2* mutants suggests, however, that proteins required for anaphase may also function as anaphase inhibitors (Funabiki *et al.*, 1996).

3.3. Other Proteins Degraded in Mitosis

We presently know of two APC functions in mitosis: the initiation of anaphase, regulated by the ubiquitination of proteins such as Pds1p, and the exit from mitosis, mediated by the ubiquitination of mitotic cyclins (Fig. 4). However, there are several other proteins that are degraded specifically in mitosis, and at least some of these may be substrates of APC (see Table V). This raises the interesting possibility that the execution of other mitotic events may depend on the proteolysis of specific proteins. For example, the budding yeast protein kinase Cdc5p, a homologue of polo in *Drosophila* and Plk-1 in vertebrate cells, is degraded in an APC-dependent manner (M. Shirayama and K. Nasmyth, personal communication). Several microtubule-binding proteins (MAPs) are degraded at the end of mitosis, including CENP-E in mammalian cells (Brown *et al.*, 1994) and Ase1p in budding yeast (Pellman *et al.*, 1995). Degradation of Ase1p requires both *CDC23* function and a sequence element in the Ase1p C-terminus that resembles the cyclin B3 D box, indicating that it is a substrate of the APC pathway (Juang *et al.*, 1997). Overexpression of a nondegradable Ase1p mutant in telophase delays exit from mitosis, whereas overexpression in G1 arrests cells at G2/M. This arrest is dependent on *MAD2*, a gene essential for the spindle assembly checkpoint, suggesting that expression of Ase1p in G1 leads to formation of an abnormal mitotic spindle. Degradation of Ase1p and perhaps other MAPs may therefore prevent premature binding of MAPs to microtubules during G1 and also facilitate spindle disassembly at the end of mitosis.

Yet another APC substrate, called geminin, has been identified in *Xenopus* based on its instability in mitotic egg extracts (T. McGarry and M. Kirschner, personal communication). Geminin contains a D box that is required for its mitosis-specific ubiquitination and degradation. *Xenopus* embryos microinjected with a D box mutant develop into cellularized embryos that lack nuclei, suggesting that the degradation of geminin may be involved in some aspect of nuclear division, such as DNA replication or chromosome segregation.

To date, all known substrates of APC contain a D box and are degraded in mitosis and G1. However, given the multisubunit nature of APC, it is conceivable

Table V
Substrates of the APC Ubiquitination Pathway

Protein	APC-dependent degradation *in vivo*?	APC-dependent ubiquitination *in vitro*?	References
B-type cyclins			
X. laevis B1 and B2	—[a]	Yes	Yu *et al.* (1996)
A. punctulata B	—[a]	Yes	Peters *et al.* (1996), King *et al.* (1995), Sudakin *et al.* (1995)
S. cerevisiae Clb2p	Yes	Yes	Zachariae and Nasmyth (1996), Irniger *et al.* (1995)
S. cerevisiae Clb3p	Yes	Not tested	S. Irniger and K. Nasmyth (personal communication)
S. cerevisiae Clb5p	Yes	Not tested	S. Irniger and K. Nasmyth (personal communication)
A-type cyclins			
S. solidissima A	—	Yes	Sudakin *et al.* (1995)
Other APC substrates			
X. laevis Geminin	Not tested	Yes	T. McGarry, J.-M. Peters, and M. Kirschner (unpublished results)
S. cerevisiae Pds1p	Yes	Yes	Cohen-Fix *et al.* (1996)
S. pombe Cut2	Yes	Not tested	Funabiki *et al.* (1996)
S. cerevisiae Ase1p	Yes	Yes	Juang *et al.* (1997), J.-M. Peters and D. Pellman (unpublished results)
S. cerevisiae Cdc5p	Yes	Not tested	M. Shirayama and K. Nasmyth (personal communication)

[a]Degradation of A- and B-type cyclins *in vivo* has been observed in higher eukaryotes (e.g., Evans *et al.*, 1983; Hunt *et al.*, 1992), but APC dependence *in vivo* has not been demonstrated.

that the complex has additional substrates that could be recognized through different substrate binding sites, perhaps allowing ubiquitination at other points in the cell cycle. This speculation arises from three reports in the literature that cannot be easily explained by the known mitotic function of APC. First, *cdc16* and *cdc27* mutants of budding yeast have been reported to overreplicate their DNA, suggesting that APC function may be required to restrict DNA replication to once per cell cycle (Heichman and Roberts, 1996). In another study it has been reported, however, that most of the DNA synthesis that occurs in APC mutants is the result of replication of mitochondrial DNA (Pichler *et al.*, 1997). Second, *Aspergillus bimE* mutants can override the interphase arrest induced by certain drugs and mutations (Ye *et al.*, 1996; James *et al.*, 1995; Osmani *et al.*, 1991), suggesting that in these cells APC may normally restrain entry into mitosis. Third, fission yeast cells that contain a mutated version of the APC subunit Cut4 are hypersensitive to heavy metals (Yamashita *et al.*, 1996). This suggests that APC may be required for the ubiquitin-dependent degradation of misfolded proteins that are induced by heavy metal treatment. However, this phenotype is not shared by other fission yeast APC mutants such as *cut9* or *nuc2*. It is important to note that in all three cases it remains unclear whether the observed phenotypes can be attributed to novel interphase functions of APC, or whether they result indirectly from interference with the mitotic ubiquitination activity of the complex.

3.4. Regulation of APC-Dependent Ubiquitination

3.4.1. Mitotic Regulation of the APC Pathway

Mitotic cyclin stability could in principle be regulated at the substrate level, as is the case for substrates of the *CDC34* pathway, or through the periodic activation of components of the APC pathway. Unlike the G1 cyclins or Sic1p, there is no evidence that phosphorylation of mitotic cyclins targets these proteins for degradation during mitosis (Li *et al.*, 1995; Izumi and Maller, 1991). Instead, the activity of the mitotic cyclin ubiquitination system is regulated. Regulation of this pathway is apparently achieved by modulating the intrinsic activity of cyclosome/APC. Whereas UBC4 and UBCx/E2-C are equally active during interphase and mitosis, the cyclosome/APC is at least fivefold more active during mitosis (King *et al.*, 1995; Sudakin *et al.*, 1995). Curiously, purified interphase APC consistently exhibits a low level of cyclin ubiquitination activity, whereas little or no mitotic cyclin ubiquitination occurs in crude interphase extracts (Peters *et al.*, 1996). Suppression of this basal activity to the low levels seen in crude interphase extracts is related to the presence of inhibitors that are removed during purification of APC (J.-M. Peters and M. W. Kirschner, unpublished results). Posttranslational modification of APC during mitosis could render the complex insensitive to such an inhibitor, enabling the complex to be fully active in mitotic extracts.

Biochemical experiments in clam and *Xenopus* suggest that phosphorylation of the cyclosome/APC plays an important role in its activation during mitosis. At least four subunits of *Xenopus* APC become phosphorylated in mitosis (Peters *et al.*, 1996), and partially purified clam cyclosome and affinity-purified APC are inhibited by incubation with protein phosphatases (Peters *et al.*, 1996; Lahav-Baratz *et al.*, 1995). Although the kinase(s) that phosphorylate APC in mitosis are unknown, purified CDC2-cyclin B kinase, which can activate cyclin degradation when added to crude interphase extracts (Félix *et al.*, 1990), can partially activate interphase (Sudakin *et al.*, 1995) or phosphatase-treated forms of the clam cyclosome (Lahav-Baratz *et al.*, 1995) and interphase *Xenopus* APC (J.-M. Peters, P. T. Stukenberg, and M. W. Kirschner, unpublished data). However, CDC2 kinase is activated early in mitosis, well before B-type cyclins become unstable. *In vitro* experiments indicate that CDC2 kinase activates cyclosome fractions with a lag period that may account for the delayed activation of cyclin B destruction observed *in vivo* (Sudakin *et al.*, 1995). Alternatively, additional kinases that regulate APC activity may become activated later in mitosis. One candidate is protein kinase A, which is activated following CDC2 kinase, and whose activity is essential for cyclin degradation in *Xenopus* extracts (Grieco *et al.*, 1996). Mitotic APC is recognized by the MPM-2 monoclonal antibody (King *et al.*, 1995), which reacts with a subset of mitotic phosphoproteins (Davis *et al.*, 1983), indicating that APC may be phosphorylated by the recently identified MPM-2 epitope kinase PLX-1 (Kumagai and Dunphy, 1996; Kuang and Ashorn, 1993).

Although CDC2-cyclin B kinase is a well-established activator of the cyclin degradation system, the ability of cyclin A to activate the degradation machinery is somewhat controversial. Initial reports demonstrated that cyclin A is incapable of activating the degradation machinery, despite the fact that it could activate histone-H1 kinase activity when added to interphase extracts (Grieco *et al.*, 1996; Lorca *et al.*, 1992b; Luca *et al.*, 1991). However, a recent study demonstrated that cyclin A, when added at higher concentrations sufficient to activate nuclear envelope breakdown, could also initiate both cyclin A and cyclin B degradation (Jones and Smythe, 1996). Whether cyclin A plays a physiological role in activating the destruction system remains to be determined.

Dephosphorylation has also been discussed as a possible mechanism for activating mitotic cyclin proteolysis (Ishii *et al.*, 1996). Protein phosphatase-1 (PP1) activity is essential for the metaphase–anaphase transition in fungi (Hisamoto *et al.*, 1994; Doonan and Morris, 1989; Ohkura *et al.*, 1989), *Drosophila* (Axton *et al.*, 1990), and mammalian cells (Ghosh and Paweletz, 1992). A multicopy suppressor of fission yeast PP1 mutants, *sds22+*, is also required for sister chromatid segregation (Ohkura and Yanagida, 1991), whereas another suppressor, *sds23+*, can also suppress mutations in APC subunits (Ishii *et al.*, 1996). However, it is not known whether PP1 directly regulates APC, or whether mutation of

PP1 stabilizes mitotic cyclins and inhibits anaphase indirectly through activation of the spindle assembly checkpoint.

Another potential regulator of cyclin proteolysis is the budding yeast gene *CDC20* (Hartwell *et al.*, 1973). Although Cdc20p activity is not essential for cyclin ubiquitination in extracts of G1-arrested cells (Zachariae and Nasmyth, 1996), *cdc20* mutants arrest prior to anaphase (Sethi *et al.*, 1991). Mutation of a related gene in *Drosophila*, *fizzy*, produces a similar arrest and stabilizes both A- and B-type cyclins (Dawson *et al.*, 1995, 1993; Sigrist *et al.*, 1995). Recent data suggest that *CDC20/fizzy* are substrate specific activators of the APC pathway (see Note Added in Proof).

Genetic studies in budding yeast suggest that additional mechanisms may regulate mitotic cyclin stability. In this organism the proteins Cdc5p, Cdc15p, Cdc14p, Lte1p, Dbf2p/Dbf20p, Ras1p/Ras2p/Rsr1p, Spo12p, and Tem1p are required for exit from mitosis and the complete degradation of mitotic cyclins but not for the initiation of anaphase (Shirayama *et al.*, 1996, 1994a,b; Morishita *et al.*, 1995; Toyn and Johnston *et al.*, 1994, 1993; Kitada *et al.*, 1993; Surana *et al.*, 1993; Johnston *et al.*, 1990). The molecular functions of these proteins can be inferred from their sequences that show similarities with either protein kinases (Cdc5p, Cdc15p, Dbf2p/Dbf20p), protein phosphatases (Cdc14p), GTPases (Ras1p/Ras2p/Rsr1p, Tem1p), or guanine nucleotide release factors (Lte1p), suggesting that these proteins may be components of a signaling pathway that somehow regulates Clb2p stability.

The findings that these proteins are required for Clb2p proteolysis but not for chromosome segregation suggest that the degradation of various APC substrates may be differentially regulated, perhaps at the level of substrate recognition. Further support for this hypothesis comes from studies of the CDK inhibitor Sic1p (Mendenhall, 1993), which, when overexpressed, is sufficient to induce Clb2p proteolysis in *cdc15* mutants or in cells arrested in nocodazole (Amon, 1997). Sic1p is expressed late in mitosis, when B-type cyclins become unstable, and its expression is dependent on Dbf2p and Cdc14p (Donovan *et al.*, 1994). Sic1p may serve as a targeting factor to direct cyclin B for ubiquitination by APC; alternatively, it has been proposed that CDC2-cyclin B activity negatively regulates APC in metaphase, and that Sic1p induces cyclin proteolysis by inactivating the kinase (Amon, 1997). However, if this is a universal mechanism for regulating cyclin stability, it would not explain why the cyclin B degradation system remains activated in *Xenopus* extracts containing nondegradable cyclin B and consequently high levels of CDC2 kinase activity (Glotzer *et al.*, 1991). Furthermore, although cells deleted for *SIC1* show a delay in exiting mitosis, they are nevertheless viable (Donovan *et al.*, 1994), suggesting that Sic1p-independent mechanisms for turning on cyclin proteolysis must exist.

Following exit from mitosis, the cyclin degradation machinery must be inactivated to allow the accumulation of new cyclins for the next mitosis. In the budding yeast cell cycle, the mitotic cyclin degradation system remains active through early G1 until G1 cyclin/Cdc28p kinase activity appears (Amon *et al.*, 1994). This suggests a simple model for maintaining the orderly progression of cell cycle phases in yeast: Mitotic cyclins cannot accumulate and thus mitosis cannot occur until the G1-Cdc28p kinase switches off the mitotic cyclin degradation system. It has been observed that the ability of G1 kinases to inhibit mitotic cyclin degradation remains reversible until the *CDC4*-dependent step (Schwob *et al.*, 1996; Amon *et al.*, 1994), suggesting that inactivation of APC may occur in two distinct steps. Recent experiments with extracts prepared from synchronized mammalian cells indicate that mitotic cyclins are also degraded during the G1 phase of higher eukaryotic somatic cells (Brandeis and Hunt, 1996). It will be interesting to determine whether CDKs that are active at the end of G1 are required for the inactivation of the cyclin degradation in metazoans. Experiments in *Drosophila* suggest that this may be the case because expression of cyclin E is necessary for the accumulation of mitotic cyclins (Knoblich *et al.*, 1994). Although G1-cyclin-dependent kinases may play a universal role in extinguishing cyclin proteolysis, the mechanisms through which these kinases inactivate cyclin degradation remain obscure.

3.4.2. Additional Mechanisms that Regulate APC-Dependent Ubiquitination

The spindle assembly checkpoint is a control mechanism that delays anaphase until every chromosome has become properly attached to the spindle, thus ensuring accurate segregation of sister chromatids (for a review, see Rudner and Murray, 1996). This mechanism is also activated by microtubule-depolymerizing or stabilizing agents such as nocodazole and taxol. Elegant micromanipulation experiments in grasshopper spermatocytes suggest that unequal tension across the meiotic spindle can generate an anaphase-inhibiting signal (Li and Nicklas, 1995). This signal may originate from the kinetochore of misattached sister chromatids where phosphoepitopes have been observed that disappear as soon as chromosomes become properly linked to the spindle (Nicklas *et al.*, 1995). Genetic studies in budding yeast have identified several proteins that are required for the spindle assembly checkpoint (Hoyt *et al.*, 1991; Li and Murray, 1991). Reconstitution of spindle assembly checkpoint conditions in *Xenopus* extracts has shown that the checkpoint-induced inhibition of mitotic cyclin proteolysis depends on the activation of MAP kinases (Minshull *et al.*, 1994). The next crucial step is to learn how the components of the spindle assembly checkpoint stabilize substrates of the APC-dependent ubiquitination system, such as Cut2, Pds1p, and the mitotic cyclins. Whether stabilization is achieved through inhibition of APC or through protection of its substrates is unknown.

Anaphase is also inhibited physiologically during the meiotic maturation of vertebrate eggs (for a review, see Sagata, 1996). Cell cycle arrest occurs at metaphase of meiosis II, and is overcome by the calcium influx into the ooplasm induced by fertilization. Metaphase-arrested eggs contain high levels of CDC2 kinase activity and mitotic cyclins, but do not enter anaphase because of the activity of cytostatic factor (CSF), a component of which is the protein kinase c-mos (Sagata *et al.*, 1989). CSF must somehow inhibit APC-dependent proteolysis. The CSF-induced arrest may depend on mechanisms similar to those activated by the spindle assembly checkpoint because activated MAP kinase is required in both systems (reviewed in Sagata, 1996). Interestingly, APC is active when purified from metaphase-II arrested extracts (R. W. King, J.-M. Peters, and M. W. Kirschner, unpublished data), suggesting that additional meiotic factors either block active APC, or protect its substrates from ubiquitination.

4. PERSPECTIVES

The recent dissection of the *CDC34* and APC pathways has changed our understanding of how cell cycle progression is controlled. Although protein phosphorylation was initially thought to be the major mechanism by which cell division was regulated, it is now clear that proteolysis is an integral part of the cell cycle control machinery (King *et al.*, 1996a, 1994; Nasmyth, 1996a,b; Deshaies, 1995a).

Although the existence of the *CDC34* and the APC pathways and their physiological importance is becoming widely recognized, many questions remain unanswered. For example, our present inventory of known *CDC34* pathway components may not be complete, and our understanding of how this pathway functions in higher eukaryotes is rudimentary. There may also be yet unidentified components in the APC pathway. Even though the minimal set of proteins that is required for the reconstitution of APC-dependent ubiquitination reactions is known, little is known about the factors that turn APC on and off during the cell cycle. In both pathways, the mechanisms of substrate recognition and ubiquitination remain poorly understood and may be different from the mechanisms proposed for other ubiquitination reactions, such as the E6-AP-dependent ubiquitination of p53.

For about a decade mitotic cyclins were the only cell cycle regulators that were known to be controlled by proteolysis. Within only a few years this has changed and we now know of several CDK inhibitors, kinases, anaphase inhibitors, and MAPs that are subject to cell cycle-regulated proteolytic control. Presently there is no reason to believe that the list of APC and Cdc34p substrates is near completion. Thus, an important goal for the future is to identify new physiological substrates for these pathways, and to assess how their degradation contributes to cell cycle progression.

Note Added in Proof

Since this article was written a number of discoveries have been made in the field of cell cycle-regulated proteolysis. We apologize to those authors whose recent papers we cannot mention in this added note due to space constraints.

Two recent papers (Feldman *et al.*, 1997; Skowyra *et al.*, 1997) have established that Cdc53p, Cdc4p, and Skp1p assemble into a ubiquitin ligase complex (dubbed SCF^{Cdc4} for Skp1p–Cdc53p or Cullin–F box receptor subunit) that interacts with Cdc34p to catalyze ubiquitination of phosphorylated Sic1p. Phospho-Sic1p binds selectively to Cdc4p/Skp1p, which in turn bind to Cdc53p and Cdc34p. Intriguingly, the Cdc4p subunit of SCF^{Cdc4} can be replaced by the F-box protein Grr1p. SCF^{Grr1} complexes bind tightly to phospho-Cln2p, but not phospho-Sic1p, suggesting that the substrate specificity of SCF complexes is dictated by the identity of the F-box subunit. Similar ubiquitin ligase complexes appear to be present in human cells, since human SKP1, CUL1, and SKP2 proteins assemble into complexes *in vivo* and *in vitro*, and human CUL1 associates with ubiquitin-conjugating activity in HeLa cell extracts (Lisztwan *et al.*, 1998, *EMBO J.* **17**:368–383; Lyapina *et al.*, submitted).

The cloning of the human APC subunits APC2, APC4, APC5, APC7 and APC8 (CDC23) has been completed (Yu *et al.*, 1998, *Science*, in press). Homologs of APC2, APC4 and APC5 and three additional subunits of low molecular mass (Apc9p, Apc10p/Doc1p, Apc11p) have been identified in budding yeast (Zachariae *et al.*, 1998, *Science*, in press; Kramer *et al.*, 1998, *EMBO J.* **17**:498–450; Hwang and Murray, 1997, *Mol. Biol. Cell* **8**:1877–1887).

Genetic experiments in budding yeast and flies have recently identified two related WD40 repeat proteins as essential components of the APC pathway. In yeast Cdc20p is required for the degradation of Pds1p and Clb5p, whereas Hct1p/Cdh1p is essential for Clb2p and Ase1p proteolysis (Schwab *et al.*, 1997, *Cell* **90**:683–693; Visintin *et al.*, *Science* **278**:460–463, 1997). In flies fizzy and fizzy-related may perform similar functions (Sigrist and Lehner, 1997, *Cell* **90**:671–681). Overexpression studies indicate that these WD40 proteins may be substrate specific activators of APC.

5. REFERENCES

Amon, A., 1997, Regulation of B-type cyclin proteolysis by Cdc28-associated kinases in budding yeast, *EMBO J.* **16**:2693–2702.

Amon, A., Irniger, S., and Nasmyth, K., 1994, Closing the cell cycle circle in yeast: G2 cyclin proteolysis initiated at mitosis persists until the activation of G1 cyclins in the next cycle, *Cell* **77**:1037–1050.

Arion, D., Meijer, L., Brizuela, L., and Beach, D., 1988, *cdc2* is a component of the M phase-specific histone H1 kinase: Evidence for identity with MPF, *Cell* **55:** 371–378.

Aristarkhov, A., Eytan, E., Moghe, A., Admon, A., Hershko, A., and Ruderman, J. V., 1996, E2-C, a cyclin selective ubiquitin carrier protein required for the destruction of mitotic cyclins, *Proc. Natl. Acad. Sci. USA* **93:**4294–4299.

Axton, J. M., Dombradi, V., Cohen, P. T., and Glover, D. M., 1990, One of the protein phosphatase 1 isoenzymes in *Drosophila* is essential for mitosis, *Cell* **63:**33–46.

Bai, C., Sen, P., Hofmann, K., Ma, L., Goebl, M., Harper, J. W., and Elledge, S. J., 1996, SKP1 connects cell cycle regulators to the ubiquitin proteolysis machinery through a novel motif, the F-box, *Cell* **86:**263–274.

Barral, Y., Jentsch, S., Mann, C., Coats, S., Flanagan, W. M., Nourse, J., Roberts, J. M., Cross, F. R., and Blake, C. M., 1995, G1 cyclin turnover and nutrient uptake are controlled by a common pathway in yeast, *Genes Dev.* **9:**399–409.

Basi, G., and Draetta, G., 1995, $p13^{suc1}$ of *Schizosaccharomyces pombe* regulates two distinct forms of the mitotic Cdc2 kinase, *Mol. Cell. Biol.* **15:**2028–2036.

Blondel, M., and Mann, C., 1996, G2 cyclins are required for the degradation of G1 cyclins in yeast, *Nature* **384:**279–282.

Brandeis, M., and Hunt, T., 1996, The proteolysis of mitotic cyclins in mammalian cells persists from the end of mitosis until the onset of S phase, *EMBO J.* **15:**5280–5289.

Brinkmann, U., Brinkmann, E., Gallo, M., and Pastan, I., 1995, Cloning and characterization of a cellular apoptosis susceptibility gene, the human homologue to the yeast chromosome segregation gene CSE1, *Proc. Natl. Acad. Sci. USA* **92:**10427–10431.

Brown, K. D., Coulson, R. M., Yen, T. J., and Cleveland, D. W., 1994, Cyclin-like accumulation and loss of the putative kinetochore motor CENP-E results from coupling continuous synthesis with specific degradation at the end of mitosis, *J. Cell Biol.* **125:**1303–1312.

Cho, J. W., and Sauer, H. W., 1994, A non-cycling mitotic cyclin in the naturally synchronous cell cycle of *Physarum polycephalum Eur. J. Cell Biol.* **65:** 94–102.

Clurman, B., Sheaff, R. J., Thress, K., Groudine, M., and Roberts, J. M., 1996, Turnover of cyclin E by the ubiquitin-proteasome pathway is regulated by cdk2 binding and cyclin phosphorylation, *Genes Dev.* **10:**1979–1990.

Coats, S., Flanagan, W. M., Nourse, J., and Roberts, J. M., 1996, Requirement of p27Kip1 for restriction point control of the fibroblast cell cycle, *Science* **272:**877–880.

Cohen-Fix, O., Peters, J.-M., Kirschner, M. W., and Koshland, D., 1996, Anaphase initiation in *Saccharomyces cerevisiae* is controlled by the APC-dependent degradation of the anaphase inhibitor, Pds1p, *Genes Dev.* **10:**3077–3080.

Connelly, C., and Hieter, P., 1996, SKP1 is an evolutionarily conserved kinetochore protein required for cell cycle progression, *Cell* **86:**275–285.

Cross, F. R., and Blake, C. M., 1993, The yeast Cln3 protein is an unstable activator of Cdc28, *Mol. Cell. Biol.* **13:**3266–3271.

D'Andrea, R. J., Stratmann, R., Lehner, C. F., John, U. P., and Saint, R., 1993, The three rows gene of *Drosophila melanogaster* encodes a novel protein that is required for chromosome disjunction during mitosis, *Mol. Biol. Cell* **4:**1161–1174.

Davis, F. M., Tsao, T. Y., Fowler, S. K., and Rao, P. N., 1983, Monoclonal antibodies to mitotic cells, *Proc. Natl. Acad. Sci. USA* **80:**2926–2930.

Dawson, I. A., Roth, S., Akam, M., and Artavanis, T. S., 1993, Mutations of the fizzy locus cause metaphase arrest in *Drosophila melanogaster* embryos, *Development* **117:**359–376.

Dawson, I. A., Roth, S., and Artavanis, T. S., 1995, The Drosophila cell cycle gene fizzy is required for normal degradation of cyclins A and B during mitosis and has homology to the CDC20 gene of *Saccharomyces cerevisiae*, *J. Cell Biol.* **129:**725–737.

Deshaies, R. J., 1995a, The self-destructive personality of a cell cycle in transition, *Curr. Opin. Cell Biol.* **7:**781–789.

Deshaies, R. J., 1995b, Make it or break it: The role of ubiquitin-dependent proteolysis in cellular regulation, *Trends Cell Biol.* **5:**428–434.

Deshaies, R. J., Chau, V., and Kirschner, M., 1995, Ubiquitination of the G1 cyclin Cln2p by a Cdc34p-dependent pathway, *EMBO J.* **14:**303–312.
Dirick, L., Böhm, T., and Nasmyth, K., 1995, Roles and regulation of Cln-Cdc28 kinases at the start of the cell cycle of *Saccharomyces cerevisiae*, *EMBO J.* **14:**4803–4813.
Donovan, J. D., Toyn, J. H., Johnson, A. L., and Johnston, L. H., 1994, P40SDB25, a putative CDK inhibitor, has a role in the M/G1 transition in *Saccharomyces cerevisiae*, *Genes Dev.* **8:**1640–1653.
Doonan, J. H., and Morris, N. R., 1989, The bimG gene of *Aspergillus nidulans*, required for completion of anaphase, encodes a homolog of mammalian phosphoprotein phosphatase-1, *Cell* **57:**987–996.
Dunphy, W. G., Brizuela, L., Beach, D., and Newport, J., 1988, The *Xenopus cdc2* protein is a component of MPF, a cytoplasmic regulator of mitosis, *Cell* **54:**423–431.
Edgar, B. A., Sprenger, F., Duronio, R. J., Leopold, P., and O'Farrell, P. H., 1994, Distinct molecular mechanisms regulate cell cycle timing at successive stages of *Drosophila* embryogenesis, *Genes Dev.* **8:**440–452.
Engle, D. B., Osmani, S. A., Osmani, A. H., Rosborough, S., Xin, X. N., and Morris, N. R., 1990, A negative regulator of mitosis in *Aspergillus* is a putative membrane-spanning protein, *J. Biol. Chem.* **265:**16132–16137.
Evans, T., Rosenthal, E. T., Youngblom, J., Distel, D., and Hunt, T., 1983, Cyclin: A protein specified by maternal mRNA in sea urchin eggs that is destroyed at each cleavage division, *Cell* **33:** 389–396.
Feldman, R. M. R., Correll, C. C., Kaplan, K. B., and Deshaies, R. J., 1997, A complex of Cdc4p, Skp1p, and Cdc53p/cullin catalyzes ubiquitination of the phosphorylated CDK inhibitor Sic1p, *Cell* **91:**221–230.
Félix, M. A., Labbé, J. C., Dorée, M., Hunt, T., and Karsenti, E., 1990, Triggering of cyclin degradation in interphase extracts of amphibian eggs by cdc2 kinase, *Nature* **346:**379–382.
Friedman, H., and Snyder, M., 1994, Mutations in PRG1, a yeast proteasome-related gene, cause defects in nuclear division and are suppressed by deletion of a mitotic cyclin gene, *Proc. Natl. Acad. Sci. USA* **91:**2031–2035.
Funabiki, H., Yamano, H., Kumada, K., Nagao, K., Hunt, T., and Yanagida, M., 1996, Cut2 proteolysis required for sister-chromatic separation in fission yeast, *Nature* **381:**438–441.
Gallant, P., and Nigg, E. A., 1992, Cyclin B2 undergoes cell cycle-dependent nuclear translocation and, when expressed as a non-destructible mutant, causes mitotic arrest in HeLa cells, *J. Cell Biol.* **117:**213–224.
Ghiara, J. B., Richardson, H. E., Sugimoto, K., Henze, M., Lew, D. J., Wittenberg, C., and Reed, S. I., 1991, A cyclin B homolog in *S. cerevisiae*: Chronic activation of the Cdc28 protein kinase by cyclin prevents exit from mitosis, *Cell* **65:**163–174.
Ghislain, M., Udvardy, A., and Mann, C., 1993, S. cerevisiae 26S protease mutants arrest cell division in G2/metaphase, *Nature* **366:**358–362.
Ghosh, S., and Paweletz, N., 1992, Okadaic acid inhibits sister chromatid separation in mammalian cells, *Exp. Cell Res.* **200:**215–217.
Glotzer, M., Murray, A. W., and Kirschner, M. W., 1991, Cyclin is degraded by the ubiquitin pathway, *Nature* **349:**132–138.
Goebl, M. G., Yochem, J., Jentsch, S., McGrath, J. P., Varshavsky, A., and Byers, B., 1988, The yeast cell cycle gene CDC34 encodes a ubiquitin-conjugating enzyme, *Science* **241:**1331–1335.
Goebl, M. G., Goetsch, L., and Byers, B., 1994, The Ubc3 (Cdc34) ubiquitin-conjugating enzyme is ubiquitinated and phosphorylated *in vivo*, *Mol. Cell. Biol.* **14:**3022–3029.
Gordon, C., McGurk, G., Dillon, P., Rosen, C., and Hastie, N. D., 1993, Defective mitosis due to a mutation in the gene for a fission yeast 26S protease subunit, *Nature* **366:**355–357.
Gordon, C., McGurk, G., Wallace, M., and Hastie, N. D., 1996, A conditional lethal mutant in the fission yeast 26 S proteasome mts3+ is defective in metaphase to anaphase transition, *J. Biol. Chem.* **271:**5704–5711.

Grieco, D., Porcellini, A., Avvedimento, E. V., and Gottesman, M. E., 1996, Requirement for cAMP-PKA pathway activation by M phase-promoting factor in the transition from mitosis to interphase, *Science* **271:**1718–1723.

Hartwell, L. H., Mortimer, R. K., Culotti, J., and Culotti, M., 1973, Genetic control of the cell division cycle in yeast. V. Genetic analysis of *cdc* mutants, *Genetics* **74:**267–286.

Hateboer, G., Kerkhoven, R. M., Shvarts, A., Bernards, R., and Beijersbergen, R. L., 1996, Degradation of E2F by the ubiquitin-proteasome pathway: Regulation by retinoblastoma family proteins and adenovirus transforming proteins, *Genes Dev.* **10:**2960–2970.

Heichman, K. A., and Roberts, J. M., 1996, The yeast CDC16 and CDC27 genes restrict DNA replication to once per cell cycle, *Cell* **85:**39–48.

Henchoz, S., Chi, Y., Catarin, B., Hershkowitz, I., Deshaies, R. J., and Peter, M., 1997, Phosphorylation and ubiquitin-dependent degradation of the cyclin-dependent kinase inhibitor Farlp in budding yeast. *Genes Dev.* **11:**3046–3060.

Hershko, A., Heller, H., Elias, S., and Ciechanover, A., 1983, Components of ubiquitin-protein ligase system. Resolution, affinity purification, and role in protein breakdown, *J. Biol. Chem.* **258:**8206–8214.

Hershko, A., Ganoth, D., Pehrson, D., Palazzo, R., and Cohen, L. H., 1991, Methylated ubiquitin inhibits cyclin degradation in clam embryo extracts, *J. Biol. Chem.* **266:**16376–16379.

Hershko, A., Ganoth, D., Sudakin, V., Dahan, A., Cohen, L. H., Luca, F. C., Ruderman, J. V., and Eytan, E., 1994, Components of a system that ligates cyclin to ubiquitin and their regulation by the protein kinase cdc2, *J. Biol. Chem.* **269:**4940–4946.

Hirano, T., Hiraoka, Y., and Yanagida, M., 1988, A temperature-sensitive mutation of the *Schizosaccharomyces pombe* gene nuc2+ that encodes a nuclear scaffold-like protein blocks spindle elongation in mitotic anaphase, *J. Cell Biol.* **106:**1171–1183.

Hirano, T., Kinoshita, N., Morikawa, K., and Yanagida, M., 1990, Snap helix with knob and hole: Essential repeats in *S. pombe* nuclear protein nuc2+, *Cell* **60:**319–328.

Hisamoto, N., Sugimoto, K., and Matsumoto, K., 1994, The Glc7 type 1 protein phosphatase of *Saccharomyces cerevisiae* is required for cell cycle progression in G2/M, *Mol. Cell. Biol.* **14:** 3158–3165.

Hofmann, F., Martelli, F., Livingston, D. M., and Wang, Z., 1996, The retinoblastoma gene product protects E2F-1 from degradation by the ubiquitin-proteasome pathway, *Genes Dev.* **10:**2949–2959.

Holloway, S. L., Glotzer, A., King, R., and Murray, A. W., 1993, Anaphase is initiated by proteolysis rather than by the inactivation of maturation-promoting factor, *Cell* **73:**1393–1402.

Hoyt, M. A., Totis, L., and Roberts, B. T., 1991, *S. cerevisiae* genes required for cell cycle arrest in response to loss of microtubule function, *Cell* **66:**507–517.

Hunt, T., Luca, F. C., and Ruderman, J. V., 1992, The requirements for protein synthesis and degradation, and the control of destruction of cyclins A and B in the meiotic and mitotic cell cycles of the clam embryo, *J. Cell Biol.* **116:**707–724.

Irniger, S., Piatti, S., Michaelis, C., and Nasmyth, K., 1995, Genes involved in sister chromatid separation are needed for B-type cyclin proteolysis in budding yeast, *Cell* **81:**269–278.

Ishii, K., Kumada, K., Toda, T., and Yanagida, M., 1996, Requirement for PP1 phosphatase and 20S cyclosome/APC for the onset of anaphase is lessened by the dosage increase of a novel gene sds23+, *EMBO J.* **15:**6629–6640.

Izumi, T., and Maller, J. L., 1991, Phosphorylation of *Xenopus* cyclins B1 and B2 is not required for cell cycle transitions, *Mol. Cell. Biol.* **11:**3860–3867.

James, S. W., Mirabito, P. M., Scacheri, P. C., and Morris, N. R., 1995, The *Aspergillus nidulans bimE* (blocked-in-mitosis) gene encodes multiple cell cycle functions involved in mitotic checkpoint control and mitosis, *J. Cell Sci.* **108:**3485–3499.

Johnson, S. L., 1991, Structure and function analysis of the yeast CDC4 gene product, Ph.D dissertation, University of Washington.

Johnston, L. H., Eberly, S. L., Chapman, J. W., Araki, H., and Sugino, A., 1990, The product of the *Saccharomyces cerevisiae* cell cycle gene DBF2 has homology with protein kinases and is periodically expressed in the cell cycle, *Mol. Cell. Biol.* **10:**1358–1366.

Jones, C., and Smythe, C., 1996, Activation of the *Xenopus* cyclin degradation machinery by full-length cyclin A, *J. Cell Sci.* **109:**1071–1079.

Juang, Y.-L., Huang, J., Peters, J.-M., McLaughlin, M. E., Tai, C.-Y., and Pellman, D., 1997, APC-mediated proteolysis of Ase1 and the morphogenesis of the mitotic spindle, *Science* **275:**1311–1314.

King, R. W., Jackson, P. K., and Kirschner, M. W., 1994, Mitosis in transition, *Cell* **79:**563–571.

King, R. W., Peters, J.-M., Tugendreich, S., Rolfe, M., Hieter, P., and Kirschner, M. W., 1995, A 20S complex containing CDC27 and CDC16 catalyzes the mitosis-specific conjugation of ubiquitin to cyclin B, *Cell* **81:**279–288.

King, R. W., Deshaies, R. J., Peters, J.-M., and Kirschner, M. W., 1996a, How proteolysis drives the cell cycle, *Science* **274:**1652–1659.

King, R. W., Glotzer, M., and Kirschner, M. W., 1996b, Mutagenic analysis of the destruction signal of mitotic cyclins and structural characterization of ubiquitinated intermediates, *Mol. Biol. Cell* **7:**1343–1357.

Kipreos, E. T., Lander, L. E., Wing, J. P., He, W. W., and Hedgecock, E. M., 1996, cul-1 is required for cell cycle exit in *C. elegans* and identifies a novel gene family, *Cell* **85:**1–20.

Kitada, K., Johnson, A. L., Johnston, L. H., and Sugino, A., 1993, A multicopy suppressor gene of the *Saccharomyces cerevisiae* G1 cell cycle mutant gene dbf4 encodes a protein kinase and is identified as CDC5, *Mol. Cell. Biol.* **13:**4445–4457.

Klotzbücher, A., Stewart, E., Harrison, D., and Hunt, T., 1996, The 'destruction box' of cyclin A allows B-type cyclins to be ubiquitinated, but not efficiently destroyed, *EMBO J.* **15:**3053–3064.

Knoblich, J. A., Sauer, K., Jones, L., Richardson, H., Saint, R., and Lehner, C. F., 1994, Cyclin E controls S phase progression and its down-regulation during *Drosophila* embryogenesis is required for the arrest of cell proliferation, *Cell* **77:**107–120.

Kobayashi, H., Stewart, E., Poon, R., Adamczewski, J. P., Gannon, J., and Hunt, T., 1992, Identification of the domains in cyclin A required for binding to, and activation of, p34cdc2 and p32cdk2 protein kinase subunits, *Mol. Biol. Cell* **3:**1279–1294.

Kornitzer, D., Raboy, B., Kulka, R. G., and Fink, G. R., 1994, Regulated degradation of the transcription factor Gcn4, *EMBO J.* **13:**6021–6030.

Kuang, J., and Ashorn, C. L., 1993, At least two kinases phosphorylate the MPM-2 epitope during *Xenopus* oocyte maturation, *J. Cell Biol.* **123:**859–868.

Kumagai, A., and Dunphy, W. G., 1996, Purification and molecular cloning of Plx1, a Cdc25-regulatory kinase from *Xenopus* egg extracts, *Science* **273:**1377–1380.

Labbé, J. C., Capony, J. P., Caput, D., Cavadore, J. C., Derancourt, J., Kaghad, M., Lelias, J. M., Picard, A., and Dorée, M., 1989, MPF from starfish oocytes at first meiotic metaphase is a heterodimer containing one molecule of cdc2 and one molecule of cyclin B, *EMBO J.* **8:**3053–3058.

Lahav-Baratz, S., Sudakin, V., Ruderman, J. V., and Hershko, A., 1995, Reversible phosphorylation controls the activity of cyclosome-associated cyclin-ubiquitin ligase, *Proc. Natl. Acad. Sci. USA* **92:**9303–9307.

Lamb, J. R., Michaud, W. A., Sikorski, R. S., and Hieter, P. A., 1994, Cdc16p, Cdc23p and Cdc27p form a complex essential for mitosis, *EMBO J.* **13:**4321–4328.

Lanker, S., Valdivieso, M. H., and Wittenberg, C., 1996, Rapid degradation of the G1 cyclin Cln2 induced by CDK-dependent phosphorylation, *Science* **271:**1597–1601.

Li, J. K., Meyer, A. N., and Donoghue, D. J., 1995, Requirement for phosphorylation of cyclin B1 for *Xenopus* oocyte maturation, *Mol. Biol. Cell* **9:**1111–1124.

Li, R., and Murray, A. W., 1991, Feedback control of mitosis in budding yeast, *Cell* **66:**519–531.

Li, X., and Nicklas, R. B., 1995, Mitotic forces control a cell-cycle checkpoint, *Nature* **373:**630–632.

Lin, W. C., and Desiderio, S., 1993, Regulation of V(D)J recombination activator protein RAG-2 by phosphorylation, *Science* **260:**953–959.

Lorca, T., Devault, A., Colas, P., Van, L. A., Fesquet, D., Lazaro, J. B., and Doree, M., 1992a, Cyclin A-Cys41 does not undergo cell cycle-dependent degradation in *Xenopus* extracts, *FEBS Lett.* **306:** 90–93.
Lorca, T., Labbe, J. C., Devault, A., Fesquet, D., Strausfeld, U., Nilsson, J., Nygren, P., Uhlen, M., Cavadore, J. C., and Doree, M., 1992b, Cyclin A-cdc2 kinase does not trigger but delays cyclin degradation in interphase extracts of amphibian eggs, *J. Cell Sci.* **102:**55–62.
Lorca, T., Cruzalegui, F. H., Fesquet, D., Cavadore, J. C., Mery, J., Means, A., and Doree, M., 1993, Calmodulin-dependent protein kinase II mediates inactivation of MPF and CSF upon fertilization of *Xenopus* eggs, *Nature* **366:**270–273.
Luca, F. C., Shibuya, E. K., Dohrmann, C. E., and Ruderman, J. V., 1991, Both cyclin A delta 60 and B delta 97 are stable and arrest cells in M-phase, but only cyclin B delta 97 turns on cyclin destruction, *EMBO J.* **10:**4311–4320.
Luo, Q., Michaelis, C., and Weeks, G., 1994, Overexpression of a truncated cyclin B gene arrests *Dictyostelium* cell division during mitosis, *J. Cell Sci.* **107:**3105–3114.
Mahaffey, D., Yoo, Y., and Rechsteiner, M., 1993, Ubiquitin metabolism in cycling *Xenopus* egg extracts, *J. Biol. Chem.* **268:**21205–21211.
Mahajan, R., Delphin, C., Guan, T., Gerace, L., and Melchior, F., 1997, A small ubiquitin-related polypeptide involved in targeting RanGAP1 to nuclear pore complex protein RanBP2, *Cell* **88:** 97–107.
Mathias, N., Johnson, S. L., Winey, M., Adams, A. E., Goetsch, L., Pringle, J. R., Byers, B., and Goebl, M. G., 1996, Cdc53p acts in concert with Cdc4p and Cdc34p to control the G1-to-S-phase transition and identifies a conserved family of proteins, *Mol. Cell. Biol.* **16:**6634–6643.
Matoskova, B., Wong, W. T., Seki, N., Nagase, T., Nomura, N., Robbins, K. C., and Di Fiore, P. P., 1996a, RN-tre identifies a family of tre-related proteins displaying a novel potential protein binding domain, *Oncogene* **12:**2563–2571.
Matoskova, B., Wong, W. T., Nomura, N., Robbins, K. C., and Di Fiore, P. P., 1996b, RN-tre specifically binds to the SH3 domain of eps8 with high affinity and confers growth advantage to NIH3T3 upon carboxy-terminal truncation, *Oncogene* **12:**2679–2688.
Matsushime, H., Roussel, M. F., Ashmun, R. A., and Sherr, C. J., 1991, Colony-stimulating factor 1 regulates novel cyclins during the G1 phase of the cell cycle, *Cell* **65:**701–713.
McKinney, J. D., and Cross, F. R., 1995, Far1 and the G1 phase specificity of cell cycle arrest by mating factor in *Saccharomyces cerevisiae*, *Mol. Cell. Biol.* **15:**2509–2516.
McKinney, J. D., Chang, F., Heintz, N., and Cross, F. R., 1993, Negative regulation of FAR1 at the start of the yeast cell cycle, *Genes Dev.* **7:**833–843.
Mendenhall, M. D., 1993, An inhibitor of p34CDC28 protein kinase activity from *Saccharomyces cerevisiae*, *Science* **259:** 216–219.
Minshull, J., Sun, H., Tonks, N. K., and Murray, A. W., 1994, A MAP kinase-dependent spindle assembly checkpoint in *Xenopus* egg extracts, *Cell* **79:**475–486.
Moreno, S., Hayles, J., and Nurse, P., 1989, Regulation of $p34^{cdc2}$ protein kinase during mitosis, *Cell* **58:**361–372.
Morgan, D. O., 1995, Principles of CDK regulation, *Nature* **374:**131–134.
Morishita, T., Mitsuzawa, H., Nakafuku, M., Nakamura, S., Hattori, S., and Anraku, Y., 1995, Requirement of *Saccharomyces cerevisiae* Ras for completion of mitosis, *Science* **270:**1213–1215.
Murray, A. W., Solomon, M. J., and Kirschner, M. W., 1989, The role of cyclin synthesis and degradation in the control of maturation promoting factor activity, *Nature* **339:**280–286.
Nakamura, T., Hillova, J., Mariage-Samson, R., Onno, M., Huebner, K., Cannizzaro, L. A., Boghosian-Sell, L., Croce, C. M., and Hill, M., 1992, A novel transcriptional unit of the *tre* oncogene widely expressed in human cancer cells, *Oncogene* **7:**733–741.
Nasmyth, K., 1993, Control of the yeast cell cycle by the Cdc28 protein kinase, *Curr. Opin. Cell Biol.* **5:**166–179.
Nasmyth, K., 1996a, At the heart of the budding yeast cell cycle, *Trends Genet.* **12:**405–412.
Nasmyth, K., 1996b, Viewpoint: Putting the cell cycle in order, *Science* **274:**1643–1645.

Nicklas, R. B., Ward, S. C., and Gorbsky, G. J., 1995, Kinetochore chemistry is sensitive to tension and may link mitotic forces to a cell cycle checkpoint, *J. Cell Biol.* **130:**929–939.

Nugroho, T. T., and Mendenhall, M. D., 1994, An inhibitor of yeast cyclin-dependent protein kinase plays an important role in ensuring the genomic integrity of daughter cells, *Mol. Cell. Biol.* **14:**3320–3328.

O'Donnell, K. L., Osmani, A. H., Osmani, S. A., and Morris, N. R., 1991, BimA encodes a member of the tetratricopeptide repeat family of proteins and is required for the completion of mitosis in *Aspergillus nidulans*, *J. Cell Sci.* **99:**711–719.

Ohkura, H., and Yanagida, M., 1991, S. pombe gene sds22+ essential for a midmitotic transition encodes a leucine-rich repeat protein that positively modulates protein phosphatase-1, *Cell* **64:** 149–157.

Ohkura, H., Kinoshita, N., Miyatani, S., Toda, T., and Yanagida, M., 1989, The fission yeast dis2+ gene required for chromosome disjoining encodes one of two putative type 1 protein phosphatases, *Cell* **57:**997–1007.

Osaka, F., Seino, H., Seno, T., and Yamao, F., 1997, A ubiquitin-conjugating enzyme in fission yeast that is essential for the onset of anaphase in mitosis, *Mol. Cell. Biol.* **17:**3388–3397.

Osmani, A. H., O'Donnell, K., Pu, R. T., and Osmani, S. A., 1991, Activation of the nimA protein kinase plays a unique role during mitosis that cannot be bypassed by absence of the bimE checkpoint, *EMBO J.* **10:**2669–2679.

Osmani, S. A., Engle, D. B., Doonan, J. H., and Morris, N. R., 1988, Spindle formation and chromatin condensation in cells blocked at interphase by mutation of a negative cell cycle control gene, *Cell* **52:**241–251.

Pagano, M., Tam, S. W., Theodoras, A. M., Beer-Romero, P., Del Sal, G., Chau, V., Yew, P. R., Draetta, G. F., and Rolfe, M., 1995, Role of the ubiquitin-proteasome pathway in regulating abundance of the cyclin-dependent kinase inhibitor p27, *Science* **269:**682–685.

Papa, F. R., and Hochstrasser, M., 1993, The yeast *DOA4* gene encodes a deubiquitinating enzyme related to a product of the human *tre*-2 oncogene, *Nature* **366:**313–319.

Pardee, A. B., 1987, Molecules involved in proliferation of normal and cancer cells: Presidential address, *Cancer Res.* **47:**1488–1491.

Patra, D., and Dunphy, W. G., 1996, Xe-p9, a *Xenopus* Sucl/Cks homolog has multiple essential roles in cell cycle control, *Genes Dev.* **10:**1503–1515.

Pellman, D., Bagget, M., Tu, H., and Fink, G. R., 1995, Two microtubule-associated proteins required for anaphase spindle movement in *Saccharomyces cerevisiae*, *J. Cell Biol.* **130:**1373–1385.

Peters, J.-M., King, R. W., Höög, C., and Kirschner, M. W., 1996, Identification of BIME as a subunit of the anaphase promoting complex, *Science* **274:**1199–1201.

Peterson, T. A., Yochem, J., Byers, B., Nunn, M. F., Duesberg, P. H., Doolittle, R. F., and Reed, S. I., 1984, A relationship between the yeast cell cycle genes CDC4 and CDC36 and the ets sequence of oncogenic virus E26, *Nature* **309:**556–558.

Philp, A. V., Axton, J. M., Saunders, R. D., and Glover, D. M., 1993, Mutations in the *Drosophila melanogaster* gene three rows permit aspects of mitosis to continue in the absence of chromatid segregation, *J. Cell Sci.* **106:**87–98.

Piatti, S., Böhm, T., Cocker, J. H., Diffley, J. F. X., and Nasmyth, K., 1996, Activation of S-phase-promoting CDKs in late Gl defines a "point of no return" after which Cdc6 synthesis cannot promote DNA replication in yeast, *Genes Dev.* **10:**1516–1531.

Pichler, S., Piatti, S., and Nasmyth, K., 1997, Is the yeast anaphase promoting complex needed to prevent re-replication during G2 and M phases? *EMBO J.* **16:**5988–5997.

Plon, S. E., Leppig, K. A., Do, H. N., and Groudine, M., 1993, Cloning of the human homolog of the CDC34 cell cycle gene by complementation in yeast, *Proc. Natl. Acad. Sci. USA* **90:**10484–10488.

Prendergast, J. A., Ptak, C., Kornitzer, D., Steussy, C. N., Hodgins, R., Goebl, M., and Ellison, M. J., 1996, Identification of a positive regulator of the cell cycle ubiquitin-conjugating enzyme Cdc34 (Ubc3), *Mol. Cell. Biol.* **16:**677–684.

Ramsay, G., Evan, G. I., and Bishop, J. M., 1984, The protein encoded by the human proto-oncogene c-myc, *Proc. Natl. Acad. Sci. USA* **81:**7742–7746.

Reich, N. C., and Levine, A. J., 1984, Growth regulation of a cellular tumour antigen, p53, in nontransformed cells, *Nature* **308:**199–201.

Richter-Ruoff, B., and Wolf, D. H., 1993, Proteasome and cell cycle. Evidence for a regulatory role of the protease on mitotic cyclins in yeast, *FEBS Lett.* **336:**34–36.

Rimmington, G., Dalby, B., and Glover, D. M., 1994, Expression of N-terminally truncated cyclin B in the *Drosophila* larval brain leads to mitotic delay at late anaphase, *J. Cell Sci.* **107:**2729–2738.

Rogers, S., Wells, R., and Rechsteiner, M., 1986, Amino acid sequences common to rapidly degraded proteins: The PEST hypothesis, *Science* **234:**364–368.

Rudner, A. D., and Murray, A. W., 1996, The spindle assembly checkpoint, *Curr. Opin. Cell Biol.* **8:** 773–780.

Sagata, N., 1996, Meiotic metaphase arrest in animal oocytes: Its mechanisms and biological significance, *Trends Cell Biol.* **6:**22–28.

Sagata, N., Watanabe, N., Vande Woude, G. F., and Ikawa, Y., 1989, The c-mos proto-oncogene product is a cytostatic factor responsible for meiotic arrest in vertebrate eggs, *Nature* **342:**512–518.

Saitoh, H., Pu, R., Cavenagh, M., and Dasso, M., 1997, RanBP2 associates with Ubc9p and a modified form of RanGAP1, *Proc. Natl. Acad. Sci. USA* **94:**3736–3741.

Salama, S. R., Hendricks, K. B., and Thorner, J., 1994, G1 cyclin degradation—The PEST motif of yeast Cln2 is necessary, but not sufficient, for rapid protein turnover, *Mol. Cell. Biol.* **14:**7953–7966.

Samejima, I., and Yanagida, M., 1994, Bypassing anaphase by fission yeast cut9 mutation: Requirement of cut9+ to initiate anaphase, *J. Cell Biol.* **127:**1655–1670.

Scherf, U., Pastan, I., Willingham, M. C., and Brinkmann, U., 1996, The human CAS protein which is homologous to the CSE1 yeast chromosome segregation gene product is associated with microtubules and mitotic spindle, *Proc. Natl. Acad. Sci. USA* **83:**2670–2674.

Schneider, B. L., Yang, Q. H., and Futcher, A. B., 1996, Linkage of replication to start by the Cdk inhibitor Sic1, *Science* **272:**560–562.

Schwob, E., Böhm, T., Mendenhall, M. D., and Nasmyth, K., 1994, The B-type cyclin kinase inhibitor p40SIC1 controls the G1 to S transition in *S. cerevisiae*, *Cell* **79:**233–244.

Schwob, E., Böhm, T., Mendenhall, M. D., and Nasmyth, K., 1996, The B-type cyclin kinase inhibitor p40SIC1 controls the G1 to S transition in *S. cerevisiae* (Erratum), *Cell* **784:**U13.

Sethi, N., Monteagudo, M. C., Koshland, D., Hogan, E., and Burke, D. J., 1991, The CDC20 gene product of *Saccharomyces cerevisiae*, a beta-transducin homolog, is required for a subset of microtubule-dependent cellular processes, *Mol. Cell. Biol.* **11:**5592–5602.

Seufert, W., and Jentsch, S., 1990, Ubiquitin-conjugating enzymes UBC4 and UBC5 mediate selective degradation of short-lived and abnormal proteins, *EMBO J.* **9:**543–550.

Seufert, W., Futcher, B., and Jentsch, S., 1995, Role of a ubiquitin-conjugating enzyme in degradation of S- and M-phase cyclins, *Nature* **373:**78–81.

Shamu, C. E., and Murray, A. W., 1992, Sister chromatid separation in frog egg extracts requires DNA topoisomerase II activity during anaphase, *J. Cell Biol.* **117:**921–934.

Sherwood, S. W., Kung, A. L., Roitelman, J., Simoni, R. D., and Schimke, R. T., 1993, *In vivo* inhibition of cyclin B degradation and induction of cell-cycle arrest in mammalian cells by the neutral cysteine protease inhibitor N-acetylleucylleucylnorleucinal, *Proc. Natl. Acad. Sci. USA* **90:**3353–3357.

Shirayama, M., Matsui, Y., and Toh-E, A., 1994a, The yeast TEM1 gene, which encodes a GTP-binding protein, is involved in termination of M phase, *Mol. Cell. Biol.* **14:**7476–7482.

Shirayama, M., Matsui, Y., Tanaka, K., and Toh-E, A., 1994b, Isolation of a CDC25 family gene, MSI2/LTE1, as a multicopy suppressor of ira1, *Yeast* **10:**451–461.

Shirayama, M., Matsui, Y., and Toh-E, A., 1996, Dominant mutant alleles of yeast protein kinase gene CDC15 suppress the lte1 defect in termination of M phase and genetically interact with CDC14, *Mol. Gen. Genet.* **251:**176–185.

Sigrist, S., Jacobs, H., Stratmann, R., and Lehner, C. F., 1995, Exit from mitosis is regulated by *Drosophila fizzy* and the sequential destruction of cyclins A, B and B3, *EMBO J.* **14:**4827–4838.
Sikorski, R. S., Boguski, M. S., Goebl, M., and Hieter, P., 1990, A repeating amino acid motif in CDC23 defines a family of proteins and a new relationship among genes required for mitosis and RNA synthesis, *Cell* **60:**307–317.
Skowyra, D., Craig, K. L., Tyers, M., Elledge, S. J., and Harper, J. W., 1997, F-box proteins are receptors that recrruit phosphorylated substrates to the SCF ubiquitin-ligase complex, 1997, *Cell* **91:**209–219.
Smith, S. A., Kumar, P., Johnston, I., and Rosamond, J., 1992, SCM4, a gene that suppresses mutant cdc4 function in budding yeast, *Mol. Gen. Genet.* **235:**285–291.
Song, A., Wang, Q., Goebl, M., and Harrington, M. A., 1996, The ubiquitin pathway is required for the rapid degradation of hyperphosphorylated MyoD, *Mol. Biol. Cell.* **7:**357a.
Starborg, M., Brundell, E., Gell, K., and Höög, C., 1994, A novel murine gene encoding a 216-kDa protein is related to a mitotic checkpoint regulator previously identified in *Aspergillus nidulans*, *J. Biol. Chem.* **269:**24133–24137.
Stemmann, O., and Lechner, J., 1996, The *Saccharomyces cerevisiae* kinetochore contains a cyclin-CDK complexing homolog, as identified by *in vitro* reconstitution, *EMBO J.* **15:**3611–3620.
Stewart, E., Kobayashi, H., Harrison, D., and Hunt, T., 1994, Destruction of Xenopus cyclins A and B2, but not B1, requires binding to p34cdc2, *EMBO J.* **3:**584–594.
Stratmann, R., and Lehner, C. F., 1996, Separation of sister chromatids in mitosis requires the *Drosophila* pimples product, a protein degraded after the metaphase/anaphase transition, *Cell* **84:** 25–35.
Sudakin, V., Ganoth, D., Dahan, A., Heller, H., Hershko, J., Luca, F. C., Ruderman, J. V., and Hershko, A., 1995, The cyclosome, a large complex containing cyclin-selective ubiquitin ligase activity, targets cyclins for destruction at the end of mitosis, *Mol. Biol. Cell* **6:**185–198.
Sudakin, V., Shteinberg, M., Ganoth, D., Hershko, J., and Hershko, A., 1997, Binding of activated cyclosome to p13^{suc1}, *J. Biol. Chem.* **272:**18051–18059.
Surana, U., Amon, A., Dowzer, C., McGrew, J., Byers, B., and Nasmyth, K., 1993, Destruction of the CDC28/CLB mitotic kinase is not required for the metaphase to anaphase transition in budding yeast, *EMBO J.* **12:**1969–1978.
Townsley, F. M., Aristarkhov, A., Beck, S., Hershko, A., and Ruderman, J. V., 1997, Dominant-negative cyclin-selective ubiquitin carrier protein E2-C/UbcH10 blocks cells in metaphase, *Proc. Natl. Acad. Sci. U.S.A.* **94:**2362–2367.
Toyn, J. H., and Johnston, L. H., 1993, Spo12 is a limiting factor that interacts with the cell cycle protein kinases Dbf2 and Dbf20, which are involved in mitotic chromatin disjunction, *Genetics* **135:** 963–971.
Toyn, J. H., and Johnston, L. H., 1994, The Dbf2 and Dbf20 protein kinases of budding yeast are activated after the metaphase to anaphase cell cycle transition, *EMBO J.* **13:**1103–1113.
Tugendreich, S., Tomkiel, J., Earnshaw, W., and Hieter, P., 1995, The CDC27HS protein co-localizes with the CDC16HS protein to the centrosome and mitotic spindle and is essential for the metaphase to anaphase transtion, *Cell* **81:**261–268.
Tyers, M., 1996, The cyclin-dependent kinase inhibitor p40^{SIC1} imposes the requirement for CLN G1 cyclin function at Start, *Proc. Natl. Acad. Sci. USA* **93:**7772–7776.
Tyers, M., Tokiwa, G., Nash, R., and Futcher, B., 1992, The Cln3-Cdc28 kinase complex of *S. cerevisiae* is regulated by proteolysis and phosphorylation, *EMBO J.* **11:**1773–1784.
van der Velden, H. M., and Lohka, M. J., 1993, Mitotic arrest caused by the amino terminus of *Xenopus* cyclin B2, *Mol. Cell. Biol.* **13:**1480–1488.
van der Velden, H. M., and Lohka, M. J., 1994, Cell cycle-regulated degradation of *Xenopus* cyclin B2 requires binding to p34cdc2, *Mol. Biol. Cell* **5:**713–724.
Verma, R., Annan, K. S., Huddleston, M. J., Carr, S. A., Reynard, G., and Deshaies, R. J., 1997a, Phosphorylation of Sic1p by G1 Cdk required for its degradation and entry into S phase, *Science* **278:**455–460.

Verma, R., Feldman, R. M. R., and Deshaies, R. J., 1997b, SIC1 is ubiquitinated in vitro by a pathway that requires CDC4, CDC34, and cyclin/CDK activities. *Mol. Biol. Cell* **8:**1427–1437.

Whitfield, W. G. F., Gonzalez, C., Maldonado-Codina, G., and Glover, D. M., 1990, The A- and B-type cyclins of *Drosophila* are accumulated and destroyed in temporally distinct events that define separable phases of the G2–M transition, *EMBO J.* **9:**2563–2572.

Willems, A. R., Lanker, S., Patton, E. E., Craig, K. L., Nason, T. F., Kobayashi, R., Wittenberg, C., and Tyers, M., 1996, Cdc53 targets phosphorylated G1 cyclins for degradation by the ubiquitin proteolytic pathway, *Cell* **86:**453–463.

Won, K.-A., and Reed, S. I., 1996, Activation of cyclin E/CDK2 is coupled to site-specific autophosphorylation and ubiquitin-dependent degradation of cyclin E, *EMBO J.* **15:**4182–4193.

Xiao, Z., McGrew, J. T., Schroeder, A. J., and Fitzgerald, H. M., 1993, CSE1 and CSE2, two new genes required for accurate mitotic chromosome segregation in *Saccharomyces cerevisiae*, *Mol. Cell. Biol.* **13:**4691–4702.

Yaglom, J., Linskens, M. H., Sadis, S., Rubin, D. M., Futcher, B., and Finley, D., 1995, p34Cdc28-mediated control of Cln3 cyclin degradation, *Mol. Cell. Biol.* **15:**731–741.

Yamamoto, A., Guacci, V., and Koshland, D., 1996a, Pds1p is required for faithful execution of anaphase in the yeast, *Saccharomyces cerevisiae*, *J. Cell Biol.* **133:**85–97.

Yamamoto, A., Guacci, V., and Koshland, D., 1996b, Pds1p, an inhibitor of anaphase in budding yeast, plays a critical role in the anaphase and checkpoint pathway(s), *J. Cell Biol.* **133:**99–110.

Yamano, H., Gannon, J., and Hunt, T., 1996, The role of proteolysis in cell cycle progression in *Schizosaccharomyces pombe*, *EMBO J.* **15:**5268–5279.

Yamashita, Y. M., Nakaseko, Y., Samejima, I., Kumada, K., Yamada, H., Michaelson, D., and Yanagida, M., 1996, 20S cyclosome complex formation and proteolytic activity inhibited by the cAMP/PKA pathway, *Nature* **384:**276–279.

Ye, X. S., Fincher, R. R., Tang, A., O'Donnell, K., and Osmani, S. A., 1996, Two S-phase checkpoint systems, one involving the function of both BIME and Tyr15 phosphorylation of $p34^{cdc2}$, inhibit NIMA and prevent poremature mitosis, *EMBO J.* **15:**3599–3610.

Yew, P. R., and Kirschner, M. W., 1997, Proteolysis and DNA replication: the CDC34 requirement in the Xenopus egg cell cycle, *Science* **277:**1672–1676.

Yoon, H. J., and Carbon, J., 1995, Genetic and biochemical interactions between an essential kinetochore protein, Cbf2p/Ndc10p, and the CDC34 ubiquitin-conjugating enzyme, *Mol. Cell. Biol.* **15:**4835–4842.

Yu, H., King, R. W., Peters, J.-M., and Kirschner, M. W., 1996, Identification of a novel ubiquitin-conjugating enzyme involved in mitotic cyclin degradation, *Curr. Biol.* **6:**455–466.

Zachariae, W., and Nasmyth, K., 1996, TPR proteins required for anaphase progression mediate ubiquitination of mitotic B-type cyclins in yeast, *Mol. Biol. Cell* **7:**791–806.

Zachariae, W., Shin, T.-H., Galova, M., Obermaier, B., and Nasmyth, K., 1996, New subunits of the anaphase promoting complex of *Saccharomyces cerevisiae*, *Science* **274:**1201–1204.

Zhang, H., Kobayashi, R., Galaktionov, K., and Beach, D., 1995, p19SKP1 and p45SKP2 are essential elements of the cyclin A-CDK S-phase kinase, *Cell* **82:**915–925.

Zhu, Y., Carroll, M., Papa, F. R., Hochstrasser, M., and D'Andrea, A. D., 1996, DUB-1, a deubiquitinating enzyme with growth-suppressing activity, *Proc. Natl. Acad. Sci. USA* **93:**3275–3279.

CHAPTER 13

Ubiquitination of Integral Membrane Proteins and Proteins in the Secretory Pathway

Ron R. Kopito

1. INTRODUCTION

Covalent modification of cellular proteins with ubiquitin represents a versatile intracellular signaling pathway that is linked to such diverse phenomena as DNA repair, organelle biogenesis, and protein turnover. Best understood among these is the role of the ubiquitin system in targeting cytoplasmic proteins for proteolysis by the 26 S proteasome complex (reviewed in Chapter 6). Although most of the substrates for the ubiquitin conjugation system studied to date have been cytoplasmic or nuclear proteins, this chapter reviews recent data that suggest a role for the ubiquitin system in degradation and trafficking of proteins within the central vacuolar system. By central vacuolar system, I refer collectively to the compartments of the secretory and endocytic/lysosomal pathways including the endoplasmic reticulum, Golgi apparatus, plasma membrane, lysosomes, and the various transitional compartments and transport vesicles.

Ron R. Kopito • Department of Biological Sciences, Stanford University, Stanford, California 94305-5020.

Ubiquitin and the Biology of the Cell, edited by Peters *et al.* Plenum Press, New York, 1998.

This chapter is divided into two main sections. The first reviews the evidence for and the implications of a role for the ubiquitin–proteasome pathway in the degradation of misfolded integral membrane and soluble proteins that are retained within the endoplasmic reticulum (ER), and the relationship of this process to "quality control" or "ER degradation." The second section reviews recent reports suggesting a role for the ubiquitin pathway in regulating the trafficking of membrane receptors and other integral plasma membrane proteins.

2. QUALITY CONTROL AND PROTEIN DEGRADATION ASSOCIATED WITH THE ER

Proteins synthesized in the secretory pathway—integral membrane and secreted proteins—usually exist in their mature conformations as oligo- or multimeric complexes (Hurtley and Helenius, 1989). The assembly of these oligomers occurs in the ER and is a prerequisite for export to the Golgi apparatus (Hurtley and Helenius, 1989; Pelham, 1989). Thus, exit of proteins from the ER can be viewed as a critical "checkpoint" in the secretory pathway that regulates the deployment of misfolded or denatured proteins. Mutant, misfolded, or unassembled oligomeric proteins are retained in the ER complexes with molecular chaperones such as BiP or calnexin and are, following widely varying lag times, degraded (reviewed in Hammond and Helenius, 1995). The independence of this degradation process of lysosomal function and the apparent absence of proteasomes from the ER, together with studies localizing substrates destined for degradation to the ER region, have been interpreted to suggest the existence of a specialized proteolytic "quality control" machinery within the ER (Fra and Sitia, 1993; Klausner and Sitia, 1990; Lippincott-Schwartz *et al.*, 1988). However, attempts to identify the proteases involved in "ER degradation," or to biochemically isolate degradation intermediates were frustrated by the lack of effective and specific inhibitors of this process. The inhibition of ER degradation by peptide aldehydes such as ALLN were initially interpreted to suggest a role for calpains in ER degradation (Inoue *et al.*, 1991; Neumann *et al.*, 1996). However, the recognition that these compounds are also effective inhibitors of the 26 S proteasome (Rock *et al.*, 1994) and the discovery in 1995 that the fungal metabolite lactacystin is a potent and apparently specific inhibitor of the 20 S proteasome (Fenteany *et al.*, 1995) pointed to the proteasome as an agent of ER proteolysis and provided a tool to investigate the role of the ubiquitin–proteasome pathway in ER degradation. Several key studies that have led to this conclusion are described in the following section.

2.1. Role of the Ubiquitin–Proteasome Pathway in the Degradation of Integral Membrane Proteins

2.1.1. CFTR

CFTR is a polytopic integral membrane glycoprotein of $M_r \sim 160{,}000$ that forms a kinase-activated, ATP-gated Cl^- channel in the plasma membrane of various epithelial and nonepithelial cells (reviewed in Riordan and Chang, 1992; Riordan *et al.*, 1989). Cystic fibrosis (CF), a common recessive genetic disease, results directly from mutations in the CFTR gene, which cause the functional absence of CFTR-mediated Cl^- channel activity from the apical plasma membranes of epithelial cells (Tsui, 1992). One CFTR allele, ΔF508, is of considerable clinical importance because it is associated with a severe form of the disease (Kristidis *et al.*, 1992) and because of its prevalence in northern European and American populations, where it accounts for nearly 70% of all CF cases (Kerem *et al.*, 1989). Individuals homozygous for ΔF508 lack plasma membrane CFTR by both functional and immunological criteria (Kartner *et al.*, 1992). Pulse–chase studies reveal that, like wild-type CFTR, ΔF508 molecules are initially synthesized as 140-kDa core-glycosylated polypeptides (Cheng *et al.*, 1990; Ward and Kopito, 1994). Unlike wild-type CFTR, however, which can be chased to a 160-kDa complex glycosylated form, ΔF508 molecules are extremely unstable ($t_{1/2}$ = 15–30 min) and are rapidly degraded without detectable lag following synthesis (Lukacs *et al.*, 1993; Ward and Kopito, 1994). Although ΔF508 does not associate with the luminal HSP70 chaperone, BiP, it does appear to be complexed with cytosolic Hsc70 (Yang *et al.*, 1993) and with calnexin (Pind *et al.*, 1994) until it is degraded. These chaperone associations are consistent with the predicted topology of CFTR as a membrane-spanning glycoprotein with little mass exposed to the ER lumen (Riordan and Chang, 1992; Riordan *et al.*, 1989). Moreover, ΔF508 degradation is insensitive to inhibitors of lysosomal proteolysis and brefeldin A (Lukacs *et al.*, 1993; Ward and Kopito, 1994). In all of these respects, ΔF508 CFTR degradation resembles classically described "ER degradation."

Recent studies strongly suggest a role for the ubiquitin pathway in the degradation of ΔF508 CFTR. Treatment of ΔF508-expressing cells with inhibitors of the 20 S proteasome, including peptide aldehyde compounds MG132 and ALLN, and lactacystin, stabilize the protein from intracellular proteolysis (Jensen *et al.*, 1995; Ward *et al.*, 1995). Stabilized ΔF508 molecules that accumulate in response to proteasome inhibitors migrate on SDS–PAGE as a typical "ladder" of high-molecular-weight bands and are immunoreactive with a c-myc antibody in cells coexpressing c-myc-tagged ubiquitin (Ward *et al.*, 1995), suggesting a polyubiquitinated intermediate in ΔF508 CFTR degradation. Finally, ΔF508 CFTR degradation is inhibited in cells that are defective in protein ubiquitination. For example, ΔF508 CFTR stability is increased in El mutants (ts20 cells) and when coexpressed with myc-tagged ubiquitin in which Lys48, which is important for the formation of branched multiubiquitin chains (Chau *et al.*, 1989), is mutated to Arg (K48R) (Ward *et al.*, 1995). Together these data suggest that the ubiquitin–proteasome pathway participates in the degradation of this large integral membrane protein.

2.1.2. Sec61

The Sec61 complex, comprised of three ER-resident integral membrane polypeptide chains, Sec61p, Sbh1p, and Sss1p (corresponding to the mammalian proteins designated Sec61α,β,γ), is necessary for translocation of membrane and secretory proteins into the mammalian ER (Görlich and Rapoport, 1993; Hartmann *et al.*, 1994). Sec61p (Sec61α) is a polytopic integral membrane protein that probably comprises a major structural component of the protein conducting channel or pore through which nascent polypeptide chains are transported across the ER membrane. Consistent with its proposed function, *SEC61* is an essential gene in yeast; temperature-sensitive mutant alleles lead to the cytoplasmic accumulation of precursor polypeptides at restrictive temperatures (Deshaies and Schekman, 1987). *SSS1*, also essential, was identified as a multicopy suppressor of a conditional allele, *Sec61-2* (Esnault *et al.*, 1993). In *sec61* cells, levels of Sec61p and Sss1p decrease in parallel on shifting to the restrictive temperature, suggesting that Sec61p and Sss1p interact to form a complex that is destabilized and presumably degraded in *sec61* cells (Esnault *et al.*, 1994).

The first suggestion of involvement of the ubiquitin pathway in Sec61 degradation came from the observation that mutations in *UBC6*, a ubiquitin-conjugating enzyme (E2) bound to the cytosolic surface of the ER membrane, suppresses a *sec61* protein translocation defect (Sommer and Jentsch, 1993). Indeed, loss-of-function mutations in *UBC6* and another E2, *UBC7*, are effective suppressors of the sec61 growth deficiency, with the *ubc7* phenotype being more robust (Biederer *et al.*, 1996). Moreover, deletion of either *UBC6* or *UBC7* effectively suppresses both the short half-life (40–50 min) and the reduced steady-state levels of Sec61p in *sec61* mutants, suggesting a role for the ubiquitin-conjugating system in Sec61p degradation (Biederer *et al.*, 1996). Finally, the instability of mutant Sec61p can also be suppressed by coexpression of mutant ubiquitin (K48R) and by a mutation (*Pre1*) that disrupts proteasome function (Biederer *et al.*, 1996). These data strongly implicate the ubiquitin–proteasome system in the degradation of this ER-resident integral membrane protein. The instability of mutant Sec61p is unaffected by mutations in *SEC18*, a gene that is essential for formation of the vesicle fusion complex in ER-to-Golgi transport (Biederer *et al.*, 1996), suggesting that

Sec61p degradation, like classical "ER degradation," is independent of post-ER compartments.

2.1.3. MHC Class 1 Heavy Chains

One way by which human cytomegalovirus (CMV) evades the host immune system is by causing the selective degradation of newly synthesized class I heavy chains (HC). At least two CMV-encoded gene products, US11 (Wiertz *et al.*, 1996a) and US2 (Wiertz *et al.*, 1996b), are each sufficient to confer this effect on uninfected cells. In US2- and US11-expressing cells, HC is synthesized and core-glycosylated, but is degraded extraordinarily rapidly ($t_{1/2} < 1$ min) in a process that is blocked by proteasome inhibitors including the proteasome inhibitor lactacystin (Fenteany *et al.*, 1995; Wiertz *et al.*, 1996a,b). Inhibition of proteasome activity with lactacystin leads to the accumulation of HC degradation intermediates the nature of which suggests a novel mechanism for degradation of an integral membrane protein. In pulse-labeled, lactacystin-treated US11-expressing cells, HC chains were first detectable as microsome-associated, core high-mannose N-glycosylated chains, indicative of translocation of the glycan acceptor sites into the ER lumen. A subsequent chase revealed the accumulation of HC chains lacking N-glycans in a nonvesicular fraction, suggesting a surprising model in which HC chains are first translocated into the ER where they become core-glycosylated, and subsequently deglycosylated and released from the membrane. Deglycosylation could be mediated by protein:N-glycanases that have been identified and purified from the cytosol of animal, plant, and fungal cells (Suzuki *et al.*, 1994). Cleavage of asparagine-linked oligosaccharides by this enzyme results in conversion of the Asn acceptor to Asp and is consistent with the observed acidic shift in HC isoelectric point. These data thus suggest a model in which HC is first translocated across the ER membrane, core-glycosylated, and in the presence of US11, is "reverse translocated" or dislocated into the cytoplasm, deglycosylated, and, in the absence of lactacystin, degraded by the proteasome (Wiertz *et al.*, 1996a,b). So far, however, there is no direct evidence suggesting participation of the ubiquitin system in US11-mediated HC degradation.

2.2. Role of the Ubiquitin–Proteasome Pathway in the Degradation of Luminal (Secretory) Proteins

2.2.1. Carboxypeptidase Y

Point mutations in a conserved active site region of the *S. cerevisiae* vacuolar protease, carboxypeptidase Y (CPY), prevent the trafficking of nascent CPY chains to the vacuole and lead to rapid ($t_{1/2}$ = 15–20 min) degradation of the mutant protein (Finger *et al.*, 1993). This process is independent of the *SEC18* gene product, and occurs without apparent α1-6 or α1-3 mannosylation of outer chain glycosyls on the core N-linked oligosaccharide (Finger *et al.*, 1993), suggesting that degradation does not involve vesicular transport from the ER or transit through the Golgi apparatus. *D*egradation in *ER* (DER) mutants were identified in a screen for mutations that increase the steady-state levels of a CPY mutant, R255G, that is subject to rapid degradation (Knop *et al.*, 1996). The identity of one of these mutants, *DER2*, to the ubiquitin-conjugating enzyme *UBC7*, suggested the participation of the cytosolic ubiquitin system in the degradation of R255G (Hiller *et al.*, 1996). Indeed, R255G polypeptides are stabilized in Δ*Ubc7* cells, and, to a lesser extent, in cells bearing a deletion of the ER-associated conjugating enzyme, *UBC6* (Hiller *et al.*, 1996). As in the case of CFTR and Sec61p degradation, R255G mutants were stabilized by coexpression of K48R ubiquitin, and by mutations in the 20 S catalytic core (*PRE1*) or the 19 S cap (*CIM3*) of the proteasome (Hiller *et al.*, 1996). Ubiquitination of R255G was directly demonstrated by coexpression of epitope-tagged ubiquitin. Although ubiquitinated R255G was associated with a membrane fraction, these conjugates were susceptible to digestion with extravesicular trypsin, under conditions in which a luminal marker, Kar2p, was not. These data suggest a model in which mutant CPY molecules are first core-glycosylated in the ER lumen, then shuttled into the cytoplasmic compartment where they become ubiquitinated and subject to degradation by the 26 S proteasome. The mechanism for translocation of CPY across the ER membrane is unknown, but recent reports suggest that the yeast ER is able to export short peptides (Romisch and Schekman, 1992) and glycosylated pro-α-factor (McCracken and Brodsky, 1996) in an ATP- and cytosol-dependent manner.

2.2.2. α_1-Antitrypsin

Failure to secrete mutant forms of α_1-antitrypsin is a primary cause for homozygous PiZZ α_1-antitrypsin deficiency, a disorder associated with chronic liver and lung disease (Carrell, 1986). This disorder is caused by a point mutation in the α_1-antitrypsin gene, which results in the retention of nascent α_1-antitrypsin chains in the ER compartments and in their rapid degradation by a process that resembles classic "ER degradation" (Qu *et al.*, 1996; Sifers *et al.*, 1989). This degradation is insensitive to inhibitors of lysosomal proteolysis, but is effectively blocked by proteasome inhibitors including lactacystin (Qu *et al.*, 1996). Degradation of α_1-antitrypsin by the proteasome was also suggested by the ATP dependence of α_1-antitrypsin degradation reconstituted *in vitro* (Qu *et al.*, 1996). Interestingly, degradation of mutant α_1-antitrypsin *in vitro* appears to require N-glycosylation and association of the mutant glycoprotein with calnexin. Although there is no evidence for ubiquitination of mutant α_1-antitrypsin polypeptides, a recent report (Qu *et al.*, 1996) suggests that in the presence of mutant α_1-antitrypsin, calnexin molecules are ubiquitinated and presumably degraded together with associated α_1-antitrypsin molecules.

2.3. Is the Cytosolic Ubiquitin–Proteasome Pathway the Major Route for "ER" Degradation of Proteins?

ER degradation has been classically defined on the basis of three criteria: (1) insensitivity to lysosomotropic drugs such as NH_4Cl or chloroquine, (2) immunolocalization of degraded " substrates" to the ER, and (3) lack of maturation of N-linked oligosaccharides, characteristic of ER-retained proteins (Fra and Sitia, 1993; Klausner and Sitia, 1990; Lippincott-Schwartz *et al.*, 1988). In addition, "classical" ER degradation is unaffected by disruption of ER–Golgi trafficking either by the use of brefeldin A or in the case of yeast *Sec* mutants. Furthermore, although substrates for ER degradation display a wide range of half-lives ranging from 15 min to several hours, the classical substrates all tend to be short-lived proteins that are degraded with $t_{1/2} < 45$ min after a variable lag period following release from the ribosome (Fra and Sitia, 1993; Klausner and Sitia, 1990; Lippincott-Schwartz *et al.*, 1988). The specificity of the ER degradation apparatus for misfolded proteins (i.e., so-called off-pathway proteins, as opposed to proteins in the process of folding) suggests the existence of a sorting process by which substrates for ER degradation are segregated from nonsubstrates.

The rough ER is an organelle that is specialized for folding and assembly of proteins, a fact that is reflected in the high concentrations of molecular chaperones such as BiP and calnexin, and of folding enzymes such as protein disulfide isomerase and peptidyl-prolyl isomerase. Given the presence of high concentrations of nascent, unfolded, or partially folded polypeptide chains, there must be stringent mechanisms within the ER to segregate the proteolytic apparatus from susceptible polypeptide chains that are in the folding pathway. Because the process of translocation of nascent polypeptide chains across the ER membrane is slow (limited by the rate of protein synthesis) compared with proteolysis, and because proteins are translocated in an unfolded state, it is unlikely that kinetic partitioning between folding and degradation pathways can account for the selectivity of the system. The selectivity of the system must therefore involve either physical segregation of degradation substrates into a distinct membrane-bound compartment (by analogy to degradation of cytosolic proteins in lysosomes), or biochemical segregation by covalent attachment of a degradation "tag" (by analogy to cytoplasmic ubiquitination). Investigation of these alternative pathways has been, until recently, severely hampered by the lack of specific inhibitors and the inability to detect intermediates of the ER degradation pathway.

As attested by the rapid proliferation of recent reports of proteasomal involvement in ER degradation (Ward *et al.*, 1995; Biederer *et al.*, 1996; Hiller *et al.*,

1996; Qu *et al.*, 1996; Werner *et al.*, 1996; Wiertz *et al.*, 1996a,b), the availability of lactacystin as a specific inhibitor of the proteasome has provided a much-needed tool to begin to biochemically dissect this process. Lactacystin labels the highly conserved NH_2-terminal threonine residue on the mammalian proteasome subunit X that, from the recently solved X-ray structure of the archaebacterial proteasome (Löwe *et al.*, 1995), is likely to be the active-site nucleophile (Fenteany *et al.*, 1995). Lactacystin is also a potent inhibitor of the trypsinlike, chymotrypsinlike peptidyl-glutamyl-peptide hydrolyzing activities of purified 20 S proteasomes *in vitro* and has no detectable effect, even on extended exposure, on calpains or other cysteine proteases or on the serine proteases trypsin and chymotrypsin (Fenteany *et al.*, 1995). On the other hand, the specific labeling of proteasome subunits with lactacystin does not rule out the possibility of other cellular targets that may be present at much lower concentrations than proteasomes. Moreover, these data cannot anticipate other uncharacterized or undiscovered proteases against which lactacystin may be active. It is therefore important to emphasize that lactacystin represents one of several tools that can be used to examine the hypothesis that ER degradation is mediated by the proteasome. The participation of the ubiquitin–proteasome pathway in a diversity of cellular phenomena raises the possibility that interference with this pathway could produce indirect, pleiotropic effects. At least in the specific cases described above, the combined use of genetic mutations and chemical inhibitors of the ubiquitin–proteasome pathway, and the biochemical demonstration of direct ubiquitin conjugation to degradation intermediates strongly support the surprising conclusion that at least some ER degradation occurs in the cytosol, or at the cytosolic face of the ER membrane.

2.4. Cytoplasmic Degradation of Membrane and Secretory Proteins

The degradation of misfolded integral membrane and, in particular, secretory proteins by the cytoplasmic ubiquitin–proteasome pathway implies the existence of a mechanism to recognize substrate proteins and to translocate them across the ER membrane. Such a process would necessarily represent a reversal of the process of biosynthetic protein translocation that is usually coupled to translation. Biosynthetic translocation of nascent polypeptide chains requires, minimally, that three conditions be met: first, that the substrate proteins be in an unfolded state; second, that translocation occur through a proteinaceous pore; and third, that there be a source of energy.

For biosynthetic translocation the protein's unfolded state is maintained either by interaction with molecular chaperones such as members of the HSC70 family (Deshaies *et al.*, 1988), or by directly coupling polypeptide translocation to translational elongation (Schatz and Dobberstein, 1996). Biosynthetic translocation also appears to require the activity of an HSC70-class chaperone (i.e., Kar2p or BiP) at the *trans* (i.e., luminal) face of the ER membrane (Brodsky *et al.*, 1995;

Panzner *et al.*, 1995; Sanders *et al.*, 1992). Folding of some polypeptides in the ER (Hammond and Helenius, 1995), as in the cytoplasm (Frydman *et al.*, 1994), appears to involve sequential binding of nascent chains to and release from molecular chaperones. In the ER these include the soluble chaperones BiP, calreticulin, and GRP94 and the transmembrane chaperone calnexin (Bergeron *et al.*, 1993; Degen and Williams, 1991). Mutant polypeptides fail to fold productively and, hence, fail to be released from complexes with chaperones such as BiP (Haas and Wabl, 1984). Undegraded, ER-retained misfolded integral membrane proteins including CFTR, and secretory proteins including α_1-antitrypsin and secretory Ig, are found as complexes with one or more of these chaperones. Interestingly, degradation of two secretory proteins, human α_1-antitrypsin (Qu *et al.*, 1996) and yeast pro-α-factor (McCracken and Brodsky, 1996), has been reported to require calnexin, although the role of N-glycosylation in this process is controversial. Calnexin is a lectin that binds to glycoproteins containing monoglucosylated high-mannose oligosaccharides (Hebert *et al.*, 1995). Rapid degradation of mutant human α_1-antitrypsin reconstituted *in vitro* requires the presence of monoglucosylated oligosaccharide structures, suggesting a requirement for calnexin (Qu *et al.*, 1996). Qu *et al.* (1996) reported that expression of mutant human α_1-antitrypsin induces an increase in calnexin ubiquitination, and proposed that ubiquitination of the cytoplasmic domain of this integral membrane chaperone may be linked to the degradation of mutant α_1-antitrypsin. In contrast, McCracken and Brodsky (1996) reported that calnexin is essential for the degradation of unglycosylated yeast pro-α-factor. The reason for this discrepancy is not apparent, although it may reflect underlying differences in the function of calnexin in yeast and mammalian cells. The CMV gene products US2 and US11, which induce the retrograde translocation of MHC class I heavy chains for degradation by cytoplasmic proteasomes (Wiertz *et al.*, 1996a,b) may serve such a chaperone function, although it is not clear whether these gene products act alone to unfold substrate or act cooperatively with endogenous chaperones. The prolonged interaction with chaperones can thus be regarded as a mechanism to maintain the competence of the misfolded chains for reverse translocation. The large differences observed in the half-lives of misfolded proteins in the ER, which range from several minutes (Wiertz *et al.*, 1996a) to hours (Fra and Sitia, 1993), may reflect different rates at which the substrate–chaperone complexes are recognized and delivered to the translocation apparatus. However, the factors that influence the decision for a chaperone-substrate complex either to continue trying to fold or to be delivered to the proteolytic machinery are not known for any substrate.

Protein translocation across the ER membrane requires the presence of a suitable transmembrane pore or channel-forming component (i.e., translocon), which, in the case of biosynthetic translocation, is composed minimally of the Sec61 complex. It is possible that the same apparatus could form the channel through which degraded proteins are transported, rendering "retrograde" trans-

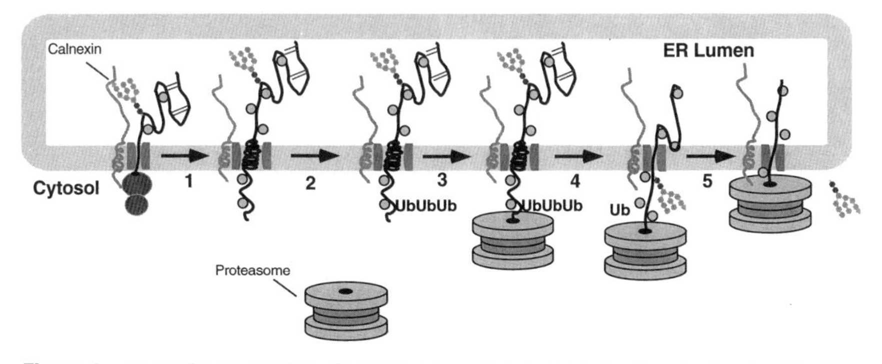

Figure 1. Model for degradation of an integral membrane protein by the cytoplasmic ubiquitin–proteasome pathway. (Step 1) Misfolded proteins fail to be released from interactions with molecular chaperones like Hsp70 (shaded circles) and calnexin. (Step 2) Some misfolded nascent chains may never be released from the translocation pore composed in part of the Sec61p complex. (Step 3) Cytoplasmic domains of proteins like CFTR and Sec61p can be multiubiquitinated. (Step 4) Proteasomes are recruited to membrane translocation (dislocation) sites by the presence of multiubiquitin chains and possibly other, yet unidentified signals. Misfolded proteins are dislocated from the membrane through a pore that may or may not be identical to the translocation pore. Dislocation may be tightly coupled to proteolysis by the proteasome. (Step 5) N-linked oligosaccharides are removed by N-glycanase proposed to be present at the cytosolic side of the ER membrane. The proteasome may form tight interaction with the dislocation site.

location (or "dislocation") a simple reversal of biosynthetic translocation (Fig. 1). In this respect, the translocon would resemble other ion channels and permeases that permit the bidirectional movement of substrates. Such a mechanism would be particularly attractive, as the translocon possesses the ability to translocate both integral membrane and secreted proteins. In cells expressing the CMV US2 gene product, which induces the retrograde translocation of MHC class I heavy chain, deglycosylated degradation intermediates were coimmunoprecipitated together with US2 protein and with Sec61β chains, consistent with the participation of the Sec61 complex in retrograde translocation (Wiertz *et al.*, 1996b). Future experiments will be needed to ascertain whether retrograde translocation occurs through the Sec61 complex, or via other, yet unidentified pathways.

Finally, removing an integral membrane protein from a lipid bilayer into an aqueous environment requires a considerable input of free energy. Protein insertion into membranes is likely to be a multistep process involving initial interaction of the stop-transfer sequence with a component of the translocation apparatus and subsequent transfer from an aqueous to a lipid environment. It is likely that formation of the correct interactions between hydrophobic membrane-spanning domains in a crowded bilayer is subject to similar types of competing off-pathway interactions that affect protein folding in aqueous environments. Thus, it is unlikely that the protein translocation apparatus transfers individual membrane-spanning domains directly into the lipid phase. Instead, these hydrophobic sequences may be stabilized against inappropriate interactions by transfer from the translocation pore to integral membrane chaperones, which may or may not be part of the translocon. Candidates for proteins that might perform such an integral membrane chaperone function could include calnexin or the TRAM complex (Görlich *et al.*, 1992).

ATPases that might participate in the process of retrograde translocation include HSP70-class chaperones, the ubiquitin conjugation/ligation apparatus, and the proteasome itself. Ubiquitination of retrograde translocated substrates could occur cotranslocationally, ensuring the irreversibility of the process, increasing the solubility of hydrophobic membrane-spanning domains, and coupling translocation directly to the proteasome. Indeed, the 19 S proteasome "cap" is thought to possess unfolding ATPase activity (Hilt and Wolf, 1995; Lupas *et al.*, 1993). It is also possible that, by analogy to biosynthetic translocation, which usually occurs cotranslationally, retrograde translocation may be directly coupled to proteasome-mediated degradation. Thus, proteins could be delivered directly to the proteasome, eliminating the problems associated with transfer of membrane polypeptides into an aqueous environment. Proteasomes have been identified both biochemically and morphologically (Palmer *et al.*, 1996; Rivett *et al.*, 1992) attached to the cytoplasmic face of the ER membrane; possibly, identification of the membrane binding structures will help to identify integral membrane components that participate in retrograde translocation.

3. THE ROLE OF MEMBRANE-PROTEIN UBIQUITINATION IN CELL-SURFACE SIGNALING

3.1. Ligand-Induced Ubiquitination of Receptor Tyrosine Kinases

The potential involvement of ubiquitination in cellular signaling was originally suggested by the presence of covalently linked ubiquitin moieties identified during the course of sequencing purified cell-surface receptors (Leung *et al.*, 1987; Siegelman *et al.*, 1986; van de Rijn *et al.*, 1990; Yarden *et al.*, 1986). Subsequent studies have established that receptor ubiquitination is a response to ligand binding/receptor activation common to many monomeric receptor tyrosine kinases (RTKs), including receptors for platelet-derived growth factor (PDGF) (Mori *et al.*, 1992), epidermal growth factor (EGF) (Mori *et al.*, 1995a), colony stimulating factor-1 (Mori *et al.*, 1995a), fibroblast growth factor (Mori *et al.*, 1995), c-kit (Miyazawa *et al.*, 1994), insulinlike growth factor 1 (IGF-1) (Sepp-Lorenzino *et al.*, 1995), and p185^{erbB2} (Mimnaugh *et al.*, 1996). Ubiquitination also appears to participate in the antigen-induced activation of the multicomponent T-cell antigen receptor complex (TCR) (Cenciarelli *et al.*, 1992, 1996; Hou *et al.*, 1994). A subset of ζ subunits of this receptor undergoes multiple tyrosine phosphorylations on receptor stimulation via a process that involves recruitment of activated members of the *src*-family of tyrosine kinases, Lck ($p56^{lck}$) and Fyn ($p59^{fyn}$) (Weissman, 1994). Activation of TCR also stimulates ubiquitination of the ζ and CD3 δ subunits (Cenciarelli *et al.*, 1992). TCR ubiquitination is increased by inhibition of tyrosine phosphatases, and is decreased in mutants lacking the tyrosine kinase $p56^{lck}$, suggesting a linkage between receptor activation and its ubiquitination (Cenciarelli *et al.*, 1996). However, there is no evidence that ubiquitination of TCR is associated with altered turnover of receptor protein.

RTKs constitute a major class of cell-surface receptors that play key roles in the regulation of cellular growth and differentiation (see Fantl *et al.*, 1993, for review). These integral membrane proteins contain extracellular ligand binding domains, a single transmembrane span, and a cytoplasmic domain that possesses intrinsic tyrosine kinase activity. Ligand binding induces dimerization of receptor monomers and consequent activation of intrinsic kinase activity. Autophosphorylation of these receptors is required for recruitment of downstream effectors and initiates a cascade of phosphorylations resulting in activation of specific cellular responses. Receptor activation also ultimately leads to downregulation via internalization through endosomes (Lund *et al.*, 1990) and delivery of the receptor–ligand complex to the lysosome for degradation (Haft *et al.*, 1994; Wiley *et al.*, 1991). However, data from several laboratories suggest that ligand binding to RTKs may also result in downregulation through a lysosome-independent pathway involving the ubiquitin–proteasome pathway.

Mori *et al.* (1992) reported that PDGF receptor ubiquitination is induced on ligand binding. Ligand-induced ubiquitination was abolished in a PDGF receptor mutant lacking a functional kinase, suggesting a link between receptor activation and ubiquitination. Moreover, mutations that decrease or abolish receptor ubiquitination inhibit the rate of ligand-induced degradation and increase mitogenic potency (Mori *et al.*, 1993). In other words, mutations that decrease receptor ubiquitination appear to decrease ligand-induced downregulation, suggesting a causal relationship between receptor ubiquitination and steady-state levels. Interestingly, the increased stability of mutant PDGF receptor is not associated with altered kinetics of receptor–ligand internalization via endocytosis, and is insensitive to lysosomotropic amines, suggesting that it does not require lysosomal function (Mori *et al.*, 1993). These data suggested the existence of a novel pathway for ligand-induced receptor downregulation, requiring ligand-induced receptor ubiquitination and degradation through nonlysosomal pathways. A role for the proteasome in RTK degradation is suggested by recent studies using lactacystin and peptide aldehyde proteasome inhibitors (Mori *et al.*, 1995b; Sepp-Lorenzino *et al.*, 1995). Those studies are limited, however, by assumptions about the specificity of the inhibitors, and moreover, shed little light on the molecular mechanisms by which receptor activation leads to ubiquitin ligation and subsequent degradation. Investigation of the mechanism of receptor downregulation by ansamycin antibiotics suggests that a crucial link in this process may be the participation of molecular chaperones of the HSP90 family.

The benzoquinonoid ansamycin antibiotics, herbimycin and geldanamycin, were originally described as inhibitors of cellular and receptor tyrosine kinases (Fukazawa *et al.*, 1990; Uehara and Fukazawa, 1991). Although these compounds can directly inhibit tyrosine kinase activity of purified RTKs at high concentrations, at low concentrations they appear to specifically promote degradation of tyrosine kinases. For example, herbimycin treatment decreased the half-life of EGF receptor and IGF-1 receptor, from > 24 hr to 6 hr, without affecting the half-lives of several other non-tyrosine kinase polypeptides (Sepp-Lorenzino *et al.*, 1995). Herbimycin-induced degradation of IGF and insulin receptors is unaffected by lysosomotropic amines or inhibitors of lysosomal cathepsins, but is blocked by peptide aldehyde proteasome inhibitors and requires functional ubiquitin-activating enzyme (El), suggesting that ansamycin-induced RTK degradation occurs via the ubiquitin–proteasome pathway (Sepp-Lorenzino *et al.*, 1995). How these drugs influence the stability of tyrosine kinases is suggested by the identification of the chaperone HSP90 (Whitesell *et al.*, 1994) as a cellular target for the drugs. Treatment of cells with geldanamycin was reported to disrupt a complex between p185^{erbB2} and GRP94, an abundant protein of the ER lumen that is homologous to HSP90 (Chavany *et al.*, 1996). Geldanamycin treatment led to p185^{erbB2} degradation following a 30-min lag, during which p185^{erbB2} was converted to a high-molecular-weight smear that was recognized by ubiquitin antibodies (Mimnaugh *et al.*, 1996). Application of lactacystin together with geldanamycin led to accumulation of high-molecular-weight polyubiquitinated forms of p185^{erbB2}, further supporting the hypothesis that ansamycins induce RTK degradation via the ubiquitin–proteasome pathway (Mimnaugh *et al.*, 1996). However, the p185^{erbB2} studies are confusing because it is not clear how p185^{erbB2} can simultaneously be present at the cell surface and also be associated with the ER chaperone, GRP94. Perhaps distinct ER and plasma membrane pools of p185^{erbB2} are subject, respectively, to degradation by distinct proteasomal and lysosomal pathways. Indeed, geldanamycin has been reported to inhibit the maturation of nascent p185^{erbB2} molecules beyond the ER, suggesting that the drug influences the folding of nascent p185^{erbB2} molecules in the secretory pathway (Mimnaugh *et al.*, 1996). Thus, although a role for the ubiquitin–proteasome pathway in RTK downregulation is suggested, its role in signaling and its relationship to the lysosomal pathway remain to be elucidated. By contrast, studies on the turnover of non-RTK cell surface receptors in yeast, discussed below, provide compelling evidence that membrane protein ubiquitination may serve as a signal in the endocytic pathway.

3.2. Ubiquitin and the Endocytic Pathway

The *STE6* gene encodes an integral membrane protein, Ste6p, an ATP-binding cassette (ABC) transporter that mediates the export of a-factor (Kuchler *et al.*, 1989; McGrath and Varshavsky, 1989). Ste6p is normally processed via the secretory pathway to the plasma membrane where it is rapidly and constitutively internalized and delivered to the vacuole for degradation (Kolling and Hollenberg, 1994). In *end4* mutants that are defective in endocytosis, ubiquitinated forms of Ste6p accumulate at the cell surface, suggesting a linkage between endocytosis and ubiquitination (Kolling and Hollenberg, 1994). Indeed, in yeast harboring mutant ubiquitin-conjugating enzymes *ubc4* and *ubc5* or double mutants thereof, Ste6 half-life is greatly extended, suggesting that ubiquitination may target Ste6p for degradation. Ste6p degradation is sensitive to mutations in *PEP4*, which encodes the vacuolar protease A, and in *pep4* cells, Ste6p accumulates in vacuolar structures, suggesting that ubiquitination of Ste6p is required for the *PEP4*-dependent turnover of Ste6p (Kolling and Hollenberg, 1994). Similar results have been reported for another ABC transporter, the multidrug transporter Pdr5p, which also exhibits a short residence at the cell surface, constitutive internalization, and rapid turnover (Egner and Kuchler, 1996). Although these studies strongly implicate a role for the ubiquitin pathway in the turnover of these yeast plasma membrane transporters by the endocytic pathway, studies on the mating pheromone receptors for a-factor (Roth and Davis, 1996) and α-factor (Hicke and Riezman, 1996) suggest that receptor ubiquitination can be a response to a signal transduction cascade initiated on ligand binding.

Binding of the mating pheromone α-factor to its G-protein-coupled receptor, Ste2p, initiates a signal transduction cascade that results in its downregulation via endocytosis (Schultz *et al.*, 1995). Ste2p is slowly internalized constitutively, but the rate of internalization is stimulated 5- to 10-fold by the presence of ligand (Jenness and Spatrick, 1986). Ubiquitination of Ste2p is induced by α-factor binding and, like Ste6p, ubiquitinated forms of Ste2p accumulate at the plasma membrane in *end4*Δ cells (Hicke and Riezman, 1996). Moreover, Ste2p internalization and transport to the vacuole are severely retarded in cells harboring mutant ubiquitin-conjugating enzymes *ubc1*, *ubc4*, and *ubc5*, suggesting that the major pathway for receptor internalization is ubiquitin dependent (Hicke and Riezman, 1996). Similar results have been reported for the receptor for a-factor, Ste3p (Roth and Davis, 1996). Unlike ubiquitin-linked "ER" degradation of misfolded proteins, which targets proteins to the proteasome, ubiquitination targets yeast permeases and pheromone receptors for degradation in the vacuole. Turnover of these plasma membrane proteins is blocked by mutations in vacuolar proteases (e.g., *PEP4*), but not by mutations in either the 20 S catalytic core (*PRE1*, *PRE2*) or the regulatory ATPase subunits (*YTA1*, *YTA5*) of the proteasome (Hicke and Riezman, 1996; Roth and Davis, 1996).

3.3. Ubiquitin as a Versatile Protein Modification in Cellular Protein Traffic

The evidence reviewed in this chapter suggests that ubiquitination of proteins in the secretory pathway is a versatile process that is used both to eliminate proteins that fail to fold productively in the ER, and to control the expression of membrane receptors and transporters at the cell surface. The data also indicate that membrane protein ubiquitination may be coupled to either of the two major proteolytic systems in the cell—the proteasome and the lysosome/vacuole. Clearly there must be additional levels of regulation beyond the presence or absence of covalently bound ubiquitin, which control the intracellular fate of these proteins.

One potential control point is the structure of the substrate. The cytoplasmic tail of Ste2p contains a nine-amino-acid sequence that appears to function as a signal for both ubiquitination and internalization of the receptor (Hicke and Riezman, 1996). Mutation of the single lysine residue (K337) in this element (one of two lysines within the cytoplasmic tail of truncated Ste2p) completely eliminated both ubiquitination and ligand-stimulated internalization, suggesting that K337 may function as a ubiquitin acceptor (Hicke and Riezman, 1996). Moreover, mutation of the three Ser residues within this cytoplasmic nonamer element blocked ubiquitination and internalization, further linking ubiquitination and internalization and suggesting a role for additional, nonlysine structural elements within the receptor polypeptide. In addition to modification by ubiquitin, endocytosis of Ste2p and Ste3p is accompanied by Ser/Thr phosphorylation and ligand binding to Ste2p induces hyperphosphorylation (Zanolari *et al.*, 1992). Mutation of the Ser triplet within the Ste2p nonameric internalization motif also blocked the mobility shift associated with hyperphosphorylation, suggesting that phosphorylation and ubiquitination may be linked, as has been demonstrated for soluble ubiquitin–proteasome substrates (Hicke and Riezman, 1996).

There may be additional structural elements within the cytoplasmic domains of plasma membrane proteins that participate in the recruitment of the ubiquitination machinery. For example, mutation of an invariant arginine in a putative cyclinlike "destruction box" motif in Fur4p, a yeast plasma membrane uracil permease, significantly stabilized the protein against stress-induced degradation (Galan *et al.*, 1994). Both constitutive and stress-induced degradation of Fur4p require ubiquitination (Galan *et al.*, 1996) as well as a ubiquitin ligase (E3) encoded by the *RSP5/NPI1* gene (Galan *et al.*, 1996; Hein *et al.*, 1995).

A second control point that may be important for determining the fate of ubiquitinated membrane proteins is in the nature of the ubiquitin modification. For example, the length or degree of branching of attached ubiquitin chains could provide either a sorting signal for delivery to lysosomes or targeting to the proteasome. Significantly, both yeast mating pheromone receptors, which are degraded by a proteasome-independent, vacuolar pathway, are mono- or diubiquitinated (Hicke and Riezman, 1996; Roth and Davis, 1996). It will be important to determine whether RTK receptors like PDGF receptor and p185^{erbB2} and substrates for "quality control" degradation by the proteasome are modified by long-chain ubiquitin polymers. Although the substrate specificity of individual UBCs has yet to be determined, genetic studies in yeast suggest a correlation. For example, degradation in the vacuole of plasma membrane substrates including Ste6p, Pdr5p, Ste3p, and Ste2p requires the *UBC1*, *UBC4*, and *UBC5* genes, whereas degradation of misfolded proteins that fail to leave the ER requires a different set of UBCs—*UBC6* and *UBC7*.

Finally, accessory proteins, including E3s and molecular chaperones, are likely to contribute to the determination of the fate of the ubiquitinated substrates. Misfolded proteins exhibit prolonged interactions with molecular chaperones such as BiP, Hsc70, and calnexin; these chaperone–substrate complexes could serve as signals to recruit the appropriate ubiquitination machinery. The action of ansamycin antibiotics to promote dissociation of the molecular chaperone HSP90 from intracellular tyrosine kinases suggests a role for molecular chaperones in ubiquitin-dependent protein degradation.

4. SUMMARY AND PROSPECTS

A considerable body of evidence now implicates the ubiquitin system in the trafficking and turnover of integral membrane protein and proteins in the secretory pathway. ER quality control mechanisms ensure that misfolded integral membrane or luminal proteins remain associated with molecular chaperones, are not deployed to post-ER compartments, and are eventually degraded. The recent finding that this degradation is mediated by the ubiquitin–proteasome pathway suggests the existence of a novel mechanism by which proteins are dislocated from the ER to the cytoplasm prior to their degradation. Future studies will be needed to determine how generally this pathway operates for different substrates, and the identities of the components involved in substrate recognition and dislocation machinery.

Ubiquitination does not always target integral membrane proteins for degradation by the proteasome. Recent studies implicate ubiquitin as a signal for the downregulation of some plasma membrane receptors by the endocytic/lysosomal pathway. It is clear now that proteins within the secretory pathway and in the plasma membrane can be substrates for the ubiquitin system. The challenge ahead is to determine how this versatile protein modification can signal such diverse fates as membrane dislocation or misfolded ER proteins and ligand-stimulated receptor downregulation.

Acknowledgment. The author thanks the members of his laboratory for their critical reading of the manuscript.

5. REFERENCES

Bergeron, J. J., Brenner, M. B., Thomas, D. Y., and Williams, D. B., 1993, Calnexin: A membrane-bound chaperone of the endoplasmic reticulum, *Trends Biochem. Sci.* **339:**257–261.

Biederer, T., Volkwein, C., and Sommer, T., 1996, Degradation of subunits of the Sec61p complex, an integral component of the ER membrane, by the ubiquitin-proteasome pathway, *EMBO J.* **15:** 2069–2076.

Brodsky, J. L., Goeckeler, J., and Schekman, R., 1995, BiP and Sec63p are required for both co- and posttranslational protein translocation into the yeast endoplasmic reticulum, *Proc. Natl. Acad. Sci. USA* **92:**9643–9646.

Carrell, R. W., 1986, Alpha 1-antitrypsin: Molecular pathology, leukocytes, and tissue damage, *J. Clin. Invest.* **78:**1427–1431.

Cenciarelli, C., Hou, D., Hsu, K. C., Rellahan, B. L., Wiest, D. L., Smith, H. T., Fried, V. A., and Weissman, A. M., 1992, Activation-induced ubiquitination of the T cell antigen receptor, *Science* **257:**795–797.

Cenciarelli, C., Wilhelm, K. G., Guo, A., and Weissman, A. M., 1996, T cell antigen receptor ubiquitination is a consequence of receptor-mediated tyrosine kinase activation, *J. Biol. Chem.* **271:**8709–8713.

Chau, V., Tobias, J. W., Bachmair, A., Marriott, D., Ecker, D. J., Gonda, D. K., and Varshavsky, A., 1989, A multiubiquitin chain is confined to specific lysine in a targeted short-lived protein, *Science* **243:**1576–1583.

Chavany, C., Mimnaugh, E., Miller, P., Bitton, R., Nguyen, P., Trepel, J. L. W., Schnur, R., Moyer, J., and Neckers, L., 1996, p185^{erbB2} binds to GRP94 in vivo, *J. Biol. Chem.* **271:**4974–4977.

Cheng, S. H., Gregory, R. J., Marshall, J., Paul, S., Souza, D. W., White, G. A., O'Riordan, C. R., and Smith, A. E., 1990, Defective intracellular transport and processing of CFTR is the molecular basis of most cystic fibrosis, *Cell* **63:**827–834.

Degen, E., and Williams, D. B., 1991, Participation of a novel 88-kD protein in the biogenesis of murine class I histocompatibility molecules, *J. Cell Biol.* **112:**1099–1115.

Deshaies, R. J., and Schekman, R., 1987, A yeast mutant defective at an early stage in import of secretory protein precursors into the endoplasmic reticulum, *J. Cell Biol.* **105:**633–645.

Deshaies, R. J., Koch, B. D., Werner Washburne, M., Craig, E. A., and Schekman, R., 1988, A subfamily of stress proteins facilitates traslocation of secretory and mitochondrial precursor polypeptides, *Nature* **332**:800–805.

Egner, R., and Kuchler, K., 1996, The yeast multidrug transporter Pdr5 of the plasma membrane is ubiquitinated prior to endocytosis and degradation in the vacuole, *FEBS Lett.* **378:**177–181.

Esnault, Y., Blondel, M. O., Deshaies, R. J., Schekman, R., and Kepes, F., 1993, The yeast SSS1 gene is essential for secretory protein translocation and encodes a conserved protein of the endoplasmic reticulum, *EMBO J.* **12:**4083–4093.

Esnault, Y., Feldheim, D., Blondel, M. O., Schekman, R., and Kepes, F., 1994, SSS1 encodes a stabilizing component of the Sec61 subcomplex of the yeast protein translocation apparatus, *J. Biol. Chem.* **269:**27478–27485.

Fantl, W. J., Johnson, D. E., and Williams, L. T., 1993, Signalling by receptor tyrosine kinases, *Annu. Rev. Biochem.* **62:**453–481.

Fenteany, G., Standaert, R. F., Lane, W. S., Chois, S., Corey, E. J., and Schreiber, S. L., 1995, Inhibition of proteasome activities and subunit-specific amino-terminal threonine modification by lactacystin, *Science* **268:**726–731.

Finger, A., Knop, M., and Wolf, D. H., 1993, Analysis of two mutated vacuolar proteins reveals a degradation pathway in the endoplasmic reticulum or a related compartment of yeast, *Eur. J. Biochem.* **218:**565–574.

Fra, A., and Sitia, R., 1993, The endoplasmic reticulum as a site of protein degradation, *Subcell. Biochem.* **21:**143–168.

Frydman, J., Nimmesgern, E., Ohtsuka, K., and Hartl, F. U., 1994, Folding of nascent polypeptide chains in a high molecular mass assembly with molecular chaperones [see comments], *Nature* **370:**111–117.

Fukazawa, H., Mizuno, S., and Uehara, Y., 1990, Effects of herbimycin A and various SH-reagents on p60v-src kinase activity in vitro, *Biochem. Biophys. Res. Commun.* **173:**276–282.

Galan, J. M., Volland, C., Urban-Grimal, D., and Haguenauer-Tsapis, R., 1994, The yeast plasma membrane uracil permease is stabilized against stress induced degradation by a point mutation in a cyclin-like "destruction box," *Biochem. Biophys. Res. Commun.* **201:**769–775.

Galan, J. M., Moreau, V., Andre, B., Volland, C., and Haguenauer-Tsapis, R., 1996, Ubiquitination mediated by the Npi1p/Rsp5p ubiquitin-protein ligase is required for endocytosis of the yeast uracil permease, *J. Biol. Chem.* **271:**10946–10952.

Görlich, D., and Rapoport, T. A., 1993, Protein translocation into proteoliposomes reconstituted from purified components of the endoplasmic reticulum membrane [see comments], *Cell* **75:**615–630.

Görlich, D., Hartmann, E., Prehn, S., and Rapoport, T. A., 1992, A protein of the endoplasmic reticulum involved early in polypeptide translocation [see comments], *Nature* **357:**47–52.

Haas, I. G., and Wabl, M., 1984, Immunoglobulin heavy chain binding protein, *Nature* **306:**387–389.

Haft, C. R., Klausner, R. D., and Taylor, S. I., 1994, Involvement of dileucine motifs in the internalization and degradation of the insulin receptor, *J. Biol. Chem.* **269:**26286–26294.

Hammond, C., and Helenius, A., 1995, Quality control in the secretory pathway, *Curr. Opin. Cell Biol.* **7:**523–529.

Hartmann, E., Sommer, T., Prehn, S., Görlich, D., Jentsch, S., and Rapoport, T. A., 1994, Evolutionary conservation of components of the protein translocation complex, *Nature* **367:**654–657.

Hebert, D. N., Foellmer, B., and Helenius, A., 1995, Glucose trimming and reglucosylation determine glycoprotein association with calnexin in the endoplasmic reticulum, *Cell* **81:**425–433.

Hein, C., Springael, J. Y., Volland, C., Haguenauer-Tsapis, R., and Andre, B., 1995, NPI1, an essential yeast gene involved in induced degradation of Gap1 and Fur4 permeases, encodes the Rsp5 ubiquitin-protein ligase, *Mol. Microbiol.* **18:**77–87.

Hicke, L., and Riezman, H., 1996, Ubiquitination of a yeast plasma membrane receptor signals its ligand-stimulated endocytosis, *Cell* **84:**277–287.

Hiller, M. M., Finger, A., Schweiger, M., and Wolf, D. H., 1996, ER degradation of a misfolded luminal protein by the cytosolic ubiquitin-proteasome pathway, *Science* **273:**1725–1728.

Hilt, W., and Wolf, D. H., 1995, Proteasomes of the yeast S. cerevisiae: Genes, structure and functions, *Mol. Biol. Rep.* **21:**3–10.

Hou, D., Cenciarelli, C., Jensen, J. P., Nguygen, H. B., and Weissman, A. M., 1994, Activation-dependent ubiquitination of a T cell antigen receptor subunit on multiple intracellular lysines, *J. Biol. Chem.* **269:**14244–14247.

Hurtley, S. M., and Helenius, A., 1989, Protein oligomerization in the endoplasmic reticulum, *Annu. Rev. Cell Biol.* **5:**277–307.

Inoue, S., Bar-Nun, S., Roitelman, J., and Simoni, R. D., 1991, Inhibition of degradation of 3-hydroxy-3-methylglutaryl-coenzyme A reductase in vivo by cysteine protease inhibitors, *J. Biol. Chem.* **266:**13311–13317.

Jenness, D. D., and Spatrick, P., 1986, Down regulation of the alpha-factor pheromone receptor in S. cerevisiae, *Cell* **46:**345–353.

Jensen, T. J., Loo, M. A., Pind, S., Williams, D. B., Goldberg, A. L., and Riordan, J. R., 1995, Multiple proteolytic systems, including the proteasome, contribute to CFTR processing, *Cell* **83:**129–135.

Kartner, N., Augustinas, O., Jensen, T. J., Naismith, A. L., and Riordan, J. R., 1992, Mislocalization of delta F508 CFTR in cystic fibrosis sweat gland, *Nature Genet.* **1:**321–327.

Kerem, B.-S., Rommens, J. M., Buchanan, J. A., Markiewicz, D., Cox, T. K., Chakravarti, A., Buchwald, M., and Tsui, L.-C., 1989, Identification of the cystic fibrosis gene: Genetic analysis, *Science* **245:**1073–1080.

Klausner, R. D., and Sitia, R., 1990, Protein degradation in the endoplasmic reticulum, *Cell* **62:** 611–614.

Knop, M., Finger, A., Braun, T., Hellmuth, K., and Wolf, D. H., 1996, Der1, a novel protein specifically required for endoplasmic reticulum degradation in yeast, *EMBO J.* **15:**753–763.

Kolling, R., and Hollenberg, C. P., 1994, The ABC-transporter Ste6 accumulates in the plasma membrane in a ubiquitinated form in endocytosis mutants, *EMBO J.* **13:**3261–3271.

Kristidis, P., Bozon, D., Corey, M., Markiewicz, D., Rommens, J., Tsui, L. C., and Durie, P., 1992, Genetic determination of exocrine pancreatic function in cystic fibrosis, *Am. J. Hum. Genet.* **50:** 1178–1184.

Kuchler, K., Sterne, R. E., and Thorner, J., 1989, Saccharomyces cerevisiae STE6 gene product: A novel pathway for protein export in eukaryotic cells, *EMBO J.* **8:**3973–3984.

Leung, D. W., Spencer, S. A., Cachianes, G., Hammonds, R. G., Collins, C., Henzel, W. J., Barnard, R., Waters, M. J., and Wood, W. I., 1987, Growth hormone receptor and serum binding protein: Purification, cloning and expression, *Nature* **330:**537–543.

Lippincott-Schwartz, J., Bonifacino, J. S., Yuan, L. C., and Klausner, R. D., 1988, Degradation from the endoplasmic reticulum: Disposing of newly synthesized proteins, *Cell* **54:**209–220.

Löwe, J., Stock, D., Jap, B., Zwickl, P., Baumeister, W., and Huber, R., 1995, Crystal structure of the 20S proteasome from the archaeon *T. acidophilum* at 3.4 Å resolution, *Science* **268:**533–539.

Lukacs, G. L., Chang, X. B., Bear, C., Kartner, N., Mohamed, A., Riordan, J. R., and Grinstein, S., 1993, The delta F508 mutation decreases the stability of cystic fibrosis transmembrane conductance regulator in the plasma membrane. Determination of functional half-lives on transfected cells, *J. Biol. Chem.* **268:**21592–21598.

Lund, K. A., Opresko, L. K., Starbuck, C., Walsh, B. J., and Wiley, H. S., 1990, Quantitative analysis of the endocytic system involved in hormone-induced receptor internalization, *J. Biol. Chem.* **265:** 15713–15723.

Lupas, A., Koster, A. J., and Baumeister, W., 1993, Structural features of 26S and 20S proteasomes, *Enzyme Protein* **47:**252–273.

McCracken, A. A., and Brodsky, J. L., 1996, Assembly of ER-associated protein degradation in vitro: Dependence on cytosol, calnexin, and ATP, *J. Cell Biol.* **132:**291–298.

McGrath, J. P., and Varshavsky, A., 1989, The yeast STE6 gene encodes a homologue of the mammalian multidrug resistance P-glycoprotein, *Nature* **340:**400–404.

Mimnaugh, E. G., Chavany, C., and Neckers, L., 1996, Polyubiquitination and proteasomal degradation of the p185$^{c\text{-}erbB\text{-}2}$ receptor protein-tyrosine kinase induced by geldanamycin, *J. Biol. Chem.* **271**:22796–22801.

Miyazawa, K., Toyama, K., Gotoh, A., Hendrie, P. C., Mantel, C., and Broxmeyer, H. E., 1994, Ligand-dependent polyubiquitination of c-kit gene product: A possible mechanism of receptor down modulation in M07e cells, *Blood* **83:**137–145.

Mori, S., Heldin, C. H., and Claesson-Welsh, L., 1992, Ligand-induced polyubiquitination of the platelet-derived growth factor beta-receptor, *J. Biol. Chem.* **267:**6429–6434.

Mori, S., Heldin, C. H., and Claesson-Welsh, L., 1993, Ligand-induced ubiquitination of the platelet-derived growth factor beta-receptor plays a negative regulatory role in its mitogenic signaling, *J. Biol. Chem.* **268:**577–583.

Mori, S., Claesson-Welsh, L., Okuyama, Y., and Saito, Y., 1995a, Ligand-induced polyubiquitination of receptor tyrosine kinases, *Biochem. Biophys. Res. Commun.* **213:**32–39.

Mori, S., Kanaki, H., Tanaka, K., Morisaki, N., and Saito, Y., 1995b, Ligand-activated platelet-derived growth factor beta-receptor is degraded through proteasome-dependent proteolytic pathway, *Biochem. Biophys. Res. Commun.* **217**:224–229.

Neumann, D., Yuk, M. H., Lodish, H. F., and Lederkremer, G. Z., 1996, Blocking intracellular degradation of the erythropoietin and asialoglycoprotein receptors by calpain inhibitors does not result in the same increase in the levels of their membrane and secreted forms, *Biochem. J.* **313**:391–399.

Palmer, A., Rivett, A. J., Thomson, S., Hendil, K. B., Butcher, G. W., Fuertes, G., and Knecht, E., 1996, Subpopulations of proteasomes in rat liver nuclei, microsomes and cytosol, *Biochem. J.* **316:** 401–407.

Panzner, S., Dreier, L., Hartmann, E., Kostka, S., and Rapoport, T. A., 1995, Posttranslational protein transport in yeast reconstituted with a purified complex of Sec proteins and Kar2p, *Cell* **81:** 561–570.

Pelham, H. R. B., 1989, Control of protein exit from the endoplasmic reticulum, *Annu. Rev. Cell Biol.* **5**:1–23.

Pind, S., Riordan, J. R., and Williams, D. B., 1994, Participation of the endoplasmic reticulum chaperone calnexin (p88, IP90) in the biogenesis of the cystic fibrosis transmembrane conductance regulator, *J. Biol. Chem.* **269**:12784–12788.

Qu, D., Teckman, J. H., Omura, S., and Perlmutter, D. H., 1996, Degradation of a mutant secretory protein, α_1-antitrypsin Z, in the endoplasmic reticulum requires proteasome activity, *J. Biol. Chem.* **271**:22791–22795.

Riordan, J. R., and Chang, X. B., 1992, CFTR, a channel with the structure of a transporter, *Biochim. Biophys. Acta* **1101**:221–222.

Riordan, J. R., Rommens, J. M., Kerem, B.-S., Alon, N., Rozmahel, R., Grzelczak, Z., Zielenski, J., Lok, S., Plavsic, N., Chou, J.-L., Drumm, M. L., Iannuzzi, M. C., Collins, F. S., and Tsui, L.-C., 1989, Identification of the cystic fibrosis gene: Cloning and characterization of complementary DNA, *Science* **245**:1066–1073.

Rivett, A. J., Palmer, A., and Knecht, E., 1992, Electron microscopic localization of the multicatalytic proteinase complex in rat liver and in cultured cells, *J. Histochem. Cytochem.* **40**:1165–1172.

Rock, K. L., Gramm, C., Rothstein, L., Clark, K., Stein, R., Dick, L., Hwang, D., and Goldberg, A. L., 1994, Inhibitors of the proteasome block the degradation of most cell proteins and the generation of peptides presented on MHC class I molecules, *Cell* **78**:761–771.

Romisch, K., and Schekman, R., 1992, Distinct processes mediate glycoprotein and glycopeptide export from the endoplasmic reticulum in Saccharomyces cerevisiae, *Proc. Natl. Acad. Sci. USA* **89**:7227–7231.

Roth, A. F., and Davis, N. G., 1996, Ubiquitination of the yeast a-factor receptor, *J. Cell Biol.* **134:** 661–674.

Sanders, S. L., Whitfield, K. M., Vogel, J. P., Rose, M. D., and Schekman, R. W., 1992, Sec61p and BiP directly facilitate polypeptide translocation into the ER, *Cell* **69**:353–365.

Schatz, G., and Dobberstein, B., 1996, Common principles of protein translocation across membranes, *Science* **271**:1519–1526.

Schultz, J., Ferguson, B., and Sprague, G. F., Jr., 1995, Signal transduction and growth control in yeast, *Curr. Opin. Genet. Dev.* **5**:31–37.

Sepp-Lorenzino, L., Ma, Z., Lebwohl, D. E., Vinitsky, A., and Rosen, N., 1995, Herbimycin A induces the 20 S proteasome- and ubiquitin-dependent degradation of receptor tyrosine kinases, *J. Biol. Chem.* **270**:16580–16587.

Siegelman, M., Bond, M. W., Gallatin, W. M., St. John, T., Smith, H. T., Fried, V. A., and Weissman, I. L., 1986, Cell surface molecule associated with lymphocyte homing is a ubiquitinated branched-chain glycoprotein, *Science* **231**:823–829.

Sifers, R. N., Finegold, M. J., and Woo, S. L., 1989, Alpha-1-antitrypsin deficiency: Accumulation or degradation of mutant variants within the hepatic endoplasmic reticulum. *Am. J. Respir. Cell Mol. Biol.* **1:**341–345.

Sommer, T., and Jentsch, S., 1993, A protein translocation defect linked to ubiquitin conjugation at the endoplasmic reticulum. *Nature* **365:**176–179.

Suzuki, T., Kitajima, K., Inoue, S., and Inoue, Y., 1994, Occurrence and biological roles of 'proximal glycanases' in animal cells. *Glycobiology* **4:**777–789.

Tsui, L.-C., 1992, The spectrum of cystic fibrosis mutations. *Trends Genet.* **8:**392–398.

Uehara, Y., and Fukazawa, H., 1991, Use and selectivity of herbimycin A as inhibitor of protein-tyrosine kinases. *Methods Enzymol.* **201:**370–379.

van de Rijn, M., Weissman, I. L., and Siegelman, M., 1990, Biosynthesis pathway of gp90MEL-14, the mouse lymph node-specific homing receptor. *J. Immunol.* **145:**1477–1482.

Ward, C. L., and Kopito, R. R., 1994, Intracellular turnover of cystic fibrosis transmembrane conductance regulator. Inefficient processing and rapid degradation of wild-type and mutant proteins. *J. Biol. Chem.* **269:**25710–25718.

Ward, C. L., Omura, S., and Kopito, R. R., 1995, Degradation of CFTR by the ubiquitin proteasome pathway. *Cell* **83:**121–127.

Weissman, A. M., 1994, The T-cell antigen receptor: A multisubunit signaling complex. *Chem. Immunol.* **59:**1–18.

Werner, E. D., Brodsky, J. L., and McCracken, A. A., 1996, Proteasome-dependent endoplasmic reticulum-associated protein degradation: an unconventional route to a familiar fate. *Proc. Natl. Acad. Sci. USA* **93:**13797–13801.

Whitesell, L., Mimnaugh, E. G., De Costa, B., Myers, C. E., and Neckers, L. M., 1994, Inhibition of heat shock protein HSP90-pp60v-src heteroprotein complex formation by benzoquinone ansamycins: Essential role for stress proteins in oncogenic transformation. *Proc. Natl. Acad. Sci. USA* **91:**8324–8328.

Wiertz, E. J., Jones, T. R., Sun, L., Bogyo, M., Geuze, H. J., and Ploegh, H. L., 1996a, The human cytomegalovirus US11 gene product dislocates MHC class I heavy chains from the endoplasmic reticulum to the cytosol. *Cell* **84:**769–779.

Wiertz, E. J., Tortorella, D., Bogyo, M., Yu, J., Mothes, W., Jones, T. R., Rapoport, T. A., and Ploegh, H. L., 1996b, Sec61-mediated transfer of a membrane protein from the endoplasmic reticulum to the proteasome for destruction. *Nature* **384:**432–438.

Wiley, H. S., Herbst, J. J., Walsh, B. J., Lauffenburger, D. A., Rosenfeld, M. G., and Gill, G. N., 1991, The role of tyrosine kinase activity in endocytosis, compartmentation, and down-regulation of the epidermal growth factor receptor. *J. Biol. Chem.* **266:**11083–11094.

Yang, Y., Janich, S., Cohn, J. A., and Wilson, J. M., 1993, The common variant of cystic fibrosis transmembrane conductance regulator is recognized by hsp70 and degraded in a pre-Golgi nonlysosomal compartment. *Proc. Natl. Acad. Sci. USA* **90:**9480–9484.

Yarden, Y., Escobedo, J. A., Kuang, W. J., Yang-Feng, T. L., Daniel, T. O., Tremble, P. M., Chen, E. Y., Ando, M. E., Harkins, R. N., Francke, U., Fried, V. A., Ulrich, A., and Williams, L. T., 1986, Structure of the receptor for platelet-derived growth factor helps define a family of closely related growth factor receptors. *Nature* **323:**226–232.

Zanolari, B., Raths, S., Singer-Kruger, B., and Riezman, H., 1992, Yeast pheromone receptor endocytosis and hyperphosphorylation are independent of G protein-mediated signal transduction. *Cell* **71:**755–763.

CHAPTER 14

Degradation of Ornithine Decarboxylase

Philip Coffino

1. ORNITHINE DECARBOXYLASE AND ITS PRODUCTS

Ornithine decarboxylase (ODC) is a key enzyme in an essential metabolic pathway (Hayashi, 1989), is a growth-responsive oncoprotein (Auvinen *et al.*, 1992), and associates with an elaborately controlled second protein that directs its degradation by the 26 S protease. Although this sounds like a job for ubiquitin, it is not. Instead, the protein that prompts the destruction of ODC is antizyme (AZ). Together ODC and AZ participate in an elaborately orchestrated process that provides a unique form of feedback regulation based not on allostery but on degradation. ODC is the initial and commonly rate-limiting step in the biosynthesis of polyamines (Tabor and Tabor, 1984). These are small, ubiquitous, abundant, positively charged molecules essential for life. Putrescine (diaminopropane) is the product of ODC. Additional enzymes are required to produce higher polyamines by sequential addition of one aminopropyl group to form spermidine, and of a second to form spermine. The aminopropyl group is generated by the action of *S*-adenosylmethionine decarboxylase. This enzyme is also labile and its destruction is regulated (Svensson and Persson, 1995), but little is

Philip Coffino • Departments of Microbiology and Immunology and of Medicine, University of California, San Francisco, San Francisco, California 94143.

Ubiquitin and the Biology of the Cell, edited by Peters *et al.* Plenum Press, New York, 1998.

known about the mechanism of this process. Recent reviews have considered the role of AZ in the degradation of ODC (Hayashi and Murakami, 1995; Hayashi *et al.*, 1996).

The biochemical function of polyamines is poorly defined and likely diverse. Optimal concentrations of polyamines benefit the fidelity and efficiency of *in vitro* transcription and translation reactions. This suggests that the rapid changes in polyamine pools observed in response to cellular signals for growth or differentiation are needed to optimize these biosynthetic processes. Polyamines probably play a structural role as well. They are present in millimolar concentrations, and are largely bound to RNA and DNA. Spermidine participates in a unique posttranslational modification of the protein eIF-5A (Park *et al.*, 1993). Preventing modification by genetic or pharmacological manipulation is lethal to cells.

2. FEEDBACK REGULATION ACTS THROUGH ANTIZYME

Excess polyamines are toxic. Cells limit their levels by several means, controlling catabolism and transport as well as synthesis (Pegg, 1988; Davis *et al.*, 1992). One effect of raising polyamine levels is to sharply and quickly reduce the amount of ODC in cells (Jänne and Hölttä, 1974; Kay and Lindsay, 1973; Seely and Pegg, 1983). The induction of AZ by polyamines is an essential event in this response. Neither AZ nor ODC mRNA levels change in response to polyamines. Instead, polyamines stimulate the synthesis of AZ by improving the efficiency of translational frameshifting at a specific "shifty" site within the mRNA (Matsufuji *et al.*, 1995). Because AZ mRNA lacks an open reading frame initiated by methionine, a +1 shift is required to move from the short initial reading frame to one downstream encoding most of AZ. Although frameshifting is essential for expression of open reading frames of some animal cell viruses, the AZ gene remains for now unique among host cell genes known to have that requirement.

3. POLYAMINES CONTROL ODC TURNOVER

What is the basis for our belief that AZ destroys ODC? AZ was first described as a stoichiometric inhibitor of ODC, but this is only part of the story. It was shown more than 20 years ago that treating cells with polyamines induces an activity that inhibits the enzymatic activity of partially purified ODC (Heller *et al.*, 1976; Fong *et al.*, 1976). This inhibitory activity, after purification to homogeneity (Kitani and Fujisawa, 1984), was characterized as a protein of about 20 kDa that binds tightly to ODC. The ODC–AZ complex is enzymatically inactive. However, salt dissociation and fractionation fully regenerates ODC activity. The association of ODC and AZ is of very high affinity, but noncovalent. The reversibility of inactivation seemed to point away from AZ as an agent of destruction.

In parallel with the investigations of AZ, studies of the effect of cellular polyamines on ODC suggested that control of ODC translation rather than degradation was responsible for the loss of activity (Kahana and Nathans, 1985; Kanamoto *et al.*, 1991; Höltta and Pohjanpelto, 1986; Persson *et al.*, 1988; Kameji and Pegg, 1987). Pulse label experiments with [^{35}S]methionine demonstrated that polyamines reduced incorporation into the ODC polypeptide, but did not lower the amount of ODC mRNA. This seemed to imply a translational effect. Because polyamines bind nucleic acids and alter their conformation (Watanabe *et al.*, 1991), it seemed likely that polyamines themselves could act on the long 5′ ODC mRNA leader (Gupta and Coffino, 1985; Brabant *et al.*, 1988) to control translation. The translation hypothesis was subjected to four tests, and it failed all (van Daalen Wetters *et al.*, 1989). (1) Polyamine treatment of cells did not change the distribution of ODC mRNA on polyribosomes. (2) A chimeric protein encoded by a transcript bearing the 5′ leader of ODC mRNA was not regulated. (3) Transcripts containing the ODC coding sequence, but no other ODC-derived gene sequence, exhibited regulation. (4) Pulse label experiments were carried out to compare the incorporation of [^{35}S]methionine into ODC in polyamine-replete and polyamine-starved cells. Differential incorporation diminished progressively as pulse label times were shortened; at the shortest labeling time used, 4 min, the difference in favor of the polyamine-starved cells diminished to less than twofold. Taken together, these results implied not translational regulation but a very rapid polyamine-induced posttranslational degradation of ODC. Degradation takes place so promptly that conventional pulse label times of 20–60 min, usually an accurate measurement of synthesis, reflect instead changes in turnover.

Cloning of the gene for AZ (Matsufuji *et al.*, 1990; Miyazaki *et al.*, 1992) provided a key to investigating its action in regulating ODC degradation. Importantly, by using an inducible promoter, its expression could be made independent of cellular polyamine status. It was thus shown that AZ production *per se* is sufficient for rapid cellular degradation of ODC (Murakami *et al.*, 1992, 1994).

4. DISTINCT ELEMENTS OF ODC CONTROL BASAL AND REGULATED DEGRADATION

Degradation of ODC is very fast when polyamine levels are high, occurring within a few minutes, but it is also fairly fast even under basal conditions, when excess polyamines do not elevate AZ. Cycloheximide chase experiments (Russell and Snyder, 1969) and pulse chase experiments with [^{35}S]methionine (Fig. 1; McConlogue and Coffino, 1983a) both showed that basal degradation proceeds

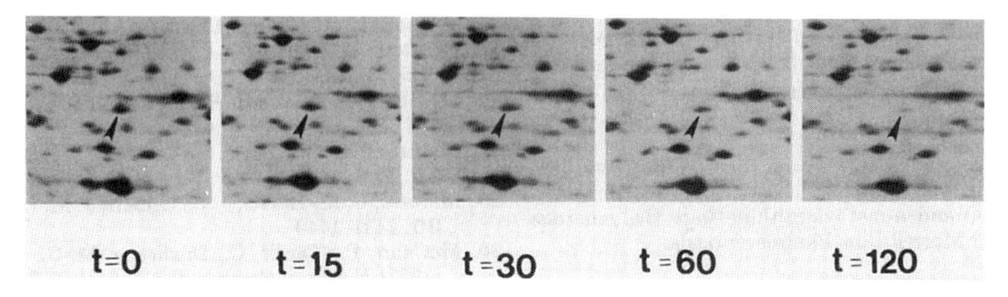

Figure 1. Turnover of ODC. S49 mouse lymphoma cells selected to overproduce ODC about 10-fold were pulse-labeled for 7 min with [^{35}S]methionine and chased for the indicated times (0–120 min). Extracts were analyzed by two-dimensional gel electrophoresis. A portion of the gel is shown; arrowheads indicate the position of ODC polypeptide. The polypeptide has a half-life of 45 min (McConlogue and Coffino, 1983a).

with a half-life of less than 1 hr, fast for an enzyme. A satisfactory explanation of lability would have to account for both basal and accelerated degradation. An obvious route to sorting out the question of how ODC is degraded would be to carry out functional studies with engineered modifications of the protein. However, when the function to be studied is degradation, this task is complicated by the fact that misfolded proteins are rapidly degraded. If specific structural elements that contribute to ODC degradation were to be revealed, engineering would need to be done with concern for preserving conformation.

Comparison of the inferred amino acid sequences of mouse and trypanosome ODCs (Gupta and Coffino, 1985; Phillips *et al.*, 1987) suggested a solution to this problem. Despite the passage of more than a billion years since the mammalian and parasite lineages separated, the proteins are 69% identical within a core region of homology (Phillips *et al.*, 1987). Chimeras between trypanosome and mouse ODC have enzymatic activity (Ghoda *et al.*, 1992) and are therefore native in conformation; domain swaps could thus be carried out with impunity. First, however, a simple comparison of the amino acid sequence of mouse and trypanosome ODC proved revealing. The mouse form had, within its 461 amino acids, about 37 extra amino acids at the C-terminus. This tail was rich in PEST amino acids (single-letter code), shown to be abundant in labile proteins (Rogers *et al.*, 1986). ODC-deficient CHO cells (Steglich and Scheffler, 1982) were available as recipients for stable transfection. In these cells, the only ODC activity present is that encoded by transfected engineered genes.

Expression of C-truncated mouse ODC lacking the last 37 amino acids of mouse ODC showed it to be stable under basal polyamine conditions (Ghoda *et al.*, 1989). The intact mouse protein when similarly transfected had, as expected, a half-life of less than 1 hr. In sharp contrast, trypanosome ODC proved to be stable in animal cells regardless of whether polyamines were high or low (Ghoda *et al.*,

1990). This finding made possible a series of experiments in which chimeras and truncated proteins were exploited to identify the protein regions that determined basal and polyamine-stimulated lability. Appending a C-terminal region of mouse ODC to trypanosome ODC sufficed to make the latter labile (Ghoda *et al.*, 1990). The C-terminus was thus shown to be necessary and sufficient for degradation under basal polyamine conditions. A series of fine-structure deletions and truncations showed that removing as few as the final 5 C-terminal amino acids (ARINV) from mouse ODC stabilized it (Ghoda *et al.*, 1992; Fig. 2). The last 5 amino acids are not the only ones in the ODC C-terminus essential for lability: Mutating a single cysteine at position 441 is also stabilizing (Miyazaki *et al.*, 1993). The 5 terminal amino acids are outside the PEST region as conventionally defined, yet are critical for degradation. Conversely, internal deletion of the PEST region within the C-terminus did not impede degradation (Ghoda *et al.*, 1992). Therefore, in the case of ODC, PEST sequences did not pass this direct test of function.

Mouse–trypanosome chimeras revealed that polyamine-dependent degradation required an additional region within mouse ODC; an essential component of the AZ binding site lay within amino acids 117–140 (Li and Coffino, 1992).

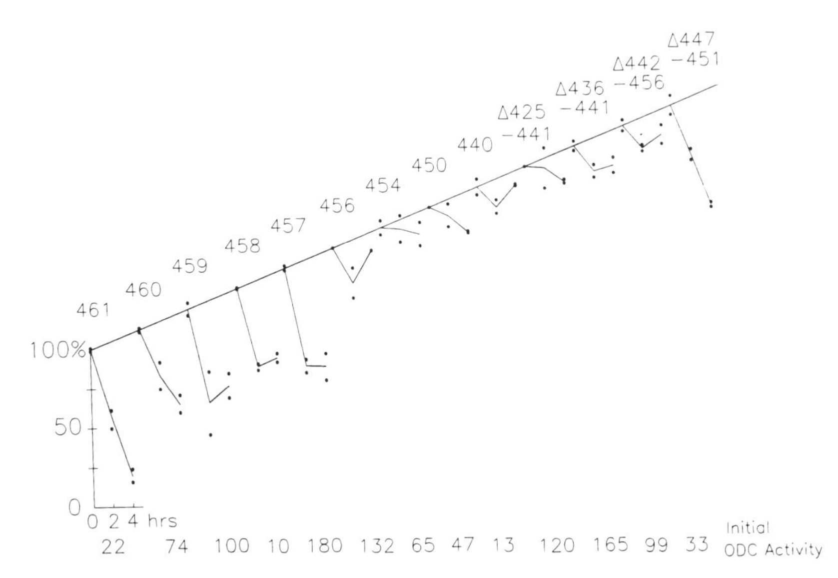

Figure 2. Turnover of wild-type, C-truncated, and deleted ODCs. ODC was truncated to the indicated residue (e.g., 457), internally deleted (e.g., Δ425–441), or left intact (461). Cells were treated with cycloheximide for 0, 2, or 4 hr to inhibit protein synthesis and extracts prepared for analysis of ODC activity. Each of the duplicate data points is shown, with line plots through the means. The origins of successive plots are displaced diagonally to display the data more clearly. Initial activity (picomoles/min/mg protein) is shown below each plot (Ghoda *et al.*, 1992).

Among a series of chimeras, a good correlation was found between AZ binding and *in vitro* AZ-dependent degradation. An *in vitro* degradation system derived from rabbit reticulocytes was utilized in which ODC degradation depends on both ATP and AZ (Li and Coffino, 1993, 1994). *In vivo* tests of critical constructs confirmed the following conclusions: The C-terminus of mouse ODC suffices for basal degradation. Truncating the C-terminus but retaining an AZ binding site allows polyamine induced regulation, but it is sluggish and attenuated. The presence of both regions is necessary and sufficient for the full response, a more than 10-fold reduction of ODC within a few hours (Fig. 3).

Figure 4 depicts some structural features of ODC. The AZ binding site, two PEST regions, and conserved residues present at the active site (amino acids 69, 169, 360) or required for dimer formation (amino acid 387) are shown. A dot matrix analysis compares the sequence of mouse ODC with that of trypanosome ODC. Note that the sequences of the two proteins are very similar, but diverge within the regions corresponding to the AZ binding site and PEST-rich sequences of mouse ODC and that trypanosome ODC has no homologue to the mouse ODC C-terminus, which contains the second PEST-rich sequence.

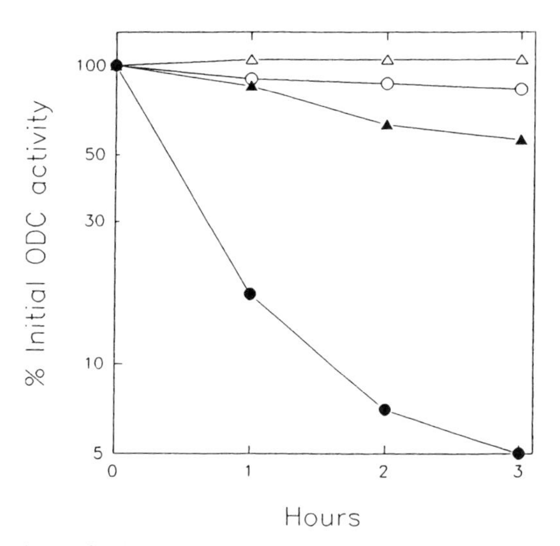

Figure 3. Dependence of polyamine regulation on AZ binding. ODC-deficient CHO cells were transfected with constructs encoding mouse ODC (circles) or a chimeric ODC in which residues 117–140 of mouse ODC were replaced with the corresponding region of trypanosome ODC (triangles). Mouse ODC binds to AZ; the chimeric ODC does not. Cells were treated to increase polyamine pools (filled symbols) or were untreated (open symbols) and extracts prepared for analysis of ODC activity (Li and Coffino, 1992).

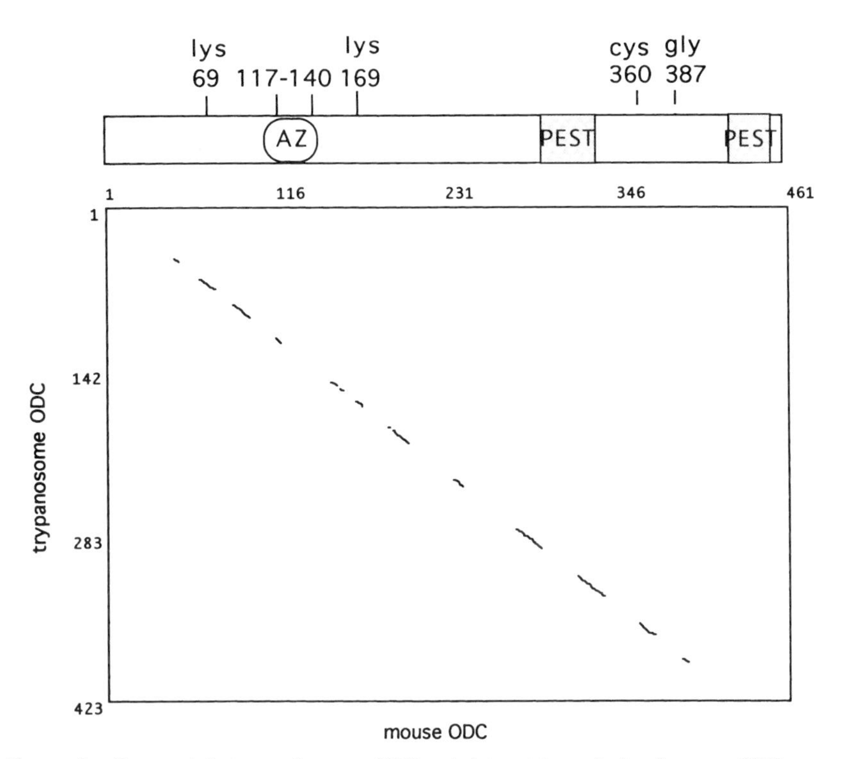

Figure 4. Structural features of mouse ODC and dot matrix analysis of mouse ODC versus trypanosome ODC. Residues 69, 169, and 360 are in the active site (Poulin *et al.*, 1992; Tsirka and Coffino, 1992) and residue 387 is required for dimer formation (Tobias *et al.*, 1993). Dot matrix analysis marks as positive a moving window of 8 contiguous amino acids if 7 or more are identical.

5. INDUCTION OF CONFORMATIONAL CHANGE IN ODC BY AZ

A natural conjecture is that AZ acts by improving the efficiency of basal degradation, perhaps by exposing a C-terminal domain that is otherwise partially occluded. ODC is a homodimer, but becomes a heterodimer in association with AZ (Mitchell and Chen, 1990). It is easy to imagine that within the dimer the C-terminus is tucked in, but becomes more exposed when the dimer is induced to dissociate. This was confirmed by demonstrating that AZ induces a change of conformation that makes the C-terminus more accessible to antibody (Fig. 5), establishing a functional connection between constitutive lability, dependent only on the C-terminus, and AZ binding (Li and Coffino, 1993). *In vitro* degradation

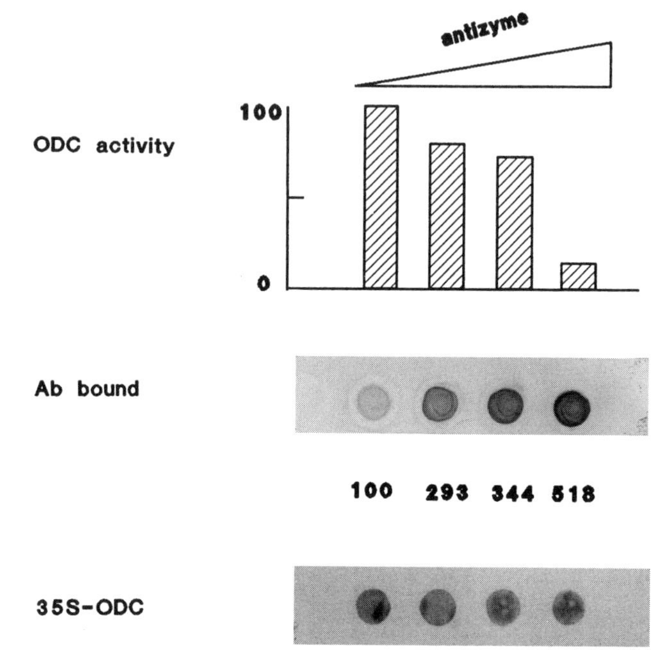

Figure 5. Effect of AZ on the reaction of mouse ODC with antibody against the ODC C-terminus. Mouse ODC and increasing amounts of AZ (0, 0.5, 2.0, and 10 μl of *in vitro* translation mixture) were incubated together. ODC activities (histogram) are shown as a percentage of that present without AZ. Identical complexes were dot-blotted to a nitrocellulose filter, immunoreacted with antibody against the C-terminus of ODC, and bound antibody determined by immunoblot (middle; intensity of immunoreaction product in arbitrary units). Reactions and blotting were carried out as above, but with ^{35}S-labeled ODC and the association of ODC with the filter determined by autoradiography (bottom) (Li and Coffino, 1993).

was prevented by an antibody specific for the C-terminus, substantiating its importance in degradation.

To further test whether mobility of the C-terminus is important, a circularly permuted form of ODC was created (Li and Coffino, 1993). The normal C- and N-termini are therein connected through a histidine bridge, and the peptide chain broken between residues 307 and 308. Tethering the C-terminus in this fashion presumably reduces its mobility. This construct, much like wild-type ODC, is enzymatically active, binds to AZ, and is inhibited by it. However, in contrast to wild-type ODC, the circularly permuted form of ODC is not degraded when associated with AZ (Li and Coffino, 1993). It might be imagined that dissociating the ODC dimer to monomers is itself sufficient to expose the C-terminus and produce degradation, but this is not the case. Elevated salt concentrations induce monomerization, but cannot promote degradation without AZ.

6. DOMAINS OF AZ REQUIRED FOR ODC BINDING AND DEGRADATION

Analysis of truncated forms of AZ revealed it to have two functional domains: an ODC binding region and a portable domain that can direct degradation of diverse proteins (Li and Coffino, 1994). The C-terminal half of AZ (amino acids 121–227) binds to ODC and exposes the C-terminal region of that protein, but this action is inadequate to direct degradation. Consequently, the model that AZ accelerates degradation of ODC merely by exposing the C-terminal domain proved insufficient as an explanation of AZ action. A region within the N-terminus of AZ (NAZ), between amino acids 70 and 121, is required for this (Li and Coffino, 1994; Ichiba *et al.*, 1994). As will be described, NAZ can function as a portable modular signal for degradation of diverse proteins.

7. UBIQUITIN SYSTEM NOT INVOLVED IN AZ-REGULATED DEGRADATION

Remarkably, the process of AZ-directed ODC degradation requires no involvement of ubiquitin, but does make use of the 26 S protease. Both *in vivo* and *in vitro* evidence support the conclusion that ubiquitin is not needed for degradation. In ts85 cells with a temperature-sensitive E1 (Finley *et al.*, 1984), polyamine-induced degradation of ODC is unimpeded at the nonpermissive temperature (Glass and Gerner, 1987; Rosenberg-Hasson *et al.*, 1989). The *in vitro* degradation properties of ODC have been characterized in rabbit reticulocyte extracts, both crude and fractionated. Depletion of components critical for ubiquitination, i.e., ubiquitin itself or E1, did not prevent AZ-dependent degradation (Bercovich *et al.*, 1989; Rosenberg-Hasson *et al.*, 1989). Fractionation of cellular extracts demonstrated that the 26 S protease, but not the 20 S protease, degraded ODC in an ATP-dependent reaction (Murakami *et al.*, 1992a). Figure 6 depicts the process of AZ induction, action, and recycling.

The products of degradation were found to be peptides (Tokunaga *et al.*, 1994); no larger intermediates of degradation were detectable. This is consistent with the view that the rate of the degradation process is limited by an early step. *In vivo* evidence supports this as well: In mutant lymphoma cells selected to devote as much as 15% of protein synthesis to ODC, no lower-molecular-mass degradation intermediates are detected (McConlogue and Coffino, 1983b). (Also not found are higher-molecular-mass forms characteristic of a ubiquitination ladder.)

Two technical characteristics of *in vitro* degradation systems in wide use should be mentioned. First, crude reticulocyte lysates degrade ODC when AZ is present, but fail to do so if AZ is absent or if its action is suppressed by AZ

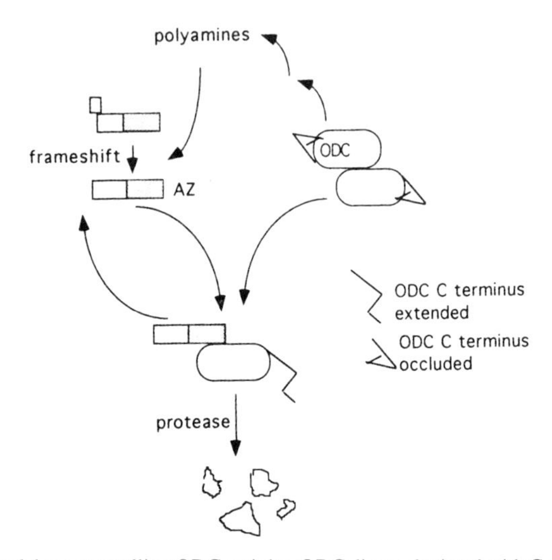

Figure 6. Feedback loop controlling ODC activity. ODC dimer, depicted with C-terminal tail tucked in, generates putrescine, the direct polyamine precursor. Polyamines promote translational frameshifting of the AZ mRNA, resulting in an increase of AZ protein, depicted with N-terminus unhatched and C-terminus hatched. The AZ C-terminus binds to ODC, exposing ODC's C-terminal tail, which, together with the AZ N-terminus, acts as a signal for degradation of ODC by the 26 S protease. AZ is recycled.

inhibitor (Murakami *et al.*, 1993). This may reflect missing components specific for basal degradation, but more likely is related to a general slowness of action, such that only the far more efficient AZ-dependent process can be observed. Second, fractionated systems are much less active than the crude reticulocyte lysates from which they are derived. Purified 26 S protease plus ATP does degrade ODC in the presence of AZ, but with much less efficiency than the equivalent amount of the protease within the crude extract (Li and Coffino, unpublished). Something is apparently excluded or made moribund by fractionation schemes in common use.

Although AZ-mediated degradation of ODC bypasses ubiquitin, this is less clear for basal degradation. Basal degradation, defined as proteolysis of ODC that does not demand interaction with AZ, is not readily observed *in vitro*. [The ODC degradation described in cell lysates not experimentally supplemented with AZ has been shown to result from traces of endogenous AZ (Murakami *et al.*, 1993).] Therefore, experiments involving depletion of ubiquitin-specific components *in vitro* cannot readily resolve the question of whether basal degradation does or does not depend on the ubiquitin system. Basal degradation must be largely independent of AZ, not simply the residual consequence of low-level AZ action, because, so long as the ODC C-terminus is present, mutant or chimeric ODCs without an AZ binding site are still degraded *in vivo*. Studying the stability of such proteins in a cell with a temperature-sensitive El could help resolve whether ubiquitin is involved in basal degradation.

AZ, like ODC, contains a PEST region, amino acids 100–113. Deletion of this part of the protein does not impede ODC degradation (Li and Coffino, 1994). ODC, as mentioned, has a PEST sequence near its C-terminus that is not required for ODC degradation. Mouse ODC also has a second PEST sequence at amino acids 298–333, not present in the trypanosome protein (Fig. 4). Analysis of mouse–trypanosome chimeras revealed that this PEST region is also not needed for degradation (Ghoda *et al.*, 1992). PEST sequences are rare among proteins, but mouse ODC and AZ together contain three. PEST regions may be plausible suspects and reliable witnesses, but have here been exonerated as perpetrators.

8. FUNCTION OF AZ IN DEGRADATION

Returning now to a discussion of how AZ acts, mention should be made of its catalytic property. One AZ molecule can direct the degradation of many molecules of ODC (Murakami *et al.*, 1992b; Mamroud-Kidron *et al.*, 1994). As described, the N-terminal region of AZ does something more than perturb ODC's conformation. It is reasonable to assume that AZ accompanies ODC to the protease, perhaps increasing the affinity of ODC for the protease or acting as an activator of proteolytic activity. AZ departs intact; ODC is destroyed. Can NAZ direct degradation if it is instead covalently grafted to ODC? It can. Fusion of NAZ to the N-terminus of ODC creates an unstable protein (Li *et al.*, 1996; Fig. 7). As is true for the normal noncovalent complex, degradation of the NAZ-ODC fusion protein requires the presence of the C-terminal region of ODC. Fusion protein degradation is not dependent on ubiquitination. In the presence of ATPγS, which supports ubiquitination but not degradation, no ladder of higher-molecular-mass forms is generated (Li *et al.*, 1996). NAZ, regardless of whether it is delivered to ODC by the normal mode of association or by molecular engineering, follows the same rules: Its function depends on the ODC C-terminus but not on ubiquitin.

Appending NAZ to other labile proteins can promote their degradation as well, and divert that process to a pathway that does not depend on ubiquitin (Li *et al.*, 1996). Cyclin B degradation is normally ubiquitin-dependent and occurs specifically in mitosis, but not in interphase cells or in rabbit reticulocyte extracts (Glotzer *et al.*, 1991). A fusion protein containing NAZ and a fragment of cyclin B previously shown to be degraded specifically in M-phase extracts can be degraded in a rabbit reticulocyte lysate. No ubiquitin conjugates can be detected in this process. Similarly, degradation of the tumor suppressor protein p53 is accelerated

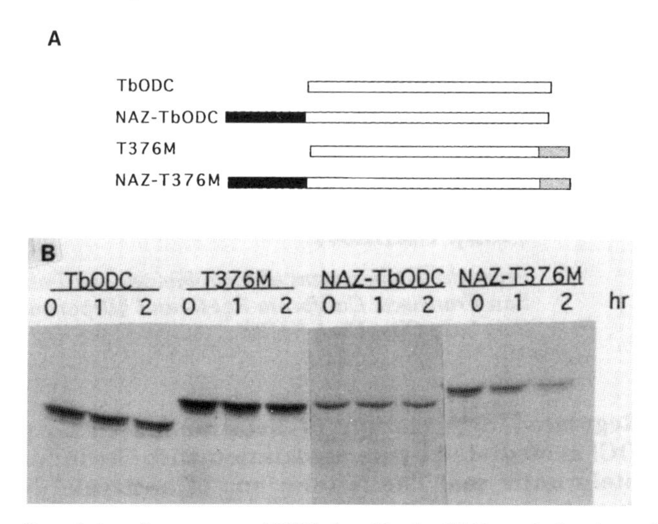

Figure 7. Degradation of trypanosome ODC induced by the AZ N-terminal region and C-terminus of mouse ODC. (Top) Structure of trypanosome ODC (open bar) and fusion to AZ N-terminal region (filled bar) or mouse ODC C-terminus (cross-hatched bar), with junction at amino acid 376. (Bottom) *In vitro* degradation of each construct. Labeled proteins were subjected to degradation in a reticulocyte lysate for 0, 1, or 2 hr and the protein remaining undegraded was evaluated by SDS–PAGE and autoradiography (Li *et al.*, 1996).

and made independent of ubiquitin by fusion to NAZ. Just as ODC requires that its C-terminal degradation domain cooperate with NAZ, cyclin and p53 also contain degradation domains. Deletion analysis revealed that, in the context of the NAZ–p53 fusion, the degradation domain of p53 lies within amino acids 100 to 150 (Li and Coffino, 1996). Appending this portion of p53 to the C-terminus of a mouse–trypanosome chimera containing an AZ binding site conferred polyamine-dependent regulation *in vivo*.

Does our understanding of the mechanism of ubiquitin-directed degradation suggest how AZ might work? Similar properties of AZ and ubiquitin are readily apparent. Both participate in a controlled process directed at specific labile proteins, and both accompany their target proteins to the 26 S protease. (However, one AZ molecule is enough, whereas multiple ubiquitins are needed.) Both are spared degradation, and can therefore act repeatedly. How does ubiquitin or AZ promote degradation? Direct evidence is lacking, but two actions seem plausible. They may act as a delivery system, by linking target to proteasome. This implies a direct or indirect affinity of AZ or ubiquitin for the proteasome. Polyubiquitin chains interact directly with a single polypeptide, subunit 5, of the 26 S protease (Deveraux *et al.*, 1994) and AZ interacts with the native 26 S protease of mammalian origin (Li *et al.*, 1996). The functional significance of these associations, however, requires scrutiny. A second potential function, presumably occurring together with delivery, is proteasome activation. Evidence for this is lacking, but could be sought by determining whether ubiquitin or AZ can alter the activity of the proteasome for a protein substrate to which they are not physically linked.

Ubiquitin and AZ differ in one important respect: The former acts autonomously in that its presence within a protein seems to constitute a sufficient signal for degradation. Not so for AZ. The presence of NAZ is necessary but insufficient for degradation. A degradation domain, for lack of a better term, must be present as well. Examples of such domains have been cited: the C-terminus of ODC, the degradation box of cyclin B, and a specific region of p53.

9. CONSERVATION OF REGULATION OF ODC

To what degree is regulation of ODC conserved in evolution? AZ is widely distributed in vertebrates, but has not been detected in lower eukaryotes (Hayashi and Murakami, 1995). However, both *Neurospora* (Davis *et al.*, 1985; DiGangi *et al.*, 1987) and *S. cerevisiae* (Fonzi and Sypherd, 1987; Fonzi, 1989) are attentive to their polyamine status, reducing ODC activity in response to excess polyamines. As in animal cells, ODC of lower eukaryotes is labile; genetic and biochemical data indicate participation of the proteasome (Bercovich and Kahana, 1992; Elias *et al.*, 1995). It is not known whether this control invokes an AZ-like interaction, ubiquitin, or some other mechanism. ODC of animal origin is also labile when expressed in yeast (Mamroud-Kidron and Kahana, 1994). This implies that both the yeast and animal proteases recognize common stigmata of lability. In sharp contrast, polyamines have no effect on ODC in trypanosomes (Bass *et al.*, 1992), which presumably use some other means to limit their toxicity. Not only is trypanosome ODC stable in its native context, so too is mouse ODC when expressed in the parasite (Bass *et al.*, 1992). Even coexpressing AZ together with mouse ODC in trypanosomes fails to elicit degradation (Hua *et al.*, 1995). The trypanosome protease has only recently been described (Hua *et al.*, 1996), and fragmentary evidence suggests it to differ in size and subunit structure from the proteases of distantly related taxa. Its substrate specificity remains to be determined, but it appears not to act on ODC.

10. WHY NOT UBIQUITIN?

Why does ODC make use of a unique means for controlling turnover when the ubiquitin system is available and suffices for other proteins? One reason may be that AZ not only degrades ODC, but also directly inhibits its enzymatic activity.

It can begin work without delay. Newly synthesized ODC, or even the ribosome-bound nascent chain, could become associated with AZ before it assumes a homodimeric enzymatically active form. Second, the means of induction of AZ, translational frameshifting, ensures that feedback is controlled by the most biochemically relevant pool of polyamines, those free to change their interactions. The bulk of cellular polyamines within cells are unavailable for this, because they are bound, predominantly to RNA and DNA (Watanabe *et al.*, 1991). Third, AZ may have additional targets related to polyamines or their action. AZ has been found to inhibit polyamine uptake by cells (Mitchell *et al.*, 1994). Other key enzymes in the regulation of polyamine biosynthesis and catabolism are under posttranscriptional control (Kameji and Pegg, 1987; Pajunen *et al.*, 1988; Pegg *et al.*, 1990), e.g., *S*-adenosylmethionine decarboxylase, required for synthesis of spermidine and spermine from putrescine, and spermine/spermine acetyltransferase, which is rate-limiting for polyamine catabolism. Posttranscriptional events are intrinsically fast.

Further regulatory complexity is suggested by the presence in cells of a protein inhibitor of AZ (Fujita *et al.*, 1982), the gene for which has recently been cloned (Murakami *et al.*, 1996). AZ has greater avidity for antizyme inhibitor than for ODC. The amino acid sequence of antizyme inhibitor is very similar to that of ODC, but it has no known enzymatic activity. Finally, gutfeeling, a protein with structural homology to the C-terminal half of AZ (the ODC binding portion) but no homology to the N-terminal half, was described in *Drosophila* (Salzberg *et al.*, 1996). It was identified in a screen for mutant flies with defects in differentiation of the peripheral nervous system. Null mutants fail to develop a normal peripheral nervous system or musculature. These findings suggest that ODC and AZ may constitute members of two mutually interacting protein families that have evolved to handle specialized tasks within a locally coordinated network concerned at least in part with polyamines and protein turnover.

Acknowledgment. This work was supported by NIH grant RO1 GM45335.

REFERENCES

Auvinen, M., Paasinen, A., Andersson, L. C., and Höltta, E., 1992, Ornithine decarboxylase activity is critical for cell transformation, *Nature* **360:**355–358.

Bass, K. E., Sommer, J. M., Cheng, Q. L., and Wang, C. C., 19922, Mouse ornithine decarboxylase is stable in Trypanosoma brucei, *J. Biol. Chem.* **267:**11034–11037.

Bercovich, Z., and Kahana, C., 1992, Involvement of the 20S proteasome in the degradation of ornithine decarboxylase, *Eur. J. Biochem.* **213:**205–210.

Bercovich, Z., Rosenberg-Hasson, Y., Ciechanover, A., and Kahana, C., 1989, Degradation of ornithine decarboxylase in reticulocyte lysate is ATP-dependent but ubiquitin-independent, *J. Biol. Chem.* **264:**15949–15952.

Brabant, M., McConlogue, L., van Daalen Wetters, T., and Coffino, P., 1988, Mouse ornithine decarboxylase gene: Cloning, structure, and expression, *Proc. Natl. Acad. Sci. USA* **85:**2200–2204.

Davis, R. H., Krasner, G. N., DiGangi, J. J., and Ristow, J. L., 1985, Distinct roles of putrescine and spermidine in the regulation of ornithine decarboxylase in Neurospora crassa, *Proc. Natl. Acad. Sci. USA* **82:**4105–4109.

Davis, R. H., Morris, D. R., and Coffino, P., 1992, Sequestered end products and enzyme regulation: The case of ornithine decarboxylase, *Microbiol. Rev.* **56:**280–290.

Deveraux, Q., Ustrell, V., Pickart, C., and Rechsteiner, M., 1994, A 26S protease subunit that binds ubiquitin conjugates, *J. Biol. Chem.* **269:**7059–7061.

DiGangi, J. J., Seyfzadeh, M., and Davis, R. H., 1987, Ornithine decarboxylase from Neurospora crassa, *J. Biol. Chem.* **262:**7889–7893.

Elias, S., Bercovich, B., Kahana, C., Coffino, P., Fischer, M., Hilt, W., Wolf, D. H., and Ciechanover, A., 1995, Degradation of ornithine decarboxylase by the mammalian and yeast 26S proteasome complexes requires all the components of the protease, *Eur. J. Biochem.* **229:**276–283.

Finley, D., Ciechanover, A., and Varshavsky, A., 1984, Thermolability of ubiquitin-activating enzyme from the mammalian cell cycle mutant ts85, *Cell* **37:**43–55.

Fong, W. F., Heller, J. S., and Canellakis, E. S., 1976, The appearance of an ornithine decarboxylase inhibitory protein upon the addition of putrescine to cell cultures, *Biochim. Biophys. Acta* **428:** 456–465.

Fonzi, W. A., 1989, Regulation of Saccharomyces cerevisiae ornithine decarboxylase expression in response to polyamine, *J. Biol. Chem.* **264:**18110–18118.

Fonzi, W. A., and Sypherd, P. S., 1987, The gene and the primary structure of ornithine decarboxylase from Saccharomyces cerevisiae, *J. Biol. Chem.* **262:**10127–10133.

Fujita, K., Murakami, Y., and Hayashi, S., 1982, A macromolecular inhibitor to the antizyme to ornithine decarboxylase, *Biochem. J.* **204:**647–652.

Ghoda, L., van Daalen Wetters, T., Macrae, M., Ascherman, D., and Coffino, P., 1989, Prevention of rapid intracellular degradation of ODC by a carboxyl-terminal truncation, *Science* **243:**1493–1495.

Ghoda, L., Phillips, M. A., Bass, K. E., Wang, C. C., and Coffino, P., 1990, Trypanosome ornithine decarboxylase is stable because it lacks sequences found in the carboxyl terminus of the mouse enzyme which target the latter for intracellular degradation, *J. Biol. Chem.* **265:**11823–11826.

Ghoda, L., Sidney, D., Macrae, M., and Coffino, P., 1992, Structural elements of ornithine decarboxylase required for intracellular degradation and polyamine-dependent regulation, *Mol. Cell. Biol.* **12:**2178–2185.

Glass, J. R., and Gerner, E. W., 1987, Spermidine mediates degradation of ornithine decarboxylase by a non-lysosomal, ubiquitin-independent mechanism, *J. Cell. Physiol.* **130:**133–141.

Glotzer, M., Murray, A. M., and Kirschner, M. W., 1991, Cyclin is degraded by the ubiquitin pathway, *Nature* **349:**132–138.

Gupta, M., and Coffino, P., 1985, Mouse ornithine decarboxylase: Complete amino acid sequence deduced from cDNA, *J. Biol. Chem.* **260:**2941–2944.

Hayashi, S. I., 1989, *Ornithine Decarboxylase: Biology, Enzymology and Molecular Genetics*, Pergamon Press, New York.

Hayashi, S., and Murakami, Y., 1995, Rapid and regulated degradation of ornithine decarboxylase, *Biochem. J.* **306:**1–10.

Hayashi, S., Murakami, Y., and Matsufuji, S., 1996, Ornithine decarboxylase antizyme: A novel type of regulatory protein, *Trends Biol. Sci.* **21:**27–30.

Heller, J. S., Fong, W. F., and Canellakis, E. S., 1976, Induction of a protein inhibitor to ornithine decarboxylase by the end products of its reaction, *Proc. Natl. Acad. Sci. USA* **73:**1858–1862.

Höltta, E., and Pohjanpelto, P., 1986, Control of ornithine decarboxylase in Chinese hamster ovary cells by polyamines, *J. Biol. Chem.* **261:**9502–9508.

Hua, S., Li, X., Coffino, P., and Wang, C. C., 1995, Rat antizyme inhibits the activity but does not promote the degradation of mouse ornithine decarboxylase in Trypanosoma brucei, *J. Biol. Chem.* **270:**10264–10271.

Hua, S., Nguyen, T. T., Wong, M.-L., and Wang, C. C., 1996, Purification and characterization of proteasomes from Trypanosoma brucei, *Mol. Biochem. Parasitol.* **78:**33–46.

Ichiba, T., Matsufuji, S., Miyazaki, Y., Murakami, Y., Tanaka, K., Ichihara, A., and Hayashi, S. I., 1994, Functional regions of ornithine decarboxylase antizyme, *Biochem. Biophys. Res. Commun.* **200:** 1721–1727.

Jänne, J., and Hölttä, E., 1974, Regulation of ornithine decarboxylase activity by putrescine and spermidine in rat liver, *Biochem. Biophys. Res. Commun.* **61:**449–456.

Kahana, C., and Nathans, D., 1985, Translational regulation of mammalian ornithine decarboxylase by polyamines, *J. Biol. Chem.* **260:**15390–15393.

Kameji, T., and Pegg, A. E., 1987, Inhibition of translation of mRNAs for ornithine decarboxylase and S-adenosylmethionine decarboxylase by polyamines, *J. Biol. Chem.* **262:**2427–2430.

Kanamoto, R., Nishiyama, M., Matsufuji, S., and Hayashi, S., 1991, Translational control mechanism of ornithine decarboxylase by asparagine and putrescine in primary cultured hepatocytes, *Arch. Biochem. Biophys.* **291:**247–254.

Kay, J. E., and Lindsay, V. J., 1973, Control of ornithine decarboxylase activity in stimulated human lymphocytes by putrescine and spermidine, *Biochem. J.* **132:**791–796.

Kitani, T., and Fujisawa, H., 1984, Purification and some properties of a protein inhibitor antizyme, of ornithine decarboxylase from rat liver, *J. Biol. Chem.* **259:**10036–10040.

Li, X., and Coffino, P., 1992, Regulated degradation of ornithine decarboxylase requires interaction with the polyamine-inducible protein antizyme, *Mol. Cell. Biol.* **12:**3556–3562.

Li, X., and Coffino, P., 1993, Degradation of ornithine decarboxylase: Exposure of the C-terminal target by a polyamine-inducible inhibitory protein, *Mol. Cell. Biol.* **13:**2377–2383.

Li, X., and Coffino, P., 1994, Distinct domains of antizyme required for binding and proteolysis of ornithine decarboxylase, *Mol. Cell. Biol.* **14:**87–92.

Li, X., and Coffino, P., 1996, Identification of a region of p53 that confers lability, *J. Biol. Chem.* **271:**4447–4451.

Li, X., Stebbins, B., Hoffman, L., Pratt, G., Rechsteiner, M., and Coffino, P., 1996, The N terminus of antizyme promotes degradation of heterologous proteins, *J. Biol. Chem.* **271:**4441–4446.

Mamroud-Kidron, E., and Kahana, C., 1994, The 26S proteasome degrades mouse and yeast ornithine decarboxylase in yeast cells, *FEBS Lett.* **356:**162–164.

Mamroud-Kidron, E., Omer-Itsicovich, M., Bercovich, Z., Tobias, K. E., Rom, E., and Kahana, C., 1994, A unified pathway for the degradation of ornithine decarboxylase in reticulocyte lysate requires interaction with the polyamine-induced protein, ornithine decarboxylase antizyme, *Eur. J. Biochem.* **226:**547–554.

Matsufuji, S., Miyazaki, Y., Kanamoto, R., Kameji, T., Murakami, Y., Baby, T. G., Fujita, K., Ohno, T., and Hayashi, S., 1990, Analyses of ornithine decarboxylase antizyme mRNA with a cDNA cloned from rat liver, *J. Biochem.* **108:**365–371.

Matsufuji, S., Matsufuji, T., Miyazaki, Y., Murakami, Y., Atkins, J. F., Gesteland, R. F., and Hayashi, S. I., 1995, Autoregulatory frameshifting in decoding mammalian ornithine decarboxylase antizyme, *Cell* **80:**51–60.

McConlogue, L., and Coffino, P., 1983a, Ornithine decarboxylase in difluoromethylornithine-resistant mouse lymphoma cells, *J. Biol. Chem.* **258:**8384–8388.

McConlogue, L. C., and Coffino, P., 1983b, A mouse lymphoma cell mutant whose major protein product is ornithine decarboxylase, *J. Biol. Chem.* **258:**12083–12086.

Mitchell, J. L. A., and Chen, H. J., 1990, Conformational changes in ornithine decarboxylase enable recognition by antizyme, *Biochim. Biophys. Acta* **1037:**115–121.

Mitchell, J. L. A., Judd, G. G., Bareyal-Leyser, A., and Ling, S. Y., 1994, Feedback repression of polyamine transport is mediated by antizyme in mammalian tissue-culture cells, *Biochem. J.* **299:**19–22.

Miyazaki, Y., Matsufuji, S., and Hayashi, S., 1992, Cloning and characterization of a rat gene encoding ornithine decarboxylase antizyme, *Gene* **113:**191–197.

Miyazaki, Y., Matsufuji, S., Murakami, Y., and Hayashi, S., 1993, Single amino-acid replacement is responsible for the stabilization of ornithine decarboxylase in HMOA cells, *Eur. J. Biochem.* **214:** 837–844.

Murakami, Y., Matsufuji, S., Kameji, T., Hayashi, S., Igarashi, K., Tamura, T., Tanaka, K., and Ichihara, A., 1992a, Ornithine decarboxylase is degraded by the 26S proteasome without ubiquitination, *Nature* **360:**597–599.

Murakami, Y., Matsufuji, S., Miyazaki, Y., and Hayashi, S., 1992b, Destabilization of ornithine decarboxylase by transfected antizyme gene expression in hepatoma tissue culture cells, *J. Biol. Chem.* **267:** 13138–13141.

Murakami, Y., Tanaka, K., Matsufuji, S., Miyazaki, Y., and Hayashi, S., 1992c, Antizyme, a protein induced by polyamines, accelerates the degradation of ornithine decarboxylase in Chinese-hamster ovary-cell extracts, *Biochem. J.* **283:**661–664.

Murakami, Y., Matsufuji, S., Tanaka, K., Ichihara, A., and Hayashi, S., 1993, Involvement of the proteasome and antizyme in ornithine decarboxylase degradation by a reticulocyte lysate, *Biochem. J.* **295:**305–308.

Murakami, Y., Matsufuji, S., Miyazaki, Y., and Hayashi, S., 1994, Forced expression of antizyme abolishes ornithine decarboxylase activity, suppresses cellular levels of polyamines and inhibits cell growth, *Biochem. J.* **304:**183–187.

Murakami, Y., Ichiba, T., Matsufuji, S., and Hayashi, S., 1996, Cloning of antizyme inhibitor, a highly homologous protein to ornithine decarboxylase, *J. Biol. Chem.* **271:**3340–3342.

Pajunen, A., Crozat, A., Jänne, O. A., Ihalainen, R., Laitinen, P. H., Stanley, B., Madhubala, R., and Pegg, A. E., 1988, Structure and regulation of mammalian S-adenosylmethionine decarboxylase, *J. Biol. Chem.* **263:**17040–17049.

Park, M. H., Wolff, E. C., and Folk, J. E., 1993, Is hypusine essential for eukaryotic cell proliferation? *Trends Biol. Sci.* **18:**475–479.

Pegg, A. E., 1988, Polyamine metabolism and its importance in neoplastic growth and as a target for chemotherapy, *Cancer Res.* **48:**759–774.

Pegg, A. E., Pakala, R., and Bergeron, R. J., 1990, Induction of spermidine/spermine N1-acetyltransferase activity in Chinese-hamster ovary cells by N1 N11-bis ethyl,norspermidine and related compounds, *Biochem. J.* **267:**331–338.

Persson, L., Holm, I., and Heby, O., 1988, Regulation of ornithine decarboxylase mRNA translation by polyamines, *J. Biol. Chem.* **263:**3528–3533.

Phillips, M. A., Coffino, P., and Wang, C. C., 1987, Cloning and sequencing of the ornithine decarboxylase gene from Trypanosoma brucei: Implications for enzyme turnover and selective difluoromethylornithine inhibition, *J. Biol. Chem.* **262:**8721–8727.

Poulin, R., Lu, L., Ackermann, B., Bey, P., and Pegg, A. E., 1992, Mechanism of the irreversible inactivation of mouse ornithine decarboxylase by alpha-difluromethylornithine. Characterization of sequences at the inhibitor and coenzyme binding sites. *J. Biol. Chem.* **267:**150–158.

Rogers, S., Wells, R., and Rechsteiner, M., 1986, Amino acid sequences common to rapidly degraded proteins: The PEST hypothesis, *Science* **234:**364–368.

Rosenberg-Hasson, Y., Bercovich, Z., Ciechanover, A., and Kahana, C., 1989, Degradation of ornithine decarboxylase in mammalian cells is ATP dependent but ubiquitin independent, *Eur. J. Biochem.* **185:**469–474.

Russell, D. H., and Snyder, S. H., 1969, Amine synthesis in regenerating rat liver: Extremely rapid turnover of ornithine decarboxylase, *Mol. Pharmacol.* **5:**253–262.

Salzberg, A., Golden, K., Bodmer, R., and Bellen, H., 1996, gutfeeling, a Drosophila gene encoding an antizyme-like protein, is required for late differentiation of neurons and muscles, *Genetics* **144:**183–196.

Seely, J. E., and Pegg, A. E., 1983, Effect of 1,3-diaminopropane on ornithine decarboxylase enzyme protein in thioacetamide-treated rat liver, *Biochem. J.* **216:**710–707.

Steglich, C., and Scheffler, I. E., 1982, An ornithine decarboxylase-deficient mutant of Chinese hamster ovary cells, *J. Biol. Chem.* **257:**4603–4609.

Svensson, F., and Persson, L., 1995, Regulation of mammalian S-adenosylmethionine decarboxylase as studied in a transient expression system, *Biochim. Biophys. Acta* **1260:**21–26.

Tabor, C. W., and Tabor, H., 1984, Polyamines, *Annu. Rev. Biochem.* **53:**749–790.

Tobias, K. E., Mamroud-Kidron, E., and Kahana, C., 1993, Gly 387 of murine ornithine decarboxylase is essential for the formation of stable homodimers, *Eur. J. Biochem.* **218:**245–250.

Tokunaga, F., Goto, T., Koide, T., Murakami, Y., Hayashi, S. I., Tamura, T., Tanaka, K., and Ichihara, A,. 1994, ATP- and antizyme-dependent endoproteolysis of ornithine decarboxylase to oligopeptides by the 26S proteasome, *J. Biol. Chem.* **269:**17382–17385.

Tsirka, S., and Coffino, P., 1992, Dominant negative mutants of ornithine decarboxylase, *J. Biol. Chem.* **267:**23057–23062.

van Daalen Wetters, T., Macrae, M., Brabant, M., Sittler, A., and Coffino, P., 1989, Polyamine-mediated regulation of mouse ornithine decarboxylase is posttranslational, *Mol. Cell. Biol.* **9:** 5484–5490.

Watanabe, S., Kusama-Eguchi, K., Kobayashi, H., and Igarashi, K., 1991, Estimation of polyamine binding to macromolecules and ATP in bovine lymphocytes and rat liver, *J. Biol. Chem.* **266:** 20803–20809.

CHAPTER 15

Ubiquitin and the Molecular Pathology of Human Disease

R. John Mayer, Michael Landon, and James Lowe

1. INTRODUCTION

The role of ubiquitin in cell stress and its central involvement in eliminating short-lived, abnormal or damaged proteins make it an important molecule for investigations of pathological cellular processes. Several areas of human pathology have been linked with the ubiquitin system:

- Chronic degenerative diseases of the nervous system and muscle
- Acute cellular injury associated with a cell stress response
- Cell atrophy
- Programmed cell death
- Tumor biology and turnover of oncogene products
- Processing of peptides for antigen presentation to the immune system

This overview sets out to catalogue and discuss the pathological cellular processes that involve ubiquitin and the possible role of ubiquitin in human

R. John Mayer and Michael Landon • Department of Biochemistry, University of Nottingham Medical School, Queen's Medical Centre, Nottingham NG7 2UH, United Kingdom. **James Lowe** • Department of Clinical Laboratory Sciences, University of Nottingham Medical School, Queen's Medical Centre, Nottingham NG7 2UH, United Kingdom.

Ubiquitin and the Biology of the Cell, edited by Peters *et al.* Plenum Press, New York, 1998.

disease. The role of the ubiquitin system in relation to tumor biology and the immune system is discussed elsewhere in this volume.

1.1. Disorders Involving the Ubiquitin System

The main human chronic degenerative disease mostly involve the nervous system, are very common, and include Alzheimer's disease, dementia with Lewy bodies (also called diffuse Lewy body disease, Lewy variant of Alzheimer's disease, senile dementia of the Lewy body type or Lewy body dementia), Parkinson's disease, amyotrophic lateral sclerosis (also called Lou Gehrig's disease or motor neuron disease), and Creutzfeldt–Jakob disease (a prion encephalopathy). In most of these disorders, insights into pathogenesis have been derived from studies of abnormal proteins that accumulate, as part of the disease, either within nerve cells or extracellularly. Proteins that accumulate within cells generally form aggregates termed *inclusion bodies* and these have characteristic appearances on histological examination. With emerging detailed knowledge of protein abnormalities in these diseases, it has become important to understand the mechanisms underlying the association of abnormal protein processing with disease. Ubiquitin appears to be incorporated into many of the inclusion bodies that characterize neurodegenerative diseases and investigation of the roles of ubiquitin is thus of great importance in extending our understanding of these disorders (Mayer *et al.*, 1991).

Reduction in the functional cell mass of a tissue, termed *atrophy*, is a very common response induced in many disease processes. It has become apparent that although several protein catabolic systems may be involved, the ubiquitin system is particularly important in bringing about cellular atrophy. The ubiquitin system is also involved in programmed cell death, another process that leads to reduction in functional cell mass of a tissue or organ. Finally, the ubiquitin system has been shown to have a role in the immediate response to acute cell damage during ischemia and oxidative stress.

1.2. Probes Used in the Study of Ubiquitin in Human Disease

In the past 10 years, antibodies to ubiquitin have become essential tools for the investigation of human disease. The accumulation of abnormal proteins in cells as inclusion bodies has long been recognized by histologists and is used to diagnose specific diseases. Immunohistochemical techniques have revolutionized the understanding of such inclusion bodies; the constituent proteins are now largely characterized and a hallmark is that many are ubiquitinated. With the development of new immunohistochemical probes for ubiquitin, enzymes of the ubiquitination and deubiquitination pathways, and proteasome-related proteins, it is becoming possible to perform studies that yield clues for the probable biogenesis of such inclusions. Given that these inclusions are presently regarded as the defining feature of certain diseases, it is anticipated that further studies of the ubiquitin system will give insight into the pathogenesis of disease, with implications for therapy.

1.2.1. Antibodies to Ubiquitin

Different treatments of ubiquitin prior to conjugation to carrier molecules and alternative conjugation methods have been used to produce polyclonal and monoclonal antibodies of differing specificities to free ubiquitin or ubiquitin–protein conjugates. The most widely used rabbit polyclonal antibodies are those raised against SDS-treated ubiquitin attached to keyhole limpet hemocyanin (Haas and Bright, 1985) and most of the work described here in relation to the detection of ubiquitin–protein conjugates involves the use of such antibodies, which have a high affinity for ubiquitin conjugates and are much less sensitive in detecting free ubiquitin. There is a range of commercially available antisera that are equivalent in the detection of ubiquitin conjugates (e.g., as supplied by Dako, Copenhagen). Our experience is largely based on the Dako product, which is suitable for use on formalin-fixed, wax-embedded tissue sections and in Western blotting. Many of the monoclonal antibodies raised against native ubiquitin or synthetic peptides are poorly reactive against conjugates but have a high affinity for free ubiquitin. Certain monoclonal antibodies to ubiquitin can be used in immunoprecipitation reactions, for example as recently described in the detection of a novel type of ubiquitinated phospholipid in viral membranes (Guarino *et al.*, 1995).

1.2.2. Antibodies with Specificity for Ubiquitin Chains

A key unanswered question related to the accumulation of ubiquitinated inclusions in pathological conditions is whether persistence occurs because proteins have been monoubiquitinated as a stabilizing signal (c.f. histones), or whether proteins are multiubiquitinated, but resistant to degradation. Recently, antibodies have been raised that detect ubiquitin–ubiquitin branches in multiubiquitin chains but not free ubiquitin (Fujimuro *et al.*, 1994). In addition, a sandwich ELISA has been developed to measure intracellular levels of multiubiquitin chains; the assay shows no significant cross-reactivity for mono-, di-, and triubiquitin, or mono- and diubiquitinated proteins, thus allowing the study of multiubiquitin chains in cell extracts (Takada *et al.*, 1995).

1.2.3. Ubiquitin-Cross-Reactive Protein

Certain antibodies, particularly those raised against ubiquitin–protein conjugates, also detect ubiquitin-cross-reactive protein (UCRP), a 15-kDa protein that has sequence homology with ubiquitin (Ahrens *et al.*, 1990) and is inducible by interferon (Loeb and Haas, 1992). UCRP is a functional ubiquitin homologue that can form stable conjugates analogous to those formed by ubiquitin (Loeb and

Haas, 1992). Interferon-responsive cells have been shown to contain a pathway for UCRP ligation that is parallel with, but distinct from, that of ubiquitin (Narasimhan *et al.*, 1996). Cross-reactivity with UCRP must therefore be taken into account in studies using ubiquitin antibodies as a detection system (Lowe *et al.*, 1995). Antibodies to recombinant UCRP are available that do not detect ubiquitin (Loeb and Haas, 1994).

1.2.4. Antibodies to Components of the Ubiquitin Pathways

Although many components of ubiquitin-dependent metabolism have been characterized at the gene and protein level, relatively few good immunoreagents applicable to human tissues are available.

PGP9.5 is an abundant protein, now known to be a ubiquitin carboxyl-terminal hydrolase (Day *et al.*, 1990; Wilkinson *et al.*, 1989), that is found at highest levels in neuronal tissues, where it has been used as a neuronal phenotypic marker (Ermisch and Schweccheimer, 1995; Harris *et al.*, 1990; Wilson *et al.*, 1988; Thompson *et al.*, 1983). Immunohistochemical studies show that PGP9.5 is enriched in several ubiquitinated inclusion bodies (Lowe *et al.*, 1990b).

In work carried out in our laboratory, we have found immunoreactivity to ubiquitin-activating enzyme, E1, localized to neurofibrillary tangles (see Section 2.2) and in hippocampal neurons (Fig. 2e): In particular, tangles within neurons of the pre-α-cell clusters of the entorhinal cortex are E1 positive (Layfield *et al.*, 1994); it has been suggested that this area is the earliest site exhibiting Alzheimer pathology (Braak and Braak, 1991). This observation may become more relevant as our understanding of the process of the pathological progression of the disease unfolds.

A failure of the 26 S proteasome could be one reason for the accumulation of ubiquitin conjugates in neurodegenerative diseases. For example, *SEN3* encodes a 945-amino-acid protein that is a subunit of the 26 S proteasome: Cells expressing mutant Sen3 accumulate ubiquitin–protein conjugates and exhibit defects in degradation of ubiquitinated proteins (De Marini *et al.*, 1995). Exposure of cells from a mouse neuronal cell line to an inhibitor of the chymotrypsinlike activity of the proteasome also results in accumulation of ubiquitinated proteins (Figueiredo-Pereira *et al.*, 1994). Recent work indicates that there may be differential expression of regulatory subunits of the proteasome in human brain tissue affected by Alzheimer's disease (see Section 2.4.2).

2. UBIQUITIN IN CHRONIC DEGENERATIVE DISEASES

Study of involvement of the ubiquitin system in the chronic degenerative diseases has led to the view that there is a close association of ubiquitin–protein conjugates with the endosome–lysosome system and with cytoskeletal abnormalities in afflicted cells. Most of the diseases that comprise this category are neurodegenerative diseases, but some are diseases of other relatively long-lived cells such as skeletal muscle and hepatocytes.

The most consistent association of ubiquitin is with a diverse set of inclusion bodies in neuronal cells, a finding with great practical importance in histological identification using immunochemical techniques. Immunostaining to detect the presence of ubiquitin–protein conjugates is now a routine part of clinical neuropathological diagnostic practice.

2.1. Ubiquitin and the Normal Endosome–Lysosome System

Evidence from several sources associates ubiquitin with the endosome–lysosome system of normal cells. Ubiquitin–protein conjugates are found in the primary (azurophilic) lysosome-related granules but not in the secondary (specific) granules in polymorphonuclear neutrophils (Laszlo *et al.*, 1991a). Enrichment of ubiquitin–protein conjugates has also been shown in multivesicular bodies and light vacuoles with membrane complexes in insect Sf9 cells (Low *et al.*, 1995, 1993). The ubiquitin–protein conjugates are localized to electron-dense areas of the former, but associated with the membranes of the latter. It has been proposed that there may be two routes for entry of ubiquitin–protein conjugates into these organelles, one via the cell surface and one via primary lysosomes, i.e., ubiquitin may have a role in endosome–lysosome biogenesis by functioning in their formation or translocation of transport vesicles (containing lysosomal enzymes) from the Golgi (Low *et al.*, 1995). Immunogold electron microscopy shows that ubiquitin–protein conjugates are approximately 12-fold enriched in the lysosomal compartment of 3T3-L1 fibroblasts (Laszlo *et al.*, 1990). Further, treatment of cells with the cysteine protease inhibitors E-64 or leupeptin leads to an expansion of the lysosomal compartment and a consequent increase in the cellular content of ubiquitin–protein conjugates, with no change in the specific enrichment of ubiquitin–protein conjugates in this compartment. This has led to the proposal that certain ubiquitin–protein conjugates are normally degraded in the lysosomal system and suggests that protein ubiquitination may also be a signal for protein uptake into lysosomes (Okada *et al.*, 1994; Cavanagh *et al.*, 1993; Laszlo *et al.*, 1990). It has been suggested that the ubiquitin-activating enzyme, E1, is associated with maturation of autophagic vacuoles: CHO ts20 cells are unable to generate ubiquitin–protein conjugates at the nonpermissive temperature, because of heat inactivation of a mutant thermolabile E1. In these cells the formation and maturation of autophagosomes into autolysosomes is normal, but the maturation of autolysosomes into residual bodies is disrupted at the nonpermissive temperature (Lenk *et al.*, 1992; Schwartz *et al.*, 1992; Gropper *et al.*, 1991).

Recent studies demonstrate that ubiquitination of the cytosolic tails of singlespan transmembrane receptors leads to their degradation (Hurtley, 1996). Ligand binding to pheromone receptor in yeast induces ubiquitination of the receptor cytoplasmic tail; this ubiquitination is required for stimulated endocytosis of the receptor. On this basis, it has been suggested that ubiquitination brings about degradation of the receptor–ligand complex by stimulating endocytosis and causing degradation within the lysosome/vacuole (Hicke and Riezman, 1996). This notion fits with the involvement of the ubiquitin system in lysosome-related degradation, in addition to its known roles in targeting proteins for degradation via the proteasome.

2.2. Disease-Related Lysosomal Accumulation of Ubiquitin

Ubiquitin immunoreactivity has been noted in lysosome-related vesicles in association with several processes (Table I). Ubiquitin immunoreactivity can be seen in the normal aging brain in a variety of species as dotlike bodies in both cortex and white matter. Ultrastructural studies have shown that there are two types of ubiquitin-reactive dotlike bodies; one results from the dilatation of nerve axons that accumulate lysosomal dense bodies, and the other is caused by myelin degeneration (Migheli *et al.*, 1992; Dickson *et al.*, 1990b). Dotlike bodies that correspond to dystrophic neurites (abnormally dilated and swollen nerve cell processes—axons or dendrites; hence the term *neurites*) are detectable in the brains of young adults, increase in number with age, and are most numerous in the oldest brains (Dickson *et al.*, 1990b; Pappolla *et al.*, 1989). The reason for age-related accumulation of ubiquitinated proteins in nerve cell processes and myelin is not yet known.

In Alzheimer's disease a key pathological feature is extracellular accumulation of amyloid, made up of fragments (Aβ peptides) of a normal transmembrane protein [the Alzheimer amyloid precursor protein (APP)], in a variety of patterns termed *plaques* (Maury, 1995; Sisodia and Price, 1995; Trojanowski *et al.*, 1995). In addition, neuronal cells accumulate ubiquitinated tau (a microtubule-binding protein) within the cytosol in the form of bundles of filaments called neurofibrillary tangles and ubiquitin–protein conjugates within a membrane-limited compartment [where the nature of the ubiquitinated protein(s) is unknown] in a process termed *granulovacuolar degeneration*. Immunohistochemical staining of nerve cell processes around plaques demonstrates the presence of APP, as well as

Table I
Lysosomal Accumulations of Ubiquitin

Form of accumulation	Organ system	Associated process/disease
Dotlike bodies	Central nervous system	Aging
Dystrophic neurites	Central nervous system	Alzheimer's disease
Dotlike bodies	Central nervous system	Prion disease
Rimmed vacuoles	Skeletal muscle	Inclusion body myositis

ubiquitin, in lysosome-related dense bodies (Yasuhara *et al.*, 1994; Dickson *et al.*, 1992, 1990b, 1989; Cochran *et al.*, 1991; Cras *et al.*, 1991).

The prion encephalopathies are associated with accumulation in the nervous system of a conformationally abnormal form of the prion protein, which leads to nerve cell death (DeArmond and Prusiner, 1995a,b; Goldfarb and Brown, 1995). The most important of these diseases are Creutzfeldt–Jakob disease in man, scrapie in sheep, and bovine spongiform encephalopathy in cattle. A histological characteristic of the affected brain tissue is the development of small vacuoles in affected areas, which were earlier termed *spongiform encephalopathy*. Disease can be transmitted by inoculation of prion proteins into the brains of experimental animals (DeArmond and Prusiner, 1995; Goldfarb and Brown, 1995).

Immunostaining brain tissue from scrapie-infected mice reveals ubiquitin immunoreactivity in lysosome-related vesicles. In experimentally induced disease in mice, this is detectable very early in the course of disease (Laszlo *et al.*, 1992; Lowe *et al.*, 1992, 1990b). Similar staining is seen in both the genetic and sporadic forms of human prion disease (Ironside *et al.*, 1993; Migheli *et al.*, 1991). Expression of the polyubiquitin C gene is increased 1.8- to 2.5-fold in the brains of scrapie-infected mice in the later stages of disease progression, with similar increases in the expression of an hsp 70 gene (Kenward *et al.*, 1994). The disease-related abnormal form of the prion protein accumulates in late endosomes (containing the cation-independent mannose 6-phosphate receptor) together with ubiquitin–protein conjugates (Arnold *et al.*, 1995). It has been proposed that conversion of normal to abnormal prion protein takes place in the endosome–lysosome system and that interference with protein processing in lysosomes might be a candidate target for therapy in these diseases (Mayer *et al.*, 1992).

Inclusion body myositis is a recently described disease of skeletal muscle that is characterized by insidious degeneration of muscle fibers leading to progressive weakness and death. Pathology of the affected muscle shows development of autophagic vacuoles (rimmed vacuoles) associated with intracellular filamentous inclusions (Askanas *et al.*, 1994). Immunohistochemical studies have shown that the rimmed vacuoles contain ubiquitin as well as the prion protein (Askanas and Engel, 1995; Askanas *et al.*, 1994, 1992, 1991; Albrecht and Bilbao, 1993; Leclerc *et al.*, 1993); however, the presence of prion protein immunoreactivity has been disputed (Sherriff *et al.*, 1995). A similar phenomenon can be induced in rat muscle by administration of chloroquine presumably as a result of the expansion of the endosome–lysosome system by this weak base (Tsuzuki *et al.*, 1995).

2.3. Ubiquitin and the Role of Endosomes and Lysosomes in Disease Pathogenesis

Several examples of the immunohistochemical and immunogold electron microscopical localization of ubiquitin immunoreactivity in endosomes and lysosomes have been described in neurodegenerative diseases. Although it is beyond doubt that some ubiquitinated proteins become internalized into the lysosomal system, the details of the mechanism(s) for uptake of ubiquitinated proteins are still uncertain. It should be noted that ubiquitin may be relatively stable within the lysosome and that immunostaining in histological sections may be enhanced with ubiquitin antisera, because of the presence of partly degraded multiubiquitinated proteins (Wilkinson, 1994). Understanding the precise role of ubiquitin in lysosomal structure and function will be of special importance in prion diseases and Alzheimer's disease where generation of the abnormal form of the prion protein or amyloidogenic fragments of APP may be taking place in the acid-vesicle system.

2.4. Inclusion Bodies in Neurodegenerative Diseases

Inclusion bodies are abnormal structures that develop within neurons in a variety of degenerative diseases, appearing to be linked to the pathogenesis of disease. Insight into inclusion body biogenesis has come from histological, ultrastructural, and immunohistochemical studies. Although the inclusions are composed of filament-forming proteins in association with other proteins, the low volume fraction of inclusions in diseased tissue has in most cases precluded their isolation and characterization by biochemical techniques.

2.4.1. Intermediate Filament/αB Crystallin/Ubiquitin Inclusions

There is a family of related inclusion bodies based on the common features of aggregation of intermediate filaments in association with ubiquitin, αB crystallin, certain enzymes of the ubiquitin-dependent proteolytic pathway, and proteasomes (Table II) (Lowe *et al.*, 1993, 1992b; Lowe and Mayer, 1990, 1989; Manetto *et al.*, 1989). For diagnostic pathologists, an important spinoff has been that immunochemical detection of ubiquitin can be used to locate the intraneuronal inclusions, called *cortical Lewy bodies*, which are based on neurofilament proteins (Lennox *et al.*, 1989). Dementia with Lewy bodies is now recognized as the second commonest cause of dementia (Crystal *et al.*, 1990; Hansen *et al.*, 1990, 1989; Perry *et al.*, 1990; Byrne *et al.*, 1989; Gibb *et al.*, 1985).

The concept that ubiquitination may target cytosolic proteins for degradation has led to suggestions that isolating inclusions and characterizing the ubiquitinated proteins might reveal the cellular systems that are damaged in these diseases. Most attempts to perform these experiments have been largely unsuccessful because of an inability to separate the inclusions, which are commonly few in number in the tissue, from other cellular materials, some of which are also ubiquitin-containing structures. One inclusion found in astrocytes and based on aggregated glial fibrillary acidic protein, termed the *Rosenthal fiber*, has been isolated from brain tissue in the rare disorder Alexander's disease; investigation of highly enriched inclusion material has shown that αB crystallin is ubiquitinated (Goldman and Corbin, 1991). It has been proposed that ubiquitination of αB crystallin might be a common factor in each of the intermediate filament/ubiquitin

Table II
Inclusion Bodies Associated with Ubiquitin Immunoreactivity

Inclusion	Main associated disease	Major constituent proteins	Cell type and subcellular location	References (where not in text)
Ubiquitin/intermediate filament inclusions				
Lewy body	Parkinson's disease	Neurofilament protein, αB crystallin, proteasomes	Neuron cytoplasm	
Cortical Lewy body	Dementia with Lewy bodies	Neurofilament protein, αB crystallin, proteasomes	Neuron cytoplasm	
Rosenthal fiber	Alexander's disease (degenerative disease of childhood)	Glial fibrillary acidic protein, αB crystallin	Astrocyte cytoplasm	
Cytoplasmic body	Myopathy	Desmin, αB crystallin	Skeletal muscle cell cytoplasm	
Mallory body	Alcoholic liver disease	Cytokeratin, αB crystallin	Hepatocyte cytoplasm	
Crook's hyaline	Cushing's disease	Cytokeratin	Endocrine cell cytoplasm	
Hyaline inclusions	Lichen amyloidosis (skin disease)	Cytokeratin	Skin epithelium cell cytoplasm	
Tau-associated inclusions				
Glial cytoplasmic inclusions	Multiple system atrophy (degenerative disease)	Tau, αB crystallin	Oligodendrocyte cytoplasm	
Neurofibrillary tangle	Alzheimer's disease Progressive supranuclear palsy	Tau protein	Neuron cytoplasm	
Pick body	Pick's disease	Tau protein, neurofilament protein	Neuron cytoplasm	Murayama *et al.* (1990b), Love *et al.* (1988), Lowe *et al.* (1988a)
Inclusions of uncertain nature				
Skein inclusions	Amyotrophic lateral sclerosis	Unknown (filamentous ultrastructure)	Neuron cytoplasm	
Marinesco bodies	Aging	Unknown (filamentous ultrastructure)	Neuron nucleus	Dickson *et al.* (1990a), Funata *et al.* (1990)
Hyaline inclusions	Rare neurodegenerative diseases	Unknown (filamentous ultrastructure)	Neuron nucleus	

inclusions, but this remains unproven (Lowe *et al.*, 1992b). Recently, ubiquitination of αB crystallin has also been shown for a related ubiquitinated structure in glial cells, the glial cytoplasmic inclusion (seen in multiple system atrophy), which is based on accumulation of the microtubule-associated protein tau (Tamaoka *et al.*, 1995).

The role of αB crystallin in this type of inclusion is likely to be structural and part of a pathological mechanism for filament aggregation in inclusion biogenesis. The α crystallins are homologous to the small heat shock proteins and have been shown to be molecular chaperones, capable of preventing the heat-induced aggregation of proteins. They are also involved in intermediate filament assembly and remodeling (Nicholl and Quinlan, 1994). In contrast to several other cell stress proteins, α crystallins do not seem to protect proteins from aggregation during refolding reactions and may only recognize nonnative intermediates that are formed on a denaturation pathway, having little or no affinity for intermediates formed in refolding pathways (Das and Surewicz, 1995). In the context of intermediate filament inclusions, ubiquitination of αB crystallin might relate to the elimination of chaperone-bound denatured proteins. Alternatively, αB crystallin may protect cells against cell death by facilitating the removal of reactive oxygen species (Mehlen *et al.*, 1996).

We have previously suggested, based on certain common features such as structural organization, association with endosome- and lysosome-related organelles, and similarities with cell-stress-related phenomena in other cells, e.g., in Epstein–Barr virus-transformed lymphoblastoid cells (Laszlo *et al.*, 1991b), that this type of inclusion is the morphological manifestation of a cell stress response (Lowe *et al.*, 1993). The dynamic nature of such inclusions is suggested by finding immunoreactivity for the ubiquitin C-terminal hydrolase, PGP9.5 (Lowe *et al.*, 1990b), and proteasomes (Masaki *et al.*, 1994; Ito *et al.*, 1991; Kwak *et al.*, 1991) within inclusions.

The formation of ubiquitin/intermediate filament/αB crystallin inclusions is probably a common cellular response to chronic sublethal injury and possibly cytoprotective (Lowe *et al.*, 1993, 1990a, 1988a; Mayer *et al.*, 1991, 1989). This type of inclusion can be experimentally induced; Mallory bodies form in hepatocytes in the livers of mice fed a diet containing griseofulvin, and this has been used in our laboratory as a model in which the relationship between the formation of a ubiquitinated intermediate-filament-containing cytoplasmic inclusion and the expression of ubiquitin genes can be explored. Histological observation of the presence of drug-induced hyaline intermediate filament inclusions coincides with the detection of ubiquitin immunoreactivity in the inclusions. Ubiquitin immunocytochemistry generally reveals intracellular inclusions with an amorphous core surrounded by a halo of intense ubiquitin immunoreactivity (Fig. 1b), corresponding to the ultrastructural distribution of the intermediate filaments. Immunogold electron microscopy with an antibody to ubiquitin–protein conjugates reveals a core zone to inclusions, composed of thick filaments labeled by relatively few gold particles, merging with peripheral 10-nm intermediate filaments associated with numerous gold particles (Fig. 1c). The inclusions are circumscribed by dense bodies reminiscent of lysosomes. Northern analysis of RNA prepared from the livers of griseofulvin and control animals detects increased expression of ubiquitin genes, particularly the Ub C polyubiquitin gene, in livers containing Mallory bodies (Fig. 1a).

2.4.2. Neuronal and Glial tau-Containing Inclusion Bodies

Several neurodegenerative diseases are characterized by abnormal neuronal accumulation of tau, a normal microtubule binding protein (Table II). The most important neuronal inclusions to be recognized in association with ubiquitin are the tau-containing neurofibrillary tangles of Alzheimer's disease (Cole and Timiras, 1987; Mori *et al.*, 1987; Perry *et al.*, 1987), in which abnormally hyperphosphorylated tau (Grundke-Iqbal *et al.*, 1986) accumulates inside neurons, prior to ubiquitination (Bancher *et al.*, 1989). Tau is then predominantly monoubiquitinated in its microtubule-binding region (Morishima-Kawashima and Ihara, 1994). The amyloid that also accumulates in Alzheimer's disease is derived from Aβ peptide, which has been found to decrease ubiquitin-dependent protein degradation *in vitro* by selectively inhibiting the chymotrypsinlike activity of the proteasome. This has been proposed as a possible mechanism for the intracellular accumulation of ubiquitinated tau in Alzheimer's disease (Gregori *et al.*, 1995).

In contrast to the intermediate-filament-containing inclusions, the vast majority of neurofibrillary tangles and Pick bodies contain neither the ubiquitin C-terminal hydrolase, PGP9.5, nor αB crystallin. A small subset of tangles do show immunoreactivity for the hydrolase, suggesting that ubiquitination/deubiquitination (see Section 2.6) may be taking place only at certain phases of inclusion body biogenesis, after which tangles may become permanently ubiquitinated (Lowe *et al.*, 1992b, 1990b). As noted above, in the disease termed *multiple system atrophy*, some glial cells also accumulate tau protein in the form of inclusions (Papp and Lantos, 1992; Papp *et al.*, 1989), which additionally contain ubiquitinated αB crystallin (Tamaoka *et al.*, 1995).

Components of the 20 S proteasome have been localized to certain types of neurofibrillary tangles as well as cortical Lewy bodies (see Section 2.4.1) (Masaki *et al.*, 1994; Ito *et al.*, 1991; Kwak *et al.*, 1991), and work has been performed to extend these observations with a range of antisera that detect regulatory components of the 26 S proteasome. Of particular interest are results obtained with an antiserum reactive against a regulatory ATPase (designated subunit 6 or TBP7), which immunostains all of the pathological features of Alzheimer's disease that involve abnormal tau protein (Fergusson *et al.*, 1996). Thus, neurofibrillary tan-

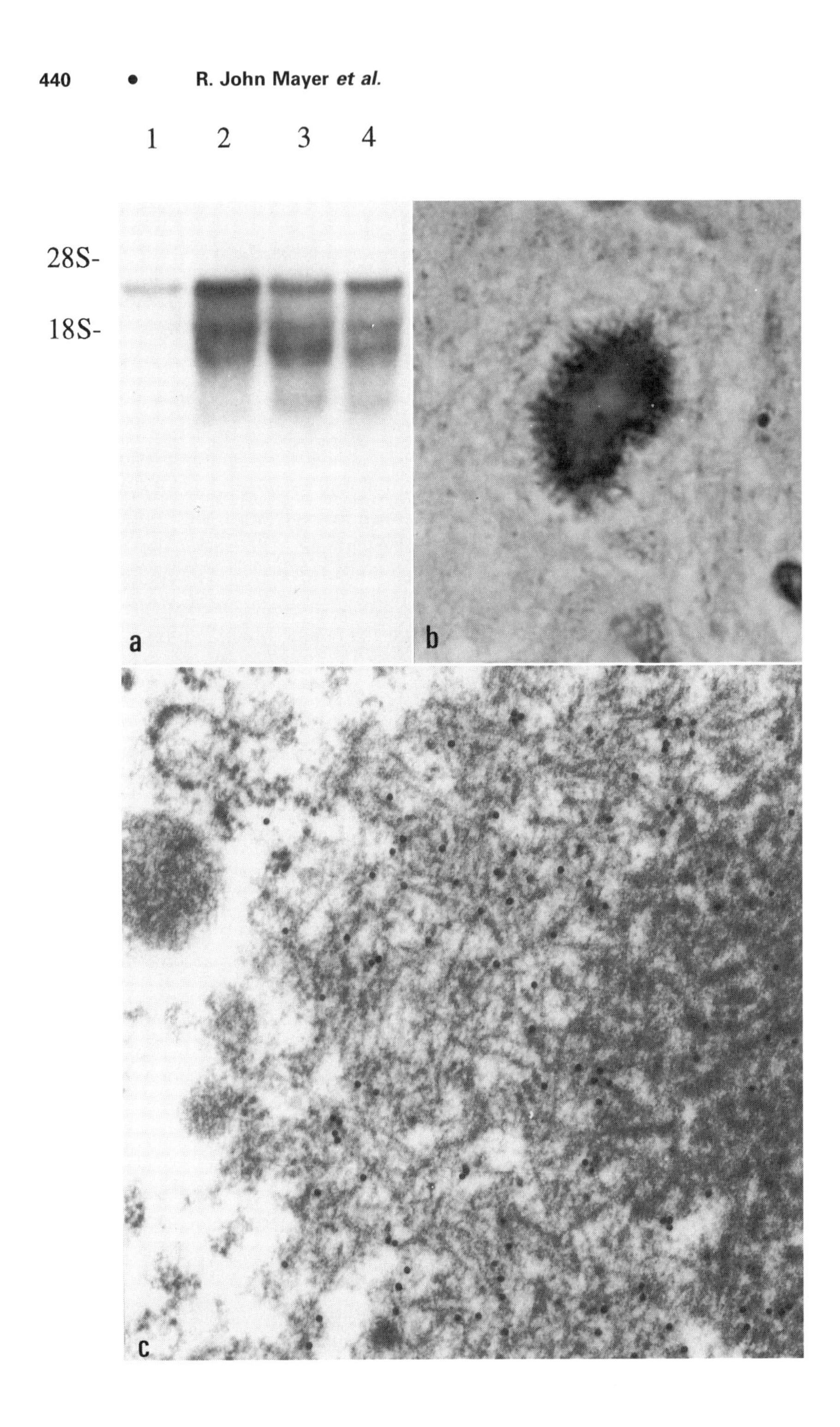
1
2
3
4
28S-
18S-
a
b
c

gles, plaque neurites, and neuropil threads are all positive for this ATPase (Fig. 2a,d), as are regions of cortical Lewy bodies (Fig. 2b,c), structures that are not visualized with antitau. These unexpected findings suggest that further investigation of relationships between proteasomes and neurodegenerative diseases might be fruitful.

2.4.3. Neuronal Filamentous Inclusions in Amyotrophic Lateral Sclerosis

Amyotrophic lateral sclerosis is a disease of unknown cause in which there is progressive death of motor neurons in the brain and spinal cord, leading to increasingly severe paralysis and typically causing death in 2–5 years (Lowe, 1994). A histological characteristic of the disease is the presence in motor neurons of ubiquitin-immunoreactive filamentous inclusions that are not easily detected by other techniques. In addition, increased levels of the polyubiquitin C-gene transcripts have been found in spinal anterior horn cells in this disease (Heggie *et al.*, 1989). The inclusions appear to be specific for both sporadic and familial amyotrophic lateral sclerosis (Schiffer *et al.*, 1991; Migheli *et al.*, 1990; Murayama *et al.*, 1990a; Lowe *et al.*, 1989, 1988b; Leigh *et al.*, 1988). The molecular nature of the filamentous inclusions in amyotrophic lateral sclerosis remains uncertain, as they have been found to be unreactive to a wide range of antisera to cytoskeletal proteins (Mather *et al.*, 1993). Inclusions with the same structural and immunohistochemical features also appear to be characteristic of frontal lobe dementia, an increasingly recognized form of dementing condition (Jackson and Lowe, 1996).

2.5. Ubiquitin Accumulation in Nerve Cell Processes

2.5.1. Neurites in Alzheimer's Disease, Dementia with Lewy Bodies, Amyotrophic Lateral Sclerosis, and Huntington's Disease

Several neurodegenerative diseases have been shown to have dystrophic neurites that are detected by immunohistochemistry for ubiquitin as well as APP (Table III) (Bacci *et al.*, 1994). In normal axons, ubiquitin, PGP9.5, and hsps 70 are transported with the slow component that is known to carry cytoskeletal and

←

Figure 1. (a) Northern analysis of RNA prepared from livers of control and griseofulvin-fed mice probed with a fragment of the human ubiquitin C gene. Control (lane 1); griseofulvin diet durations: 12 weeks (lane 2); 19 weeks (lane 3); 28 weeks (lane 4).

Mallory bodies from the liver of griseofulvin-fed mice. (b) Photomicrograph showing a halo of intense ubiquitin–protein conjugate immunoreactivity. This section was developed using immunoperoxidase/hematoxylin. (c) Immunogold electron micrograph showing a core zone composed of thick filaments merging with peripheral intermediate filaments. Immunostaining with anti-ubiquitin–protein conjugates shows a lesser number of gold particles associated with the core than the periphery.

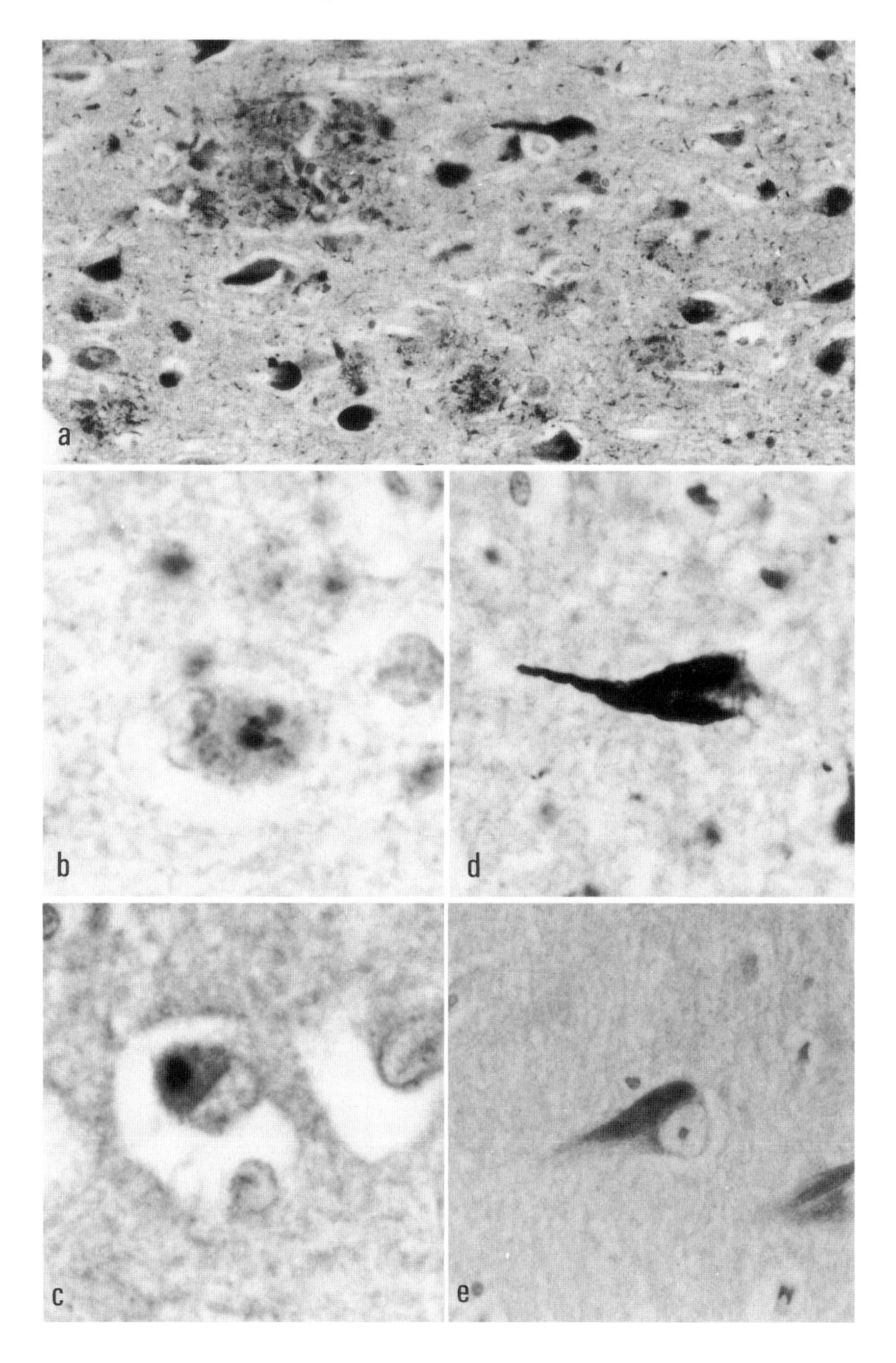
a
b
d
c
e

Table III
Accumulations of Ubiquitin in Nerve Cell Processes

Disease	Location	Association
Alzheimer's disease	Around amyloid plaques (dystrophic neurites)	Lysosome-related Complexed to tau protein
Dementia with Lewy bodies	Areas of neuronal loss (dystrophic neurites)	Associated with filamentous structures
Huntington's disease	Cerebral cortex (dystrophic neurites)	Associated with filamentous structures
Frontal dementia	Cerebral cortex (dystrophic neurites)	Uncertain
Trauma	At site of axon tearing (spheroids)	Lysosome-related

cytoplasmic proteins (Bizzi *et al.*, 1991) and therefore may accumulate in dystrophic neurites.

In Alzheimer's disease, in addition to the abnormal nerve cell processes surrounding amyloid plaques, there is widespread accumulation of tau protein associated with ubiquitin to form dilated nerve cell processes in the cerebral cortex in structures called *neuropil threads* (Perry *et al.*, 1991). Immunohistochemistry has shown that the ends of threads are devoid of ubiquitin immunoreactivity, suggesting that ubiquitination of tau protein is a late phenomenon (Iwatsubo *et al.*, 1992). It is widely believed that these accumulated protein aggregates in neuritic processes prevent normal axoplasmic transport and therefore contribute to cortical dysfunction (Schmidt *et al.*, 1993; Perry *et al.*, 1991).

A second example of abnormal neurites is the ubiquitin-immunoreactive swollen neuronal processes that are seen in the CA2–3 region of the hippocampus and in other subcortical nuclei in Parkinson's disease and dementia with Lewy bodies (Gai *et al.*, 1995; Kim *et al.*, 1995; Dickson *et al.*, 1994, 1991). Such neurites contain neither tau protein nor αB crystallin, but show immunoreactivity for the ubiquitin C-terminal hydrolase, PGP9.5; ultrastructural examination has shown

←

Figure 2. Immunoreactive lesions in brain sections with antisera: (a–d) raised against the *Manduca sexta* equivalent (the immunochemical reactivity of individual proteasomal ATPases is well conserved across species) of human S6/TBP7 ATPase component of the 19 S regulator of the 26 S proteasome (Dawson *et al.*, 1995); and (e) against the ubiquitin-activating enzyme, El (Layfield *et al.*, 1994). Development was with immunoperoxidase/hematoxylin. Hippocampal sections from Alzheimer's disease show: (a) positive staining of plaques, neurofibrillary tangles, and neuropil threads; (d, e) individual neurofibrillary tangles at higher power. Cingulate gyrus sections from cases of dementia with Lewy bodies showing regions of staining within cortical Lewy bodies (b, c).

that the neurites contain packed filaments that are believed to be altered neurofilaments (Gai *et al.*, 1995).

In amyotrophic lateral sclerosis associated with extramotor cortical involvement, abnormal neurites can be detected in superficial cortical layers using ubiquitin immunohistochemistry (J. Lowe and M. Jackson, unpublished observations). Increasingly, this extramotor involvement is becoming recognized as being associated with the development of frontal lobe dementia. The abnormal neurites cannot be detected by other conventional histological techniques (Tolnay and Probst, 1995). In Huntington's disease, a genetically determined disorder caused by expansion of a triplet repeat region in a gene coding for the protein huntingtin (Huntington's Disease Collaborative Research Group, 1993), there is degeneration of nerve cells in the caudate nucleus and putamen of the brain that is associated with loss of cerebral cortical volume (Vonsattel *et al.*, 1985). Recently, abnormal neurites demonstrable using ubiquitin immunohistochemistry have been found in the cerebral cortex of affected patients (Jackson *et al.*, 1995; Cammarata *et al.*, 1993). The nature of the underlying abnormality and any relationship with abnormal huntingtin remain uncertain.

Axonal accumulation of ubiquitinated material is not unique to these conditions, being seen in other disorders of the nervous system (Cochran *et al.*, 1991). The importance of these abnormalities is that they imply a severe disruption to the cytoskeleton in nerve cell processes, preventing normal axoplasmic flow and thus interfering with nerve cell function.

2.5.2. Traumatic Damage to Axons and Axonal Spheroids

Normal axons do not exhibit ubiquitin-conjugation immunoreactivity but show very high levels of the protein PGP9.5 (Wilkinson *et al.*, 1989; Wilson *et al.*, 1988; Thompson *et al.*, 1983). The functional significance of this abundant axonal, ubiquitin C-terminal hydrolase remains unexplained. Following traumatic severance of nerves and in some neurodegenerative diseases, axons develop swelling of their proximal terminal portion, forming *spheroids*. Axonal spheroids are also a feature of rare inherited degenerative diseases of the nervous system, ubiquitin immunoreactivity being seen in these structures in both Hallervorden–Spatz disease and infantile neuroaxonal dystrophy (Malandrini *et al.*, 1995). Such spheroids contain microtubules, abundant neurofilament protein, synaptic vesicles, mitochondria, and numerous lysosome-related bodies that accumulate in the dilated end, probably as a result of impaired axonal transport systems. Immunohistochemical studies indicate that spheroids, in addition to being immunoreactive for ubiquitin, exhibit high levels of PGP9.5 (Li and Farooque, 1996; Schweitzer *et al.*, 1993); it is likely that much of the former immunoreactivity may be associated with lysosome-related vesicles. Recent studies have proposed that the ubiquitin system may be involved in degradation of neurofilaments: antiubiquitin labels the neurofilament M subunit and, to a lesser degree, the two other subunits of the neurofilament triplet; experiments also show that addition of ATP and ubiquitin can stimulate neurofilament proteolysis by crude soluble fractions from nervous tissues (Gou and Leterrier, 1995).

In experimental crush injury to rat sciatic nerves, there is a 10-fold increase in the posttranslational arginylation of proteins; protein N-terminal arginylation leads to ubiquitination and proteolysis (Jack *et al.*, 1992). Work suggests that proteins that are arginylated following injury to axons become candidates for ubiquitin-mediated proteolysis (Chakraborty and Ingoglia, 1993; see also Chapter 8).

2.6. Proposed Generalized Role for Ubiquitin in Neurodegenerative Disease

In certain forms of axonal pathology, notably where the axon becomes ballooned and there is continued axonal transport, ubiquitin immunoreactivity appears to be related to the accumulation of lysosome-related dense bodies. In dementia with Lewy bodies, Huntington's disease, and amyotrophic lateral sclerosis, dilated neuronal processes become filled with ubiquitin-reactive material that appears filamentous. Although not seen with several antibodies to neurofilament proteins, it still seems likely that these changes are associated with alteration and ubiquitination of neurofilaments or associated proteins; ubiquitination of neurofilament protein has been suggested in one study (Gou and Leterrier, 1995). Given the abundance of PGP9.5, it is still uncertain why ubiquitin–protein conjugates accumulate in this manner, unless this hydrolase is specific for other types of conjugates. It is possible that the development of focal accumulation of ubiquitinated cytoskeletal proteins is the result of impaired PGP9.5 activity, but there is as yet no direct evidence for this proposal.

Finally, there is a more general question on the role of the ubiquitin system in neurodegeneration. While ubiquitin was only thought of as a cofactor for nonlysosomal proteolysis, there was good reason to focus on ubiquitinated proteins in filamentous inclusions. For example, some abnormality, malfunction, overload, or substrate recognition problem of the ubiquitin–proteasome system could result in the accumulation of ubiquitinated proteins in filamentous inclusions. However, now that there is increasing evidence that protein ubiquitination is involved in the endocytic process and that proteasomes may degrade the cytosolic tails of single-span membrane proteins and cytosolic loops of multispan membrane proteins, it is reasonable to believe that the ubiquitin–proteasome system may have a more general role in neurodegeneration, i.e., in the process of amyloid biogenesis in the endocytic pathway. Further progress will rely on the production of transgenic animal models with deletions and mutations in components of the ubiquitin–26 S proteasome pathway of intracellular proteolysis targeted, with appropriate promoters, e.g., nerve-specific enolase to the nervous system.

3. UBIQUITIN IN ACUTE CELL INJURY: THE CELL STRESS RESPONSE

In addition to involvement in the chronic neurodegenerative diseases, ubiquitin is involved in the response to acute cell injury. Transcription of ubiquitin mRNA is induced by heat shock and other stresses in mammalian cells and ubiquitin is therefore a cell stress protein (Bond and Schlesinger, 1986, 1985). Stress induction of polyubiquitin genes is also important for cell survival in yeast (Finley *et al.*, 1987). In human cells, heat stress induces the 2.5-kb Ub C and 1.0-kb Ub B transcripts, with coordinate induction of hsp 27 mRNA and hsp 70 mRNA (Fornace *et al.*, 1989). There are suggested roles for ubiquitin both during the immediate stress response and in recovery (Bond *et al.*, 1988; Parag *et al.*, 1987). The conservation of this response is a reflection of the need for degradation of damaged proteins in cells that have undergone stress. One proposal is that the rate of macromolecular degradation in stressed cells is the key determinant that sets an upper time limit for survival during chronic environmental stress (Hand and Hardewig, 1996).

Although it might be expected that overexpression of ubiquitin should always confer additional protection against cellular damage, for certain stimuli this has been shown not to be the case. In yeast, ubiquitin overexpression markedly increases tolerance to the amino acid analogue canavanine, marginally increases resistance to ethanol and osmotic stress, but has no effect on tolerance to heat, and is associated with reduced tolerance to arsenite, cadmium, and paromomycin (Chen and Piper, 1995).

3.1. Ischemia

Cerebrovascular disease is one of the most important contributors to morbidity and mortality in Western societies. There is great interest in understanding both the selective vulnerability of neurons to ischemic damage and also mechanisms that might be manipulated to afford cytoprotection.

Hippocampal neurons are extremely vulnerable to short periods of ischemia; affected cells develop delayed neuronal necrosis. After neuronal ischemic damage, there is a substantial selective increase in ubiquitin conjugates (Hayashi *et al.*, 1993), which is also seen after experimentally induced excitotoxic damage by the glutamate agonist NMDA (Yee *et al.*, 1993). In experimental models of cerebral ischemia, cells that recover exhibit increased ubiquitin immunoreactivity over a 72-h period (Magnusson and Wieloch, 1989). Changes in cellular ubiquitin with ischemia have been investigated in gerbil hippocampal neurons, using a panel of antibodies to ubiquitin, with reactivities either for free ubiquitin or ubiquitin–protein conjugates. A short period of ischemia causes depletion of free ubiquitin in those neurons that are most vulnerable to ischemic damage; this reduction may be the result of an inability to degrade ubiquitin conjugates already formed or a failure of *de novo* ubiquitin synthesis (Morimoto *et al.*, 1996).

Ubiquitin gene expression following ischemia/reperfusion has been studied in rat brain and is initially decreased after reperfusion but then increases before returning to normal levels after 24–48 h of restoration of blood flow; Ub C polyubiquitin gene expression is significantly upregulated, suggesting that this might be useful as an indicator of ischemic stress (Noga and Hayashi, 1996). Degradation of proteins via the ubiquitin system appears to be part of the mechanism of neuronal recovery following cerebral ischemia (Dewar *et al.*, 1994; Hayashi *et al.*, 1991).

The activities of both 20 and 26 S proteasomes have been examined in gerbil cerebral cortex after transient ischemia and following restoration of blood supply. Results in this model suggest that proteasome activity is not irreversibly impaired after the ischemic episode and that transient inhibition of ATP-dependent conversion of 20 S to 26 S proteasomes might underlie the accumulation of ubiquitin–protein conjugates in the early reperfusion period (Kamikubo and Hayashi, 1996).

3.2. Oxidative Stress

Oxidative stress has been implicated in several human disease processes, particularly in the etiology of neurodegenerative diseases (Agundez *et al.*, 1995; Dexter *et al.*, 1994; Tsuda *et al.*, 1994; Bowling *et al.*, 1993; Fahn and Cohen, 1992; Jenner *et al.*, 1992; Olanow, 1990), where it has been postulated that oxidative damage to proteins may be caused by exposure of genetically susceptible individuals to environmental agents. There are several lines of evidence that point to a role for the ubiquitin system in the response to oxidative stress. Polyubiquitin gene expression contributes to oxidative stress resistance in *Saccharomyces cerevisiae* (Cheng *et al.*, 1994). In mouse-derived C1300 N2A neuroblastoma cells, both heat shock and exposure to iron result in increased levels of mRNA coding for the inducible form of hsp 70 and for ubiquitin, suggesting that there is a cytoprotective cell stress response to iron-mediated cell damage (Uney *et al.*, 1993). Exposure of bovine lens epithelial cells to oxidative stress brings about inactivation of the ubiquitin conjugating system, with coordinate reduction of proteolysis; during recovery from stress, enhancement of ubiquitin conjugation and proteolytic activities appears to be important in the restoration of normal function in these cells (Shang and Taylor, 1995). In view of experiments in yeast on levels of ubiquitin and resistance to oxidative stress (Cheng *et al.*, 1994), any human genetic polymorphism that affects the levels of ubiquitin in cells would be a candidate molecular risk factor for increased susceptibility to environmental oxidative stress and hence of great interest in relation to particular neurodegenerative diseases.

Two important human diseases have been particularly related to abnormal susceptibility and response to oxidative stress, namely, amyotrophic lateral sclerosis and Parkinson's disease. In amyotrophic lateral sclerosis, great interest has been shown in the role of mutant cytosolic Cu/Zn-binding superoxide dismutase (SOD1), a homodimeric metalloenzyme that catalyzes dismutation of the toxic superoxide anion O_2^- to O_2 and H_2O_2 (Wong *et al.*, 1995; Jones *et al.*, 1994; Rosen *et al.*, 1993). As previously mentioned, one of the characteristics of this disease is accumulation of ubiquitinated filamentous inclusions in affected neurons (Lowe *et al.*, 1989, 1988; Leigh *et al.*, 1988a), and there is a twofold increase in the expression of the Ub C polyubiquitin gene in motor cortex and spinal cord from patients with this disease compared with control cases (Heggie *et al.*, 1989). These features suggest that the increased transcription of this stress-inducible polyubiquitin gene is related to the generation of ubiquitinated inclusion bodies and, in some familial cases, might be related to expression of mutant SOD1.

Several studies suggest a role for oxidant stress in the etiology of Parkinson's disease. There are increased levels of the early components of the peroxidation chain in brains of patients with Parkinson's disease, suggesting continued free radical generation (Dexter *et al.*, 1994). The activity of the reduced form of nicotinamide adenine dinucleotide coenzyme Q reductase (complex I) is reduced in the main brain area affected in patients dying of Parkinson's disease; this may be related to a genetic abnormality of complex I acting as a susceptibility factor for the development of the disease (Schapira, 1994, 1993). A key feature of Parkinson's disease is the development of ubiquitin-immunoreactive inclusions (Lewy bodies) in affected neurons (Lowe *et al.*, 1988a).

4. UBIQUITIN INVOLVEMENT IN CELL ATROPHY AND PROGRAMMED CELL DEATH

Cellular atrophy can be caused by a variety of processes, most often those that result in removal of a trophic factor. A characteristic of cellular atrophy is protein degradation and reduction in cell size leading to abnormal cellular function. There is currently great interest in the control of atrophy and in devising strategies that might be used to slow or prevent protein degradation. The ubiquitin-mediated degradation system is the major cytosolic mechanism to eliminate proteins from cells and may, therefore, have a key role in atrophy. Most studies have investigated mechanisms involved in atrophy of skeletal muscle.

4.1. Skeletal Muscle Atrophy

Although the most common causes of skeletal muscle atrophy are disuse and denervation, systemic metabolic disturbances, notably glucocorticoid excess, systemic acidosis, starvation, conditions in which there are elevated levels of cytokines such as tumor necrosis factor α, can also cause atrophy. In such pathological situations, rapid atrophy of skeletal muscles is pictured as largely the result of an increased rate of protein degradation mediated by the ubiquitin system, for blocking the lysosomal or the Ca^{2+}-dependent proteolytic pathways does not prevent this breakdown (Medina *et al.*, 1991). Exercise-induced damage (Thompson and Scordilis, 1994) and systemic sepsis are also able to selectively induce ubiquitin-mediated proteolysis of muscle (Voisin *et al.*, 1996; Hasselgren, 1995; Tiao *et al.*, 1994). In contrast to other model systems, burn injury in rats stimulates multiple proteolytic pathways in skeletal muscle, including the lysosomal, calcium-activated, and ubiquitin systems (Fang *et al.*, 1995). The observations from animal models have been supported by work on muscle from hospitalized patients with severe head trauma. Parameters of active muscle breakdown were mirrored by increased muscle mRNA levels for cathepsin D, m-calpain, ubiquitin, a ubiquitin carrier protein, and proteasome subunits indicating that multiple proteolytic mechanisms account for muscle wasting (Mansoor *et al.*, 1996).

As well as being induced in atrophy, ubiquitin is found in the Z-bands of normal mammalian skeletal and cardiac muscle, as well as in intercalated discs in the latter (Hilenski *et al.*, 1992), and in insect muscle, the protein arthrin is ubiquitinated actin (Ball *et al.*, 1987); however, the roles of ubiquitin at these sites remain uncertain at present.

In starvation, nonlysosomal ATP-dependent proteolysis has been shown to increase 2.5- to 4.5-fold in association with a parallel increase in polyubiquitin transcripts, whereas no change was detected in the level of mRNAs for enzymes of other proteolytic systems, such as cathepsin L, cathepsin D, and calpain-1. As expected, the muscle content of ubiquitin and ubiquitin–protein conjugates also becomes greater, with the ubiquitin concentration increasing by 60–90%; similar changes are seen following denervation (Medina *et al.*, 1995, 1991). It appears that, in starvation, glucocorticoids are essential for enhanced muscle protein breakdown (Wing and Goldberg, 1993), a feature also exhibited in muscle atrophy associated with metabolic acidosis (Mitch, 1995; Mitch *et al.*, 1994; Price *et al.*, 1994) or following sepsis (Tiao *et al.*, 1996). *In vitro*, acidification alone can induce increased expression of mRNAs of the ubiquitin system, but enhancement of protein degradation requires interaction of both acidosis and glucocorticoids (Isozaki *et al.*, 1996). This interaction has also been suggested *in vivo* (Bailey *et al.*, 1996; Mitch, 1996; Price *et al.*, 1996; England and Price, 1995).

Ubiquitin conjugates increase by 1.5- to 3.5-fold in muscle after food deprivation. It has been suggested that the accumulation of ubiquitin–protein conjugates in muscle undergoing atrophy is a consequence of degradation by the 26 S proteasome becoming rate-limiting (Wing *et al.*, 1995), even though mRNAs for several proteasome subunits are also increased 2- to 4-fold in such muscles (Medina *et al.*, 1995). Other cell stress genes in addition to polyubiquitin genes do not appear to be activated in muscle in association with food deprivation (Medina *et al.*, 1995).

Similar findings have been obtained in rodent models of muscle wasting associated with malignancy (Temparis *et al.*, 1994). In rats bearing transplanted hepatoma, muscle atrophy is associated with increased levels of mRNAs for ubiquitin (7- to 10-fold) and for multiple subunits of the proteasome (2- to 3-fold); there is also a small increase (30–40%) in mRNA for cathepsins B and D, but not for calpain 1 or hsp 70 (Baracos *et al.*, 1995). In such models, plasma levels of tumor necrosis factor α are elevated along with other cytokine and endocrine disturbances, which have together been implicated in triggering protein catabolism (Llovera *et al.*, 1994; Garcia-Martinez *et al.*, 1994).

Possibilities for Therapy in Muscle Atrophy

Metabolic disturbances leading to respiratory or metabolic acidosis are very common in patients in critical care facilities and the loss of muscle during illness is a major part of the requirement for rehabilitation and recovery (Mitch, 1996).

Great interest is therefore being shown in the development of drugs that specifically inhibit the proteasome system (Fenteany *et al.*, 1995; Iqbal *et al.*, 1995; Rock *et al.*, 1994), as they might well be effective in the treatment of both acute and chronic wasting disease, e.g., in association with muscle disuse after bone fracture, in cachexia associated with infection (including HIV-1), in cachexia related to malignancy, or in wasting associated with metabolic disturbances and acidosis.

Increased muscle proteolysis may be prevented not only directly with proteasome inhibitors but also indirectly by modulation of signal transduction pathways. For example, clenbuterol largely inhibits skeletal muscle atrophy in tumor-bearing rats; protein breakdown was normalized through a decrease in the activity of the ubiquitin-dependent proteolytic pathway (Costelli *et al.*, 1995). The muscles of interleukin-6 transgenic mice undergo atrophy involving an activation of the ubiquitin system that can be blocked by treatment with an antibody against mouse interleukin-6 receptor, suggesting that this might be effective in preventing muscle wasting in sepsis and malignacy-related cachexia (Tsujinaka *et al.*, 1996).

4.2. Programmed Cell Death

Mechanisms involved in the process of programmed cell death are currently very much in focus in the context of a wide range of pathological processes, including abnormal development, tumor biology, inflammation, response to infection, and neurodegenerative diseases: The most studied form of this process is apoptosis (Fraser and Evan, 1996; Bredesen, 1995; Orrenius, 1995; Thompson, 1995). The ubiquitin system is only one of a number of protease systems that are active in apoptosis, the most important being ICE-like proteases (Kumar and Harvey, 1995; Martin and Green, 1995).

In several examples of programmed cell death, the ubiquitin system is activated to facilitate protein catabolism. For example, apoptosis induced in interphase human lymphocytes by gamma radiation results in an increase in ubiquitin mRNA 15 to 90 min after initiation of radiation. Concomitantly, there is ubiquitination of nuclear proteins (Delic *et al.*, 1993). However, induction of the ubiquitin system may not be universally required for apoptosis, as it was not observed in at least one neuronal model (D'Mello and Galli, 1993). Additionally, studies on a model of programmed neuromuscular cell death have again shown the importance of the ubiquitin system. In the development of the tobacco hawkmoth (*Manduca sexta*), the intersegmental abdominal muscles are programmed to die as levels of the insect molting hormone, 20-hydroxyecdysone, decline. At the same time there is a 10-fold induction of a polyubiquitin gene and 20-hydroxyecdysone delays degeneration in this system, by preventing an increase in this polyubiquitin mRNA (Schwartz *et al.*, 1990). There are also increases in total ubiquitin, ubiquitin-activating enzyme (E1), ubiquitin carrier proteins (E2s), ubiquitin ligases (E3s), and ubiquitin conjugates in this model (Haas *et al.*, 1995). These developmentally programmed changes in the ubiquitin-related enzyme system are accompanied by an increase in the concentration of the core 20 S proteasome, together with differential reprogramming of the ATPase regulators of the 26 S proteasome, presumably to bring about destruction of the intersegmental muscles (Dawson *et al.*, 1995); these proteasomal changes are not seen in flight muscles that are needed by the moth after eclosion (Takayanagi *et al.*, 1996).

5. SUMMARY

The investigation of ubiquitin in pathological processes has been very rewarding. The immediate practical benefit has been that antibodies directed to ubiquitin label a whole series of cellular abnormalities that are either very difficult or impossible to visualize in histological preparations by any other technique. Three main areas have emerged where further understanding of the ubiquitin system has potential for the development of therapy in disease.

- Ubiquitin is associated with abnormal accumulations of lysosome-related vesicles, both in normal aging as well as in disease. This association is now being underpinned by observations on the likely role of the ubiquitin system in endosome–lysosome biogenesis.
- Ubiquitin has a consistent association with abnormal filaments in inclusion bodies in several different cell types, especially in the degenerative diseases of the nervous system, which are both socially and economically

important. In some cases, the association appears to be part of a cell stress response to eliminate abnormal proteins; in others, the reason for involvement of ubiquitin remains uncertain.

- Processes in which there is a requirement for elimination of cellular proteins, such as atrophy and programmed cell death, are associated with activation of the ubiquitin–proteasome degradation pathway.

In other chapters of this book, descriptions of the involvement of the ubiquitin system in the degradation of oncoproteins, involvement in immunity and in viral pathogenesis serve to emphasize the importance of this system in human disease.

Acknowledgments. We wish to thank Kevin Bailey and Phyllis Heggie for work on griseofulvin-induced Mallory bodies and Jill Fergusson for help in the preparation of the manuscript. We wish to acknowledge financial support for the work from the Wellcome Trust, MRC, BBSRC, European Union (Copernicus Programme), Parkinson's Disease Society, Research into Ageing and the Neurodegenerative Support Group at the Queen's Medical Centre.

6. REFERENCES

Agundez, J. A., Jimenez, F., Luengo, A., Bernal, M. L., Molina, J. A., Ayuso, L., Vazquez, A., Parra, J., Duarte, J., Coria, F., and Ladero, J. M., 1995, Association between the oxidative polymorphism and early onset of Parkinson's disease, *Clin. Pharmacol. Ther.* **57**:291–298.

Ahrens, P. B., Besancon, F., Memet, S., and Ankel, H., 1990, Tumour necrosis factor enhances induction by beta-interferon of a ubiquitin cross-reactive protein, *J. Gen. Virol.* **71**:1675–1682.

Albrecht, S., and Bilbao, J. M., 1993, Ubiquitin expression in inclusion body myositis. An immunohistochemical study, *Arch. Pathol. Lab. Med.* **117**:789–793.

Arnold, J. E., Tipler, C., Laszlo, L., Hope, J., Landon, M., and Mayer, R. J., 1995, The abnormal isoform of the prion protein accumulates in late-endosome-like organelles in scrapie-infected mouse brain, *J. Pathol.* **176**:403–411.

Askanas, V., and Engel, W. K., 1995, New advances in the understanding of sporadic inclusion-body myositis and hereditary inclusion-body myopathies, *Curr. Opin. Rheumatol.* **7**:486–496.

Askanas, V., Serdaroglu, P., King Engel, W., and Alvarez, R. B., 1991, Immunolocalization of ubiquitin in muscle biopsies of patients with inclusion body myositis and oculopharyngeal muscular dystrophy, *Neurosci. Lett.* **130**:73–76.

Askanas, V., Serdaroglu, P., Engel, W. K., and Alvarez, R. B., 1992, Immunocytochemical localization of ubiquitin in inclusion body myositis allows its light-microscopic distinction from polymyositis, *Neurology* **42**:460–461.

Askanas, V., Engel, W. K., Bilak, M., Alvarez, R. B., and Selkoe, D. J., 1994, Twisted tubulofilaments of inclusion body myositis muscle resemble paired helical filaments of Alzheimer brain and contain hyperphosphorylated tau, *Am. J. Pathol.* **144**:177–187.

Bacci, B., Cochran, E., Nunzi, M. G., Izeki, E., Mizutani, T., Patton, A., Hite, S., Sayre, L. M., Autilio-Gambetti, L., and Gambetti, P., 1994, Amyloid beta precursor protein and ubiquitin epitopes in human and experimental dystrophic axons. Ultrastructural localization, *Am. J. Pathol.* **144:** 702–710.

Bailey, J. L., Wang, X., England, B. K., Price, S. R., Ding, X., and Mitch, W. E., 1996, The acidosis of chronic renal failure activates muscle proteolysis in rats by augmenting transcription of genes encoding proteins of the ATP-dependent ubiquitin-proteasome pathway, *J. Clin. Invest.* **97:** 1447–1453.

Ball, E., Karlik, C. C., Beall, C. J., Saville, D. L., Sparrow, J. C., Bullard, B., and Fyrberg, E. A., 1987, Arthrin, a myofibrillar protein of insect flight muscle, is an actin–ubiquitin conjugate, *Cell* **51:** 221–228

Bancher, C., Brunner, C., Lassmann, H., Budka, H., Jellinger, K., Wiche, G., Seitelberger, F., Grundke Iqbal, I., and Wisniewski, H. M., 1989, Accumulation of abnormally phosphorylated tau precedes the formation of neurofibrillary tangles in Alzheimer's disease, *Brain Res.* **477:**90–99.

Baracos, V. E., De Vivo, C., Hoyle, D. H., and Goldberg, A. L., 1995, Activation of the ATP-ubiquitin-proteasome pathway in skeletal muscle of cachectic rats bearing a hepatoma, *Am. J. Physiol.* **268:** E996–1006.

Bizzi, A., Schaetzle, B., Patton, A., Gambetti, P., and Autilio-Gambetti, L., 1991, Axonal transport of two major components of the ubiquitin system: Free ubiquitin and ubiquitin carboxyl-terminal hydrolase PGP 9.5 [published erratum in *Brain Res.* 1991, **557:**359], *Brain Res.* **548:**292–299.

Bond, U., and Schlesinger, M. J., 1985, Ubiquitin is a heat shock protein in chicken embryo fibroblasts, *Mol. Cell. Biol.* **5:**949–956.

Bond, U., and Schlesinger, M. J., 1986, The chicken ubiquitin gene contains a heat shock promoter and expresses an unstable mRNA in heat-shocked cells, *Mol. Cell. Biol.* **6:**4602–4610.

Bond, U., Agell, N., Haas, A. L., Redman, K., and Schlesinger, M. J., 1988, Ubiquitin in stressed chicken embryo fibroblasts, *J. Biol. Chem.* **263:**2384–2388.

Bowling, A. C., Schulz, J. B., Brown, R. J., and Beal, M. F., 1993, Superoxide dismutase activity, oxidative damage, and mitochondrial energy metabolism in familial and sporadic amyotrophic lateral sclerosis, *J. Neurochem.* **61:**2322–2325.

Braak, H., and Braak, E., 1991, Neuropathological staging of Alzheimer-related changes, *Acta Neuropathol.* **82:**239–259.

Bredesen, D. E., 1995, Neural apoptosis, *Ann. Neurol.* **38:**839–851.

Byrne, E., Lennox, G., Lowe, J., and Godwin-Austen, R. B., 1989, Diffuse Lewy body disease: Clinical features in 15 cases, *J. Neurol. Neurosurg. Psychiatry* **52:**709–717.

Cammarata, S., Caponnetto, C., and Tabaton, M., 1993, Ubiquitin-reactive neurites in cerebral cortex of subjects with Huntington's chorea: A pathological correlate of dementia? *Neurosci. Lett.* **156:** 96–98.

Cavanagh, J. B., Nolan, C. C., Seville, M. P., Anderson, V. E., and Leigh, P. N., 1993, Routes of excretion of neuronal lysosomal dense bodies after ventricular infusion of leupeptin in the rat: A study using ubiquitin and PGP 9.5 immunocytochemistry, *J. Neurocytol.* **22:**779–791.

Chakraborty, G., and Ingoglia, N. A., 1993, N-terminal arginylation and ubiquitin-mediated proteolysis in nerve regeneration, *Brain Res. Bull.* **30:**439–445.

Chen, Y., and Piper, P. W., 1995, Consequences of the overexpression of ubiquitin in yeast: Elevated tolerances of osmostress, ethanol and canavanine, yet reduced tolerances of cadmium, arsenite and paromomycin, *Biochim. Biophys. Acta* **1268:**59–64.

Cheng, L., Watt, R., and Piper, P. W., 1994, Polyubiquitin gene expression contributes to oxidative stress resistance in respiratory yeast (Saccharomyces cerevisiae), *Mol. Gen. Genet.* **243:**358–362.

Cochran, E., Bacci, B., Chen, Y., Patton, A., Gambetti, P., and Autilio-Gambetti, L., 1991, Amyloid precursor protein and ubiquitin immunoreactivity in dystrophic axons is not unique to Alzheimer's disease, *Am. J. Pathol.* **139:**485–489.

Cole, G. M., and Timiras, P. S., 1987, Ubiquitin–protein conjugates in Alzheimer's lesions, *Neurosci. Lett.* **79:**207–212.

Costelli, P., Garcia-Martinez, C., Llovera, M., Carbo, N., Lopez-Soriano, F. J., Agell, N., Tessitore, L.,

Baccino, F. M., and Argiles, J. M., 1995, Muscle protein wasting in tumor-bearing rats is effectively antagonized by a beta 2-adrenergic agonist (clenbuterol). Role of the ATP-ubiquitin-dependent proteolytic pathway, *J. Clin. Invest.* **95:**2367–2372.

Cras, P., Kawai, M., Lowery, D., Gonzalez, D. P., Greenberg, B., and Perry, G., 1991, Senile plaque neurites in Alzheimer disease accumulate amyloid precursor protein, *Proc. Natl. Acad. Sci. USA* **88:**7552–7556.

Crystal, H., Dickson, D., Lizardi, J., Davies, P., and Wolfson, L., 1990, Antemortem diagnosis of diffuse Lewy body disease, *Neurology* **40:**1523–1528.

Das, K. P., and Surewicz, W. K., 1995, On the substrate specificity of alpha-crystallin as a molecular chaperone, *Biochem. J.* **311:**367–370.

Dawson, S., Arnold, J., Mayer, N. J., Reynolds, S., Billett, M. A., Kloetzel, P., Tanaka, K., and Mayer, R. J., 1995, Developmental changes of the 26S proteasome in abdominal intersegmental muscles of Manduca sexta during programmed cell death, *J. Biol. Chem.* **270:**1850–1858.

Day, I. N., Hinks, L. J., and Thompson, R. J., 1990, The structure of the human gene encoding protein gene product 9.5 (PGP9.5), a neuron-specific ubiquitin C-terminal hydrolase, *Biochem. J.* **268:**521–524.

DeArmond, S. J., and Prusiner, S. B., 1995a, Etiology and pathogenesis of prion diseases, *Am. J. Pathol.* **146:**785–811.

DeArmond, S. J., and Prusiner, S. B., 1995b, Prion protein transgenes and the neuropathology in prion diseases, *Brain Pathol.* **5:**77–89.

Delic, J., Morange, M., and Magdelenat, H., 1993, Ubiquitin pathway involvement in human lymphocyte gamma-irradiation-induced apoptosis, *Mol. Cell. Biol.* **13:**4875–4883.

De Marini, D. J., Papa, F. R., Swaminathan, S., Ursic, D., Rasmussen, T. P., Culbertson, M. R., and Hochstrasser, M., 1995, The yeast SEN3 gene encodes a regulatory subunit of the 26S proteasome complex required for ubiquitin-dependent protein degradation in vivo, *Mol. Cell. Biol.* **15:**6311–6321.

Dewar, D., Graham, D. I., Teasdale, G. M., and McCulloch, J., 1994, Cerebral ischemia induces alterations in tau and ubiquitin proteins, *Dementia* **5:**168–173.

Dexter, D. T., Holley, A. E., Flitter, W. D., Slater, T. F., Wells, F. R., Daniel, S. E., Lees, A. J., Jenner, P., and Marsden, C. D., 1994, Increased levels of lipid hydroperoxides in the parkinsonian substantia nigra: An HPLC and ESR study, *Mov. Disord.* **9:**92–97.

Dickson, D. W., Crystal, H., Mattiace, L. A., Kress, Y., Schwagerl, A., Ksiezak-Reding, H., Davies, P., and Yen, S. H., 1989, Diffuse Lewy body disease: Light and electron microscopic immunocytochemistry of senile plaques, *Acta Neuropathol.* **78:**572–584.

Dickson, D. W., Wertkin, A., Kress, Y., Ksiezak-Reding, H., and Yen, S. H., 1990a, Ubiquitin immunoreactive structures in normal human brains. Distribution and developmental aspects, *Lab. Invest.* **63:**87–99.

Dickson, D. W., Wertkin, A., Mattiace, L. A., Fier, E., Kress, Y., Davies, P., and Yen, S. H., 1990b, Ubiquitin immunoelectron microscopy of dystrophic neurites in cerebellar senile plaques of Alzheimer's disease, *Acta Neuropathol.* **79:**486–493.

Dickson, D. W., Ruan, D., Crystal, H., Mark, M. H., Davies, P., Kress, Y., and Yen, S. H., 1991, Hippocampal degeneration differentiates diffuse Lewy body disease (DLBD) from Alzheimer's disease: Light and electron microscopic immunocytochemistry of CA2–3 neurites specific to DLBD, *Neurology* **41:**1402–1409.

Dickson, D. W., Crystal, H. A., Mattiace, L. A., Masur, D. M., Blau, A. D., Davies, P., Yen, S. H., and Aronson, M. K., 1992, Identification of normal and pathological aging in prospectively studied nondemented elderly humans, *Neurobiol. Aging* **13:**179–189.

Dickson, D. W., Schmidt, M. L., Lee, V. M., Zhao, M. L., Yen, S. H., and Trojanowski, J. Q., 1994, Immunoreactivity profile of hippocampal CA2/3 neurites in diffuse Lewy body disease, *Acta Neuropathol.* **87:**269–276.

D'Mello, S. R., and Galli, C., 1993, SGP2, ubiquitin, 14K lectin and RP8 mRNAs are not induced in neuronal apoptosis, *Neuroreports* **4:**355–358.

England, B. K., and Price, S. R., 1995, Acidosis and glucocorticoids interact to provoke muscle protein and amino acid catabolism, *Blood Purification* **13:**147–152.

Ermisch, B., and Schweccheimer, K., 1995, Protein gene product (PGP) 9.5 in diagnostic (neuro-) oncology. An immunomorphological study, *Clin. Neuropathol.* **14:**130–136.

Fahn, S., and Cohen, G., 1992, The oxidant stress hypothesis in Parkinson's disease: Evidence supporting it, *Ann. Neurol.* **32:**804–812.

Fang, C. H., Tiao, G., James, H., Ogle, C., Fischer, J. E., and Hasselgren, P. O., 1995, Burn injury stimulates multiple proteolytic pathways in skeletal muscle, including the ubiquitin-energy-dependent pathway, *J. Am. Coll. Surg.* **180:**161–170.

Fenteany, G., Standaert, R. F., Lane, W. S., Choi, S., Corey, E. J., and Schreiber, S. L., 1995, Inhibition of proteasome activities and subunit-specific amino-terminal threonine modification by lactacystin, *Science* **268:**726–731.

Fergusson, J., Landon, M., Lowe, J., Dawson, S. P., Layfield, R., Hanger, D. P., and Mayer, R. J., 1996, *Neurosci. Letts.* **219:**167–170.

Figueiredo-Pereira, P. M., Berg, K. A., and Wilk, S., 1994, A new inhibitor of the chymotrypsin-like activity of the multicatalytic proteinase complex (20S proteasome) induces accumulation of ubiquitin–protein conjugates in a neuronal cell, *J. Neurochem.* **63:**1578–1581.

Finley, D., Ozkaynak, E., and Varshavsky, A., 1987, The yeast polyubiquitin gene is essential for resistance to high temperatures, starvation, and other stresses, *Cell* **48:**1035–1046.

Fornace, A. J., Jr., Alamo, I., Jr., Hollander, M. C., and Lamoreaux, E., 1989, Ubiquitin mRNA is a major stress-induced transcript in mammalian cells, *Nucleic Acids Res.* **17:**1215–1230.

Fraser, A., and Evan, G., 1996, A license to kill, *Cell* **85:**781–784.

Fujimoro, M., Sawada, H., and Yokosawa, H., 1994, Production and characterization of monoclonal antibodies specific to multi-ubiquitin chains of polyubiquitinated proteins, *FEBS Lett.* **349:** 173–180.

Funata, N., Maeda, Y., Koike, Y., Yano, M., Kaseda, M., Muro, T., Okeda, R., Iwata, M., and Uokoji, M., 1990, Review article: Neuronal intranuclear hyaline inclusion disease: Report of a case and review of the literature, *Clin. Neuropathol.* **9:**89–96.

Gai, W. P., Blessing, W. W., and Blumbergs, P. C., 1995, Ubiquitin-positive degenerating neurites in the brainstem in Parkinson's disease, *Brain* **118:**1447–1459.

Garcia-Martinez, C., Llovera, M., Agell, N., Lopez-Soriano, F. J., and Argiles, J. M., 1994, Ubiquitin gene expression in skeletal muscle is increased by tumour necrosis factor-alpha, *Biochem. Biophys. Res. Commun.* **201:**682–686.

Gibb, W., Esiri, M., and Lees, A., 1985, Clinical and pathological features of diffuse cortical Lewy body disease (Lewy body dementia), *Brain* **110:**1131–1153.

Goldfarb, L. G., and Brown, P., 1995, The transmissible spongiform encephalopathies, *Annu. Rev. Med.* **46:**57–65.

Goldman, J. E., and Corbin, E., 1991, Rosenthal fibers contain ubiquitinated alpha B-crystallin, *Am. J. Pathol.* **139:**933–938.

Gou, J. P., and Leterrier, J. F., 1995, Possible involvement of ubiquitination in neurofilament degradation, *Biochem. Biophys. Res. Commun.* **217:**529–538.

Gregori, L., Fuchs, C., Figueiredo-Pereira, M. E., Van Nostrand, W. E., and Goldgaber, D., 1995, Amyloid beta-protein inhibits ubiquitin-dependent protein degradation in vitro, *J. Biol. Chem.* **270:**19702–19708.

Gropper, R., Brandt, R. A., Elias, S., Bearer, C. F., Mayer, A., Schwartz, A. L., and Ciechanover, A., 1991, The ubiquitin-activating enzyme, E1, is required for stress-induced lysosomal degradation of cellular proteins, *J. Biol. Chem.* **266:**3602–3610.

Grundke-Iqbal, I., Iqbal, K., Tung, Y. C. H., Quinlan, M., Wisniewski, H. M., and Binder, L. I., 1986, Abnormal phosphorylation of the microtubule-associated protein tau in Alzheimer-cytoskeletal pathology, *Proc. Natl. Acad. Sci. USA* **83:**4913–4917.

Guarino, L. A., Smith, G., and Wen, D., 1995, Ubiquitin is attached to membranes of baculovirus particles by a novel type of phospholipid anchor, *Cell* **80:**301–309.

Haas, A. L., and Bright, P. M., 1985, The immunochemical detection and quantitation of intracellular ubiquitin–protein conjugates, *J. Biol. Chem.* **260:**12464–12473.

Haas, A. L., Baboshina, O., Williams, B., and Schwartz, L. M., 1995, Coordinated induction of the ubiquitin conjugation pathway accompanies the developmentally programmed death of insect skeletal muscle, *J. Biol. Chem.* **270:**9407–9412.

Hand, S. C., and Hardewig, I., 1996, Downregulation of cellular metabolism during environmental stress: Mechanisms and implications, *Annu. Rev. Physiol.* 539–563.

Hansen, L. A., Masliah, E., Terry, R. D., and Mirra, S. S., 1989, A neuropathological subset of Alzheimer's disease with concomitant Lewy body disease and spongiform change, *Acta Neuropathol.* **78:**194–201.

Hansen, L., Salmon, D., Galasko, D., Masliah, E., Katzman, R., DeTeresa, R., Thal, L., Pay, M., Hofsteter, R., Klauber, M., Rice, V., Butters, N., and Alford, M., 1990, The Lewy body variant of Alzheimer's disease, *Neurology* **40:**1–8.

Harris, M. D., Moore, I. E., Steart, P. V., and Weller, R. O., 1990, Protein gene product (PGP) 9.5 as a reliable marker in primitive neuroectodermal tumours—An immunohistochemical study of 21 childhood cases, *Histopathology* **16:**271–277.

Hasselgren, P. O., 1995, Muscle protein metabolism during sepsis, *Biochem. Soc. Trans.* **23:**1019–1025.

Hayashi, T., Takada, K., and Matsuda, M., 1991, Changes in ubiquitin and ubiquitin–protein conjugates in the CA1 neurons after transient sublethal ischemia, *Mol. Chem. Neuropathol.* **15:**75–82.

Hayashi, T., Tanaka, J., Kamikubo, T., Takada, K., and Matsuda, M., 1993, Increase in ubiquitin conjugates dependent on ischemic damage, *Brain Res.* **620:**171–173.

Heggie, P., Burdon, T., Lowe, J., Landon, M., Lennox, G., Jefferson, D., and Mayer, R. J., 1989, Ubiquitin gene expression in brain and spinal cord in motor neurone disease, *Neurosci. Lett.* **102:**2–3.

Hicke, L., and Riezman, H., 1996, Ubiquitination of a yeast plasma membrane receptor signals its ligand-stimulated endocytosis, *Cell* **84:**277–287.

Hilenski, L. L., Terracio, L., Haas, A. L., and Borg, T. K., 1992, Immunolocalization of ubiquitin conjugates at Z-bands and intercalated discs of rat cardiomyocytes in vitro and in vivo, *J. Histochem. Cytochem.* **40:**1037–1042.

Huntington's Disease Collaborative Research Group, 1993, A novel gene containing a trinucleotide repeat that is expanded and unstable on Huntington's disease chromosomes, *Cell* **72:**971–983.

Hurtley, S. M., 1996, Lysosomal degradation of ubiquitin-tagged receptors, *Science* **271:**617.

Iqbal, M., Chatterjee, S., Kauer, J. C., Das, M., Messina, P., Freed, B., Biazzo, W., and Siman, R., 1995, Potent inhibitors of proteasome, *J. Med. Chem.* **38:**2276–2277.

Ironside, J. W., McCardle, L., Hayward, P. A., and Bell, J. E., 1993, Ubiquitin immunocytochemistry in human spongiform encephalopathies, *Neuropathol. Appl. Neurobiol.* **19:**134–140.

Isozaki, Y., Mitch, W. E., England, B. K., and Price, S. R., 1996, Protein degradation and increased mRNAs encoding proteins of the ubiquitin-proteasome proteolytic pathway in BC3H1 myocytes require an interaction between glucocorticoids and acidification, *Proc. Natl. Acad. Sci. USA* **93:** 1967–1971.

Ito, H., Ii, K., Hirano, A., and Dickson, D., 1991, Immunohistochemical identification of the proteasome in ubiquitin reactive abnormal structures of the nervous system, *J. Neuropathol. Exp. Neurol.* **50:**360.

Iwatsubo, T., Hasegawa, M., Esaki, Y., and Ihara, Y., 1992, Lack of ubiquitin immunoreactivities at both ends of neuropil threads. Possible bidirectional growth of neuropil threads, *Am. J. Pathol.* **140:**277–282.

Jack, D. L., Chakraborty, G., and Ingoglia, N. A., 1992, Ubiquitin is associated with aggregates of arginine modified proteins in injured nerves, *Neuroreports* **3:**47–50.

Jackson, M., and lowe, J., 1996, The new neuropathology of degenerative front of temporal dementias, *Acta Neuropathol.* **91:**127–134.

Jackson, M., Gentleman, S., Lennox, G., Ward, L., Gray, T., Randall, K., Morrell, K., and Lowe, J., 1995, The cortical neuritic pathology of Huntington's disease, *Neuropathol. Appl. Neurobiol.* **21:**18–26.

Jenner, P., Dexter, D. T., Sian, J., Schapira, A. H., and Marsden, C. D., 1992, Oxidative stress as a cause of nigral cell death in Parkinson's disease and incidental Lewy body disease. The Royal Kings and Queens Parkinson's Disease Research Group, *Ann. Neurol.* **32:**7.

Jones, C. T., Swingler, R. J., and Brock, D. J., 1994, Identification of a novel SOD1 mutation in an apparently sporadic amyotrophic lateral sclerosis patient and the detection of Ile113Thr in three others, *Hum. Mol. Genet.* **3:**649–650.

Kamikubo, T., and Hayashi, T., 1996, Changes in proteasome activity following transient ischemia, *Neurochem. Int.* **28:**209–212.

Kenward, N., Hope, J., Landon, M., and Mayer, R. J., 1994, Expression of polyubiquitin and heat-shock protein 70 genes increases in the later stages of disease progression in scrapie-infected mouse brain, *J. Neurochem.* **62:**1870–1877.

Kim, H., Gearing, M., and Mirra, S. S., 1995, Ubiquitin-positive CA2/3 neurites in hippocampus coexist with cortical Lewy bodies, *Neurology* **45:**1768–1770.

Kumar, S., and Harvey, N. L., 1995, Role of multiple cellular proteases in the execution of programmed cell death, *FEBS Lett.* **375:**169–173.

Kwak, S., Masaki, T., Ishiura, S., and Sugita, H., 1991, Multicatalytic proteinase is present in Lewy bodies and neurofibrillary tangles in diffuse Lewy body disease brains, *Neurosci. Lett.* **128:**21–24.

Laszlo, L., Doherty, F. J., Osborn, N. U., and Mayer, R. J., 1990, Ubiquitinated protein conjugates are specifically enriched in the lysosomal system of fibroblasts, *FEBS Lett.* **261:**365–368.

Laszlo, L., Doherty, F. J., Watson, A., Self, T., Landon, M., Lowe, J., and Mayer, R. J., 1991a, Immunogold localisation of ubiquitin–protein conjugates in primary (azurophilic) granules of polymorphonuclear neutrophils, *FEBS Lett.* **279:**175–178.

Laszlo, L., Tuckwell, J., Self, T., Lowe, J., Landon, M., Smith, S., Hawthorne, J. N., and Mayer, R. J., 1991b, The latent membrane protein-1 in Epstein–Barr virus-transformed lymphoblastoid cells is found with ubiquitin–protein conjugates and heat-shock protein 70 in lysosomes oriented around the microtubule organizing centre, *J. Pathol.* **164:**203–214.

Laszlo, L., Lowe, J., Self, T., Kenward, N., Landon, M., McBride, T., Farquhar, C., McConnell, I., Brown, J., Hope, J., 1992, Lysosomes as key organelles in the pathogenesis of prion encephalopathies, *J. Pathol.* **166:**333–341.

Layfield, R., Rog, D., Arnold, J., Lowe, J., Mayer, R. J., and Landon, M., 1994, Immunoreactivity to ubiquitin-activating enzyme (E1) is detectable in some neurofibrillary tangles of Alzheimer's disease, *Neuropathol. Appl. Neurobiol.* **20:**504.

Leclerc, A., Tome, F. M., and Fardeau, M., 1993, Ubiquitin and beta-amyloid-protein in inclusion body myositis (IBM), familial IBM-like disorder and oculopharyngeal muscular dystrophy: An immunocytochemical study, *Neuromusc. Disord.* **3:**283–291.

Leigh, P. N., Anderton, B. H., Dodson, A., Gallo, J. M., Swash, M., and Power, D. M., 1988, Ubiquitin deposits in anterior horn cells in motor neurone disease, *Neurosci. Lett.* **93:**2–3.

Lenk, S. E., Dunn, W. A., Jr., Trausch, J. S., Ciechanover, A., and Schwartz, A. L., 1992, Ubiquitin-activating enzyme, E1, is associated with maturation of autophagic vacuoles, *J. Cell Biol.* **118:** 301–308.

Lennox, G., Lowe, J., Morrell, K., Landon, M., and Mayer, R. J., 1989, Anti-ubiquitin immunocytochemistry is more sensitive than conventional techniques in the detection of diffuse Lewy body disease, *J. Neurol. Neurosurg. Psychiatry* **52:**67–71.

Li, G. L., and Farooque, M., 1996, Expression of ubiquitin-like immunoreactivity in axons after compression trauma to rat spinal cord, *Acta Neuropathol.* **91:**155–160.

Llovera, M., Garcia-Martinez, C., Agell, N., Marzabal, M., Lopez-Soriano, F. J., and Argiles, J. M., 1994, Ubiquitin gene expression is increased in skeletal muscle of tumour-bearing rats, *FEBS Lett.* **338:**311–318.

Loeb, K. R., and Haas, A. L., 1992, The interferon-inducible 15-kDa ubiquitin homolog conjugates to intracellular proteins, *J. Biol. Chem.* **267:**7806–7813.

Loeb, K. R., and Haas, A. L., 1994, Conjugates of ubiquitin cross-reactive protein distribute in a cytoskeletal pattern, *Mol. Cell. Biol.* **14:**8408–8419.

Love, S., Saitoh, T., Quijada, S., Cole, G. M., and Terry, R. D., 1988, Alz-50, ubiquitin and tau immunoreactivity of neurofibrillary tangles, Pick bodies and Lewy bodies, *J. Neuropathol. Exp. Neurol.* **47**:393–405.

Low, P., Doherty, F. J., Sass, M., Kovacs, J., Mayer, R. J., and Laszlo, L., 1993, Immunogold localisation of ubiquitin–protein conjugates in Sf9 insect cells. Implications for the biogenesis of lysosome-related organelles, *FEBS Lett.* **316**:152–156.

Low, P., Doherty, F. J., Fellinger, E., Sass, M., Mayer, R. J., and Laszlo, L., 1995, Related organelles of the endosome–lysosome system contain a different repertoire of ubiquitinated proteins in Sf9 insect cells, *FEBS Lett.* **368**:125–131.

Lowe, J., 1994, New pathological findings in amyotrophic lateral sclerosis, *J. Neurol. Sci.* **124**:38–51.

Lowe, J., and Mayer, R. J., 1989, Ubiquitin: New insights into chronic degenerative diseases, *Br. J. Hosp. Med.* **42**:462–466.

Lowe, J., and Mayer, R. J., 1990, Ubiquitin, cell stress and diseases of the nervous system, *Neuropathol. Appl. Neurobiol.* **16**:281–291.

Lowe, J., Blanchard, A., Morrell, K., Lennox, G., Reynolds, L., Billett, M., Landon, M., and Mayer, R. J., 1988a, Ubiquitin is a common factor in intermediate filament inclusion bodies of diverse type in man, including those of Parkinson's disease, Pick's disease, and Alzheimer's disease, as well as Rosenthal fibres in cerebellar astrocytomas, cytoplasmic bodies in muscle, and Mallory bodies in alcoholic liver disease, *J. Pathol.* **155**:9–15.

Lowe, J., Lennox, G., Jefferson, D., Morrell, K., McQuire, D., Gray, T., Landon, M., Doherty, F. J., and Mayer, R. J., 1988b, A filamentous inclusion body within anterior horn neurones in motor neurone disease defined by immunocytochemical localisation of ubiquitin, *Neurosci. Lett.* **94**:203–210.

Lowe, J., Aldridge, F., Lennox, G., Doherty, F. J., Jefferson, D., Landon, M., and Mayer, R. J., 1989, Inclusion bodies in motor cortex and brainstem of patients with motor neurone disease are detected by immunocytochemical localisation of ubiquitin, *Neurosci. Lett.* **105**:7–13.

Lowe, J., McDermott, H., Kenward, N., Landon, M., Mayer, R. J., Bruce, M., McBride, P., Somerville, R. A., and Hope, J., 1990a, Ubiquitin conjugate immunoreactivity in the brains of scrapie infected mice, *J. Pathol.* **162**:61–66.

Lowe, J., McDermott, H., Landon, M., Mayer, R. J., and Wilkinson, K. D., 1990b, Ubiquitin carboxyl-terminal hydrolase (PGP 9.5) is selectively present in ubiquitinated inclusion bodies characteristic of human neurodegenerative diseases, *J. Pathol.* **161**:153–160.

Lowe, J., Fergusson, J., Kenward, N., Laszlo, L., Landon, M., Farquhar, C., Brown, J., Hope, J., and Mayer, R. J., 1992a, Immunoreactivity to ubiquitin–protein conjugates is present early in the disease process in the brains of scrapie-infected mice, *J. Pathol.* **168**:169–177.

Lowe, J., McDermott, H., Pike, I., Spendlove, I., Landon, M., and Mayer, R. J., 1992b, Alpha B crystallin expression in non-lenticular tissues and selective presence in ubiquitinated inclusion bodies in human disease, *J. Pathol.* **166**:61–68.

Lowe, J., Mayer, R. J., and Landon, M., 1993, Ubiquitin in neurodegenerative diseases, *Brain Pathol.* **3**:55–65.

Lowe, J., McDermott, H., Loeb, K., Landon, M., Haas, A. L., and Mayer, R. J., 1995, Immunohistochemical localization of ubiquitin cross-reactive protein in human tissues, *J. Pathol.* **177**: 163–169.

Magnusson, K., and Wieloch, T., 1989, Impairment of protein ubiquitination may cause delayed neuronal death, *Neurosci. Lett.* **96**:264–270.

Malandrini, A., Cavallaro, T., Fabrizi, G. M., Berti, G., Salvestroni, R., Salvadori, C., and Guazzi, G. C., 1995, Ultrastructure and immunoreactivity of dystrophic axons indicate a different pathogenesis of Hallervorden–Spatz disease and infantile neuroaxonal dystrophy, *Virchows Arch.* **427**: 415–421.

Manetto, V., Abdul-Karim, F. W., Perry, G., Tabaton, M., Autilio-Gambetti, L., and Gambetti, P., 1989, Selective presence of ubiquitin in intracellular inclusions, *Am. J. Pathol.* **134**:505–513.

Mansoor, O., Beaufrere, B., Boirie, Y., Ralliere, C., Taillandier, D., Aurousseau, E., Schoeffler,

P., Arnal, M., and Attaix, D., 1996, Increased mRNA levels for components of the lysosomal, Ca^{2+}-activated, and ATP-ubiquitin-dependent proteolytic pathways in skeletal muscle from head trauma patients, *Proc. Natl. Acad. Sci. USA* **93:**2714–2718.

Martin, S. J., and Green, D. R., 1995, Protease activation during apoptosis: Death by a thousand cuts? *Cell* **82:**349–352.

Masaki, T., Ishiura, S., Sugita, H., and Kwak, S., 1994, Multicatalytic proteinase is associated with characteristic oval structures in cortical Lewy bodies: An immunocytochemical study with light and electron microscopy, *J. Neurol. Sci.* **122:**127–134.

Mather, K., Martin, J. E., Swash, M., Vowles, G., Brown, A., and Leigh, P. N., 1993, Histochemical and immunocytochemical study of ubiquitinated neuronal inclusions in amyotrophic lateral sclerosis, *Neuropathol. Appl. Neurobiol.* **19:**141–145.

Maury, C. P., 1995, Molecular pathogenesis of beta-amyloidosis in Alzheimer's disease and other cerebral amyloidoses, *Lab. Invest.* **72:**4–16.

Mayer, R. J., Lowe, J., Lennox, G., Landon, M., MacLennan, K., and Doherty, F. J., 1989, Intermediate filament-ubiquitin diseases: Implications for cell sanitization, *Biochem. Soc. Symp.* **55:**193–201.

Mayer, R. J., Arnold, J., Laszlo, L., Landon, M., and Lowe, J., 1991, Ubiquitin in health and disease, *Biochim. Biophys. Acta* **1089:**141–157.

Mayer, R. J., Landon, M., Laszlo, L., Lennox, G., and Lowe, J., 1992, Protein processing in lysosomes: The new therapeutic target in neurodegenerative disease, *Lancet* **340:**156–159.

Medina, R., Wing, S. S., Haas, A., and Goldberg, A. L., 1991, Activation of the ubiquitin-ATP-dependent proteolytic system in skeletal muscle during fasting and denervation atrophy, *Biomed. Biochim. Acta* **50:**347–356.

Medina, R., Wing, S. S., and Goldberg, A. L., 1995, Increase in levels of polyubiquitin and proteasome mRNA in skeletal muscle during starvation and denervation atrophy, *Biochem. J.* **307:**631–637.

Mehlen, P., Kretz-Remy, C., Preville, X., and Arrigo, A.-P., 1996, Human hsp27, Drosophila hsp27 and human alphaB-crystallin expression-mediated increase in glutathione is essential for the protective activity of these proteins against TNFalpha-induced cell death, *EMBO J.* **15:**2695–2706.

Migheli, A., Autilio-Gambetti, L., Gambetti, P., Mocellini, C., Vigliani, M. C., and Schiffer, D., 1990, Ubiquitinated filamentous inclusions in spinal cord of patients with motor neuron disease, *Neurosci. Lett.* **114:**5–10.

Migheli, A., Attanasio, A., Vigliani, M. C., and Schiffer, D., 1991, Dystrophic neurites around amyloid plaques of human patients with Gerstmann-Straussler-Scheinker disease contain ubiquitinated inclusions, *Neurosci. Lett.* **121:**55–58.

Migheli, A., Attanasio, A., Pezzulo, T., Gullotta, F., Giordana, M. T., and Schiffer, D., 1992, Age-related ubiquitin deposits in dystrophic neurites: An immunoelectron microscopic study, *Neuropathol. Appl. Neurobiol.* **18:**3–11.

Mitch, W. E., 1995, Cellular mechanisms of catabolism activated by metabolic acidosis, *Blood Purification* **13:**368–374.

Mitch, W. E., 1996, Metabolic acidosis stimulates protein metabolism in uremiae, *Miner Electrol. Metab.* **22:**62–65.

Mitch, W. E., Medina, R., Grieber, S., May, R. C., England, B. K., Price, S. R., Bailey, J. L., and Goldberg, A. L., 1994, Metabolic acidosis stimulates muscle protein degradation by activating the adenosine triphosphate-dependent pathway involving ubiquitin and proteasomes, *J. Clin. Invest.* **93:**2127–2133.

Mori, H., Kondo, J., and Ihara, Y., 1987, Ubiquitin is a component of paired helical filaments in Alzheimer's disease, *Science* **235:**1641–1644.

Morimoto, T., Ide, T., Ihara, Y., Tamura, A., and Kirino, T., 1996, Transient ischemia depletes free ubiquitin in the gerbil hippocampal CA1 neurons, *Am. J. Pathol.* **148:**249–257.

Morishima-Kawashima, M., and Ihara, Y., 1994, Posttranslational modifications of tau in paired helical filaments, *Dementia* **5:**282–288.

Murayama, S., Mori, H., Ihara, Y., Bouldin, T. W., Suzuki, K., and Tomonaga, M., 1990a, Immunocytochemical and ultrastructural studies of lower motor neurons in amyotrophic lateral sclerosis, *Ann. Neurol.* **27:**137–148.

Murayama, S., Mori, H., Ihara, Y., and Tomonaga, M., 1990b, Immunocytochemical and ultrastructural studies of Pick's disease, *Ann. Neurol.* **27:**394–405.

Narasimhan, J., Potter, J. L., and Haas, A. L., 1996, Conjugation of the 15-kDa interferon-induced ubiquitin homolog is distinct from that of ubiquitin, *J. Biol. Chem.* **271:**324–330.

Nicholl, I. D., and Quinlan, R. A., 1994, Chaperone activity of alpha-crystallins modulates intermediate filament assembly, *EMBO J.* **13:**945–953.

Noga, M., and Hayashi, T., 1996, Ubiquitin gene expression following transient forebrain ischemia, *Mol. Brain Res.* **36:**261–267.

Okada, M., Miyake, T., Kitamura, T., Kawasaki, K., and Mizushima, Y., 1994, Anti-ubiquitin immunoreactivity associates with pyramidal cell death induced by intraventricular infusion of leupeptin in rat hippocampus, *Neurosci. Res.* **19:**59–66.

Olanow, C., 1990, Oxidative reactions in Parkinson's disease, *Neurology* **40**(Suppl. 3)**:**32–37.

Orrenius, S., 1995, Apoptosis: Molecular mechanisms and implications for human disease, *J. Intern. Med.* **237:**529–536.

Papp, M. I., and Lantos, P. L., 1992, Accumulation of tubular structures in oligodendroglial and neuronal cells as the basic alteration in multiple system atrophy, *J. Neurol. Sci.* **107:**172–182.

Papp, M. I., Kahn, J. E., and Lantos, P. L., 1989, Glial cytoplasmic inclusions in the CNS of patients with multiple system atrophy (striatonigral degeneration, olivopontocerebellar atrophy and Shy-Drager syndrome), *J. Neurol. Sci.* **94:**79–100.

Pappolla, M. A., Omar, R., and Saran, B., 1989, The 'normal' brain. 'Abnormal' ubiquitinilated deposits highlight an age-related protein change, *Am. J. Pathol.* **135:**585–591.

Parag, H. A., Raboy, B., and Kulka, R. G., 1987, Effect of heat shock on protein degradation in mammalian cells: Involvement of the ubiquitin system, *EMBO J.* **6:**55–61.

Perry, G., Friedman, R., Shaw, G., and Chau, V., 1987, Ubiquitin is detected in neurofibrillary tangles and senile plaque neurites of Alzheimer disease brains, *Proc. Natl. Acad. Sci. USA* **84:**3033–3036.

Perry, G., Kawai, M., Tabaton, M., Onorato, M., Mulvihill, P., Richey, P., Morandi, A., Connolly, J. A., and Gambetti, P., 1991, Neuropil threads of Alzheimer's disease show a marked alteration of the normal cytoskeleton, *J. Neurosci.* **11:**1748–1755.

Perry, R. H., Irving, D., Blessed, G., Fairburn, A., and Perry, K., 1990, Senile dementia of the Lewy body type: A clinically and neuropathologically distinct form of dementia in the elderly, *J. Neurol. Sci.* **95:**119–139.

Price, S. R., England, B. K., Bailey, J. L., Van Vreede, K., and Mitch, W. E., 1994, Acidosis and glucocorticoids concomitantly increase ubiquitin and proteasome subunit mRNAs in rat muscle, *Am. J. Physiol.* **267:**C955–960.

Price, S. R., Bailey, J. L., and England, B. K., 1996, Necessary but not sufficient: The role of glucocorticoids in the acidosis-induced increase in levels of mRNAs encoding proteins of the ATP-dependent proteolytic pathway in rat muscle, *Miner. Electrol. Metab.* **22:**72–75.

Rock, K. L., Rothstein, L., Clark, K., Stein, R., Dick, L., Hwang, D., and Goldberg, A. L., 1994, Inhibitors of the proteasome block the degradation of most cell proteins and the generation of peptides presented on MHC class 1 molecules, *Cell* **78:**761–771.

Rosen, D. R., Siddique, T., Patterson, D., Figlewicz, D. A., Sapp, P., Hentati, A., Donaldson, D., Goto, J., O'Regan, J. P., and Deng, H. X., 1993, Mutations in Cu/Zn superoxide dismutase gene are associated with familial amyotrophic lateral sclerosis, *Nature* **362:**59–62.

Schapira, A. H., 1993, Mitochondrial complex I deficiency in Parkinson's disease, *Adv. Neurol.* **60:** 288–291.

Schapira, A. H., 1994, Evidence for mitochondrial dysfunction in Parkinson's disease—A critical appraisal, *Mov. Disord.* **9:**125–138.

Schiffer, D., Autilio-Gambetti, L., Chio, A., Gambetti, P., Giordana, M. T., Gullotta, F., Migheli, A., and Vigliani, M. C., 1991, Ubiquitin in motor neuron disease: Study at the light and electron microscope, *J. Neuropathol. Exp. Neurol.* **50:**463–473.

Schmidt, M. L., Murray, J. M., and Trojanowski, J. Q., 1993, Continuity of neuropil threads with tangle-bearing and tangle-free neurons in Alzheimer disease cortex. A confocal laser scanning microscopy study, *Mol. Chem. Neuropathol.* **18:**299–312.

Schwartz, A. L., Brandt, R. A., Geuze, H., and Ciechanover, A., 1992, Stress-induced alterations in autophagic pathway: Relationship to ubiquitin system, *Am. J. Physiol.* **262:**C1031–1038.

Schwartz, L. M., Myer, A., Kosz, L., Engelstein, M., and Maier, C., 1990, Activation of polyubiquitin gene expression during developmentally programmed cell death, *Neuron* **5:**411–409.

Schweitzer, J. B., Park, M. R., Einhaus, S. L., and Robertson, J. T., 1993, Ubiquitin marks the reactive swellings of diffuse axonal injury, *Acta Neuropathol.* **85:**503–507.

Shang, F., and Taylor, A., 1995, Oxidative stress and recovery from oxidative stress are associated with altered ubiquitin conjugating and proteolytic activities in bovine lens epithelial cells, *Biochem. J.* **307:**297–303.

Sherriff, F. E., Joachim, C. L., Squier, M. V., and Esiri, M. M., 1995, Ubiquitinated inclusions in inclusion-body myositis patients are immunoreactive for cathepsin D but not beta-amyloid, *Neurosci. Lett.* **194:**37–40.

Sisodia, S. S., and Price, D. L., 1995, Role of the beta-amyloid protein in Alzheimer's disease, *FASEB J.* **9:**366–370.

Takada, K., Nasu, H., Hibi, N., Tsukada, Y., Ohkawa, K., Fujimuro, M., Sawada, H., and Yokosawa, H., 1995, Immunoassay for the quantification of intracellular multi-ubiquitin chains, *Eur. J. Biochem.* **233:**42–47.

Takayanagi, K., Dawson, S., Reynolds, S. E., and Mayer, R. J., 1996, Specific developmental changes in the regulatory subunits of the 26S proteasome in intersegmental muscles preceding eclosion in Manduca sexta, *Biochem. Biophys. Res. Commun.*, **228:**517–523.

Tamaoka, A., Mizusawa, H., Mori, H., and Shoji, S., 1995, Ubiquitinated alpha B-crystallin in glial cytoplasmic inclusions from the brain of a patient with multiple system atrophy, *J. Neurol. Sci.* **129:**192–198.

Temparis, S., Asensi, M., Taillandier, D., Aurousseau, E., Larbaud, D., Obled, A., Bechet, D., Ferrara, M., Estrela, J. M., and Attaix, D., 1994, Increased ATP-ubiquitin-dependent proteolysis in skeletal muscles of tumor-bearing rats, *Cancer Res.* **54:**5568–5573.

Thompson, C. B., 1995, Apoptosis in the pathogenesis and treatment of disease, *Science* **267:**1456–1462.

Thompson, H. S., and Scordilis, S. P., 1994, Ubiquitin changes in human biceps muscle following exercise-induced damage, *Biochem. Biophys. Res. Commun.* **204:**1193–1198.

Thompson, R. J., Doran, J. F., Jackson, P., Dhillon, A. P., and Rode, J., 1983, PGP 9.5—A new marker for vertebrate neurons and neuroendocrine cells, *Brain Res.* **278:**224–228.

Tiao, G., Fagan, J. M., Samuels, N., James, J. H., Hudson, K., Lieberman, M., Fischer, J. E., and Hasselgren, P. O., 1994, Sepsis stimulates nonlysosomal, energy-dependent proteolysis and increases ubiquitin mRNA levels in rat skeletal muscle, *J. Clin. Invest.* **94:**2255–2264.

Tiao, G., Fagan, J., Roegner, V., Lieberman, M., Wang, J. J., Fischer, J. E., and Hasselgren, P. O., 1996, Energy-ubiquitin-dependent muscle proteolysis during sepsis in rats is regulated by glucocorticoids, *J. Clin. Invest.* **97:**339–348.

Tolnay, M., and Probst, A., 1995, Frontal lobe degeneration: Novel ubiquitin-immunoreactive neurites within frontotemporal cortex, *Neuropathol. Appl. Neurobiol.* **21:**492–497.

Trojanowski, J. Q., Shin, R. W., Schmidt, M. L., and Lee, V. M., 1995, Relationship between plaques, tangles, and dystrophic processes in Alzheimer's disease, *Neurobiol. Aging* **16:**335–340; discussion 341–345.

Tsuda, T., Munthasser, S., Fraser, P. E., Percy, M. E., Rainero, I., Vaula, G., Pinessi, L., Bergamini, L., Vignocchi, G., and McLachlan, D. R., 1994, Analysis of the functional effects of a mutation in SOD1 associated with familial amyotrophic lateral sclerosis, *Neuron* **13:**727–736.

Tsujinaka, T., Fujita, J., Ebisui, C., Yano, M., Kominami, E., Suzuki, K., Tanaka, K., Katsume, A., Ohsugi, Y., Shiozaki, H., and Monden, M., 1996, Interleukin 6 receptor antibody inhibits muscle atrophy and modulates proteolytic systems in interleukin 6 transgenic mice, *J. Clin. Invest.* **97:**244–249.

Tsuzuki, K., Fukatsu, R., Takamaru, Y., Yoshida, T., Mafune, N., Kobayashi, K., Fujii, N., and Takahata, N., 1995, Co-localization of amyloid-associated proteins with amyloid beta in rat soleus muscle in chloroquine-induced myopathy: A possible model for amyloid beta formation in Alzheimer's disease, *Brain Res.* **699:**260–265.

Uney, J. B., Anderton, B. H., and Thomas, S. M., 1993, Changes in heat shock protein 70 and ubiquitin mRNA levels in C1300 N2A mouse neuroblastoma cells following treatment with iron, *J. Neurochem.* **60:**659–665.

Voisin, L., Breuille, D., Combaret, L., Pouyet, C., Taillandier, D., Aurousseau, E., Obled, C., and Attaix, D., 1996, Muscle wasting in a rat model of long-lasting sepsis results from the activation of lysosomal, Ca^{2+}-activated, and ubiquitin-proteasome proteolytic pathways, *J. Clin. Invest.* **97:** 1610–1617.

Vonsattel, J. P., Myers, R. H., Stevens, T. J., Ferrante, R. J., Bird, E. D., and Richardson, E. J., 1985, Neuropathological classification of Huntington's disease, *J. Neuropathol. Exp. Neurol.* **44:**559–577.

Wilkinson, K. D., 1994, Cellular roles of ubiquitin, in *Heat Shock Proteins in the Nervous System* (R. J. Mayer and I. Brown, eds.), pp. 192–234, Academic Press, San Diego.

Wilkinson, K. D., Lee, K. M., Deshpande, S., Duerksen, H. P., Boss, J. M., and Pohl, J., 1989, The neuron-specific protein PGP 9.5 is a ubiquitin carboxyl-terminal hydrolase, *Science* **246:** 670–673.

Wilson, P. O., Barber, P. C., Hamid, Q. A., Power, B. F., Dhillon, A. P., Rode, J., Day, I. N., Thompson, R. J., and Polak, J. M., 1988, The immunolocalization of protein gene product 9.5 using rabbit polyclonal and mouse monoclonal antibodies, *Br. J. Exp. Pathol.* **69:**91–104.

Wing, S. S., and Goldberg, A. L., 1993, Glucocorticoids activate the ATP-ubiquitin-dependent proteolytic system in skeletal muscle during fasting, *Am. J. Physiol.* **264:**E668–676.

Wing, S. S., Haas, A. L., and Goldberg, A. L., 1995, Increase in ubiquitin–protein conjugates concomitant with the increase in proteolysis in rat skeletal muscle during starvation and atrophy denervation, *Biochem. J.* **307:**639–645.

Wong, P., Pardo, C., Borchelt, D., Lee, M., Copeland, N., Jenkins, N., Sisodia, S., Cleveland, D., and Price, D., 1995, An adverse property of a familial ALS-linked SOD1 mutation causes motor neuron disease characterized by vacuolar degeneration of mitochondria, *Neuron* **14:**1105–1116.

Yasuhara, O., Kawamata, T., Aimi, Y., McGeer, E. G., and McGeer, P. L., 1994, Two types of dystrophic neurites in senile plaques of Alzheimer disease and elderly non-demented cases, *Neurosci. Lett.* **171:**73–76.

Yee, W. M., Frim, D. M., and Isacson, O., 1993, Relationships between stress protein induction and NMDA-mediated neuronal death in the entorhinal cortex, *Exp. Brain Res.* **94:**193–202.

Index